Handbook of Research on Blockchain Technology

Handbook of Research on Blockchain Technology

Edited by

Saravanan Krishnan
Valentina E. Balas
E. Golden Julie
Y. Harold Robinson
S. Balaji
Raghvendra Kumar

Academic Press is an imprint of Elsevier
125 London Wall, London EC2Y 5AS, United Kingdom
525 B Street, Suite 1650, San Diego, CA 92101, United States
50 Hampshire Street, 5th Floor, Cambridge, MA 02139, United States
The Boulevard, Langford Lane, Kidlington, Oxford OX5 1GB, United Kingdom

Copyright © 2020 Elsevier Inc. All rights reserved.

No part of this publication may be reproduced or transmitted in any form or by any means, electronic or mechanical, including photocopying, recording, or any information storage and retrieval system, without permission in writing from the publisher. Details on how to seek permission, further information about the Publisher's permissions policies and our arrangements with organizations such as the Copyright Clearance Center and the Copyright Licensing Agency, can be found at our website: www.elsevier.com/permissions.

This book and the individual contributions contained in it are protected under copyright by the Publisher (other than as may be noted herein).

Notices
Knowledge and best practice in this field are constantly changing. As new research and experience broaden our understanding, changes in research methods, professional practices, or medical treatment may become necessary.

Practitioners and researchers must always rely on their own experience and knowledge in evaluating and using any information, methods, compounds, or experiments described herein. In using such information or methods they should be mindful of their own safety and the safety of others, including parties for whom they have a professional responsibility.

To the fullest extent of the law, neither the Publisher nor the authors, contributors, or editors, assume any liability for any injury and/or damage to persons or property as a matter of products liability, negligence or otherwise, or from any use or operation of any methods, products, instructions, or ideas contained in the material herein.

British Library Cataloguing-in-Publication Data
A catalogue record for this book is available from the British Library

Library of Congress Cataloging-in-Publication Data
A catalog record for this book is available from the Library of Congress

ISBN: 978-0-12-819816-2

For Information on all Academic Press publications
visit our website at https://www.elsevier.com/books-and-journals

Publisher: Mara Conner
Acquisitions Editor: Chris Katsaropoulos
Editorial Project Manager: Ana Claudia A. Garcia
Production Project Manager: Punithavathy Govindaradjane
Cover Designer: Mark Rogers

Typeset by MPS Limited, Chennai, India

Contents

List of Contributors xvii
About the Editors... xxi
Preface xxv
About the Book... xxix
Key Features xxxi

CHAPTER 1 Ethical Considerations and Issues of Blockchain Technology-Based Systems in War Zones: A Case Study Approach 1
Ramnath Reghunadhan

1.1 Introduction 1
1.2 Methodology... 2
1.3 Blockchain Technology.. 2
1.4 The Process Involved 3
1.5 Syria... 5
1.6 Syria: The Beginning of Modern Nation-State 6
1.7 Syria: Under Bashar al-Assad Regime 8
1.8 Case Study 1: Refugees and Internally Displaced Persons in Syria Since 2011 10
1.9 Arab Spring and the Emergence of the Syrian Civil War ... 11
1.10 Major Actors in the Conflict... 12
1.11 Ethical Considerations and Issues in "De-escalation Zones" for Refugees and Internally Displaced Persons.................... 14
1.12 Blockchain Technology for Refugees and Internally Displaced Persons............... 15
1.13 Case Study 2: US Defense Forces and Industry................ 17
1.14 Issues and Challenges 21
1.15 Future of Blockchain Technology for Global Governance ... 25
1.16 Conclusion 28
References... 29
Further Reading 33

CHAPTER 2 Blockchain-Based Framework for Data Storage in Peer-to-Peer Scheme Using Interplanetary File System 35
Randhir Kumar and Rakesh Tripathi

2.1 Introduction 35
2.1.1 Motivation 36
2.2 Related Work... 36

v

vi Contents

2.3	Working Model of Framework	37
2.4	Record Storage Using Interplanetary File System and Blockchain Over Centralized Service Provider	39
	2.4.1 Interplanetary File System–Based Decentralized Storage System	39
	2.4.2 Blockchain Structure to Maintain Availability	41
2.5	Working of Interplanetary File System	45
	2.5.1 Installation of Interplanetary File System	45
	2.5.2 Interplanetary File System Commands	45
	2.5.3 Interplanetary File System Connection and Deployment	46
2.6	Implementation of Framework Using Interplanetary File System and Blockchain	48
	2.6.1 Uploading Transactions to the Interplanetary File System and Blockchain Network	49
	2.6.2 Validating the Transaction Using Mining Process	50
	2.6.3 Adding Transaction Into Blockchain Network	50
	2.6.4 Access of Transaction Using Content-Addressed Technique	51
	2.6.5 Adding Peers Into Blockchain Network	54
	2.6.6 Use of Consensus in Blockchain Network	55
2.7	Conclusion	57
	References	57

CHAPTER 3 Integration of Blockchain and Internet of Things **61**
S. Porkodi and D. Kesavaraja

3.1	Introduction	61
	3.1.1 Integrity	62
	3.1.2 Decentralization	62
	3.1.3 Anonymity	62
3.2	Research Survey	63
3.3	Limitations in Security of Internet of Things	64
	3.3.1 Internet of Things Characteristics	64
	3.3.2 Analysis of Security on Internet of Things Applications	66
3.4	Existing Technologies of Blockchain	68
	3.4.1 Data Structure	69
	3.4.2 Consensus Protocol and Byzantine General Problem	72
	3.4.3 Analysis of Blockchain Security	74
3.5	Applications of Blockchain in Internet of Things	75
	3.5.1 Malicious Behaviors in Internet of Things Devices	76
	3.5.2 Sensor Data Correctness	76
	3.5.3 Integration of Blockchain and Internet of Things Projects and Applications	77
	3.5.4 Structure of Blockchain Integrated With Internet of Things Applications	77

Contents **vii**

3.6 Security in Blockchain With Ciphertext Policy and Hybridized Weighted Attribute-Based Encryption.. 79
    3.6.1 Experimental Setup.. 80
    3.6.2 Working of Ciphertext Policy and Hybridized Weighted Attribute-Based Encryption Algorithm 80
    3.6.3 Experimental Results... 81
    3.6.4 Result Analysis 86
3.7 Conclusion.. 89
References.. 90
Further Reading 94

CHAPTER 4 A Deep Dive Into Security and Privacy Issues of Blockchain Technologies **95**
Neha Gupta

4.1 Introduction 95
4.2 Blockchain Aspects for Consideration 96
    4.2.1 Security of Blockchain.. 96
    4.2.2 Privacy of Blockchains 97
4.3 Security Issues of Blockchain Technology 97
    4.3.1 51% Vulnerability 98
    4.3.2 Double Spending.. 98
    4.3.3 Mining Pool Attacks 99
    4.3.4 Client Side Security Threats 99
    4.3.5 Forking 99
    4.3.6 Criminal Activity 100
    4.3.7 Private Key Security 101
    4.3.8 Transaction Privacy Leakage 101
4.4 Privacy Issues of Blockchain Technology... 102
    4.4.1 Physical–Cyberspace Boundary 102
    4.4.2 Information Storage and Inference 103
    4.4.3 Nature of the Blockchain—Eternal Records 104
4.5 Types of Attacks 104
    4.5.1 Selfish Mining Attack.. 104
    4.5.2 DAO Attack 105
    4.5.3 BGP Hijacking Attack 106
    4.5.4 Eclipse Attack 106
    4.5.5 Liveness Attack 107
4.6 Security Enhancement to Blockchain Systems 107
    4.6.1 SmartPool 107
    4.6.2 Quantitative Framework 108
    4.6.3 Oyente 108
    4.6.4 Hawk 110

viii Contents

4.7 Future Directions 110
4.8 Conclusion 111
References 111
Further Reading 111

CHAPTER 5 Blockchain Implementation for Internet of Things Applications 113
Vibha Nehra, Ajay K. Sharma and Rajiv K. Tripathi
5.1 Introduction 113
5.1.1 Bitcoin 113
5.1.2 Bitcoin Energy Consumption Index 114
5.1.3 Blockchain 114
5.2 Literature Work 117
5.3 Blockchain for Internet of Things 119
5.3.1 Challenges 119
5.3.2 Research Opportunities 122
5.4 Energy-Efficient Blockchain for Internet of Things 123
5.4.1 Experimental Setup 125
5.4.2 Energy Consumption in Attached Machines 126
5.5 Results 128
5.6 Conclusion 129
5.7 Future Work 130
Acknowledgments 130
References 130

CHAPTER 6 Blockchain Implementation Using Smart Grid-Based Smart City ... 133
M. Afshar Alam and Sapna Jain
6.1 Introduction 133
6.2 Background 135
6.3 Blockchain Concept 136
6.4 Blockchain Types 138
6.5 Center Components of Blockchain Architecture: How Does It Work 141
6.6 Carbon Credit 142
6.7 The Paris Agreement 143
6.8 Carbon Credit Exchange 145
6.8.1 Carbon Credits and Carbon Markets 146
6.9 Carbon Emission Effect 147
6.9.1 Blockchain Application for Distributed Energy Resources
Management in Smart City 151
6.10 Blockchain Implementation in Power Generation, Transmission and
Distribution 152
6.10.1 Smart Energy Grids 153

Contents ix

6.11 Smartgrid Blockchain Implementation in India—A Case Study of Puducherry Project 153
    6.11.1 System Functionality 155
6.12 Blockchain Implemented Projects Worldwide in Energy Conservation 161
6.13 Conclusion and Recommendations 164
    Glossary 165
    Discussion Questions 166
    References 166
    Further Reading 167

CHAPTER 7 Cloud-Based Blockchaining for Enhanced Security **171**
D. Jeyabharathi, D. Kesavaraja and D. Sasireka
7.1 Introduction 172
    7.1.1 Blockchaining 172
    7.1.2 Transparent 173
    7.1.3 Autonomy 173
    7.1.4 Immutable 173
    7.1.5 Anonymity 174
7.2 Block Chain Structure 174
    7.2.1 Main Data 174
    7.2.2 Timestamp 174
    7.2.3 Types of Blockchain 174
    7.2.4 Blockchain Security Process 175
7.3 Related Works 176
7.4 Incorporation of Blockchaining With Cloud Environment 177
    7.4.1 Layer 1: Access Policy Checking 177
    7.4.2 Layer 2: Virtual Machine Measurement Security Checking Process 178
    7.4.3 Layer 3: Sensitive and Nonsensitive Data Separation and Data Package Security Checking Process 178
7.5 Conclusion 179
    References 180
    Further Reading 180

CHAPTER 8 Blockchain Integration With Low-Power Internet of Things Devices **183**
Sandeep B. Kadam and Shajimon K. John
8.1 Introduction 183
    8.1.1 The Concepts of Internet of Things and Blockchain 184
    8.1.2 The Concept of Blockchain 187
    8.1.3 Features of Blockchain 189
    8.1.4 Structure of Blockchain 190

Contents

	8.1.5 Elements in a Blockchain	190
	8.1.6 Types of Blockchain	192
	8.1.7 Consensus	195
8.2	Blockchain for Internet of Things Applications	198
	8.2.1 Hyperledger	198
	8.2.2 IOTA	199
	8.2.3 Ethereum	199
	8.2.4 Challenges in Integrating Blockchain With Internet of Things	200
	8.2.5 Risks in Internet of Things Blockchain Integration	201
8.3	Proposed Architecture	201
	8.3.1 Smart Contract	203
	8.3.2 Installation of Lamp Server	204
	8.3.3 Deploying Smart Contract on Network	205
	8.3.4 Interfacing ESP32 With Ropsten Network	205
8.4	Results and Discussion	206
	References	209

CHAPTER 9 Applications of Blockchain Technology **213**

Chetna Laroiya, Deepika Saxena and C. Komalavalli

9.1	Introduction to Blockchain Technology Applications	213
	9.1.1 Relevance of Blockchain Technology Applications	214
9.2	Background of Blockchain	215
	9.2.1 What Is Bitcoin?	215
	9.2.2 Emergence of Bitcoin	216
	9.2.3 Working of Bitcoin	217
	9.2.4 Risk in Bitcoin	218
	9.2.5 Legal Issues in Bitcoin	219
9.3	Applications of Blockchain Technology	220
	9.3.1 Applications of Blockchain Technology in Financial Services	220
	9.3.2 Blockchain Applications in Insurance Sector	223
	9.3.3 Blockchain Technology and Financial Technology	225
	9.3.4 Application of Blockchain in Healthcare	225
	9.3.5 Application of Blockchain in Voting	229
	9.3.6 Application of Blockchain in Real Estate	230
	9.3.7 Application of Blockchain Technology in Supply Chain	232
	9.3.8 Application of Blockchain Technology in Music Industry	235
	9.3.9 Application of Blockchain Technology in Identity Management	236
9.4	Blockchain Challenges and Concern	237
9.5	Future Cases	238
	9.5.1 Blockchain Technology in Military	238
	9.5.2 Blockchain Technology in Internet of Things	239

Contents   **xi**

9.6 Categories of Blockchain Development 240
    9.6.1 Application of Blockchain Technology as a Development Platform 240
    9.6.2 Application of Blockchain Technology as a Smart Contract 240
    9.6.3 Application of Blockchain Technology as a Marketplace 240
    9.6.4 Application of Blockchain Technology as Trusted Service
    Application 241
9.7 Summary 241
    Reference 241
    Further Reading 241

CHAPTER 10 Blockchain-Powered Smart Healthcare System **245**
    Rashmi G. Shukla, Anuja Agarwal and Shekhar Shukla
10.1 Introduction 245
10.2 Healthcare System 247
    10.2.1 Insurance Industry 248
    10.2.2 Pharmaceutical Industry 249
    10.2.3 Physicians 250
    10.2.4 Patients 250
    10.2.5 Government 250
10.3 Issues and Challenges for a Healthcare System 251
10.4 The Blockchain Technology: Concept and Application Areas 253
10.5 Blockchain Applications in Healthcare 256
10.6 Challenges to Blockchain Applications in Healthcare 259
10.7 An Integrated Blockchain-Powered Smart Healthcare Framework:
    Conceptual Model and Analysis 261
10.8 Conclusion and Implications 264
10.9 Future Research Directions 264
    References 266

**CHAPTER 11 Internet of Things and Blockchain: Integration, Need,
Challenges, Applications, and Future Scope** **271**
    *Deepak Kumar Sharma, Ajay Kumar Kaushik, Aarti Goel and
Saakshi Bhargava*
11.1 Introduction: Blockchain 271
    11.1.1 Basics of Blockchain 272
    11.1.2 Why Do We Need Blockchain? 272
    11.1.3 Crypto Currency 273
    11.1.4 Types of Blockchain 273
    11.1.5 Applications of Blockchain 275
    11.1.6 Potential Areas of Research in Blockchain 277
    11.1.7 Challenges With Blockchain 278

xii Contents

11.2	Internet of Things	279
	11.2.1 Architecture of Internet of Things	280
	11.2.2 How a Device Becomes Smart Device?	281
	11.2.3 Applications of Internet of Things	281
	11.2.4 Advantages of Using Internet of Things Technology	285
	11.2.5 Challenges With Internet of Things Systems	286
11.3	Integration of Internet of Things and Blockchain	286
	11.3.1 Need for Integration of Internet of Things and Blockchain	287
	11.3.2 Challenges in Internet of Things and Blockchain Integration	288
	11.3.3 Benifits of Integration	288
	11.3.4 Industrial Internet of Things	288
	11.3.5 Internet of Things on Cloud Platform	289
	11.3.6 Creating an Environment	289
11.4	Discussion	290
11.5	Future Work	291
	References	292
	Further Reading	294

CHAPTER 12 Blockchain and Internet of Things: An Overview **295**

A. Sherly Alphonse and M.S. Starvin

12.1	Introduction	295
12.2	Overview of Internet of Things	297
	12.2.1 Application of Internet of Things	297
	12.2.2 Architecture, Challenges, and Existing Security Technologies Used for Internet of Things	298
12.3	Overview of Blockchain Technology	303
12.4	Properties of Blockchain	305
	12.4.1 Transparent	307
12.5	Challenges	308
12.6	Research on Blockchain	308
12.7	Scope for Research	309
12.8	Some Recent Advances in the Technology	310
	12.8.1 Payment Networks With Bitcoin	310
12.9	Blockchain and Internet of Things	311
	12.9.1 Components	313
	12.9.2 Use Cases	315
	12.9.3 Static Sensors and Autonomous Machines	319
12.10	The Convergence of Blockchain, Internet of Things, and Artificial intelligence	319
12.11	Conclusion	320
	References	321

Contents **xiii**

CHAPTER 13 Cryptocurrency Mechanisms for Blockchains: Models, Characteristics, Challenges, and Applications **323**

Deepak Kumar Sharma, Shrid Pant, Mehul Sharma and Shikha Brahmachari

13.1 Introduction 323
    13.1.1 Introduction to Cryptocurrency 323
    13.1.2 Contemporaneous Advancements in Blockchains 323
    13.1.3 Motivations for the Use of Blockchains in Cryptocurrencies 324
    13.1.4 Introduction to Cryptocurrency Mechanisms for Blockchains 324
13.2 Principal Disciplines of Cryptocurrency Mechanisms for Blockchains 325
    13.2.1 Blockchain Architecture 325
    13.2.2 Blockchain Consensus Model 329
    13.2.3 Blockchain Challenges and Opportunities 335
    13.2.4 Case Study: Bitcoin 339
13.3 Comparative Analysis With Other Mechanisms 340
    13.3.1 The IOTA Tangle 340
    13.3.2 Hedera Hashgraph 342
13.4 Further Advancements 344
13.5 Conclusion 345
    References 345

CHAPTER 14 Overview of Blockchain Technology Concepts **349**

C. Komalavalli, Deepika Saxena and Chetna Laroiya

14.1 Introduction 349
    14.1.1 Network Architectures 349
14.2 Motivations Behind Blockchain 350
    14.2.1 Foundation of Blockchain 352
    14.2.2 Need for Blockchain 352
14.3 Blockchain Concepts 353
    14.3.1 History of Blockchain 353
    14.3.2 Network View 353
    14.3.3 Database Versus Blockchain 353
    14.3.4 Driving Concepts for Business 354
14.4 Key Terms Related to Blockchain 355
    14.4.1 Block Structure 356
    14.4.2 Genesis Block 357
    14.4.3 Block Time 358
    14.4.4 Block Size 358
    14.4.5 Cryptography in Blockchain 358

xiv Contents

14.5 Formation of Chain in Blockchain 359
14.6 Consenus Algorithm 361
    14.6.1 Proof of Work 361
    14.6.2 Proof of Stake 361
    14.6.3 Practical Byzantine Fault Tolerance 361
    14.6.4 Proof of Activity 361
    14.6.5 Proof of Burn Time 362
    14.6.6 Proof of Capacity 362
    14.6.7 Proof of Importance 362
14.7 Blockchain Versions 362
    14.7.1 Blockchain 1.0 Currency 362
    14.7.2 Blockchain 2.0 Smart Contracts 362
    14.7.3 Blockchain 3.0 DApps 362
    14.7.4 Blockchain 4.0 363
    14.7.5 Different Users of Blockchain Technology 363
14.8 Benefits of Blockchain Technology 363
14.9 Types of Blockchain 364
    14.9.1 Public 364
    14.9.2 Private 364
    14.9.3 Hybrid 364
14.10 Working of Blockchain 365
14.11 Platforms 365
    14.11.1 Criteria for Selecting Blockchain Platform 365
    14.11.2 Different Platforms 366
14.12 Challenges in Blockchain Technology 368
14.13 Strengths of Technology 369
14.14 Applications 369
14.15 Summary 370
    Further Reading 370

CHAPTER 15 Scalability in Blockchain: Challenges and Solutions 373
    Gagandeep Kaur and Charu Gandhi
15.1 Introduction 373
15.2 Core Concepts 375
15.3 Scalability Challenges in Blockchain 377
15.4 Scalability Solutions in Blockchain 379
15.5 Future Directions 403
15.6 Conclusion 404
    References 404
    Further Reading 406

Contents **xv**

CHAPTER 16 A Vital Role of Blockchain Technology Toward Internet of Vehicles.. **407**
Vikram Puri, Raghvendra Kumar, Chung Van Le, Rohit Sharma and Ishaani Priyadarshini

16.1 Introduction 407
16.2 Related Studies 409
16.3 Architecture 409
16.4 Application of Blockchain With Internet of Vehicles 411
16.5 Security and Privacy Issues 413
    16.5.1 Security.. 413
    16.5.2 Privacy... 414
16.6 Conclusion 415
    References... 415

CHAPTER 17 Need to Know About Combined Technologies of Blockchain and Machine Learning **417**
X. Alphonse Inbaraj and T. Rama Chaitanya

17.1 Introduction 417
17.2 Literature Survey 417
17.3 Bitcoin Methodology 418
    17.3.1 A Decentralized Chain 419
17.4 Blockchain Implementations in Various Industries 419
    17.4.1 Blockchain Transaction in Banking 419
    17.4.2 Education... 419
    17.4.3 Energy 420
    17.4.4 Legal, Government, and Insurance... 420
    17.4.5 Internet of Things and Artificial Intelligence 420
    17.4.6 Health Care and Medicine 420
17.5 Process Stages of Implementation of Blockchain Technology............. 420
17.6 Creation of Bitcoin Transaction and Mining.. 421
    17.6.1 Transaction of Data.. 423
    17.6.2 Chaining the Blocks (With Hash) 424
    17.6.3 Chaining 424
    17.6.4 Final Chaining Process 424
    17.6.5 Alteration in First Block 425
    17.6.6 Hash Signature Created 425
17.7 Combining Blockchain With Artificial Intelligence/Machine Learning 427
    17.7.1 Energy Consumption and Efficiency... 427
    17.7.2 Blockchain Can Help Artificial Intelligence/Machine-Learning Systems to Better Explain Themselves 427
    17.7.3 More Data 427

xvi Contents

        17.7.4 Smart Contracts 427

        17.7.5 Reduce Instances of 51% Attacks 428

17.8 Startup With Blockchain and Machine Learning 428

17.9 Census Types and Pros and Cons 428

17.10 Artificial Intelligence Versus Blockchain 428

17.11 Future Directions 431

17.12 Conclusion 431

        References 431

Index 433

List of Contributors

Anuja Agarwal
Technology Management Area, MPSTME, NMIMS, Mumbai, India

M. Afshar Alam
Department of Computer Science and Engineering, Jamia Hamdard University, New Delhi, India

A. Sherly Alphonse
DMI Engineering College, Aralvaimozhi, India

Saakshi Bhargava
Department of Electronics and Communication, Banasthali Vidyapith, Vanasthali, India

Shikha Brahmachari
CAITFS, Department of Information Technology, Netaji Subhas University of Technology (Formerly known as Netaji Subhas Institute of Technology), New Delhi, India

T. Rama Chaitanya
Professor, Department of Computer Science and Engineering, PACE Institute of Technology and Sciences, Ongole, India

Charu Gandhi
Department of CSE&IT, Jaypee Institute of Information Technology, Noida, India

Aarti Goel
CAITFS, Department of Information Technology, Netaji Subhas University of Technology (Formerly known as Netaji Subhas Institute of Technology), New Delhi, India

Neha Gupta
MRIIRS, Faridabad, India

X. Alphonse Inbaraj
Research Scholar, Department of Information Engineering, I-Shou University, Taiwan, R.O.C

Sapna Jain
Department of Computer Science and Engineering, Jamia Hamdard University, New Delhi, India

D. Jeyabharathi
Department of IT, Sri Krishna College of Technology, Coimbatore, India

Shajimon K. John
Saintgits College of Engineering, Kottukulam Hills, Pathamuttam, Kerala

Sandeep B. Kadam
Department of Electronics & Communication Engineering, Saintgits College of Engineering, Kottukulam Hills, Pathamuttam, Kerala

Gagandeep Kaur
Department of CSE&IT, Jaypee Institute of Information Technology, Noida, India

xviii List of Contributors

Ajay Kumar Kaushik
CAITFS, Department of Information Technology, Netaji Subhas University of Technology (Formerly known as Netaji Subhas Institute of Technology), New Delhi, India

D. Kesavaraja
Department of Computer Science and Engineering, Dr. Sivanthi Aditanar College of Engineering, Tiruchendur, India

C. Komalavalli
Jagan Institute of Management Studies, Rohini, New Delhi, India

Raghvendra Kumar
Department of Computer Science and Engineering, GIET University, Gunupur, India

Randhir Kumar
Department of Information Technology, National Institute of Technology, Raipur, India

Chetna Laroiya
Jagan Institute of Management Studies, Rohini, New Delhi, India

Chung Van Le
Duy Tan University, Danang, Vietnam

Vibha Nehra
Department of Computer Science and Engineering, National Institute of Technology Delhi, New Delhi, India

Shrid Pant
CAITFS, Department of Information Technology, Netaji Subhas University of Technology (Formerly known as Netaji Subhas Institute of Technology), New Delhi, India

S. Porkodi
Department of Computer Science and Engineering, Dr. Sivanthi Aditanar College of Engineering, Tiruchendur, India

Ishaani Priyadarshini
University of Delaware, Newark, DE, United States

Vikram Puri
Duy Tan University, Danang, Vietnam

Ramnath Reghunadhan
Research Scholar, Department of Humanities and Social Sciences, Indian Institute of Technology Madras (IITM), Chennai, India

D. Sasireka
VV College of Engineering, Tisaiyanvillai, India

Deepika Saxena
Jagan Institute of Management Studies, Rohini, New Delhi, India

Ajay K. Sharma
Department of Computer Science and Engineering, National Institute of Technology Jalandhar, Jalandhar, India

List of Contributors   **xix**

Deepak Kumar Sharma
CAITFS, Department of Information Technology, Netaji Subhas University of Technology
(Formerly known as Netaji Subhas Institute of Technology), New Delhi, India

Mehul Sharma
CAITFS, Department of Information Technology, Netaji Subhas University of Technology
(Formerly known as Netaji Subhas Institute of Technology), New Delhi, India

Rohit Sharma
Department of Electronics & Communication Engineering, SRM Institute of Science and
Technology, Ghaziabad, India

Rashmi G. Shukla
Technology Management Area, MPSTME, NMIMS, Mumbai, India

Shekhar Shukla
Information Management Area, S.P. Jain Institute of Management and Research, Mumbai, India

M.S. Starvin
University College of Engineering, Nagercoil, India

Rajiv K. Tripathi
Department of Electronics and Communication Engineering, National Institute of Technology
Delhi, New Delhi, India

Rakesh Tripathi
Department of Information Technology, National Institute of Technology, Raipur, India

About the Editors

Saravanan Krishnan, PhD, is currently working as a Senior Assistant Professor, Department of Computer Science and Engineering at Anna University, Regional Campus, Tirunelveli, Tamil Nadu, India. He has published papers in 14 international conferences and 24 international journals. He has also written six book chapters and three edited books with international publishers. He has done four research projects and two consultancy projects with the total worth of Rs. 70 Lakhs. He is an active researcher and academician. Also, he is a reviewer for many reputed journals in Elsevier, Springer, IEEE, etc. He has also received an outstanding reviewer certificate from Elsevier, Inc. He is Mentor of Change for Atal Tinkering Lab of NITI Aayog. He has professional membership with ISTE, ACM, IEI, and ISCA. He has 14 years of experience in the academic and IT industries. He has previously worked in Cognizant Technology Solutions, Pvt. Ltd. as a software associate. He has done PhD in the year 2015 and completed ME (Software Engineering) in 2007.

Valentina E. Balas, PhD, is currently a Full Professor in the Department of Automatics and Applied Software at the Faculty of Engineering, "Aurel Vlaicu" University of Arad, Romania. She holds a PhD in Applied Electronics and Telecommunications from Polytechnic University of Timisoara. She is the author of more than 270 research papers in refereed journals and international conferences. Her research interests are in intelligent systems, fuzzy control, soft computing, smart sensors, information fusion, modeling, and simulation. She is the Editor-in-Chief to *International Journal of Advanced Intelligence Paradigms* (*IJAIP*) and *International Journal of Computational Systems Engineering* (*IJCSysE*), member of the Editorial Board of several national and international journals, and evaluator expert for national and international projects. She served as General Chair of the International Workshop Soft Computing and Applications in seven editions 2005−16 held in Romania and Hungary. She participated in many international conferences as organizer, session chair, and member in International Program Committee. Now she is working in a national project with EU funding support from BioCell-NanoART = Novel Bio-inspired Cellular Nano-Architectures—For Digital Integrated Circuits, a 2M Euro project from National Authority for Scientific Research and Innovation. She is a member of EUSFLAT and ACM, senior member of IEEE, member in TC—Fuzzy Systems (IEEE CIS), member in TC—Emergent Technologies (IEEE CIS), and member in TC—Soft Computing (IEEE SMCS). She was Vice President (Awards) of IFSA International Fuzzy Systems Association Council (2013−15) and is a Joint Secretary of the Governing Council of Forum for Interdisciplinary Mathematics (FIM), a multidisciplinary academic body, India.

E. Golden Julie, PhD, is currently working as a Senior Assistant Professor in the Department of Computer Science and Engineering, Anna University, Regional Campus, Tirunelveli, Tamil Nadu, India. She has received PhD degree in Information and Communication Engineering from Anna University, Chennai, India, in the year 2017. She has completed ME degree in Computer Science and Engineering in Nandha Engineering College, Tamil Nadu, India, in the year 2008 and completed BE Computer Science and Engineering in Tamil Nadu College of Engineering, Coimbatore, Tamil Nadu, India. She is having more than 12 years of experience in teaching. She has published more than 34 papers in various international journals and presented more than 20 papers in both national and international conferences. She has written six book chapters by Springer, IGI global Publication. She is acting as an editor for a book titled *Successful Implementation and Deployment of IoT Projects in Smart Cities*. IGI Global in the Advances in Environmental Engineering and Green Technologies (AEEGT) book series. She is one of the editors for the book titled *Handbook of Research on Blockchain Technology: Trend and Technologies* published by Elsevier. She is acting as a reviewer for many journals such as *Computers & Electrical Engineering* published by Elsevier, and received best reviewer certificate, *Wireless Personal Communication* by Springer Publication. She has given many guest lecturers in various subjects such as multicore architecture, operating system, compiler design in Premier Institutions. She has also given an invited talk in one-day state-level technical symposium SPACES 2k18 organized by department of MCA, PET Engineering College, Vellore. She has acted as a jury at national level and international IEEE conferences, project fair, and symposium. She has attended various seminars workshops and faculty development programs to enhance the knowledge of the student's community. Her research area includes wireless sensor adhoc networks, soft computing, blockchain, Internet of Things, and image processing. She is also an active lifetime member of Indian Society of Technical Education.

Y. Harold Robinson is currently working as an Associate Professor and Head in the Department of Computer Science and Engineering, SCAD College of Engineering and Technology, Tirunelveli, Tamil Nadu, India. He has received PhD degree in Information and Communication Engineering from Anna University, Chennai, India, in the year 2016. He is having more than 15 years of experience in teaching. He has published more than 50 papers in various international journals and presented more than 45 papers in both national and international conferences. He has written four book chapters by Springer, IGI Global Publication. He is acting as an editor for a book titled *Successful Implementation and Deployment of IoT Projects in Smart Cities*. IGI Global in the Advances in Environmental Engineering and Green Technologies (AEEGT) book series. He is one of the editors for the book titled *Handbook of Research on Blockchain Technology: Trend and Technologies* published by Elsevier. He is acting as a reviewer for many journals such as *Multimedia Tools and Applications*, *Wireless Personal Communication* by Springer Publication. He has given many guest lecturers in various subjects such as pointer, operating system, compiler design in Premier Institutions. He has also given an invited talk in the technical symposium. He has acted as a convenor, coordinator, and jury at national level and international IEEE conferences, project fair, and symposium. He has attended various seminars workshops and faculty development programs to enhance the knowledge of the student's community. His research area includes wireless sensor adhoc networks, soft computing, blockchain, Internet of Things, and image processing. He is also an active lifetime member of Indian Society of Technical Education.

S. Balaji is currently working as a Professor in the Department of Computer Science and Engineering, Francis Xavier Engineering College, Tirunelveli, Tamil Nadu, India. He has received PhD degree in Information and Communication Engineering from Anna University, Chennai, Tamil Nadu, India, in the year 2016. He is having more than 17 years of experience in teaching. He has published more than 30 papers in various international journals and presented more than 25 papers in both national and international conferences. He has given many guest lecturers in various subjects in Premier Institutions. He has acted as a convenor, coordinator, and jury at national level and international conferences, and symposium. He has attended various seminars workshops and faculty development programs. His research area includes wireless sensor, mobile computing, cloud computing, adhoc networks, network security. His methodology of teaching about TCP&UDP is hosted on Wipro Mission 10X portal. He is also an active lifetime member of Indian Society of Technical Education.

Raghvendra Kumar, PhD, is working as an Associate Professor in Computer Science and Engineering Department at GIET University, India. He received BTech, MTech, and PhD in Computer Science and Engineering, India, and Postdoc Fellow from Institute of Information Technology, Virtual Reality, and Multimedia, Vietnam. He serves as series editor *Internet of Everything (IOE): Security and Privacy Paradigm, Green Engineering and Technology: Concepts and Applications*, published by CRC Press, Taylor & Francis Group, United States, and *Bio-Medical Engineering: Techniques and Applications*, published by Apple Academic Press, CRC Press, Taylor & Francis Group, United States. He also serves as acquisition editor for *Computer Science* by Apple Academic Press, CRC Press, Taylor & Francis Group, United States. He has published number of research papers in international journal (SCI/SCIE/ESCI/Scopus) and conferences including IEEE and Springer as well as serve as organizing chair (RICE-2019, 2020), volume editor (RICE-2018), Keynote speaker, session chair, co-chair, publicity chair, publication chair, advisory board, technical program committee members in many international and national conferences and serve as guest editors in many special issues from reputed journals (indexed by Scopus, ESCI, SCI). He also published 13 chapters in edited book published by IGI Global, Springer, and Elsevier. His researches areas are computer networks, data mining, cloud computing and secure multiparty computations, theory of computer science, and design of algorithms. He authored and edited 23 computer science books in field of Internet of Things, data mining, biomedical engineering, big data, robotics by IGI Global Publication, United States, IOS Press Netherland, Springer, Elsevier, CRC Press, United States. He is Managing Editor in *International Journal of Machine Learning and Networked Collaborative Engineering (IJMLNCE)*, ISSN 2581-3242.

Preface

Blockchain is one of the recent advanced technologies that exploits the decentralized networks and used in many sectors for reliable, cost-effective, and rapid business transactions. Financial services, retail, insurance, logistics, supply chain, and public sectors are now investing in blockchain for their business growth. For example, block chain prevents the double spending in financial transactions without the need of a trusted authority or central server. It is a decentralized ledger platform that facilitates the verifiable transactions between the parties in smart way. Smart contract and cryptocurrency are the main applications of blockchain. Smart contract is a self-executing contract, in which the terms of agreement between the buyer and seller is directly written into the lines of code. Foundation for blockchain concept lies in Markel tree having the leaf nodes, intermediate node, and root nodes. Every individual node contains the hash of its left child and right child, except the leaf node. The root contains the combined hash of its left child and a right child. Every transaction in blockchain is securely communicated using the cryptographic mechanisms such as digital signature, public and private key, and hashing algorithms. Here, system privacy is provided tamper proof, which is extremely hard to make a change in the blockchain architecture. These transactions constitute a block that is immutable and highly secure. Intensive computation should be performed by the miners to solve the hash functions in the block transactions. Miners get reward in the form of digital currency for the construction of consecutive blocks in the network. These blocks are added sequentially in the existing chain, and each user will receive the updated block using peer-to-peer network. Enterprise blockchain requires the properties such as consensus, provenance, immutability, and finality in executing the smart contracts between the different parties in the business chain. Enterprises can implement new business models with reduced cost, time, and risks.

There were many cryptocurrencies used in blockchain transactions such as Bitcoin, Litecoin, Bytecoin, Peercoin, Emercoin, Gridcoin, Ripple, and so on. Currently, three blockchain platforms are used extensively: (1) Hyperledger Fabric (permissioned private blockchain, which allows only the predefined users), (2) Bitcoin, and (3) Ethereum (both are permissionless public blockchains, which mean no restriction for miners). Currently, many platforms exist in the scalable blockchain development such as Ethereum, Hyperledger, Multichain, HydraChain, Open Chain, and IBM Bluemix Blockchain. These platforms help the developers in rapid prototyping and programing using the scripting languages such as python/ruby/javascript. Built-in scripting function languages are used in the implementation of decentralized blockchain applications. Technologies such as decentralized peer-to-peer networking, Internet of Things (IoT), cryptography, consensus protocols are involved in the blockchain ecosystem. Though it offers many advantages to the current business applications, many issues concerning the adaption of blockchain in the real-world scenarios. The core issues in implementing blockchain are scalability and interoperability; governing protocols and security; and trust and privacy. Several use cases and architectural implementation are necessary for the successful deployment in the business sectors. This edited book aimed to explore the blockchain technology (BCT) from fundamental concepts to the research solutions. The book is organized into 17 chapters.

Chapter 1, Ethical Considerations and Issues of Blockchain Technology-Based Systems in War Zones: A Case Study Approach, discusses about the number of actors who have taken initiatives to record, authenticate, and verify identities, but the documentation processes and mechanisms have often been found practically lacking. This issue could be dealt with by providing verifiable universal identity (a global passport) to the population, particularly the refugees and internally displaced persons (IDPs). Blockchain technology-based system (BTBS) has been increasingly used by various actors to identify and provide support, funding, aid, and food to refugees and IDPs. At the same time, the disruptive nature of BTBS as a foundational technology into the defense sector has also been dealt with. Interestingly, the pros and cons of the system have been one that is debatable. In this context, the chapter seeks to discuss the features, deployment, technical innovations, ethical considerations, and issues related to the global use and implementation of BTBS.

Chapter 2, Blockchain-Based Framework for Data Storage in Peer-to-Peer Scheme Using Interplanetary File System, proposes scheme can effectively solve the problem of availability, reliability, and storage of a centralized service provider. In this proposed framework, we are storing the academic record of the student (jpg format) on IPFS (peer-to-peer scheme). To implement the blockchain framework, we have used python (anaconda), python ask, postman, and IPFS.

Chapter 3, Integration of Blockchain and Internet of Things, discusses about security to the data in the internet of things, a lot of challenges arise in the blockchain, such as large set of device management, usage of computational power, network structure, and bandwidth for communication. So, existing blockchain technologies are to be analyzed with importance to the internet of things. The BCT that suits the internet of things applications are adapted and enhanced. The future work will be for integrating the BCT into the network of internet of things in an effective manner.

Chapter 4, A Deep Dive Into Security and Privacy Issues of Blockchain Technologies, discusses the security and the privacy of blockchain along with their impact with regard to different trends and applications. The chapter is intended to discuss key security attacks and the enhancements that will help to develop better blockchain systems.

Chapter 5, Blockchain Implementation for Internet of Things Applications, proposes a modified energy-efficient technique for computation of proof of work, followed by blockchain implementation on resource-constrained IoT for ambient living or ambient assisted living applications for elderly care. This application gathers data from different sensors on a single-board computer for a big hall, which mines for up to a convincing level of difficulty. Later, the computation task is appointed to a few compute-efficient machines to reduce the computation time. The emulation results show that the proposed solution is energy efficient and more suitable to resource-limited IoT applications as compared to standard blockchain implementation. The chapter concludes with the answers to potential research questions aroused in recent works in the domain on efficacy of blockchain for IoT.

Chapter 6, Blockchain Implementation Using Smart Grid-Based Smart City, exhibits the usage and investigation of burden the executives by utilizing the stage of Smart Grids for private buyer setup, a case study of utility in Puducherry under open joint effort. The smart meters introduced in the utility, which have highlights of two-way correspondence, surveying interim and information estimation, stockpiling and show of utilization information to customers, support for variable duty. The chapter focuses on the blockchain smart grid framework in smart cities with its analysis and features.

Chapter 7, Cloud-Based Blockchaining for Enhanced Security, discusses about the growth of social technology and cloud-based access, which has gained more attention. The lack of security is a main problem for accessing the data. To avoid this security issues, cloud-based authentication is done with the help of blockchaining.

Chapter 8, Blockchain Integration With Low-Power Internet of Things Devices, discusses about the missing association toward structure a really decentralized, trustless, and secure condition for the IoT, and in this thesis, we discuss about the current IoT architecture, IoT security threats, brief about blockchain, and correlation of major blockchain innovations that can be coordinated with IoT gadgets and propose an ethereum-based IoT architecture for low-power IoT devices.

Chapter 9, Applications of Blockchain Technology, aims to discuss existing real-time applications across the globe by various organizations and also covers various future use cases, which are under implementation. The chapter emphasizes the benefits this technology provides to various industries, organizations, and society at large, and how this technology can transform the lives. It also aims to explore various concerns and challenges for the implementation of BCT. The chapter also puts these business applications into separate categories to help the developers and practitioners for their future research, development, and application.

Chapter 10, Blockchain-Powered Smart Healthcare System, presents a conceptual model along with research-based implementation instances to define and illustrate a smart healthcare system powered with blockchain. This work provides a holistic perspective in terms of applicability and understanding of blockchain to healthcare and helps in planning and strategizing the usage of BCT.

Chapter 11, Internet of Things and Blockchain: Integration, Need, Challenges, Applications, and Future Scope, evaluates how blockchain helps in strengthening the security in the IoT. After unfolding the basics of blockchain, it explains the most pertinent BIoT applications that focus on the influence of customary Cloud-centering IoT uses.

Chapter 12, Blockchain and Internet of Things: An Overview, gives an overview of IoT, its applications and challenges, BCT, its pros and cons, the convergence of blockchain technology with IoT, the different architectures, evaluation techniques, the advanced scenarios, and use cases.

Chapter 13, Cryptocurrency Mechanisms for Blockchains: Models, Characteristics, Challenges, and Applications, focuses on cryptocurrency mechanisms for blockchains. Cryptocurrencies are a new form of virtual currency, first introduced to the masses with the creation of Bitcoin, developed by Satoshi Nakamoto. They are a more secure and verified form of transaction, which uses cryptographic protocols in a peer-to-peer system generating a distributed ledger. With the growing popularity of blockchains, a large number of cryptocurrencies have been introduced, greatly changing the prospects of blockchains in the domain of finance. Greater attention is now being paid by stakeholders around the world to various mechanisms of using blockchains in cryptocurrencies.

Chapter 14, Overview of Blockchain Technology Concepts, focuses on the perspective of software architecture, and understanding the working model of blockchain helps in the implementation of technology in different domains. In the working model of blockchain, one party initiates the transaction by creating a block. Computers in the network validate the block, and finally, block is added to the chain and stored across the network. Even though BCT gained momentum in the future Internet systems, a number of challenges are to be addressed carefully. Expertise in BCT technology is much needed since we are in nasal stage of this technology. Adoption of BCT offers promising features in various domains, but huge initial cost for the infrastructure is a major concern for the industries. Privacy and security factors also play an important role in the implementation of blockchain technology. Scalability and legal regulations are also posing a major challenge for the adoption of blockchain technology.

xxviii Preface

Chapter 15, Scalability in Blockchain: Challenges and Solutions, covered chain partitioning-based scalability, directed acyclic graphs-based scalability, and horizontal scalability through sharding and have discussed future directives.

Chapter 16, A Vital Role of Blockchain Technology Toward Internet of Vehicles, discusses about blockchain-enabled Internet of vehicle architecture as well as illustrates the application of architecture in various domains. Moreover, the chapter discusses about how it becomes beneficial for the future research. Analysis of security and privacy issues is also discussed in this chapter.

Chapter 17, Need to Know About Combined Technologies of Block Chain and Machine Learning, explains about the combination of machine learning and blockchain, together with issuing revolutionary applications in different ways. These are revolutionizing modern industries by transforming business models, user experiences, and behaviors. In this chapter, we will see how the two are making strides in major industries.

Saravanan Krishnan, Valentina E. Balas, E. Golden Julie,
Y. Harold Robinson, S. Balaji and Raghvendra Kumar

About the Book

The purpose of this edited book is to present the detailed exploration of adaption and implementation of blockchain technologies in the real-world business applications. Blockchain is getting momentum in all the sectors that transact huge numbers every day. This edited book covers all the aspects of the blockchain with a complete 360-degree view spectrum. It can be used at basic and intermediate level for computer science postgraduate students, researchers, and practitioners. It presents the rapid advancement in the existing business model by applying the blockchain techniques. Novel architectural solutions in the deployment of blockchain are the core aspects of this book. Several use cases with Internet of Things and smart city will also be incorporated. The wide variety of topics it presents offers readers multiple perspectives on a variety of disciplines including number of chapters in the edited book.

Key Features

1. Covers the evolution of blockchain from fundamental theories to present forms
2. Diversified business applications of blockchain with use cases
3. Contributors from different parts of the world
4. Cryptocurrency mechanisms for blockchain
5. Presents multiple perspectives such as academia, industry, and research fields
6. Successful implementation of blockchain in cloud/edge computing, smart city, and Internet of Things
7. Emphasizes the advances and cutting-edge technologies throughout
8. Focused on different tools, platforms, and techniques
9. Smart contracts tools and consensus algorithms
10. It facilitates supply chain management and tracking of goods and delivery

CHAPTER 1

ETHICAL CONSIDERATIONS AND ISSUES OF BLOCKCHAIN TECHNOLOGY-BASED SYSTEMS IN WAR ZONES: A CASE STUDY APPROACH

Ramnath Reghunadhan

Research Scholar, Department of Humanities and Social Sciences, Indian Institute of Technology Madras (IITM), Chennai, India

1.1 INTRODUCTION

Blockchain technology is considered to provide a revolutionary technological paradigm, a shift in every form of transactional relations and even behaviors as well. The main factor entailed to be creating this transitional, transactional, and transparent nature in the blockchain technology is the creation of blocks of information (or data) that is accessible to all stakeholders in a decentralized manner. A number of actors have taken initiatives to record, authenticate, and verify identities, but the documentation processes and mechanisms have often been found practically lacking. This issue could be dealt with by providing verifiable universal identity (a global passport) to the population, particularly the refugees and internally displaced persons (IDPs). Blockchain Technology-Based System (BTBS) has been increasingly used by various actors to identify, provide support, funding, aid, and food to refugees and IDPs. At the same time, the disruptive nature of blockchain as a foundational technology into the defense sector has also been dealt with. Interestingly, the pros and cons of the system have been one that is debatable. In this context, the chapter seeks to discuss the features, deployment, technical innovations, ethical considerations, and issues related to the global use and implementation of blockchain. The use of blockchain technology, one which have been increasingly used by various actors to identify, from providing support, funding, aid, and food to refugees and IDPs to improving weapon systems reliance and enabling more secure approach to national security, blockchain technology has the potential to be a major disruptor in the human society [1–4].

A specific focus on the disruptive nature of blockchain (both positive and negative) as a foundational technology for governance in resolving issues related to refugee and IDPs as well as the defense sector has also been dealt with in this chapter. The focus is both on the positive and negative disrupting tendencies toward governance in the context of Syria and for the US Defense industry. Interestingly, the pros and cons of the BTBS have been one that is debatable, one which has been dealt in much detail in relation to technological aggregation; ethical and moral dilemmas as

Handbook of Research on Blockchain Technology. DOI: https://doi.org/10.1016/B978-0-12-819816-2.00001-0
© 2020 Elsevier Inc. All rights reserved.

1

2 CHAPTER 1 ETHICAL CONSIDERATIONS AND ISSUES OF BLOCKCHAIN

well. In this context, the chapter seeks to discuss the features, deployment, technical innovations, ethical considerations, and issues related to the global use and implementation of the blockchain technology and related system. The digital technologies have transformed warfare, particularly during the last decade of the 20th century. The chapter undertakes qualitative research by undertaking two case studies: first, BTBSs for dealing with the issues and crisis related to refugee and IDPs in Syria and second, BTBSs in the US Defense industry in War Zones.

1.2 METHODOLOGY

The study undertakes qualitative research methodology, entailing on the case study approach, and further on content analysis as well as historiographic approach for contextualizing the research and origins of blockchain technology, Syria and the Civil War. This adheres to the overall framework of the chapter. Moreover, an extensive literature review, as well as intensive discussions and interview of various experts, scholars, and academicians and other stakeholders in this field, has been undertaken.

1.3 BLOCKCHAIN TECHNOLOGY

Blockchain technology is an emerging technology with a lot of promises in various sectors of the society that is continuously connected in this period of "digitalized globalization" or "Globalization 2.0." It has created various cost-effective solutions to address the various aspects of security and privacy but enabling decentralized access to the individual end user as well. It is also an underpinning technology that provides functional characteristics to Bitcoin and other cyptocurrencies. It could generally be considered to have "a network of computers that all have the same history of transactions." Further, the "transactions are recorded as blocks on network held... [And accessible to] multiple people at the same time. [Moreover, every] ten minutes of transactions become [a single] block." Thus, the database is decentralized, accessible to, and validated by each of the stakeholders within the network of computers. Moreover, accessible does not ensue in options to alter the history (of transaction), information or data on the blocks, thus rendering it near impossible to defraud any of the stakeholders. Thus, blockchain acts more like an interface for various software codes, computer, and technological applications to run. The absence of any central authorizing entity is the beauty of the technology of blockchain [5].

For the first time in scholarly academic literature, the functionality of chain of blocks was explained by research scientists in Bell Communication Research (Bellcore), Stuart Haber and Scott Stornetta through a paper titled *How to Time-Stamp a Digital Document*, 3 years before they founded their own company, called Surety. The authors focused on various aspects of privacy, bandwidth, storage, competence, and trust of the time-stamping service (TSS). In a trusted TSS, the paper suggested two improvements, first, the use of hash, by making "use of a family" of cryptographically secure *collision-free hash functions* that is, h: $\{0,1\}^*$ tending to $\{0,1\}^l$, whereby l is a fixed length. And second, the use of digital signatures, through the algorithm called *signature scheme*, though it was already written by W, Diffie and M.E. Hellman in the 1976 article

"New directions in cryptography" published in *IEEE Transactions on Information Theory*, and M. O. Rabin's article "Digitalized signatures" published as part of the edited book titled *Foundations of Secure Computation* in 1978 [6, pp. 99–111].

The company Surety "utilized the Linked Time-Stamping Authority (TSA) framework," was considered to be the first commercial blockchain service provider in the world [7]. These events had a greater influence on the publication of the Whitepaper (in 2008) on the working of Bitcoin, by (a reportedly anonymous) person or a group of people known as Satoshi Nakamoto. In 2008, 1 year before the release of the Bitcoin software, the Whitepaper explained the workings of a peer-to-peer (P2P) electronic cash system for a digital currency called Bitcoin, based on the digital platform provided by the blockchain technology. It can be considered as a data structure that used the method of organizing information and/or knowledge through a distributed ledger that is shared among a multitude of people and places at the same time. It enables transaction to be made that avoids the need to have a traditional concept of "legitimate" central supervisory authority; through reliance on "innovative combination of distributed consensus protocols, cryptography" and "in-built economic incentives based on game theory," thereby ensuring greater reliance, transparency as well as "integrity" to the transactions being conducted. Thus, blockchain technology is different from the traditional concept of "legitimate" central supervisory authority. It is based on "innovative combination of distributed consensus protocols" and "cryptography," enhancing reliance, transparency, and integrity to the digital transactions [8–10].

1.4 THE PROCESS INVOLVED

The first transaction that is added to a block is known as "generation transaction or coin-base transaction." More importantly, the extent of success in regard to the execution of the transaction enables the formation of a "peer-to-peer Bitcoin network," and once it undergoes verification through the process known as "mining" it becomes permanently imprinted in the ledger. This is created by a miner or network of "miners," who runs a "mining rig" specialized computers and related hardware used for mining bitcoins [11]. Fig. 1.1 shows sample model of transactions taking place in blockchain network.

The miner then competes with each other approving the unconfirmed transactions (which are temporarily stored but not validated until confirmed) and later adds them as a "new block." In a blockchain, each block comprises a unique hash-based proof-of-work, which is attached to each transaction and can be identified as a hash value. This hash is produced after each transaction and is reminiscent of a digital fingerprint. Once the verification is done, it becomes permanently imprinted through the process known as "mining." This is created by a "miner" or "network of miners," which is only confirmed when validated. The key security feature of blockchain technology is that thus preventing any "adjustment" or malpractice to take place without informing of all the stakeholders. Now each computer or electronic device connected to the blockchain's network is called a "node." Each node has a copy of the entire ledger and works in coordination with other nodes. If any node disappears or the server goes down, the information is still retained [8,9,11,12]. There are various models of networks in blockchain like the Merkle tree is a network in tree form, "constructed by hashing paired data (the leaves), then pairing and hashing the results until a single

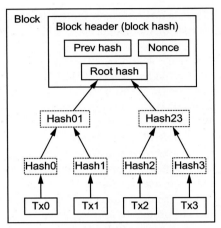 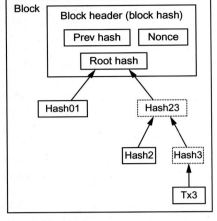

FIGURE 1.1

Transactions hashed in Merkle tree in a block and after pruning the first three transactions from the block.

Courtesy: S. Nakamoto, Bitcoin: A Peer-to-Peer Electronic Cash System, Available from: <https://bitcoin.org/bitcoin.pdf>, 2008.

hash remains, [which is] the Merkle root." A valid Merkle root is considered to be "a descendant of all the hashed pairs in the tree," that is, the root node of the Merkle tree. In cryptocurrencies like the Bitcoin, the paired data "are almost always transactions from a single block" [13]. Fig. 1.2 shows an illustration of Merkle tree, which simulates the creation of a chain of blocks as the inherent process within blockchain technology.

The later development and inclusion of a new method of storing data in a more accurate and secure manner, one which was entailed through the Merkle trees. It was invented and coined by Ralph Merkle, an American computer scientist, who is also considered to be among "the father of modern public-key cryptography." This transformed the functionality of cryptocurrencies including Bitcoin, in a much more effective manner. As in Fig. 1.2, the primary (bottom) transactions *L1*, *L2*, *L3*, and *L4* are known as leaves. Each leaf gets a transaction id or "hash" (by miners), which later is paired and "hashed again" as *Hash0-0 (hash L1)*, *Hash0-1 (hash L1)*, *Hash 1-0 (hash L1)*, and *Hash 1-1 (hash L1)*, thereby creating another level on the Merkle tree. Further, *hash {Hash 0-0 and Hash 0-1}* becomes *Hash 0*, while *hash {Hash 1-0 and Hash 1-1}* becomes *Hash 1*. The process keeps on going whereby elimination of hash takes place until only one hash remains "at the top of the tree," that is, Top Hash, which is *hash {Hash 0 and Hash 1}*. The Top Hash that remains at the top of the Merkle tree is also known as Merkle root. These have ensued for the modern-day nature of "cryptocurrencies to be open-sourced, independently verifiable, secure, and highly accurate" [14].

The key security feature of blockchain is that any modification of a parent block causes a change in all the subsequent blocks of the chain, and thus would automatically force recalculation/change of all "subsequent blocks to be recalculated and for each subsequent block new proof-of-work would have to be provided." This would require computational power that would surpass the

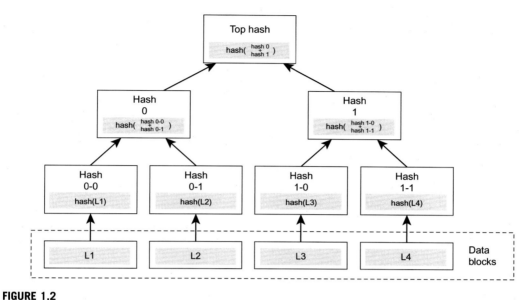

FIGURE 1.2

Illustration of a Merkle tree simulating the creation of a chain of blocks as in the blockchain technology.

Compiled from Wikipedia, Merkle tree, Available from: <https://en.wikipedia.org/wiki/Merkle_tree, 2019>.

capability of an individual node even though they (could) work together. Thus, manipulation or changes by an individual node or a small group is prevented through the "consensus mechanism," whereby consent of the majority of stakeholders is required for any permanent change to be stored. The consensus mechanism (set of rules) based on which the network uses to verify each transaction and agree on the current state of the blockchain is ensured. Once a block of data is recorded on the blockchain ledger, it becomes extremely difficult to change or move, unless the majority of the nodes reach a consensus to examine and modify. Through the use of cryptography, each participant (stakeholder with verifiable access) is able to change the ledger, which is only temporary unless verified by a majority of the stakeholders, thus providing for the decentralization of decision making in governing matters [8,9,11,12]. The focus is on two case studies: first, BTBSs for dealing with the issues and crisis related to refugee and IDPs in Syria and second, BTBSs in the US Defense industry in War Zones.

1.5 SYRIA

Until the year 1516, the Ottoman Empire ruled over the region that encompassed Syria. This changed in the earlier of the 20th century when the Ottoman became part of the First World War. To the end of the First World War, on October 1, 1918, the Desert Mounted Corps of Emir Feisal, with the support of the British, attacked the retreating Ottomans, and in the process, occupied the city of Damascus (modern-day capital of Syria) [15, p. 69]. The end of the First World War in the

6 CHAPTER 1 ETHICAL CONSIDERATIONS AND ISSUES OF BLOCKCHAIN

month of November saw the later institutionalization of the Paris Peace Conference in June 1919. Following the end of World War and the defeat of Germany and the Ottoman Empire, Emir Faisal and the Allied Powers entailed the self-rule for the Arabs and the creation of Syria. This ensued through the Versailles Peace Conference, and further meetings in this regard went on for 1 year. This was formalized and later agreed upon within the League of Nations in January 1920 [16].

On July 2, 1919, the General Syrian Congress comprising of 85 members passed resolutions in the city of Damascus that sought the independence of Syria and Palestine, and recognition under the sovereign rule of Faisal. The British and the French viewed this as a repudiation of the (secret) Sykes—Picot Agreement (between the English and the French), the Balfour Declaration (between the English and the Arabs), and a huge bottleneck toward the plan of major powers to partition Syria and possible potential issues with regard to the Jewish leaders on promise of the formation of a Jewish State [17]. Further discussions and treaties went on for another 4 years, with the conclusion of treaties with Turkey (successor of the Ottoman Empire), Hungary [18]. Two months later, on March 8, 1920, Emir Feisal was proclaimed as the ruler of Syria by the National Congress, located in the city of Damascus. Thus, the Syrian Arab Republic, which was earlier home to the four-century humungous empire of Ottomans, came into existence within the modern Westphalian nation-state form. It covered an area of 185,000 km^2, was the end product of the First World War [19]. Fig. 1.3 shows the division of autonomous regions during 1922.

During the period of formation Syria extended from the Taurus Mountains (also called *TorosDağlari* in Turkish) that borders Turkey to the North [20], the Mediterranean Sea (a sea of the Atlantic Ocean) that divides Eurasia and Africa to the West [21], and the region near the Sinai Peninsula (also called *ShibhJazīratSīnā* in Arabic) that borders political boundary of Egypt to the South [22]. Interestingly, just 3 months after the coronation of Faisal, in the San Remo Conference in June 1920, the Syria—Lebanon region is taken over by the French, with the support of the British, and Faisal goes to exile. In the next few years, under the French Mandate, Syria is divided into different autonomous regions. Accordingly, separate areas like the Alawite State (with Latakia as capital) and Greater Lebanon (with Beirut as capital) respectively as West and South—West regions were created. Further, the State of Aleppo (with Aleppo as capital) in the North, the State of Damascus (with Damascus as capital), and Jabal al-Druze State (with As-Suwayda as capital) in the South were created. Within the next few years, there began various uprisings against the French rule, which became further entrenched since 1925. In the next 3 years, for the first time, elections were held for the constituent assembly that led to the drafting of the Syrian Constitution, all of which was rejected by the French. The French position changed within almost a decade, and in 1936 they agreed to Syrian independence and dissolution of autonomous regions, though military was maintained and economic relations of Syria was still retained by the French in 1936, also controlling Lebanon [23].

1.6 SYRIA: THE BEGINNING OF MODERN NATION-STATE

In 1943 Shukri al-Kuwatli becomes the first elected President of Syria, though full independence of the Syrian is only achieved in 1946. One year later, the Arab Socialist Baath Party is founded by Michel Aflaq and Salah-al-Din al Bitar and was part of repeated coups within the civilian

1.6 SYRIA: THE BEGINNING OF MODERN NATION-STATE

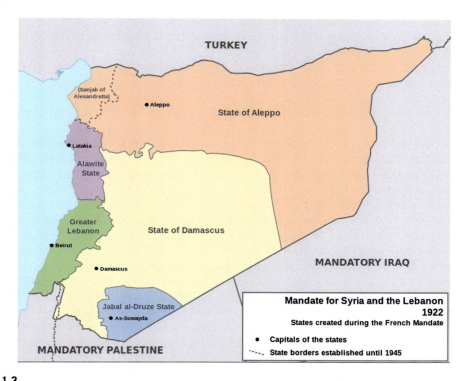

FIGURE 1.3

Division of autonomous regions created during the French control of Mandate for Syria and Lebanon.
University of Central Arkansas, French Syria (1919-1946), Available from: <https://uca.edu/politicalscience/dadm-project/middle-eastnorth-africapersian-gulf-region/french-syria-1919-1946/>, 2019.

administration and government. Almost a decade later, al-Kuwatli returns to power in the year 1955, trying to solidify Syria's bilateral relations with its southern neighbor Egypt. Three years later, the United Arab Republic (UAR) (*Al-Jumhūrīyah al- Arabīyah al-Mutta idah* in Arabic) was formed under the leadership of Gamal Abdel Nasser, the President of Egypt in collaboration of Syria on February 1, 1958. This union ended on September 28, 1961, when a military coup in Damascus, by a group of Syrian army officers fragments UAR and leads to its dissolution [24]. Almost 2 years later, in the month of March, the Baathist military group seize reign of the Syrian State. Further, in February 1966, an internal coup against the Baath leadership took place under the leadership of Salah Jadid. Under the new leadership, Hafez al-Assad takes over the defense portfolio.

In June 1967, the Golan Heights, a strategic location, was captured from Syria by Israel as a consequence of the Six-Day War or June War, one which entails a formative event in the history of Syria. Almost 3 years later, Assad imprisons the Baathist leader Salah Jadid, and through *coup d'etat* becomes President of Syria by overthrowing the then President Nur al-Din al-Atasi. Further, he amends the constitutional requirement for the President of Syria to be from the Muslim

8 **CHAPTER 1** ETHICAL CONSIDERATIONS AND ISSUES OF BLOCKCHAIN

community. The enmity between Syria and Israel persisted too much more into the future, with Assad feeling the urge to avenge the loss of prestige and Golan Heights as well [25]. Almost 10 months after a secret agreement with Egypt, in October 1973 Syria (along with Egypt and Jordan) was again engaged in war with Israel, but was defeated again as well [1]. Three years later, the Syrian army becomes an active party to the Civil War in Lebanon, supporting the Lebanese Maronite Christians, who are one of the largest Eastern-rite communities of the Roman Catholic Church [4], and who along with other Christians constitute up to 40% of total population in Lebanon [26].

At the beginning of the 1980s, the rise of the Muslim Brotherhood, in the aftermath of Islamic Revolution in Iran changed the societal dynamics in Syria. This was further exacerbated with conflict between Iran and Iraq, with Syrian backing toward Iran. Interestingly, the tensions in the region deteriorated when in December 1981, Israel formally annexed Golan Heights. Within the next 2 months, in city of Hama, a place located on the banks of River Orontes 305 km north of Damascus, 10,000−20,000 were reportedly killed by the army. Further, the situation in the region further deteriorates, when Israel in the aftermath of invading Lebanon, attacks Syrian Army as well. Two years later Rifaar al-Assad, the older brother of Hafez al-Assad, who played a central role in the repression of 1980s Islamic insurgency takes charge as one of the three Vice Presidents of Syria on March 11, 1984. In 1990 Syria for the first time joins the US-led coalition acting against Iraq's invasion of Kuwait, thereby improving bilateral relations with the United States and even Egypt. In October, Syria was a participant in the first post−Cold War conference jointly organized by the United States and USSR on the issue of peace in Middle East (modern-day conception of West Asia) at Madrid, Spain, though its initiatives to get back Golan Heights did not bear fruit. Three years after the conference, there was a thaw in Syria's international relations, while the internal turmoil was deteriorating. The President's likely successor, the elder of the two sons, was reportedly killed in a car accident. A few years later, the brother of the President was dismissed from the post of Vice President as well, on the grounds of treason and undertaking activities with the intent to overthrowing the President [27].

1.7 SYRIA: UNDER BASHAR AL-ASSAD REGIME

In the aftermath of the death of Hafiz al-Assad (on 10th June), the Syrian Parliament voted to lower the minimum age for presidential candidates, so as to enable Bashar to become eligible for the office of the President. Within 1 month the younger son of the former president, Bashar al-Assad was elected with more than 97% of the vote, as the President of Syria at the age of 34 on July 11, 2000. Concomitantly, he was also selected as the leader of the Ba'ath Party and the commander-in-chief of the Armed Forces in Syria [28]. He declared looking at options towards "creative thinking" entailing the need for "constructive criticism" inducing grassroots movements towards "transparency" and "democracy" in Syria. Five months into the succession of the London-educated younger Assad ordered the release of hundreds of political prisoners. This was perceived as a move away from totalitarian regime, especially by many of the human rights activists, both foreign and domestic, outside and inside the country. This reversed the next year when Assad gradually undertook repressive measures to clampdown on protests. In April 2001, he bans the Muslim Brotherhood,

1.7 SYRIA: UNDER BASHAR AL-ASSAD REGIME

9

while exiling many of its leaders. In August 2001, the release of political prisoners was completely stopped and more people were jailed in the process. Human Rights Watch has often in its reports alleged that since ascension of Bashar al-Assad, the Syrian prison was filled "with political prisoners, journalists, and human rights activists (Annex 1 lists 92 political [as well as] human rights activists detained...)" [29]. Five months later, many of the members of the People's Council of Syria are detained, along with other political activists, though intermittent release and amnesties were given from time to time [30].

On September 22, 2001, a *Decree No.50/2001* (Press Law) was enacted to provide the Syrian Administration with greater teeth to control content in print and online media [29]. Around 4 months after George W. Bush's State of the Union address on the *Axis of Evil* [14], in May 2002, the Under Secretary of State from US Mr. John Bolton in his speech *Beyond the Axis of Evil*, included Syria along with Cuba and Libya in the list of nations to be targeted by the US Administration. Syria was thus categorized as among the "rogue states" that was reportedly "active [in] attempting to develop weapons of mass destruction [WMD]" and are considered as an existential threat to the United States [31]. In January 2004, the Syrian President visits Turkey for the first time, thereby thawing the bilateral relations between the two countries. Four months later, the US Administration imposes economic sanctions on Syria, in line with the sanctions entailed to the *axis of evil* list of countries to prevent activities related to acquisition, production, and/or distribution of weapons of mass destruction as well as support to terrorism. The relations with the United States further worsens in 2005, when the Prime Minister of Lebanon Rafik al-Hariri was killed in a car bombing incident in Beirut, the capital and the largest city in Lebanon. The suspicions against Syrian administration diminishes the image of Syria in the international stage, deteriorating its relations with even neighboring countries in West Asia and international organizations like the UN. In April 2005, Syria completely withdrew its army from Lebanon, ending what many considers as the nearly three-decade military occupation of Lebanon [17]. The regional peace initiatives in West Asia were once again bottlenecked when Israel carried out an aerial strike on Al-Kubar (a nuclear facility), near the Deir al-Zor region (eastern Syria) on September 6, 2007.

The relations with Iraq were thawed for the first time since the Gulf War, when in November 2006 the diplomatic relations were restored. This was mainly because Syrian Administration had good relations with then leaders of Iraq including Jalal Talabani, Nouri al-Maliki, other Iraqi Baathist leaders, militia and jihadist groups, and interestingly even Shia groups that opposed the later Maliki Administration [32, pp. 3–24]. Almost 5 months later, the European Union (EU) restarted the dialogue process with Syria, through diplomatic channels. One month later, the US House of Representatives Speaker met the Syrian President for the first time in history. This was followed by the meeting between the US Secretary of State, Condoleeza Rice, and the Syrian Foreign Minister Walid al-Muallem at a regional conference on May 4 in Egypt [33]. The relations were further exacerbated when the US administration alleged that Syria was constructing a secret nuclear reactor, which again was destroyed by Israel again in 2008 [30]. The meeting between Bashar al-Assad and the French counterpart Nicolas Sarkozy in mid-2008 signaled an end in diplomatic isolation of Syria from West Europe by bringing out a joint Franco-Syrian statement expressing the desired peace in the region. France expressed its willingness to act the role of mediating direct talks between Syria, Israel, and the United States. In the joint statement it was stated that "...France, together with the [US, wishes to contribute for]... a future peace agreement between Syria and Israel, both to the direct peace talks and to the implementation of

10 **CHAPTER 1** ETHICAL CONSIDERATIONS AND ISSUES OF BLOCKCHAIN

the peace agreement." This was a reset of relations between the European Union and Syria since the assassination of Lebanese PM Hariri 3 years prior. This thawed the relations between Syria and Lebanon as well, paving way to establishment of diplomatic relations in October 2008. Further, the process was welcomed and supported by the Sarkozy administration in the aftermath of the talks in July 2008 [10].

Under the Obama Administration's initiatives to entail outreach activities to Syria, in March 2009, the then acting Assistant US Secretary of State Jeffrey Feltman (who later on became the Assistant Secretary of State for Near Eastern Affairs) and White House National Security aide Daniel Shapiro (who later on became the Assistant Secretary of State for Political—Military Affairs) [34,35] undertook the "highest-level bilateral meeting in years... [as part of the] strategic realignment" in diplomatic relations between the United States and Syria, meeting Syrian Foreign Minister Muallem. The reciprocity was increasingly seen from the Assad regime, though the Syrian spokesperson did set preconditions like dropping of the "Syria Accountability Act [that] sanctions and... [removes] Syria from the list... [known as] State Sponsors of Terrorism" [3].

Under the Assad regime, Syria undertook measures to liberalize its economy providing more accessibility to foreign companies, one which it undertakes even earlier than the other "neo-patrimonial rentier states" in the West Asian region. Raymond Hinnebusch has argued that even in early the 1990s Syrian regime's economic policy imperatives were determined by the long-term need to entail balancing rationality, at both political and economic paradigms. Hinnebusch argued that the "systemic requisites" could be understood through the use of the intervening variables like class interests and the political participation in Syria, which was hitherto heard of in any other country in the West Asia [36, pp. 249–265]. But the situation changed in 2 months when the United States renewed the sanctions against Syria due to the alleged support to the terrorist groups like the Hezbollah in Lebanon [32, p. 23]. But the relationship contour between European Union and Syria went detour when the alleged construction of secret nuclear power reactor by the Assad regime was further attested to by the International Atomic Energy Agency in 2011, with further allegations that assistance from North Korea has been taken as well [37]. Moreover, the domestic situation has not changed much since the "state of emergency enacted in 1963, remains in place, and the government continues to rule by emergency powers" [29].

1.8 CASE STUDY 1: REFUGEES AND INTERNALLY DISPLACED PERSONS IN SYRIA SINCE 2011

In the postwar era, the Syrian Civil War has become the bloodiest, prolonged, and protracted conflict in the 21st century. It inherently includes various dimensions, increasing complexity in the pattern, both in the defense sector and in the ensuing displacement-cum-migration crises, on a scale that the world has hitherto witnessed. This has pretty much caused a huge (direct and/or indirect) paradigm shift in the socioeconomic and political patterns and processes around the World. Although the political crisis has induced large-scale migration, including mass exodus, which many have described as the "The Great Humanitarian Crisis," the defense sector is slowly imbibing various technologies including artificial intelligence and digitalization into it. The Syrian Civil War has resulted in the loss of nearly half a million Syrian lives, injuring over 1 million injured and many

1.9 ARAB SPRING AND THE EMERGENCE OF THE SYRIAN CIVIL WAR

millions who are either IDPs or have become refugees, besides degrading societal conditions and instability. An estimated 5.6 million Syrians have been forced to become refugees, around 6.2 million accounts for IDPs. Globally, almost one-sixth of the world population are without provable identity, while 20 trillion USD of unowned capital lay around. A prominent issue in resolving The Great Humanitarian Crisis is to legitimately identify the refugees and IDPs. But a large number of the refugees and IDPs lack evidence or verification documents like identity cards, property deeds, passports; most of which would have been lost while fleeing for their lives, amidst conflicts or bombardments. This creates a number of hindrances for their return; reentry; identity verification, validation, certification, and rehabilitation; repatriation and resettlement as well. This also leads to issues of increasing illegal entry points, lack of property restitution and/or compensation, along with the lack of livelihood opportunities or overdependence on aid. Moreover, issues like trust deficit, leakages in distribution and accessibility to goods and services exacerbate the situation [38,39, p. 1].

1.9 ARAB SPRING AND THE EMERGENCE OF THE SYRIAN CIVIL WAR

A series of prodemocracy movements arose in various Muslim-dominated countries in West Asia and North African (WANA) region countries. This mainly included Tunisia (Zine El Abidine Ben Ali's rule), Morocco (King Mohamed VI's rule), Syria (Bashar al-Assad's rule), Libya (Colonel MuammerGaddafi's rule), Egypt (Hosni Mubarak's rule), and Bahrain (King Hamad al-Khalifa's rule), but had smaller offshoots in other countries as well [40,41]. The beginning of the movement occurred during the spring, and since it began in Tunisia in December 2010, it was also known as Jasmine Revolution. Interestingly, these movements were considered to have intended toward an upheaval against political and social systems in each of these countries. But this movement reportedly increased political instability and societal fragmentation, though arguably increased cohesion in terms of social movements through nongovernmental channels [42].

The Syrian Civil War that began in the year 2011 is considered to be one of the most protracted conflicts in the 21st century, but with the involvement of multiple actors. In May 2017, this conflict accounted for half a million deaths, over a million injured and more than 12 million Syrian fleeing their homes. Those who fled their homes either became considered as IDPs and if somehow crossed the borders of the country they became categorized as refugees. In the post−Cold War era, the Syrian civil war has become the bloodiest, prolonged, and protracted conflict in the 21st century. It inherently includes various dimensions, increasing complexity in the pattern, both in the defense sector and in the ensuing displacement-cum-migration crises, on a scale that the world has hitherto witnessed. This has pretty much caused a huge (direct and/or indirect) paradigm shift in the socioeconomic and political patterns and processes around the World. The protests against the Assad regime were initially peaceful during the early March 2011, but later transformed into very "violent, brutal attacks, all culminating into the bloodiest conflict in the 21st century" [39, p. 1]. This is one of the rare conflicts wherein multiple State and non-State actors were perceived to be equally prominent and decider in flaring up and in dousing the intensity of the conflict. Moreover, it also had an impact on the global world order and political systems, which saw an increased tendency toward nationalism and trade protectionism.

12 **CHAPTER 1** ETHICAL CONSIDERATIONS AND ISSUES OF BLOCKCHAIN

1.10 **MAJOR ACTORS IN THE CONFLICT**

Though initially some militia and armed groups were considered to be the prominent stakeholders during the conflict, this was completely transformed when Islamic State (IS) came into the fore. Currently, it is believed to have thousands of armed militia groups with more than hundreds of thousands of fighters, including women [43]. The major rebel coalitions include:

- The Supreme Military Council of the Free Syrian Army, also called Free Syrian Army (FSA) led by Brigadier General Salim Idris, and their affiliate groups like Martyrs of Syrian Brigades, Northern Storm Brigade, and AhrarSouriya Brigade [43].
- The Islamic Front led by Ahmed al-Sheikh (Suqour al-Sham) and their affiliate groups like HarakatAhrar al-Sham al-Islamiyya, Jaysh al-Islam, Suqour al-Sham, Liwa al-Tawhid, Liwa al-Haqq, KataibAnsar al-Sham, and Kurdish Islamic Front [43].
- The Syrian Islamic Liberation Front (SILF) was formed based on alliance by more nearly two dozen rebel groups that include Farouq Brigades, the Islamic Farouq Brigades, Liwa al-Tawhid, Liwa al-Fath, Liwa al-Islam, Suqour al-Sham, and the Deir al-Zour Revolutionaries Council in 2012−13. This loose coalition contains ideological variations from ultraconservative Salafist to moderate Islamist, but have recognized the FSA as a friendly group to work together with [43].
- Other independent groups include Ahfad al-Rasoul Brigades, Asalawa al-Tanmiya Front, Durou al-Thawra Commission. TajammuAnsar al-Islam, Yarmouk Martyrs' Brigade, National Unity Brigades [43].
- The Jihadist groups include the Islamic State of Iraq and Levant (ISIS or IS), Al-Nusra Front, Jaysh al-Muhajirinwa al-Ansar, and the likes [43].
- The Kurdish groups include Popular Protection Units (YPG), an armed wing of the political party of the Kurds like the Democratic Unity Party (PYD). PYD is an affiliated party of the Kurdistan Workers' Party (PKK) that runs the de facto autonomous Kurdish-controlled region in Northeast Syria [43].
- The United States carried out its first airstrike against a Syrian-government controlled airbase only on April 7, 2017; though it was engaging with various actors well before that through arms supply, personnel training and humanitarian aid to various actors in conflict with the Assad regime. The US military action on the end of the first week of April saw nearly five dozen "missile strikes," an action it surmised the aftermath of the deadly chemical gas attack in Idlib, a province in northwestern Syria. Moreover, since 2013 Central Intelligence Agency (CIA) has reportedly spent nearly half a billion dollars to train more than dozens of fighters and groups [39,44,45, pp. 2−3].
- Russia began its "bombing campaign against" in support of Assad regime against rival militia groups in Syria on September 30, 2015. It was reported that about "444 combat sorties" targeting around "1500 targets" were undertaken in the second week of February 2016. According to Russian official position, these bombings mainly targeted ISIS or ISIL and other alleged "terrorist groups" like Al-Nusra Front and others [46]. Russia has an Air Force base in Latakia (in Western Syria) and a naval based in Tartus (along the Mediterranean coast). They have been heavily involved in peace talks with all major actors in the conflict [47].
- Turkey is one of the neighboring countries that was affected by the Syrian conflict, mainly due to the initially porous border (911 km) and the perceived threats from the Kurdish militia. Turkey

1.10 MAJOR ACTORS IN THE CONFLICT   13

has participated mainly through military and diplomatic means, being some of the few countries to engage in two-way cooperation with both the United States and Russia as well. It is a major actor that deals with terrorist organizations like ISIL, Al-Nusra, and has been a major stakeholder for the creation of the *Global Coalition Against DEASH* in September 2014. Since July 24, 2015, Turkey has officially been part of the military activities in Syria, and on August 24, 2016, it initiated the *Operation Euphrates Shield* (OES). It considers YPG/PYD as terrorist groups and thus is opposed to US support to the Kurds, both in the form of arms and funding as well [48,49].

- Iran reportedly supported the Assad regime from 2012, providing support in form of military support, training, and tactical intelligence. Being a major power in the Middle East, with nearly half a million active personnel in its defense forces, it has a history of fighting long and protracted wars, especially against Iraq and indirectly against Israel. The Islamic Revolutionary Guard Corps (IRGC) and the Shia-affiliated militia groups are Iran's assets supporting the Assad regime. Syria has reportedly supported weapons transport and transit from Iran to Hezbollah, opposing Israel's rule over Palestine as well as its military activities in Syria, and actively participating in Lebanon as well. It has been a major partner of Russia in the region, setting aside divergences with Turkey to support peace talks and processes in the region. The rivalry between Iran and Saudi Arabia in West Asia tends to be reminiscent of the old Cold War rivalry of United States—USSR [47].
- Additionally, various coalitions and alliances between 60 countries are being led by the United States, along with various militia and rebel groups, though the scenario keeps changing each year. This has prompted to breaking down of some alliances, say to the ISIL, and the coalition of rebels to new bonding with the Kurdish forces and other local militias as well [47].

A number of peace talks have been proposed so far by multiple actors and groups but has been mostly met with defiance with either or both of the parties. In February 2012, Kofi Annan was appointed as the Special Envoy for peacemaking in Syria, jointly by the UN and the Arab League [50]. In March 2012, the UN Special Envoy Kofi Annan put forward a "six-point peace plan." The proposal called for Syrian authorities to cooperate with the Envoy to provide for "inclusive Syrian-led political process," with the appointment of "an empowered interlocutor" on the invitation of the Envoy, "an effective United Nations-supervised" ceasefire in every form and by every actor with the intent toward protecting civilians and prevent destabilization of the country. Further, the "timely provision of humanitarian assistance to all areas affected," early and speedy release of detainees and (peaceful) political prisoners, "freedom of movement" in Syria especially for journalists through "nondiscriminatory visa policy," and finally, "freedom of association and the right to demonstrate peacefully" as inscribed under the laws of the country. Additionally, "a comprehensive ceasefire" within 2 days with the rebel forces and an endorsement toward "a military withdrawal" by 10th of April 2012 was agreed upon as well. It was accepted 17 days later on March 27 by the Assad regime but failed to fructify on the ground level [51]. This prompted the initial UN-supported talks on June 2012, in Geneva [52]. The United Nations Supervision Mission in Syria (UNSMIS) was put in place, but purportedly with just 300 military observers (which was headed by Major General Robert Mood of Norwegian Armed Forces) and a vague mandate, which was altogether lackadaisical in defusing the crisis [50].

In September 2013, the United States and Russia jointly announced the Geneva II conference on Syria to be convened in Switzerland, with three key objectives. These include establishment of

14 CHAPTER 1 ETHICAL CONSIDERATIONS AND ISSUES OF BLOCKCHAIN

first: a "permanent international contact group containing key regional players," second: "to prioritize urgently needed humanitarian assistance," and third: to "legitimize the idea of intra-Syrian negotiations." But it was criticized as a "hopeless exercise" due to its inability to provide a concrete proposal, in-fighting between various actors internationally and domestically as well as the largely polarized camp formation between Saudi Arabia and Iran in the conflict [53]. Later on, the Geneva III peace talks were organized to find an amicable solution to the political crisis in Syria in an inclusive manner. The Assad government and the High Negotiations Committee (HNC) representing the Syrian opposition were brought together, with the need to adhere to "Vienna Roadmap." It was established by the International Syrian Support Group (ISSG) calling for establishing a nationwide ceasefire between different parties, creation of a new constitution through a transparent and accountable electoral system and a timeline for elections to take place within one and a half years from December 18, 2015. But again, this failed as an increased stalemate in talks, failure of the ISSG to come into an agreement and further failure of the US Special Envoy to restart talks leading to the resignation of the Mohammed Alloush, the chief negotiator of the HNC [54].

In January 2017, the negotiations began in Astana (which was later renamed as Nur-sultan in March 2019), the capital of Kazakhstan. This saw the rebel groups representing themselves in a unified manner, though some influential militia and rebels groups like Ahrar al-Sham refused to come. Additionally, the Kurdish groups like the PYD/YPG, any representation from the ISIL, al-Qaeda or affiliated groups like Jabhat Fatah al-Sham (al-Nusra) were not welcomed in the negotiation process [55]. In first half of March 2017, the fourth round of talks on Syrian conflict dubbed a Geneva IV peace talks ended with a "tentative agreement" among various actors with potentiality toward "future negotiations." The 9-day-long negotiation process led by UN Special Envoy Staffan de Mistura included the Assad Government lead negotiator Bashar al-Jaafari, who locked horns with the opposition group. The initial 10 months of the negotiation, for most of the part, was "closed-door consultations." The differences in positions on "counterterrorism" clause (on the insistence of both the Assad government and Russia) as well as the lack of consensus for the "political transition" process created serious a wedge to the progress of the talks. Counterterrorism was only the last among the four core subjects that were considered on the agenda for negotiations (also called "baskets") but stalled and derided the whole peace talks [56].

1.11 ETHICAL CONSIDERATIONS AND ISSUES IN "DE-ESCALATION ZONES" FOR REFUGEES AND INTERNALLY DISPLACED PERSONS

In the first week of May 2017, Iran, Russia, and Turkey (IRT) came together to delineate areas to be considered as safe zones, also known as "de-escalation zones" at Astana. The IRT, who play a major role in the region designated themselves as "Guarantors" and pledged to "reduce the military tensions, to bring the refugees as well as the internally displaced persons (IDPs) [back safely] and [to] stabilise the political condition of Syria." Besides the pledge to deal with the militia and terrorist groups and their reaffirmation to protect "the sovereignty, independence, unity and territorial integrity of Syria" under the United Nations Security Council Resolution (UNSCR) 2254. At the end of October 2017, the "de-escalation zones" were to be demilitarized as safe zones for civilians and were divided into four areas along the eight provinces of Syria [39,57, p. 1].

The de-escalation zones that include Aleppo, Hama, Latakia, and Idlib in the first zone, Talbiseh and Al-Rastan in the second zone, eastern Ghouta in the third zone, and Quneitra and Daraa in the fourth zone. The first zone was largely dominated by affiliates of Al-Qaeda and comprises of more than a million civilian population, while the second zone is reportedly consisting of nearly 200,000 civilians. The third zone is to the north of Damascus (the capital of Syria) and was largely controlled by Jaish al-Islam, consisting of more than 700,000 civilians, while the fourth zone south of Damascus that is largely controlled by rebel groups along the border of Jordan and consists of at least 100,000 more civilians than the third zone. The agreement was expected to induce the formation of a joint working group (JWG) to securitize and regulate activities within the de-escalation zones, particularly focusing on resolution of operational and technical aspects in regard to its implementation. More importantly, the increased provision of "unhindered humanitarian access... [providing] conditions to restore basic infrastructure facilities, including provision for accessibility to water supply, medical assistance, electricity and distribution networks... security checkpoints and observation points to ensure compliance of ceasefire... [were] agreed upon by IRT" [39, pp. 1−2].

The Assad government reportedly bombarded the Idlib region (in the first zone) and eastern Ghouta (third zone), reportedly conducting aerial bombings and in the process even using chemical weapons as well. This derided the potential for Astana accord to reduce violence in de-escalation zones, although both the US President Donald Trump and the Russian President Vladimir Putin have enunciated on as well. The attacks on Atarib (neighboring province of Idlib) are largely legitimized by the Administration that terrorist groups have infiltrated the zones, though the ground reality says otherwise. The lack of legitimate, transparent, and accountable system of identification has largely hindered viable talks to take place. The negotiations and international diplomatic process are largely lacking in terms of understanding the ground reality while the civilian population are mostly affected by the violence that takes place. This includes the use of "Surrender or starve" strategy by government against opposing factions and groups. More importantly, the international agencies, organizations, and other State actors are unable to determine "legitimate act of defense" by the State as against "inhumane act of violence" and forced displacement against its own civilians [58,59]. This could only be circumvented through initiatives that enable digitization of identities for refugees and IDPs, moreover, providing access, transparency, and accountability.

1.12 BLOCKCHAIN TECHNOLOGY FOR REFUGEES AND INTERNALLY DISPLACED PERSONS

The construction of digital identity and its management is one which has been the main attribute to BTBS, specifically for the refugees and IDPs. According to Jim Yong, the President of the World Bank Group, digital identity is "the greatest poverty killer app we've ever seen." Moreover, the use of blockchain technology in the case of cryptocurrencies or other assets has "to be rendered digitally" for any transaction within BTBS to engage successfully with the owner or transacting party. The vice-chairman of Mastercard Walt Macnee has opined that of the total 2 billion "unbanked" people, and one-fourth additional new "consumers" and one-tenth "new merchants can be brought into the global economy." This according to him will provide "opportunities for greater financial inclusion around the... [globe, and] is a path to long-term sustainable economic growth." An increasing number of "new partnership models" have emerged, which saw coming together of tech companies, nongovernmental organizations, and governments. This has increased exponentially in terms of support to refugees and displaced population. This facilitates credit "payments and data protection" for domestic enterprises and business communities [60]. In relations to these aspects, the blockchain technology or the BTBS (at implementation stage) could enable greater accessibility, transparency, accountability as well as efficiency in the existing institutions of the society while providing for the creation of new institutions that often could transcend the traditional notions, perceptions as well as understandings. With the integration of blockchain technology, refugees and IDPs could control their personal data accessibility, prevent (if not reduce) data loss, interoperability, ensure greater accountability from authorities and have low-cost verification, validation as well as authentication in various circumstances, particularly for needs of accessibility to goods and services.

In countries like the United States, Finland, and Moldova, the authorities have already invited the support of start-ups to bring innovative solutions by using the BTBS. Besides being used for data storage, it supports the provision of prepaid debit card, digital identity for children, in securing authentication procedures by linking the identity of cardholders to the blockchain. There are multiple start-ups and developers, like ConsenSyswho intends to launch a biometrics-backed digital identity collaboration for Ubuntu phones and tablets, while *BlockApps* intends to deliver e-wallet and identity system. The *uPort* (digital wallet) allows third-party authorities as well as peers to validate the user's information. The three companies based in Netherlands, Tykn, FissaCoin, and CareerChain provide digital identity, payment, and even job opportunities, while Save the Children UK (collaborating with Adobe Marketing Cloud) and Doctor's Link are trying to bring out "humanitarian passport" and "medical passport" based on BTBS in East Africa. For a refugee this might be the only way to receive cash, even from international agencies or kins far away, along with legitimizing their identity before the authorities, particularly to access to goods and services. This ensures effective removal of middle-men, helps in bridging loopholes in the distribution system and plug leakages at the ground level. The service called ShoCard is currently developing solutions to overcome the issues of forged authentication, misrepresentation, human trafficking, and offers access to a trusted identity distributor, while BanQu is providing digital identity to refugees and those categorized as "unbanked." The Blockchain Emergency ID (BE-ID) undertaken by Bitnation (advocacy group) facilitates and provides identification, emergency services, humanitarian aid to refugees and better governance, while Disberse tracks from financial transactions to distribution of aid. The UN's World Food Programme (WFP) has deployed a pilot project known as "Building Blocks" and "Blockchain against Hunger" for more than ten thousands of refugees in the Azraq and Zaatari refugee camp in Jordan using BTBS to make cash-based transfers, payments, access to food, and other means of entitlements in a much faster, cheaper but more secure manner. Additionally, the concept like a "refugee bank" can supplemented by features like the e-wallets to the mainstream as well as enable institutionalization of a productive life for the refugees [2,16,47,60−64].

In 2017 the ID2020 Summit that took place in the United States saw the biggest global initiative by technology companies for digitizing identity, or what is known as "digital identity." Global multinational corporations (MNCs) like Accenture and Microsoft with support from UN and other agencies, have initiated the ID2020 for providing legal-cum-digital identity refugees around the world by 2020. It integrates biometric information such as retinal scans, fingerprints, birth date, medical records, education, travel, and bank accounts with BTBS, thus providing for the necessity to create a legal identity. This program of a universal digital ID (it was argued) could ease the issue of transit, travel by having just a smartphone app. But this will lead to centralization of identities, data, and information into the hands of selected few, one which is against the basic principle of blockchain technology. The recent ID2020 Summit that took place in New York saw technology companies, including that from developing countries vying to create a digital identification; by linking iris scans, fingerprints, birth date, medical records, education, travel, bank accounts, and the like [65]. This program of a universal digital ID could ease the issue of transit, travel by having just a smartphone app. A number of countries including Finland, Sweden, Estonia, and others have already enabled the support of start-ups have brought innovative solutions for providing identity by using the blockchain technology for data storage. This could enable support starting from citizens to refugees, asylum seekers, provide prepaid debit card, and secure authentication procedures and links the identity of cardholders to the blockchain [64].

1.13 CASE STUDY 2: US DEFENSE FORCES AND INDUSTRY

In the 1990s, with the addition of the Network Centric Operations (NCOs), digitalization of new implications and impact came onto the warfront. There was increased coordination for aerial strikes, during wars, conflicts, and even during small-scale skirmishes. Besides, air fighters and strategic bombers, drones started playing a new role, especially in terrains that are largely vulnerable to full-scale fighters and loss of human lives as well. The emergence of blockchain technologies as a standalone technology and in complementing as well as supplementing other existing and emerging technologies have transformed the defense sector. Salvador Llopis Sanchez, the Project Officer in the European Defense Matters (EDA) feels that blockchain technology will be one of the very rare disruptive defense innovations due to its attributes like "encryption... novel approach towards information security, authentication, data integrity and resilience... and others." Additionally, other aspects that include "distributed database, the peer-to-peer transmission and the computational logic will be candidate technological building blocks" [66]. But the defense sector is relatively slow imbibing the various aspects of the blockchain technology along with new and emerging paradigms, one which major powers around the world are inculcating into various theaters of conflict [67].

The US-based Defense Advanced Research Projects Agency (DARPA) and the Department of Defense (DoD) have already started looking at the use of blockchain in "battlefield autonomy and paradigm-shifting [areas]" [68]. It is being stated that the US military is becoming hugely "dependent on secure, timely [and] accurate data." This is particularly important in the context of threats from cyber warfare that is increasingly challenging the nation-states from digital space. During a vulnerability testing by the US Government Accountability Office (GAO), it was found that there were evident "mission-critical cyber vulnerabilities in 86 weapon systems" that were only in nascent stages. In 2010 the US Air Force (USAF) lost contact with the *Francis E. Warren Air Force Base* in Wyoming (United States) that created serious concerns about information warfare on the nuclear command and control networks. This airbase contained about 50 *Minuteman III*

18 **CHAPTER 1** ETHICAL CONSIDERATIONS AND ISSUES OF BLOCKCHAIN

Intercontinental Ballistic Missiles (ICBMs), thus is considered to be a critical national security blackout [67].

In 2013 the Defense Science Board (DSB) of the Pentagon, that was institutionalized as a civilian experts committee over the recommendation of the Second Hoover Commission to advice the Pentagon and the US DoD on "scientific, technical, manufacturing, acquisition process, and other matters of special interest" found out that there are severe vulnerabilities, cyber and asymmetric threats to US military systems within and outside the territorial limits. This was brought out in its various reports, and include *Basic Research* (in January 2012), *The Role of Autonomy in DoD Systems* (in July 2012), *Predicting Violent Behavior* (in August 2012), *Resilient Military Systems and the Advanced Cyber Threat* (in January 2013), *Assessment of Nuclear Monitoring and Verification Technologies* (in January 2014), *DSB Summer Study Report on Strategic Surprise* (in July 2015), *DSB Summer Study on Autonomy* (in June 2016), *Cyber Defense Management* (in September 2016), *Next-Generation Unmanned Undersea Systems* (in October 2016), *Seven Defense Priorities for the New Administration* (in December 2016), and *Cyber Deterrence* (in February 2017) [69]. Further, major regulatory documents and guidance issued by other US Government Agencies for addressing cryptocurrency included the *FIN-2013-G001* (in March 2013), *FIN-2014-R011*, and *FIN-2014-R012* (in October 2014) by FinCEN, *Notice 2014-21* (in March 2014) by IRS, *Investor Alert* (in May 2014) by Securities and Exchange Commission, *Consumer Advisory* (in August 2014) by Consumer Financial Protection Bureau [70, p. 70]. A recent survey by the US-based Foundation for Defense Democracies found that:

> A single line of malicious code could be activated and weaponized by the enemy through a back-door to destroy military weapon systems from far. This code could be inserted during the manufacturing or production process and not be discovered until it is too late. [The implications on the defense and military supply chain is very pertinent and serious in this regard.] The issue that records of these supply chains are scattered across computer systems and paper documents and thus the. . . Government does not have full visibility into what companies are involved in the manufacturing and production of. . . most sensitive weapon systems [71].

Foundation of Defense of Democracies (FDD) along with Combating Terrorism Technical Support Office (CTTSO) and Engineering Center Steyr (ECS) conducted a field-test using a smart contract and an Ethereum (blockchain) platform. The test simulated the real-world process in Microsoft's Azure platform was found to be self-executing, autonomous, transparent, fast, with automated authorization and deliberation, but stored in the immutable ledger distributed among various stakeholders. The simulation focused on deliberate alteration of programming code by a subcontractor but was ascertained by the smart contract, thereby aborting the process while pinpointing the error. Further, pairing of blockchain technology with other technologies could thus enhance the security by "conducting antitampering checks, validating geo-location information and recording who has had access to sensitive items as they are transported around the world." The Federal Aviation Administration (FAA) in United States has already incorporated the Automatic Dependent Surveillance—Broadcast (ADS-B), envisioned as a part of the Aviation Blockchain Infrastructure (ABI) ensuing a "new surveillance system" [72].

A smart contract, thus, can catch and expose alterations in weapon systems before harm is done, flagging and immediately blacklisting malicious and negligent subcontractors, and

1.13 CASE STUDY 2: US DEFENSE FORCES AND INDUSTRY   19

enabling... [Governments] to do an immediate and thorough investigation [71]. The use of blockchain has also been increasingly depended by the US Army to entail the provision of "flow of data through terrestrial and satellite networks" whereby the "trustworthy" feature of the information and data could be relied upon. It has been reported that an increasing number of State militaries globally are reliant upon the blockchain technology, mainly in communication due to its encrypted nature and security aspects inherent in cryptographic mechanisms as well. In 2016 the North Atlantic Treaty Organization Communication and Information Agency (NATO Communications and Information Agency), an agency set up by NATO in 2012, hosted the *2016 Innovation Challenge* that generally encompassed solutions involving "military-grade blockchain applications... regarding military logistics, procurement, and finance." The Space and Terrestrial Communications Directorate of the US Army (S&TCD) has also started pushing research and scientific projects addressing various aspects related to blockchain, especially security and privacy of encrypted communication data within blockchain. More features related to machine learning, Internet of Things (IoT), and artificial intelligence (AI) have been looked upon as well [73,74].

The Royal Institute of International Affairs, an organization based in the United Kingdom gave a serious warning in January 2018 of critical vulnerabilities on nuclear weapon systems of countries like the United States, United Kingdom, and other nuclear powers as well. According to the report released, the threats of "cyber-attacks on nuclear weapons systems is relatively high and increasing from advanced persistent threats from states and non-state groups" [67]. On February 14−15, 2018, the workshop organized by DARPA's Information Innovation Office (I20) examined three aspects, vis-à-vis. "Incentivizing Distributed Consensus Protocols without Money, Economic-Driven Security Models for Distributed Computation Protocols, The Centralities of Distributed Consensus Protocols" [75]. In December 2018, it invited individuals and agencies to present ideas in a workshop on utilizing "permissionless distributed consensus protocols" (generally called as blockchain technology) into the defense sector. It is being argued that DARPA is looking at options for creating ledgers using cryptographic protocol that will be enabling tracking record of transactions thus reducing incidences of falsification of improperly generated inputs. Moreover, DARPA is looking at methods to combine utilitarian economic notions with the world of computer science protocols and in categorizing each participant as "honest" or "malicious." Finally, DARPA is also looking at options to centralize the distributed protocols, to find a backdoor to the program, to understand and access all information this regard. DARPA is looking at possibilities to reduce uncertainty. DARPA feels that blockchain technology could have serious transformations in the defense sector, with "implications for the security and resilience of critical data storage and computation tasks, including the [DoD]" [68].

The application of blockchain in various other military sectors like drones, aviation, battleship control systems, and additive manufacturing (AM) or 3D printing. The use of AI and blockchain, it has recently been explained by experts and showcased as creating multiple possibilities like in autonomous robots and drones. Here blockchain collects and records data in real time, without the need for human intervention or control. In the field of AM or 3D printing, the US Navy since 2017 has been integrating the blockchain technology to provide reliable information to ensure the efficiency and prevent counterfeit component(s) or malicious code(s) from entering defense equipment. In battleships, the various parts or armaments like sensors, radars, high-speed computers with centralized systems ensue greater possibilities toward failure. Blockchain technology has already been

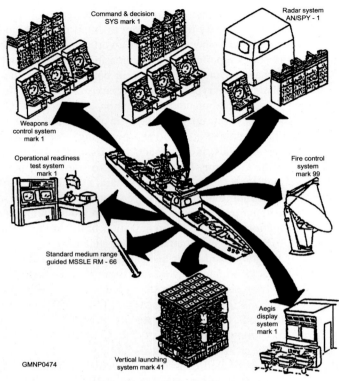

FIGURE 1.4

Illustration of the AEGIS Weapon System Mk7.

J. Pike, AEGIS Weapon System Mk 7, Available from: <https://www.globalsecurity.org/military/systems/ship/systems/aegis-core.htm>, 2019.

incorporated by Lockheed Martin, the US-based defense contractor manufacturing the Aegis Combat System, an integrated naval weapons system [76], is shown in Fig. 1.4.

The US-based Nuclear Threat Initiative (NTI) group came up with a report titled Nuclear Weapons in The *New Cyber Age: Report of the Cyber-Nuclear Weapons Study Group* in September 2018. It was jointly authored by Page O. Stoutland and Samantha Pitts-Keifer that found out that "nuclear weapons systems" are potentially vulnerable to cyber-related attacks, even from non-State actors. Samantha Pitts-Kiefer, the Senior Director of the Global Nuclear Policy Program at the NTI opined that "an adversary could insert malicious code or malware into the supply chain into a component of a nuclear weapon" [77]. But moreover, unintentional issues of parts and goods in the supply chain is the biggest challenge for the United States. The use of outdated or vulnerable equipment by manufacturers, especially the Original Equipment Manufacturers (OEMs) could attract huge threats on the ground level, especially against adversaries that rely on low-cost attrition. This has, and will have serious implications on the defense sector, including the future potential threats from a flaw, or the discovery of the flaw. The prospects of blockchain to identity, prevent, and blacklist vendors (including the OEMs) have been argued as one of the best deterrence in the field of deterrence against (un)intentional flaws in defense armament, weapon systems, and other critical infrastructure of various actors. An implication of this is inherently linked to the national security threats, one which derides the capacities and capabilities of a nation-state. Another major issue with the weapon suppliers including the United States is the tracking of flow of its weapons to its own adversaries or rival nations [78], one which blockchain could help track in an authentic manner. This includes countries like Iran, Syria, and non-State actors like the ISIS, Al-Qaeda, and other militia groups as well. Sanchez feels that the challenges that may arise from the nature of blockchain "may require further research in areas such as interoperability, network infrastructure and a thorough analysis of its regulatory framework" [66].

1.14 ISSUES AND CHALLENGES

The inclusion of blockchain technology into emerging digitization could enable the creation of a "digital economy" that is increasingly inclusive in nature, invariably secure, transparent as well as accountable to the users. It will provide the public with a robust mechanism that provides greater integration toward economic collaboration and cooperation at the transborder level. This could play "a critical role in strengthening economic resilience," and ensure relatively greater equity for the provision of accessibility and distribution of resources [8]. When complemented with other attributes like smart contracts, digital governance, and IoT, blockchain is considerably more transparent, accountable, and highly secure against data manipulation or hacking. This can greatly reduce the prospects of fraud or data mismanagement in the country. It also allows for the direct transaction between different parties, and thus removes the requirement for intermediaries like banks, reduce payment costs, and greatly increase protection to beneficiary data. It could provide identity solutions for children particularly in refugee camps or among internally displace populations (IDPs) so as to enable access to institutionalize mechanisms in any country, irrespective of the political boundaries, social hindrances, or any other physical bottlenecks. It ensures documentation, voluntary identification, that which can grant access to healthcare, social welfare services, livelihood opportunities, and so forth.

Concomitantly, the challenges lie in the creation of an architectural framework that enables the provision, procurement, and development of BTBS particularly in governance and accessibility to other public goods and services. The government, as well as companies, have the greatest opportunity in the form of blockchain but still are unequipped in dealing with the "technological turn" that will occur as technological wherewithal progresses, as it is expected to be. There are emerging challenges include the activities for manipulating and/or editing blockchain by the likes of major developers and corporates. This has manifold ramifications on the implementation of blockchain, particularly to the reliability and trust that blockchain has in the international arena. A lot of issues may arise when a "trusted administrator" or "limited key access" is given the power to record, rewrite, and verify transactions, thus possibly disintegrating the decentralized nature of blockchain to the interest of few. This contradicts the basic principle followed by blockchain that is if half the

22 **CHAPTER 1** ETHICAL CONSIDERATIONS AND ISSUES OF BLOCKCHAIN

computers on the network agreed to alter the transaction records, the records could be modified, which is unlikely to happen since the network is too decentralized. The currently proposed "chameleon" hash function could circumvent the transparency and accountability that currently is the key feature of blockchain technology [79].

Though this is, in turn, stated as increasing the security features of cryptocurrencies like Ethereum, Ripple, Litecoin, NEO, Stellar, and others, but unfortunately transforms the genuine intention with which blockchain was originally devised. The emergence of issues of the volatility of these cryptocurrencies has but become a greater threat not only to the global financial system but also to the political system as well. Though the use of blockchain technology promises a better, viable as well as more secure payment system that replaces the e-currency in the traditional bank accounts where most importantly moving toward cashless economy; many are skeptical about them. This includes suspicions of possible manipulation or abuse, even impacting the society at greater length. The means of exchange that provided for the basis of trust between the citizens as well as the governments is an important symbol of legitimacy over the people, which when manipulated will disrupt not just the economic but sociopolitical fabric as well [80].

Another prominent issue that has emerged in relation to BTBS is that of "manipulation" or "editing" of these systems by the likes of major developers, companies, and corporates. The "centralization" of blockchain technology is not only "de-decentralizes" it, but also vulnerable to hacking and exploitation. This will have manifold ramifications in reliability and trust that BTBS had in the first place. A lot of these issues are related to when a "trusted administrator" or "limited key access" is given to a particular authority or entity. This includes the power to record, rewrite, and verify transactions, thus possibly destroying the "decentralized nature" of BTBS for the interest of the "few." Basically, this contradicts the fundamental principle of BTBS, which is consensus-based distribution protocols. The currently proposed and researched "chameleon" hash function could inadvertently circumvent the transparency and accountability feature of the BTBS. The changes devised on features of cryptocurrencies, particularly in Bitcoin, Ethereum, Ripple, Litecoin, NEO, Stellar, and others could transform the genuine intention with which blockchain technology was originally devised. The major incidents that calculated in USD include Japanese-based Mt Gox, US-based BitFloor, US-based Poloniex, Slovenian-based Bitstamp, and Hong Kong-based Bitfinex [81].

Moreover, the attacks from 2017 to 2018 siphoned 882 million USD with most of the cases lacking in conviction or arrests [7]. As of 28 June 2019, the market value some of the major cryptocurrencies above 1 billion market capitalization (market cap) club are Bitcoin (BTC) with a market cap of 197 billion USD and price of 11,087.94 USD, Ethereum (ETH) with a market cap of 31 billion USD and price of 297.23 USD, Ripple (XRP) with a market cap of 17 billion USD and price at 0.406 USD, Bitcoin Cash (BCH) with a market cap of 7.3 billion USD and price at 406.1 USD, Litecoin (LTC) with a market cap of 7.2 billion USD and price at 114.97 USD, EOS.IO (EOS) with a market cap of 5.4 billion USD and price at 5.82 USD, Binance Coin (BNB) with a market cap of 4.8 billion USD and price at 33.79 USD, Bitcoin SV (BSV) with a market cap of 3.54 billion USD and price at 198.21 USD, Tether (USDT) with a market cap of 3.53 billion USD and price at 0.986 USD, Cardano (ADA) with a market cap of 2.23 billion USD and price at 0.086 USD, TRON (TRX) with a market cap of 2.17 billion USD and price at 0.032 USD, Stellar (XLM) with a market cap of 2.11 billion USD and price at 0.108 USD, UNUS SED LEO (LEO) with a market cap of 1.75 billion USD and price at 1.75 USD, Monero (XMR) with a market cap of 1.63

billion USD and price at 95.32 USD, Dash (DASH) with a market cap of 1.38 billion USD and price at 155.75 USD, NEO (NEO) with a market capitalization of 1.22 billion USD and price at 17.24 USD, IOTA (MIOTA) with a market cap of 1.1 billion USD and price at 0.394 USD as well as Cosmos (ATOM) with a market cap of 1.04 billion USD and price at 5.44 USD [82]. It is estimated that since June 2011 there have been many incidents and is shown in Table 1.1.

Table 1.1 Chronological Timeline of Crypto Hacks From 2011 to 2019.

Year	The Provider (Company)	Founded at (Country)	Amount in (USD/Cryptocurrencies)
Jun 2011	Mt. Gox	United States	2,643 BTC
Jul 2011	Bitomat	Poland	17,000 BTC
Aug 2011	MyBitCoin	United States	154,406 BTC
Oct 2011	Bitcoin7	United States	11,000 BTC
Oct 2011	Mt. Gox	United States	2,609 BTC
Mar 2012	Linode (Slush, Bitcoinica, Tradehill)	United States	46,000 + BTC
May 2012	Bitcoinica	New Zealand	18,000 BTC
Jul 2012	BTC-e	Russia	4,500 BTC
Jul 2012	Mt. Gox (Bitcoinica)	United States	40,000 BTC
Aug 2012	Bitcoin Saving and Trust	United States	265,000 BTC
Sep 2012	BitFloor	United States	24,000 BTC
Nov 2012	Bitcoin Saving and Trust	United States	200,000 BTC
May 2013	Vicurex	China	1,600 + BTC
Jun 2013	PicoStocks	Marshall Is.	1,300 BTC
Oct 2013	Silk Road Market Place	United States	1,606 BTC
Nov 2013	BitCash.cz	Czech Republic	484 BTC
Nov 2013	PicoStocks	Marshall Is.	5,896 BTC
Mar 2014	Mt. Gox	United States	850,000 BTC
Mar 2014	Poloniex	United States	97 BTC
Mar 2014	Bitcurex	Poland	Unknown
Jul 2014	Mt. Gox (Cryptsy)	United States	13,000 BTC; 300,000 LTC
Aug 2014	Bter	China	(~1 million USD) NXT
Oct 2014	KipCoin	China	~690,000 USD
Oct−Dec 2014	Mt. Gox (Mintpal)	United States (United Kingdom)	16,894 BTC
Dec 2014	BitPay	United States	5000 BTC
Jan 2015	796Exchange	China	1000 BTC
Jan 2015	Bitstamp	Slovenia	19,000 BTC
Jan 2015	LocalBitcoins	Finland	17 BTC
Feb 2015	Bter	China	7000 + BTC
Feb 2015	KipCoin	China	3000 BTC

(Continued)

24 CHAPTER 1 ETHICAL CONSIDERATIONS AND ISSUES OF BLOCKCHAIN

Table 1.1 Chronological Timeline of Crypto Hacks From 2011 to 2019. *Continued*

Year	The Provider (Company)	Founded at (Country)	Amount in (USD/Cryptocurrencies)
Mar 2016	Cointrader.net	Canada	Unknown
Apr 2016	ShapeShift	Switzerland	469 BTC; 5800 ETH; 1900 LTC
May 2016	GateCoin	Hong Kong	250 BTC; 185,000 ETH
Jun 2016	Decentralized Autonomous Organization (DAO)	Europe	3.6 million ETH
Aug 2016	Bitfinex	Hong Kong China	120,000 BTC
Oct 2016	Bitcurex	Poland	2300 BTC
Apr 2017	Bithumb	South Korea	3816 BTC
Apr–Dec 2017	YouBit (Yabizon)	South Korea	3816 BTC
Jun 2017	QuadrigaCX	Canada	67,000 ETH
Jul 2017	Coindash	China	43,500 ETH
Jul–Nov 2017	Parity	United Kingdom	153,000–514,000 ETH
Dec 2017	Nicehash	Slovenia	4000 BTC
Jan 2018	Bitstamp	Slovenia	18,000 BTC
Jan 2018	Coincheck	Japan	523 million NEM/XEM
Feb 2018	BitGrail	Italy	17 million NANO
Apr 2018	Dantang	China	~13 million USD
Apr 2018	GainBitcoin	India	~300 million USD
Apr 2018	M-Coin	South Korea	~20 million USD
Apr 2018	CoinSecure	India	438 BTC
Apr 2018	MyEtherWallet	United States	215 ETH
May 2018	Bitcoin Gold	United States	388,000 BTG
May 2018	Taylor	Estonia	2578 ETH; 659,000 TAY
Jun 2018	Coinrail	South Korea	1927 ETH; 2.6 billion NPXS; 93 million ATX; 831 million DENT
Jun 2018	Bithumb	South Korea	~31 million USD
Jul 2018	Bancor	Switzerland	~23.5 million USD
Sep 2018	Zaif	Japan	5966 BTC
Oct 2018	MapleChange	Canada	913 BTC
Dec 2018	QuadrigaCX	Canada	26,350 BTC
Jan 2019	HitBTC	Chile	Unknown
Jan 2019	Cryptopia	New Zealand	21,065 ETH
Feb 2019	Coinmama	Israel	User data compromised
Mar 2019	Dragonex	Singapore	20 types of cryptocurrencies stolen
Apr 2019	Bithumb	South Korea	20.2 million XRP
May 2019	Binance	Canada	7000 BTC
May 2019	Poloniex	United States	1800 BTC
Jun 2019	GateHub	Slovenia	(~10 million USD) XRP
Jun 2019	Bitrue	Singapore	9.3 million XRP; 2.5 million ADA

Compiled by the author as of June 2019.

1.15 FUTURE OF BLOCKCHAIN TECHNOLOGY FOR GLOBAL GOVERNANCE   25

Besides, the increasing volatility of cryptocurrencies has become a threat to the viability of financial systems that provide payments. In 2011 Chuck Schumar characterized Bitcoin as "an online form of money laundering used to disguise the source of money, and to disguise who's both selling and buying drugs." In 2013 Mythili Raman, the former US Assistant Attorney General during a Senate testimony said that "virtual currency is not necessarily synonymous with anonymity" [70, p. 80]. A pertinent challenge to the use of the new challenges to privacy, data concentration, lack of standardization, and/or internationally accepted "benchmark-ism," differential scalability (size of ledger and speed of transactions), validation, regulatory bottlenecks, expensive nature of maintaining and updating as well as the existence of unknown vulnerabilities are general challenges that exist in the system. However, there are specific challenges like consensus hijack (51% hijack), sidechains or Sybil attack (proxy tokens), Distributed Denial of Service (DDoS) (crippling server), routing attack (compromising Internet Service Provider and intercepting internet traffic), wallet management (loss of access, compromise of a wallet and/or irreversible fraudulent transactions), smart contract management issues (error from people, processes, and/or technology), and even the exploitation of permissioned blockchain (exploitation of regulator's capabilities) [11,65,79,80,83−85].

1.15 FUTURE OF BLOCKCHAIN TECHNOLOGY FOR GLOBAL GOVERNANCE

These initiatives could enable the creation of a digital economy that is inclusive in nature, increasingly secure, transparent, and accountable to all stakeholders, providing greater integration in states. This could play "a critical role in strengthening economic resilience," and ensure relatively greater equity for the provision of accessibility and distribution of resources. When complemented with other features like smart contracts, digital governance, and IoT, BTBScan greatly reduces the prospects of fraud, data, and economic mismanagement. It removes (or reduces) the requirement for intermediaries like banks, reduce payment costs, and greatly increases the protection to beneficiaries and their data. It ensures documentation, voluntary identification, which can grant access to healthcare, social welfare services, livelihood opportunities, and so forth [8,86]. The new provision of digital identity provides moreover indestructible, untamperable access to goods, services, and other provisions to the population not only outside but inside one's own country.

In Estonia, the applications have been implemented in electoral systems, digital identity, creation and for providing healthcare services. In one of the South-Atlantic states of the United States, companies like Consensys has started examining the options in insurance, stock market, and corporate trade. Additionally, the department in the United States like Homeland Security, Health, and Human Services have started pilot projects as well. In UAE, the State is looking at bringing in aspects of governance like registration for businesses, trade-related activities, and banking operations [87]. In April 2016, the Central Securities Depository of the Russian known as National Settlement Depository (NSD) started looking at options to include blockchain-based automated voting systems, entailing on the regulatory environment 3 months later in July. In September 2016, a discussion between various experts and asset managers looked explicably at the options at improving corporate governance, proxy voting processes, implementing *ISO 20022* and the challenges for cross-border corporate actions [88]. A pilot test that is been conducted in both developing countries

26 CHAPTER 1 ETHICAL CONSIDERATIONS AND ISSUES OF BLOCKCHAIN

and developed countries, categorized into three categories viz. in progress, planned and announced. In the first category, for digital currency/payments it has been implemented in countries like China, India, and Canada; for land registration the countries include Sweden and Georgia; for voting (elections) the countries include Ukraine, South Korea, Denmark, Estonia, and some districts in the United States like Illinois and New York; for identity management the country is Estonia; in supply chain traceability it has been implemented in Texas (in the United States); in healthcare it has been implemented in Estonia; for voting (proxy) it has been implemented by National Association of Securities Dealers Automated Quotations System (NASDAQ) (in the United States); for corporate registration it has been implemented in Delaware (in the United States); and in regard to entitlements management it has been so far implemented in the United Kingdom. In the second category, for digital currency/payments it has been implemented in countries Denmark, South Korea, Singapore, Senegal, and Nigeria; for land registration it has been implemented in Illinois, Ghana, and Brazil; for identity management in Switzerland; for supply chain traceability in Finland; for healthcare by US Department of Health and Human Services (USHHS), US Food and Drug Administration (USFDA), US Postal Services (USPS); for voting (proxy) in UAE, Russia, and South Africa. In the third category, it has been implemented for land registration in Texas [87].

In 2018 ArtemDuvanov (the Director of Innovations at the NSD) spoke about the need to improve the "e-voting system prototype" currently at NSD, and to improve the regulatory environment for blockchain technology in Russia, one which he opined as a "breakthrough technology." NSD went on to partner with the corporate-sector investment arm of Sberbank, one of the oldest commercial bank in Russia [74]. PawanDuggal, a Cyber Expert, and Legal Practitioner-based in India exclaimed that blockchain technology is a "game-changer." According to him, the basic principle of blockchain is that "the record of transaction is maintained simultaneously by all the stakeholders on different computers." Moreover, the use of cryptography encoding all the information provides for much tight securitization of what is known as tokens or "tokenization." While, Virag Gupta, a Cyber Expert and lawyer-based in India described how blockchain is transforming the technological ecosystem as a whole, overwhelmingly influencing all other sectors as well. He entailed on how in India, the Government of India and Reserve Bank of India (RBI) are carrying forward with options to accelerate research and development (R&D), in utilizing the blockchain technology in various sectors as well as areas in the socioeconomic sphere [5,89].

Countries could and are (increasingly) adopting blockchain technologies in "identity management, land registration and voting." First, in terms of the use of blockchain technology for identity management, the inherent base lies in the aspect of digital identity, wherein identification, rights and security converge together. Additionally, whether it is Bitcoin (cryptocurrencies in general) or digitally connected objects (Internet of Things, i.e., IoT) or even people (Internet of People, i.e., IoP), each becomes a node or asset that enters the paradigm of the blockchain and utilizes it. This is more attributable in the public sectors, wherein at least one-fifth of the population is estimated to be without an identity that can be authenticated and secured at real time. Although, the issues that need to be addressed include the standardization (international benchmarkization) of the digital identity, diverging processes for attestation procedures and "entry points" that has be to mitigated, prevented, and streamlined as well. Additionally, the value proposition inherently needs security, self-sovereignty in identity thereby enabling efficient transactions across sections, as well as stakeholders (and individuals). The inherent nature of blockchain that provides decentralized but "explicit control" overidentification is much more secure in the contemporary period [87].

1.15 FUTURE OF BLOCKCHAIN TECHNOLOGY FOR GLOBAL GOVERNANCE

Second, for land registration, the major application lies at securing "deeds and titling," particularly an issue related to real estate and land sale as well as ownership. This enhances the protection provided for buyers and even sellers, while increasingly supporting targeted investment, land development, growth as well as prosperity of both investors and the resident people. Interestingly, this is more sporadic in developing countries than among developed countries, and is reportedly considered "unique and noncorruptible." Additionally, it has greater reliability record for storage of "property record[s], [while] validating changes" among all the nodes (users or stakeholders). The issues and challenges include the processes with regard to licensing and registering, which at present is not just paper-based but fragmented as well, thereby more expensive, inefficient and tamper-prone. In 2014−15 in the United States alone a reported 800 million USD was spent by landowners for covering risks related to titles of land. The addition of blockchain could enable more decentralized nature for aspects of land registration, and digitization of records, which can reduce middle-men, increase trustworthiness in transaction, efficiency while reduces procedural bottlenecks, delays, total time period, and the processing time. It could reportedly save billions for individual stakeholders and the burden on the State as well [87].

Third, blockchain is very critical and provides greater legitimacy in terms of conducting electoral processes and as such voting. The citizens could cast their votes and make sure their votes are not "hacked" or manipulated by any attacks through the cyberspace, making sure that the inherent foundation of the democratic process is not broken as such. Additionally, the verification, authentication, and validation are secure than all existing systems, with many potentialities in solutions and features that include "digital identity management, anonymous vote-casting, individualized ballot process... and ballot casting confirmation [that is] verifiable by the voter." This BTBS system enables tokenizing the votes as such and thus making sure they are immutable to any degree. The issues and challenges include the expensive nature of implementation of BTBS, especially related to manufacturing, distribution, maintenance of the governance architecture, and the training of the personnel as well as sensitization of the citizen voters in this regard. In countries with high illiteracy rates, this could be a greater challenge. Moreover, the increasing number of security vulnerabilities, threats, and issues from cyber-related attacks could be a grave threat to the political stability and security of the State and the political systems in the country. The important challenge in this regard is the issues of transparency, especially in centralized blockchain-based electoral process, which (according to Table 1.1) is more prone to hacking and being compromised as well. The delays in voting and the related inefficiencies from "remote/absentee voting" are additional challenges that need to be pondered over as well. But the potential for blockchain to reduce the expenses, enhance security, accountability, audibility of electoral process, enhance political participation, and increase transparency as well [87].

The use of the internet and digital technologies are transforming the transactional and informational relations in society. An extensive survey by the Deutsche Bank and FT Remark in 2016 found that 39% of the financial market participants believed that blockchain will be in active use within 5−6 years, while 36% believed the time period to be much lesser at around 3−4 years. A combined three-fourths of them thus believed that BTBS will be dominant in the financial markets within the coming 3−6 years, especially at its implementation stage [90]. But the pertinent questions that still persist are *how*, *where*, and *who*. Moreover, the validity of the transaction, secure nature of the transactions, and the legal validity of the entities involved in the transaction be verified. Duggal states that the applications of blockchain extend to multiple points including banking, insurance, financial services, e-governance, digital storage, cybersecurity, real estate, education, healthcare, defense, and almost all the areas in society and individual life as well. According to him, "blockchain will completely transform in the way in which nation, entities and other related actors actually start working in doing their day-to-day operations." Frank Rausan Pereira of the Rajya Sabha Television Network (RSTV) opines that blockchain technology can even end corruption in a larger and effective way. But the blockchain technology provides a much secure, and confidential pathway toward communication through a cryptocurrency, overwhelmingly beating other contemporary technologies for security and privacy in a long way. Moreover, it enables banking services to be much safer, relatively cost-effective, and also entails faster banking transactions. Additionally, it also helps secure legal documents, healthcare data, information on notaries and personal information and other important documents as well [5,89].

Through the use of blockchain in governance, it can help in the delivery of subsidies, governmental assistance, scholarship, funding, and grants, as well as effective digitalized distribution of land records. The technology could also be used in cloud storage, biometric-cum-digital identification, smart communication, and e-voting. Duggal elucidates on how South Korea has utilized blockchain for e-governance, in maintaining records on e-health, e-land, e-storage, e-insurance, e-banking, for the financial system and many others with future possibilities. This is possible because blockchain is able to provide anytime and real-time access to the actors on the pertinent information regarding the ownership and transactional relations as well as the history of everything in this regard. This is explained as a distributed ledger system or P2P. It helps overcome issues of systemic—structural deficiencies, reduces expensive maintenance, helps in cost-reduction, and in decentralizing the access to various powers and data through a networked approach. Real estate and land leasing activities could be done in a digitalized manner, while various academic institutes and universities are utilizing the cloud hosting platform for educational purposes. Further, it provides a transparent and accountable platform for students, teaching faculties, and even the corporate sector in developed countries [5,89].

1.16 CONCLUSION

Blockchain technology is an emerging technology with a lot of promise in various sectors of the society that is continuously connected in this period of "digitalized globalization" or "Globalization 2.0." It has created various cost-effective solutions to address the various aspects of security and privacy but enabling decentralized access to the individual end user as well. Thus it could enable greater accessibility, transparency, accountability as well as efficiency in the existing institutions of the society while providing for the creation of new institutions that often could transcend the traditional notions, perceptions as well as understandings. The crisis emanating from the Syrian Civil War has created one of the largest and increasingly complex humanitarian problems in the 21st century world. Besides being unresolved, "The Great Humanitarian Crisis" has had a huge impact on international politics, setting in motion the emergence of postliberal world order. The foundations of globalization have been threatened, especially with the emergence of nationalist tendencies that is reminiscent of the interwar period. This has resulted in an increasing number of refugees and IDPs, who are in constant search for securing shelter, food as well as security.

In the current scenario, the use of BTBS enhances the chances of refugees and IDPs for access to targeted aid, grant, funding, and food in the host country. It enhances efficiency in regard to institutionalizing or even replacing existing systems, processes, and procedures to provide for refugees and IDPs. With the integration of blockchain citizens, refugees, and IDPs could control their personal data accessibility, prevent (if not reduce) data loss, interoperability, ensure greater accountability from authorities and have low-cost verification, validation as well as authentication in various circumstances, particularly for needs of accessibility to goods and services. But besides the pros of blockchain technology, it has its cons that could not only affect its implementation but ensue exploitation by "the few." This increasing tendency toward centralizing BBTS entails control and increases possibilities toward State and corporate authoritarianism. A move when implemented will inviolably create a tendency toward surveillance capitalism, and determinism left within the bounds of some MNCs. This overturns the real intention of the development of blockchain, which was decentralization. Oligopoly (or monopoly) and the centralization and monetization of the blockchain technology or BTBS by "the few" will incidentally create more disparities and ramifications for refugees and IDPs, exacerbate the already existing societal divide as well. The inherent ethical issues and inherent problems in the BTBS need to be plugged, centralization prevented and opportunities for development of concrete solutions should be enhanced, both vertically and horizontally.

REFERENCES

[1] United Nations (UN), Syria Conflict at 5 Years: The Biggest Refugee and Displacement Crisis of Our Time Demands a Huge Surge in Solidarity, Available from: < https://www.unhcr.org/news/press/2016/3/56e6e3249/syria-conflict-5-years-biggest-refugee-displacement-crisis-time-demands.html > , 2016.

[2] DuDe, Blockchain, Dive into the Refugee Camp in Jordan that Runs on Blockchain, Hackernoon, Available from: < https://hackernoon.com/dive-into-the-refugee-camp-in-jordan-that-runs-on-blockchain-924a8fde2d9d > , 2018.

[3] N. Smolenski, Blockchain Records for Refugees, Medium, Available from: < https://medium.com/learning-machine-blog/blockchain-records-for-refugees-bd27ad6e6da1 > , 2017.

[4] H. James, The Bitcoin Threat, Available from: < https://www.project-syndicate.org/commentary/bitcoin-threat-to-political-stability-by-harold-james-2018-02?utm_source = Project + Syndicate + Newsletter&utm_campaign = 4bf9c5d8e0-sunday_newsletter_4_2_2018&utm_medium = email&utm_term = 0_73bad5b7d8-4bf9c5d8e0-105499057 > , 2018.

[5] Rajya Sabha TV, In Depth—Blockchain Technology, YouTube, Available from: < https://www.youtube.com/watch?v = _CNdIUD9H2E > , 2018.

[6] S. Haber, W.S. Stornetta, How to time-stamp a digital document, J. Cryptol. 3 (2) (1991) 99−111.

[7] World Crypto Index, W. Scott Stornetta Bio, Available at < https://www.worldcryptoindex.com/creators/w-scott-stornetta/ > , 2018.

[8] Centre for International Governance Innovation (CIGI), Blockchain: A Technology With Policy Potential, Available from: < https://www.cigionline.org/interactives/2017annualreport/?gclid = EAIaIQobChMIgcS8hY6B2QIV1rrACh3tVgSlEAAYAiAAEgLOQ_D_BwE#/?slide = 13 > , 2017.

[9] European Union Agency for Network and Information Security (ENISA), Blockchain, Available from: < https://www.enisa.europa.eu/topics/csirts-in-europe/glossary/blockchain > , 2018.

30 CHAPTER 1 ETHICAL CONSIDERATIONS AND ISSUES OF BLOCKCHAIN

[10] S. Nakhoul, Sarkozy Meets Syria's Assad Ahead of Summit, Reuters, Available from: < https://uk.reuters.com/article/uk-france-syria-lebanon/sarkozy-meets-syrias-assad-ahead-of-summit-idUKL1246476620080712 > , 2008.

[11] National Settlement Depository (NSD), Media About Us: NSD Leads Blockchain Implementation in Russia, Available from: < https://www.nsd.ru/en/press/pubs/index.php?id36 = 634189 > , 2016.

[12] S. Nakamoto, Bitcoin: A Peer-to-Peer Electronic Cash System, Available from: < https://bitcoin.org/bitcoin.pdf > , 2008.

[13] A.M. Antonopoulos, Mastering Bitcoin: Programming the Open Blockchain, O'Reilly Media, Inc, Sebastopol, CA, 2017.

[14] World Crypto Index, What is a Merkle Tree?, Available at < https://www.worldcryptoindex.com/what-is-a-merkle-tree/ > , 2018.

[15] S. Khatwani, Top 5 Biggest Bitcoin Hacks Ever, Available from: < https://coinsutra.com/biggest-bitcoin-hacks/ > , 2018.

[16] The Editors of Encyclopaedia Britannica, Paris Peace Conference, Available from: < https://www.britannica.com/event/Paris-Peace-Conference > , 2019.

[17] The Regents of the University of California, Africa, the Near East and the War: Lectures Delivered Under the Auspices of the Committee on International Relations on the Los Angeles Campus of the University of California Spring 1942, University of California Press, Los Angeles, CA, 1943.

[18] The Editors of Encyclopaedia Britannica, Treaty of Lausanne, Available from: < https://www.britannica.com/event/Treaty-of-Lausanne-1923 > , 2019.

[19] B.D. Istanbul, UN Warns of 'Worst Humanitarian Disaster' of 21st Century as 30,000 Flee Syria's Idlib, Independent, Available from: < https://www.independent.co.uk/news/world/middle-east/syria-civil-war-idlib-province-battle-un-russia-air-strikes-assad-a8531976.html > , 2018.

[20] The Editors of Encyclopaedia Britannica, Taurus Mountains, Available from: < https://www.britannica.com/place/Taurus-Mountains > , 2019.

[21] The Editors of Encyclopaedia Britannica, Mediterranean Sea, Available from: < https://www.britannica.com/place/Mediterranean-Sea > , 2019.

[22] The Editors of Encyclopaedia Britannica, Sinai Peninsula, Available from: < https://www.britannica.com/place/Sinai-Peninsula > , 2019.

[23] University of Central Arkansas, French Syria (1919−1946), Available from: < https://uca.edu/politicalscience/dadm-project/middle-eastnorth-africapersian-gulf-region/french-syria-1919-1946/ > , 2019.

[24] The Editors of Encyclopaedia Britannica, United Arab Republic, Available from: < https://www.britannica.com/place/United-Arab-Republic > , 2019.

[25] E. Zisser, June 1967: Israel's capture of the Golan Heights, Israel Stud. 7 (1) (2002) 168.

[26] British Broadcasting Corporation (BBC), Who Are the Maronites?, Available from: < http://news.bbc.co.uk/2/hi/middle_east/6932786.stm > , 2007.

[27] British Broadcasting Corporation (BBC), Profile: Syrian City of Hama, Available from: < https://www.bbc.com/news/world-middle-east-17868325 > , 2012.

[28] Al-Jazeera Media Network, Syria's Civil War Explained From the Beginning, Available from: < https://www.aljazeera.com/news/2016/05/syria-civil-war-explained-160505084119966.html > , 2018.

[29] K. Wurx, President Bush Axis of Evil Speech, Youtube [Video], Available from: < https://www.youtube.com/watch?v = btkJhAM7hZw > , 2013.

[30] British Broadcasting Corporation (BBC), Syria Profile−Timeline, Available from: < https://www.bbc.com/news/world-middle-east-14703995 > , 2019.

[31] British Broadcasting Corporation (BBC), US Expands 'Axis of Evil', Available from: < http://news.bbc.co.uk/2/hi/1971852.stm > , 2002.

REFERENCES   31

[32] P. Salem, Iraq's Tangled Foreign Interests and Relations, Carnegies Middle East Center, December 2013, Available from: < https://carnegieendowment.org/files/Iraqs_Tangled_Foreign_Interests_and_Relations.pdf > , 2013.

[33] United Press International Inc., Syrian Foreign Minister Would Meet Rice, Available from: < https://www.upi.com/Top_News/2007/04/25/Syrian-foreign-minister-would-meet-Rice/56951177512218/ > , 2007.

[34] US Government Publishing Office, Administration of Barack H. Obama, 2009: Digest of Other White House Announcements, Available from: < https://www.govinfo.gov/content/pkg/DCPD-2009DIGEST/pdf/DCPD-2009DIGEST.pdf > , 2009.

[35] B.Y. Saab, On a New Footing: U.S.-Syria Relations, The Brookings Institution, Available from: < https://www.brookings.edu/articles/on-a-new-footing-u-s-syria-relations/ > , 2009.

[36] History.com Editors, Arab Spring, History, A&E Television Networks, Available from: < https://www.history.com/topics/middle-east/arab-spring > , 2019.

[37] A. France-Presse, Israel Admits 2007 Strike on Syrian Nuclear Site, The National, Available from: < https://www.thenational.ae/world/mena/israel-admits-2007-strike-on-syrian-nuclear-site-1.714815 > , 2018.

[38] World Vision, Syrian Refugee Crisis: Facts, FAQs, and How to Help, World Vision, Available from: < https://www.worldvision.org/refugees-news-stories/syrian-refugee-crisis-facts > , 2018.

[39] R. Reghunadhan, The impact of "de-escalation" zones in Syria, Centre Air Power Stud. 1 (2017) 2−3. Available from:. Available from: http://capsindia.org.managewebsiteportal.com/files/documents/CAPS_Infocus_RR_00.pdf.

[40] British Broadcasting Corporation (BBC), Arab Uprising: Country by Country − Bahrain, Available from: < https://www.bbc.com/news/world-12482295 > , 2013.

[41] J.N. Sater, Morocco's "Arab" Spring, Middle East Institute, Available from: < https://www.mei.edu/publications/moroccos-arab-spring > , 2011.

[42] Homeland Security Studies and Analysis Institute (HSSAI), Risks and Threats of Cryptocurrencies, Homeland Security Enterprise, Available from: < https://www.anser.org/docs/reports/RP14-01.03.03-02_Cryptocurrencies%20508_31Dec2014.pdf > , 2014.

[43] British Broadcasting Corporation (BBC), Guide to the Syrian Rebels, Available from: < https://www.bbc.com/news/world-middle-east-24403003 > , 2013.

[44] Al-Jazeera Media Network, US Launches Cruise Missiles on Syrian Airbase, Available from: < https://www.aljazeera.com/news/2017/04/us-missiles-syria-170407013424492.html > , 2017.

[45] A. France-Presse, Israel Admits 2007 Strike on Syrian Nuclear Site, The National, Available from: < https://www.thenational.ae/world/mena/israel-admits-2007-strike-on-syrian-nuclear-site-1.714815 > , 2018.

[46] Al-Jazeera Media Network, Syria: Under Russia's Fist, Available from: < https://www.aljazeera.com/programmes/peopleandpower/2016/02/syria-russia-fist-160225053929748.html > , 2016.

[47] C. Winter, Iran's Military Power: What You Need to Know, Deutsche Welle, Available from: < https://www.dw.com/en/irans-military-power-what-you-need-to-know/a-43756843 > , 2018.

[48] Ministry of Foreign Affairs, Relations Between Turkey−Syria, Available from: < http://www.mfa.gov.tr/relations-between-turkey%E2%80%93syria.en.mfa > , 2019.

[49] Global Coalition, Welcome to the Global Coalition Against Daesh, Available from: < https://theglobal-coalition.org/en/ > , 2019.

[50] K. Collin, 7 Years into the Syrian War, Is There a Way Out? Brookings, Available from: < https://www.brookings.edu/blog/order-from-chaos/2018/03/16/7-years-into-the-syrian-war-is-there-a-way-out/ > , 2018.

[51] R. Hinnebusch, Syria: the politics of economic liberalisation, Third World Q. 18 (2) (1997) 249−265.

[52] Al-Jazeera Media Network, Profile: Bashar al-Assad, Available from: < https://www.aljazeera.com/news/middleeast/2007/07/200852518514154964.html > , 2018.

[53] J. Barnes-Dacey, D. Levy, Three goals for making the most of Geneva II on Syria, Eur. Coun. Foreign Relat. (2014) MENA Programme, Available from:. Available from: https://www.ecfr.eu/page/-/Making_the_most_of_Geneva_II.pdf.

[54] International Institute for Democracy and Electoral Assistance (International IDEA), Constitutional History of Syria, Constitution Net, Available from: < http://constitutionnet.org/country/constitutional-history-syria >, 2016.

[55] British Broadcasting Corporation (BBC), Syria Conflict: War of Words as Peace Talks Open in Astana, < https://www.bbc.com/news/world-middle-east-38714441 >, 2017.

[56] D. Collins, Geneva 4 and the Shifting Shape of Syria Diplomacy, Al-Jazeera Media Network, Available from: < https://www.aljazeera.com/indepth/features/2017/03/geneva-4-shifting-shape-syria-diplomacy-170311081959353.html >, 2017.

[57] British Broadcasting Corporation (BBC), Nursultan: Kazakhstan Renames Capital Astana After ex-President, Available from: < https://www.bbc.com/news/world-asia-47638619 >, 2019.

[58] Amnesty International, Syria: The Biggest Humanitarian Crisis of Our Time, Available from: < https://www.amnesty.org.nz/take-action/syria-crisis >, 2017.

[59] Human Rights Watch (HRW), A Wasted Decade: Human Rights in Syria During Bashar al-Assad's First Ten Years in Power, Available from: < https://www.hrw.org/report/2010/07/16/wasted-decade/human-rights-syria-during-bashar-al-asads-first-ten-years-power >, 2010.

[60] MEHR, Refugee Bank to Provide Financial Inclusion for Refugee Crisis, Available from: < http://en.mehrnews.com/news/110981/Refugee-Bank-to-provide-financial-inclusion-for-refugee-crisis >, 2015.

[61] J. Redman, Blockchain Identity: Solving the Global Identification Crisis, Available from: < https://bitcoinist.com/blockchain-identity-solutions-solving-global-identification-crisis/ >, 2015.

[62] A. Srivastava, 5 Blockbuster Blockchain Startups from the Windmill Country, Available from: < https://siliconcanals.nl/news/startups/5-blockbuster-blockchain-startups-from-the-windmill-country/ >, 2018.

[63] World Food Programme (WFP), Blockchain Against Hunger: Harnessing Technology in Support of Syrian Refugees, Available from: < https://www.wfp.org/news/news-release/blockchain-against-hunger-harnessing-technology-support-syrian-refugees >, 2017.

[64] Middle East Intelligence Bulletin (MEIN), Dossier: Rifaat Assad, Available from: < https://www.meforum.org/meib/articles/0006_sd.htm >, 2000.

[65] Bitnation, Blockchain emergency ID (BE-ID), Available from: < https://refugees.bitnation.co/blockchain-emergency-id-be-id/ >, 2018.

[66] S.L. Sanchez, Blockchain Technology in Defence, European Defence Matters, Available from: < https://www.eda.europa.eu/webzine/issue14/cover-story/blockchain-technology-in-defence >, 2019.

[67] V. Adams, Why Military Blockchain Is Critical in the Age of Cyber Warfare, CONSENSYS, Available from: < https://media.consensys.net/why-military-blockchain-is-critical-in-the-age-of-cyber-warfare-93bea0be7619 >, 2019.

[68] K.D. Atherton, DARPA Wants to Know if It Can Do Anything With Blockchain, C4ISRNET, Available from: < https://www.c4isrnet.com/c2-comms/2018/12/06/darpa-wants-to-know-if-it-can-do-anything-with-blockchain/ >, 2018.

[69] United States Department of Defense (US DoD), Defense Science Board, Available from: < https://fas.org/irp/agency/dod/dsb/index.html >, 2017.

[70] C.E. Humud, C.M. Blanchard, M.B.D. Nikitin, Armed Conflict in Syria: Overview and U.S. Response, Congressional Research Service, RL33487, Available from: < https://crsreports.congress.gov >, 2019.

[71] Foundation of Defense of Democracies (FDD), How to Use the Blockchain to Secure a Supply Chain, YouTube, Available from: < https://www.youtube.com/watch?v = LaMlp22NQtU >, 2019.

[72] International Coalition for the Responsibility to Protect (ICRtoP), Syria & Geneva III: The Peace Process at a Glance, Available from: < http://responsibilitytoprotect.org/geneva-iii-timeline.pdf >, 2018.

[73] iHLS, Blockchain's New Civil Aviation Application, Available from: < https://i-hls.com/archives/89109 >, 2019.

[74] Nuclear Threat Initiative (NTI), Nuclear Weapons in the New Cyber Age, YouTube, Available from: < https://www.youtube.com/watch?v = uQjV8KfP6B8 >, 2018.

[75] C. Cordell, DARPA Wants to Take a Look at Blockchain's Security Rules, FEDSCOOP, Available from: < https://www.fedscoop.com/darpa-wants-take-look-blockchains-security-rules/ >, 2018.

[76] iHLS, Blockchain Solution Needed for Coping With Military Challenge, Available from: < https://i-hls.com/archives/92002 >, 2019.

[77] Nuclear Threat Initiative (NTI), The Nuclear Threat: Despite Progress, the Nuclear Threat Is More Complex and Unpredictable Than Ever, Available from: < https://www.nti.org/learn/nuclear/ >, 2019.

[78] O'Reilly Media, Inc., Chapter 8. Mining and Consensus, Chimera, Available from: < http://chimera.labs.oreilly.com/books/1234000001802/ch08.html#_proof_of_work_algorithm >, 2013.

[79] Blockchain Innovation Conference, David Birch — Tokenizing All Assets and Blockchain 3.0 | BIC18, YouTube, Available from: < https://www.youtube.com/watch?v = irh6lboEJ4U >, 2018.

[80] Accenture, Editing the Uneditable Blockchain: Why Distributed Ledger Technology Must Adapt to an Imperfect World, Available from: < https://www.accenture.com/cn-en/insight-editing-uneditable-blockchain >, 2016.

[81] E. Kedourie, The capture of Damascus1 October, 1918 Middle East. Stud. 1 (1) (1964) 70.

[82] CoinMarketCap, Top 100 Cryptocurrencies by Market Capitalization, Available from: < https://coinmarketcap.com/ >, 2019.

[83] Rokkex, A Huge List of Cryptocurrency Thefts, Hackernoon, Available from: < https://hackernoon.com/a-huge-list-of-cryptocurrency-thefts-16d6bf246389 >, 2019.

[84] J. Risberg, Yes, the Blockchain Can Be Hacked, Coincentral, Available from: < https://coincentral.com/blockchain-hacks/ >, 2018.

[85] P. Bendor-Samuel, The Primary Challenge to Blockchain Technology, Available from: < https://www.forbes.com/sites/peterbendorsamuel/2017/05/23/the-primary-challenge-to-blockchain-technology/2/#414157537a12 >, 2017.

[86] S. Ardittis, How Blockchain Could Make Refugee Programs More Transparent, Available from: < https://www.newsdeeply.com/refugees/community/2018/02/26/how-blockchain-could-make-refugee-programs-more-transparent >, 2018.

[87] M. White, J. Killmeyer, B. Chew, Will Blockchain Transform the Public Sector?, Deloitte Insights, Available from: < https://www2.deloitte.com/insights/us/en/industry/public-sector/understanding-basics-of-blockchain-in-government.html >, 2017.

[88] NATO Communications and Information Agency (NCIA), About, Available from: < https://www.ncia.nato.int/About/Pages/default.html >, 2019.

[89] E. Korkmaz, Blockchain for Refugees: Great Hopes, Deep Concerns, Oxford Department of International Development, Available from: < https://www.qeh.ox.ac.uk/content/blockchain-refugees-great-hopes-deep-concerns >, 2019.

[90] Fintechnews Switzerland, 87% of Financial Market Participants Say Blockchain Will Disrupt the Industry, Available from: < http://fintechnews.ch/blockchain_bitcoin/deutsche-bank-survey-87-of-financial-market-participants-say-blockchain-will-disrupt-the-industry/8174/ >, 2016.

FURTHER READING

C. Winter, Iran's Military Power: What You Need to Know, Deutsche Welle, Available from: <https://www.dw.com/en/irans-military-power-what-you-need-to-know/a-43756843>, 2018.

E. Zisser, June 1967: Israel's capture of the Golan Heights, Israel Stud. 7 (1) (2002) 168.

H. James, The Bitcoin Threat, Available from: <https://www.project-syndicate.org/commentary/bitcoin-threat-to-political-stability-by-harold-james-2018-02?utm_source = Project + Syndicate + Newsletter&utm_campaign = 4bf9c5d8e0-sunday_newsletter_4_2_2018&utm_medium = email&utm_term = 0_73bad5b7d8-4bf9c5d8e0-105499057>, 2018.

34 **CHAPTER 1** ETHICAL CONSIDERATIONS AND ISSUES OF BLOCKCHAIN

J. Pike, AEGIS Weapon System Mk 7, Available from: <https://www.globalsecurity.org/military/systems/ship/systems/aegis-core.htm>, 2019.

J. Young, Round-Up of Crypto Exchange Hacks So Far in 2019 — How Can They Be Stopped?, Bithoven, Available from: <https://cointelegraph.com/news/round-up-of-crypto-exchanges-hack-so-far-in-2019-how-can-it-be-stopped>, 2019.

N. Smolenski, Blockchain Records for Refugees, Medium, Available from: <https://medium.com/learning-machine-blog/blockchain-records-for-refugees-bd27ad6e6da1>, 2017.

S. Haber, W.S. Stornetta, How to time-stamp a digital document, J. Cryptol. 3 (2) (1991) 99−111.

T. Grove, F. Schwartz, U.S. dismisses Russia's ban on aircraft over Syrian safe zones, Wall Street J. (2017). Available from: <https://www.wsj.com/articles/russia-u-s-aircraft-barred-from-syria-safe-zones-1493983021>.

The Editors of Encyclopaedia Britannica, Maronite Church, Available from: <https://www.britannica.com/topic/Maronite-church>, 2019.

The Editors of Encyclopaedia Britannica, Rafiq al-Hariri, Available from: <https://www.britannica.com/biography/Rafiq-al-Hariri>, 2019.

US Government Publishing Office, Administration of Barack H. Obama, 2009: Digest of Other White House Announcements, Available from: <https://www.govinfo.gov/content/pkg/DCPD-2009DIGEST/pdf/DCPD-2009DIGEST.pdf>, 2009.

CHAPTER

BLOCKCHAIN-BASED FRAMEWORK FOR DATA STORAGE IN PEER-TO-PEER SCHEME USING INTERPLANETARY FILE SYSTEM

2

Randhir Kumar and Rakesh Tripathi

Department of Information Technology, National Institute of Technology, Raipur, India

2.1 INTRODUCTION

Recently, blockchain is an emerging technology that provides facilities of immutability, reliability, and availability of resources [1,2]. Moreover, it facilitates the decentralized platform, where data get disseminated among peers of the blockchain network. In the field of education system, blockchain technology is emerging as a promising technology that ensures the storage of different records of students such as grades, credits, and rewards [3,4]. Furthermore, blockchain maintains decorum to ensure the privacy and security of the systems.

This chapter highlights the importance of peer-to-peer storage system that is, interplanetary file system (IPFS), which provides decentralized hash table and version control scheme, where each transaction gets the hash value that is efficient to store in the blockchain network [5]. To reduce the size of the blockchain transaction, we have proposed the IPFS-based peer-to-peer decentralized storage system. The originality of the records gets verified by the hash of the content (transaction) [6].

Along with the facilities of Internet most of the education systems have initiated the attention to realizing the modernization of the higher education system. An important part of the current education system is to knowing the status of student including the details of students [7−9]. However, most of the current system is based on the centralized storage structure that has different drawback such as nontransparent, mutable, and nonavailability of the resource [10]. To overcome the existing drawback, we have proposed blockchain- and IPFS-based working model to store the status of students which ensures the immutability, availability, reliability, and decentralized storage scheme. We have also facilitated the content-based addressed access model rather than location-based access scheme using IPFS peer-to-peer distributed model [5].

In this chapter, the IPFS peer-to-peer storage is being proposed to store the record of students for the reliability and availability. Moreover, the immutability of the record gets maintained in the chain (ledger) of the blockchain network [11−13]. Furthermore, the fee structure of student which is in the format of image (jpg format) is maintained in the peer-to-peer IPFS system.

Handbook of Research on Blockchain Technology. DOI: https://doi.org/10.1016/B978-0-12-819816-2.00002-2
© 2020 Elsevier Inc. All rights reserved.

36 CHAPTER 2 BLOCKCHAIN-BASED FRAMEWORK FOR DATA STORAGE

2.1.1 MOTIVATION

Today, various applications are working to maintain the student records. However, these schemes are designed on a centralized approach that leads to numerous drawbacks, such as mutability of record, privacy of record, availability of record, and reliability of records. The various literature surveys have proposed the model of the blockchain in education system, but still most of them have not been implemented yet. In Liu et al. [2], the authors have applied the hyperledger fabric to store the record of student including credit score. However, the hyperledger scheme again works as a centralized storage with the top layer of blockchain, that is, chaincode. To overcome the issue of centralized approach of storage, we have proposed decentralized IPFS- and blockchain-based storage for student record maintenance. In our proposed working model, we have also included the access policy scheme for the storage chain (ledger) in the blockchain network.

2.2 RELATED WORK

The authors of Ref. [14] proposed blockchain-based storage to store the digital certificate of the candidate to improve the digital education system. The blockchain provides the decentralized platform, where the record can be traced easily. The blockchain is used to store the operational skill data set that enables the various matrices, which ensure the trust in digital education system. The work in Ref. [2] presented hyperledger composer-based permissioned blockchain storage for the storage of credit store of student which will be visible to employer industries. The information of students can be easily traced out on the chain. Hence, this approach becomes suitable for the industry person to identify the students.

The authors of Ref. [3] have used private and public blockchain to store the big data storage, where the public chain stores the storage address of the person and the private chain stores the data of the person. There are two parallel chains that are being used in the storage scheme. The authors have used this scheme to evaluate the learning behavior of the student and teaching behavior of the teacher. The private chain is stored using distributed storage techniques, whereas the public chain uses storage of centralized storage. Distributed chain stores the achievement, credits, and rewards, whereas the central stores the record achievement, credits, and rewards which include videos, audios, and image data.

The work in Ref. [4] presented the blockchain approach b-learning management system to store the data of students, where different smart contracts are being applied for managing the QUIZ, credits management, content sharing, and instant rewarding. The storage scheme manages the different certification data, knowledge data, and shared data

In Ref. [7], the authors have focused on the higher education system, where the same university has different campuses in the country. To manage the secure transaction that includes student profile and certification, the blockchain-based structure has been proposed.

The authors of Ref. [8] have addressed the data science skill gap between industry and data science education. To link the data science education, industry blockchain-based approach has been proposed by the authors. It is feasible for the industry to search the candidate of data science through this platform.

The work in Ref. [9] has modeled blockchain-based technique to implement the word-learning system for health-care decentralized data storage scheme. The personalized context-aware ubiquitous learning system has been used for word-learning approach that detects the content in the learning system.

In Ref. [15], the authors have addressed the limitation of blockchain technology in the field of education system. The main issue of the blockchain technology is that every peers must have the same chain (ledger), and the verification of the transaction must have to accomplished by almost every peers in the network; however, this approach leads to an issue for those peers who do not want to access the different information or set of information of the ledger or the peers who do not want to include in the verification process of transactions.

The authors of Ref. [16] have used the consortium blockchain (combination of public and private blockchain) approach to store the candidate answer in the subchain (private chain). These answers are matched against the prime chain(public chain) that already contains the list of answers of the quiz questions. The authors have used this model to evaluate the course credit system using quiz. However, the authors have not mentioned how consensus gets executed and how transactions dissemination is performed with this approach.

The authors of Ref. [17] have applied blockchain-based approach to store graduation requirement index of the university. This index will be stored by the evaluator with different types of evaluation of students. This approach also provides the facilities to evaluate the course outcome of the university. However, learning block consists of two different blocks, that is, student ability chain and course chain, where student ability chain stores the details of the student and course chain keeps the record of outcome of the course based on student performance.

The authors in Ref. [18] have implemented the course credit system of university (EduCTX) using blockchain technology, where each student can trace their record even after the completion of the course.

The author in Ref. [19] has implemented the ChainTutor application, the blockchain technology in classroom teaching. Moreover, the author has also included the wallet concept and its working in the blockchain technology.

2.3 WORKING MODEL OF FRAMEWORK

In this section, we have proposed working model of the transaction storage incorporating IPFS and blockchain. The transaction consists of record of a student with the details such as ID, name, address, course, and fee receipt. The transactions are uploaded to the IPFS network and blockchain for the availability. The ID, name, address, and course are added to the blockchain as it is, whereas fee receipt gets uploaded to the IPFS-distributed peer-to-peer network. The IPFS returns hash of the receipt during validation of the transaction. The resultant IPFS hash gets recorded in the blockchain network. The size of returned hash is fixed 46 bytes long which is much smaller than original size of the receipt. The proposed structure also reduces the size of blockchain owing to the IPFS hash. The proposed structure improves the efficiency of blockchain network because adding the original size of transaction (audio, video, image, and pdf) leads to the inefficiency of blockchain network. In this proposed structure, we have stored the fee receipt in the JPG format, which is about 20 KB, whereas the size of hash is about 46 bytes (368 bits) which is not even 1 KB in size. The blockchain network size is increasing day by day owing to the feature of append only. To overcome the drawback of append only feature of blockchain, our working model of storage can be utilized. As shown in Fig. 2.1, the scheme has been proposed for the distributed peer-to-peer storage of student record using blockchain and IPFS. The proposed model provides the availability, reliability, and scalable storage platform to store the record of student.

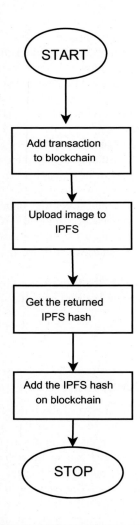

FIGURE 2.1

Proposed model of storage using interplanetary file system and blockchain.

2.4 RECORD STORAGE USING INTERPLANETARY FILE SYSTEM AND BLOCKCHAIN OVER CENTRALIZED SERVICE PROVIDER

In this section, we have discussed the significance of IPFS and blockchain over the centralized structure of data storage. The working model of student record storage is divided into two different structures such as IPFS and blockchain. The IPFS-distributed peer-to-peer system stores the image, and the blockchain structure keeps record of transactions. These two structures make the proposed working model reliable and secure.

2.4.1 INTERPLANETARY FILE SYSTEM–BASED DECENTRALIZED STORAGE SYSTEM

The IPFS is a version control system where each uploaded file gets the hash with new version [5]. Every time the new hash gets updated when any changes are made to the image (student fee receipt). The IPFS works on Merkle root, where root contains hash of the files with different versions. As shown in Fig. 2.2, the same file consists of different hash values after the modifications in file owing to the version control feature of IPFS-distributed file storage system.

These are the following features that signify the role of IPFS in our proposed working model.

2.4.1.1 Reliable and persistence

The IPFS-distributed network keeps the content (transaction) reliable and secure forever. The networks never become a failure due to the peer-to-peer decentralized structure, where each peer can

```
QmZPfgDjt6VB9kUNHd88Ur8z3udE7GxfscvzKU6mvL6Fu2
QmbYFwEDQpoqNQeHgiibn6uo8crSgWnpEFhkTWmRb2KWXu
QmbhLc9xWGGKk5VMMAsqGmrg43KYctdC7NZQDVkYqzVSQug
QmNo144Cn784WqGsJKXRbnXdoWjytgKmVyUJ65xFSf3f23
QmTJqq2FGPovXuQSNbHKGsxJbK4bghStuQ5WEF9v3X872r
QmWNGK1Nx4WPMAPJR11cqszLcbsPuqo8xcsKbxYdqAuXZg
QmfMoPaDucp5fQCqAQvLRAt4Bf3ThrJkASHtfXYGKeSbuL
QmPiXohb9oE5hEGLn6k2AdGdQAKmdTDvzVpaiHDbHQfJBb
QmPu4txvgob8zrwaa53bfF2d995KzoFzn6fmKHtmZo9oSQ
QmYV9Zvye6KNEVNJcvizdeyHfkDbsVsRj3wp27Qmkke58Z
QmctHQMBF3apFNJK7LeMzLuyjAqrmiAW2aHqWmWPHG86BJ
QmQ7MAX2wc1B47hh6bME5mkRnH7kVsX2MYbsXGSSSEYR6V
QmTh9Urge3BRQSUpMmWQxCrDLGD4SM7Sf3TXXhdUzTacGd
QmYwdsJsZZ4ZLTZT653etfgAr8BV97iv82GEKkmMpfDpUf
QmZeU8Pt8DijCDytkZnmbohfg2HSaCxEafQJtc74ECmyAu
QmerhhyfXHJwmUVM7RDZHSPmWgfWMH5uv4vDDjkkmkb7Av
QmQe1S1za3DPir5v8DaUCf7s6qRpEX1dozRLQ3ffGZfcJz
QmTXemfdV3EbtoPyTKsGq2SF9ysr3X3KB9h54j43EnF8Wc
QmXShgAE4NBVgqCDtUcU6ak9ayWqvLCDa3qBLwX5pK5NSY
QmYJXkB11VR7YvkjrEtLHr8uraDrqymXVkPkGp5VrjDh8g
```

FIGURE 2.2

Different version of hash for same file using interplanetary file system.

access the same content based on their hash value. The IPFS network is scalable and provides suitability in order to maintain larger files or set of files. If one of the peers fails in the network, then the same content (student record) can be accessed by other peers in the network [20].

2.4.1.2 Secured against distributed denial of service-style attacks
The content of IPFS-distributed peer-to-peer file storage system cannot be affected by the distributed denial of service (DDoS) attack and denial of service (DoS). The DDoS attack occurs when multiple systems from different locations synchronize to attack on a single target. This attack is different from the simple DoS attack, where attack is performed from the single location [5]. We have implemented our student record storage system using IPFS which ensures the security, attack resistance, and fault tolerant storage. In our working model, we have maintained the distributed structure where each peer contains the same chain (ledger) which ensures security against these attacks.

2.4.1.3 Previously viewed content available off-line
The caching approach of IPFS-distributed peer-to-peer system ensures the availability of the content off-line by default which is regularly viewed by the peer in the network. The dynamic content cannot be viewed off-line owing to the different versions of same content (version control), but the static content that does not get updated regularly can be viewed off-line. The static content can be viewed by their fingerprint (IPFS Hash) [5].

2.4.1.4 Decentralized hash table–based storage approach
As shown in Fig. 2.3, the file that exists on IPFS-distributed peer-to-peer storage has a unique IPFS hash to represent the file, and whenever the changes are made to the respective file, a new hash gets generated. These hashes play significant role in access of a file. The hashes of each file are distributed among all the peers of the IPFS network [20,21]. If any peer requests for the hash of a particular file, he or she can get the hash from any of the peers who are having the hash of that file. The IPFS-distributed system provides facility to make availability of the content quickly and

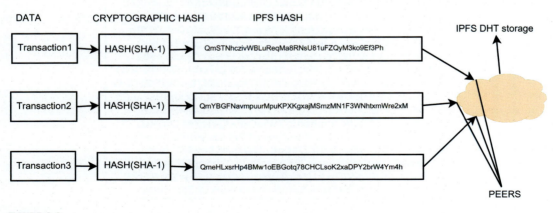

FIGURE 2.3

Interplanetary file system—distributed structure using decentralized hash table storage scheme.

2.4 RECORD STORAGE USING IPFS AND BLOCKCHAIN 41

accurately to the peers, regardless of the original host of the content. Moreover, the hashes can be checked against the correctness of the content, as a single bit changes the result in different hashes.

2.4.2 BLOCKCHAIN STRUCTURE TO MAINTAIN AVAILABILITY

The blockchain is a decentralized storage that maintains the list of growing records, called blocks. These blocks are linked using cryptographic hash of previous block, timestamp, and list of transactions. The blockchain consists of a public ledger that records transactions across many peers of the blockchain network, and these transactions cannot be altered without the modification of all subsequent blocks and the consensus of the blockchain network [22−24].

The important features of the blockchain network are presented in the following sections.

2.4.2.1 Decentralized peer-to-peer structure over centralized structure of storage

The centralized structure of data storage has various vulnerabilities:

1. In the centralized storage, all the data are stored at one place which can be easily targeted by potential hackers.
2. If any software upgradation is required in the centralized structure, then all the operations that is executing on the server must be stopped or halted.
3. If the central server goes down or some time shuts down for whatever reason, then nobody will get access of any information.
4. The biggest problem with centralized system: What if the central server gets corrupted and malicious, then all the data inside can be compromised.

In the decentralized structure, the information does not get stored in one place or single entity. Moreover, everyone in the network owns the information. The validations of transactions are being done by almost all the peers in the network. The information gets disseminated over all the peers of the network, and above all, the consistency is being maintained owing to different consensus.

The centralized and decentralized storage structures are shown in Fig. 2.4, where all the peers in the centralized storage depend on single entity to get information access. In the decentralized storage structure, every peer in the network can access the history of transaction and confirm new transaction without the need for central authority.

2.4.2.2 Transparency of record

As shown in Fig. 2.5, the blockchain storage structure provides high level of transparency ensuring that details of transactions are shared among all the participants of blockchain network [19,25]. In addition to high-level transparency built into blockchain network, there is also high level of integrity. The blockchain network includes auditable and valid chain (ledger) that is unforgeable and indelible. The transaction in the block can be added only after the validation by peers. Each entry of the transaction is visible to all the peers, which ensures transparency in the blockchain network. Fraud transaction can be easily detected owing to the feature of transparency in blockchain storage structure. The transparency removes the hurdle to check the transaction again and again, and this feature also reduces the manpower cost to verify the transaction every time requested by the peers.

42 CHAPTER 2 BLOCKCHAIN-BASED FRAMEWORK FOR DATA STORAGE

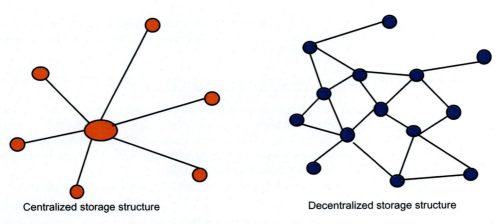

FIGURE 2.4

Centralized versus decentralized storage structure.

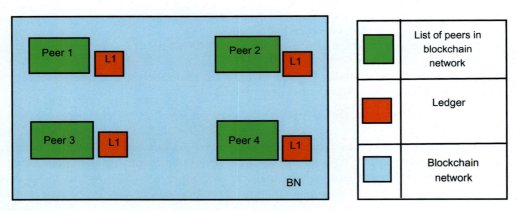

FIGURE 2.5

Peer-to-peer structure of blockchain transparency.

2.4.2.3 Immutability of record

The blockchain structure implies the immutable feature that ensures inability to make changes in data once they are recorded in the chain (ledger) [10,26]. The property is achieved owing to the feature of blockchain decentralization, where single peer of the blockchain network is responsible for the authenticity of the transaction.

As shown in Fig. 2.6, the Merkle root contains hash of all transactions hashes and keeps this into a block. The Merkle root hash provides quick verification of the block in peer-to-peer network. Each transaction associated with the blockchain network has a unique hash value. However, all

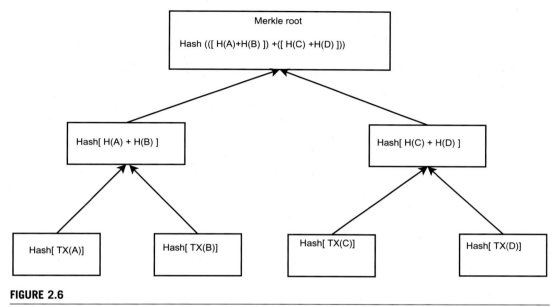

FIGURE 2.6

Merkle root hash of all transaction hashes.

these hashes do not get stored in sequential order in the block. These hashes are represented in block in the form of tree structure. The transaction hashes of the block are linked together which results into Merkle root. Modification in one transaction will also make change in the root of the block with the Merkle root hash. Similarly, modification in any block needs to make changes in all the sequential blocks of blockchain network; hence, the blockchain technology ensures immutability of the record (transactions) as well as tempering block in the blockchain structure.

2.4.2.4 Integrity of record

The manipulation of data is a serious threat to data integrity, where data might be tampered with and malicious peer might use this for their advantage. This problem generally occurs in centralized storage system. To overcome this malicious activity and data tamper, we have proposed blockchain-based structure, where hash of the block gets validated by the peers in the blockchain network [19,23]. The actual transaction of the block can be validated at any time against the hash of the block. In the proposed scheme of student storage system using blockchain, the hash gets distributed to all the peers to verify the transactions of a block. To maintain data integrity in our proposed scheme, we have designed the working model with certain access policy, where peers in the blockchain network can only verify and read the transaction. The peers are not allowed to make any modification in the chain (ledger) of the network. Moreover, the hash of each subsequent block gets attached to each other which maintains integrity in the blockchain network. Furthermore, the consensus of the blockchain network also ensures the integrity of the chain (ledger).

2.4.2.5 Availability of record

The blockchain peers contain same chain (ledger) in the network owing to the consensus algorithm that ensures consistency in the network. Moreover, the chain (ledger) gets duplicated over all the peers in the network which signifies availability of the information [19,24]. In contrast to centralized storage, the blockchain-based decentralized structure provides availability of information in such a way that if one of the peer fails in the network then the same chain (ledger) can be accessed by the other peers in the network. The drawback of centralized storage is that if the central entity gets compromised or corrupted, then it becomes impossible to get the availability of records. In our proposed working model, the designed structure of student record storage gets disseminated among all the peers in the network to access and verify the chain (ledger) of blockchain network. The peers in the network follow the consensus (proof of work) to retrieve similar chain of the network.

2.4.2.6 Access control policy

The blockchain is divided into two different parts, such as private blockchain and public blockchain. In this chapter, we are using private blockchain that is shown in Fig. 2.7, where only

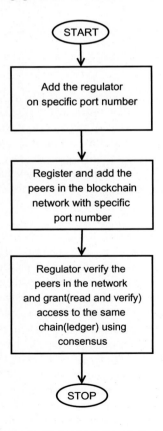

FIGURE 2.7

Access control policy for the proposed model.

authorized peers can join the network, whereas in the case of public network, anonymous peers can join the network and perform transaction sharing and access [27]. We have used private blockchain structure in which known peers can join the network and perform the transaction access. However, we have applied access control that is read-only access to peers of the network. The Regulator http://127.0.0.1:500/chain has control over the main chain (ledger) of the blockchain network. The regulator can give permission to other peers to join the network and get access of the main chain. We have applied the WORM (write once read many) policy while designing our proposed working model of student record storage [28].

2.5 WORKING OF INTERPLANETARY FILE SYSTEM

IPFS is a distributed file-sharing system that is designed to create a content-addressable and peer-to-peer model. IPFS allows user to share and receive the content without any central authority. In contrast to centralized system, the IPFS follows the decentralized structure of storage [5,11,20]. The IPFS protocols are designed with Merkle root concept which are used to maintain the different versions of file. The IPFS provides high-throughput, content-addressed block model and peer-to-peer storage system.

2.5.1 INSTALLATION OF INTERPLANETARY FILE SYSTEM

1. Install the go language from the given https://golang.ort/dl/ with the suitable operating system (click the version and download the. msi file).
2. Create the directory with name "go" in the any of the drive (c:\go).
3. Install the go language on the selected directory and set the path c:\go\bib.
4. Download go-ipfs from the given http://ipfs.io/docs/install and extract the ipfs into "c: go bin" directory.
5. Write the IPFS init on command prompt. It will generate the peer identity.
6. To interact with IPFS, use the ipfs cat/ipfs/ hash of ipfs /readme command. Now IPFS platform will get connected to the symbol of IPFS.
7. To move inside IPFS directory, type the command IPFS ls /ipfs/hash of ipfs.

2.5.2 INTERPLANETARY FILE SYSTEM COMMANDS

The following IPFS commands have been used in our proposed working model:

1. IPFS pin ls (it shows all the files that are shared by the specific peer ID). As shown in Fig. 2.8, the IPFS pin ls shows two different types of hash such as recursive and indirect. The indirect hash denotes the system file or the built-in ipfs file, whereas recursive hash is the hash of shared file by the peers on the IPFS-distributed network.
2. IPFS id(it shows all the details of the peers). As shown in Fig. 2.9, the details of the peer are shown with the peer ID and public key of the peer. The public key of the peer gets disseminated among other peers in the IPFS peer-to-peer file storage system. The peer ID and public key gets verified while sharing the file on the IPFS-distributed file storage platform.

46 CHAPTER 2 BLOCKCHAIN-BASED FRAMEWORK FOR DATA STORAGE

QmWpXCG3i14UskKD99QD2MqPuZ5dsKc3CT7SeBNxkuLhD3	indirect
QmWuPhaWLdrdDJ94yTt8jod4UZ7Rvk9n8NpN4dXZHjuqTx	indirect
QmWxG1wgUP81WekBMFBGtR3LeojubcHBouJusYExHrjoej	indirect
QmZaPMzBnk5LPsBa1oTvxm2yTyq13zb3JUavxznhdh55nn	indirect
Qmf7y8nbBhGczXXqnAP5TrM4Kt6Y4cMKvH4g1W7bHNGwzz	indirect
QmNuy8KakkVkbeUtvrFiMLHPs8bXeS1z8dif8UcH1Br5VL	indirect
QmRnRuSAm7Dsyv8A4GGVoPVs2LBgPYkKLFacKsqx2Avgv9	indirect
QmRwa4xWm6JaAFPZ2KkR1FPM6Vgu8dXmEnTi5McsGtUnuh	recursive
QmTd5FzTNkSGbxMEBcsELjJmjE5SYi1CmvDy4oiHQvfYLJ	indirect
QmVq91aerrtWtRb5ZmPRSCe1cPXwNQywh5voNVGTX5ydt6	recursive
QmeLAfTxgHbxZgAiLTAfeNk6EyTSWRV7gshPo4HN237SEP	indirect
QmNooYDeK4VnVraUXdHuDWf7vHtPwasKtJippdFtzCGeWa	indirect
QmTkaf6RUPNvS1jofxaYhuqiqnTs6oQdGiav71oMYHexAC	recursive
QmVSd7U7yXLJbPZc2hBkyPwVJJyZdeQRj3JXYzUpduwkM2	indirect
QmZhy1YzmDj5Z81hkeMUbdrjgeZpfsD9egUpBQmkCJtBYm	indirect
QmdNhX6WjCCy7HjtWACXjFDzGgZykr3Jd2Ks2DEQafzcmc	indirect
Qmf5RdXJqrvV89YTD5LjSgGxj2NFLq3L4HCaRu4HfcyjXD	indirect
QmQ5vhrL7uv6tuoN9KeVBwd4PwfQkXdVVmDLUZuTNxqgvm	indirect

FIGURE 2.8

List of files shared on interplanetary file system—distributed peer-to-peer storage system.

```
C:\Users\ITLab3_Phd>ipfs id
{
        "ID": "QmUJ2KVCBZQFYJD7YQmts6KsQYSqyrJE3PoCZdimg3A2Vn",
        "PublicKey": "CAASpgIwggEiMADGCSqGSIb3DQEBAQUAA4IBDwAwggEKAoIBAQDtxXfOV3w1QWfegWN/PKYET3qm2BWcpp6ACnudwELOhLYC2
W1rwnKIGEyxnpRhRnNA6FxDkWDkABRaYcPh8Av47VkJ1SMtEmdJSB7owFQ6TaXOV3wpaW1SE9qAAMJjch6viH9Zb1Bm7srKnRjgLGmLOYbWceJGEWWEoIWn$
4WYUMSx9NOX59NAgMBAAE=",
        "Addresses": [
                "/ip4/169.254.179.176/tcp/4001/ipfs/QmUJ2KVCBZQFYJD7YQmts6KsQYSqyrJE3PoCZdimg3A2Vn",
                "/ip4/172.22.52.231/tcp/4001/ipfs/QmUJ2KVCBZQFYJD7YQmts6KsQYSqyrJE3PoCZdimg3A2Vn",
                "/ip4/192.168.56.1/tcp/4001/ipfs/QmUJ2KVCBZQFYJD7YQmts6KsQYSqyrJE3PoCZdimg3A2Vn",
                "/ip4/127.0.0.1/tcp/4001/ipfs/QmUJ2KVCBZQFYJD7YQmts6KsQYSqyrJE3PoCZdimg3A2Vn",
                "/ip6/::1/tcp/4001/ipfs/QmUJ2KVCBZQFYJD7YQmts6KsQYSqyrJE3PoCZdimg3A2Vn",
                "/ip6/2001:0:17ca:e7a7:491:36d8:53e9:cb18/tcp/4001/ipfs/QmUJ2KVCBZQFYJD7YQmts6KsQYSqyrJE3PoCZdimg3A2Vn"
        ],
        "AgentVersion": "go-ipfs/0.4.19/",
        "ProtocolVersion": "ipfs/0.1.0"
}
```

FIGURE 2.9

Details of peer with peer ID and their public key.

3. http://127.0.0.1:8080/ipfs/ < hashoffile > (The following command access the file that is shared on the IPFS-distributed peer-to-peer file storage system.)

To get the more details of IPFS commands follow the link https://docs.ipfs.io/reference/api/cli/.

2.5.3 INTERPLANETARY FILE SYSTEM CONNECTION AND DEPLOYMENT

As shown in Fig. 2.10, the commands that we have used is to connect with IPFS-distributed peer-to-peer file system.

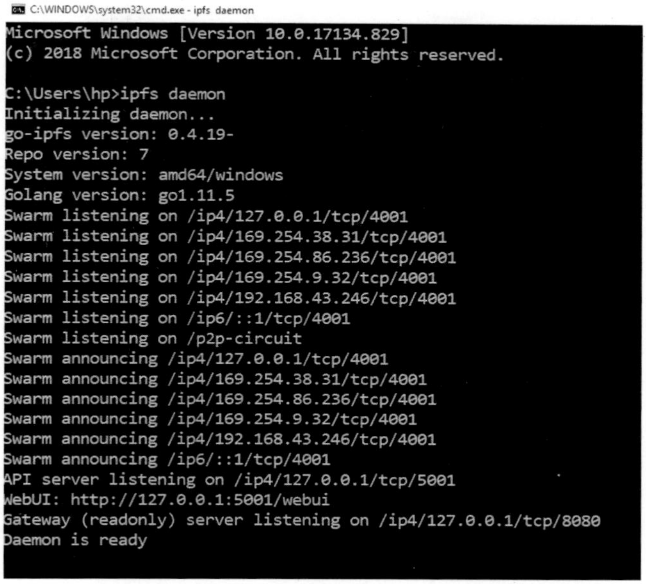

2.5 WORKING OF INTERPLANETARY FILE SYSTEM   47

```
C:\WINDOWS\system32\cmd.exe - ipfs  daemon
Microsoft Windows [Version 10.0.17134.829]
(c) 2018 Microsoft Corporation. All rights reserved.

C:\Users\hp>ipfs daemon
Initializing daemon...
go-ipfs version: 0.4.19-
Repo version: 7
System version: amd64/windows
Golang version: go1.11.5
Swarm listening on /ip4/127.0.0.1/tcp/4001
Swarm listening on /ip4/169.254.38.31/tcp/4001
Swarm listening on /ip4/169.254.86.236/tcp/4001
Swarm listening on /ip4/169.254.9.32/tcp/4001
Swarm listening on /ip4/192.168.43.246/tcp/4001
Swarm listening on /ip6/::1/tcp/4001
Swarm listening on /p2p-circuit
Swarm announcing /ip4/127.0.0.1/tcp/4001
Swarm announcing /ip4/169.254.38.31/tcp/4001
Swarm announcing /ip4/169.254.86.236/tcp/4001
Swarm announcing /ip4/169.254.9.32/tcp/4001
Swarm announcing /ip4/192.168.43.246/tcp/4001
Swarm announcing /ip6/::1/tcp/4001
API server listening on /ip4/127.0.0.1/tcp/5001
WebUI: http://127.0.0.1:5001/webui
Gateway (readonly) server listening on /ip4/127.0.0.1/tcp/8080
Daemon is ready
```

FIGURE 2.10

Daemon execution to connect interplanetary file system peer-to-peer storage network.

1. To execute the IPFS, use the command ipfs daemon (C:\Users\hp > ipfsdaemon).

   The IPFS daemon command initiates the execution of IPFS-go peer-to-peer file system. The command establishes the connection on the http://127.0.0.1:5001/webui to share and deploy various files from the same peer ID.

2. As shown in Fig. 2.11, the connection of IPFS is established on the given http://127.0.0.1:5001/webui.

   The dashboard of IPFS consists of five different options such as status, file, explore, peers, and settings. The status is used to provide the details of the peers, size of shared files by the peers, number of peers connected, and the bandwidth that is utilized by the peer. There are two types of bandwidth utilized such as IN and OUT. The IN bandwidth is used to share the files on IPFS, and OUT bandwidth is utilized to access the files from the IPFS. The file options describe the details of files with their IPFS hash (content-addressed). The explore options describe the advance setting on IPFS-distributed file-sharing system. The Peers provides list of peers with the details such as peer ID, ip address, county, etc. The setting is used to explore the IPFS configurations.

3. To add the file on IPFS, use the command ipfs add./filename.jpg (C:\Users\hp\ipfsadd./Image_Upload_1.png>).

As shown in Fig. 2.12, the file gets uploaded into IPFS peer-to-peer file storage system, and the respective hash (46-byte size) is stored on the blockchain network. The command facilitates deployment of file with the peers on the IPFS network.

48 CHAPTER 2 BLOCKCHAIN-BASED FRAMEWORK FOR DATA STORAGE

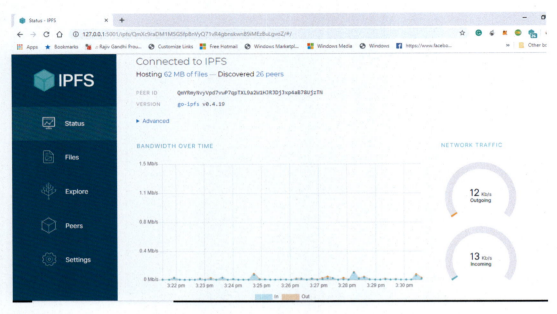

FIGURE 2.11

Execution of interplanetary file system Web-UI.

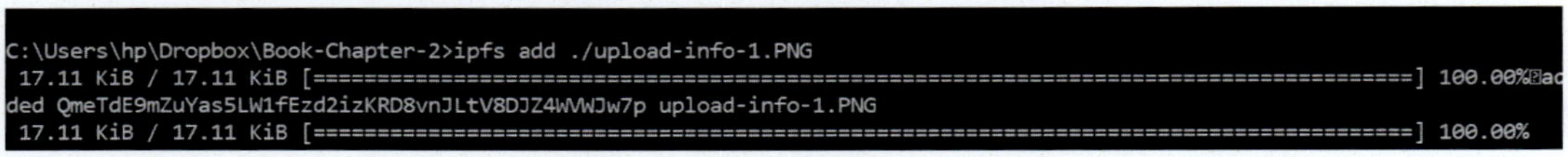

FIGURE 2.12

To add file into interplanetary file system network.

2.6 IMPLEMENTATION OF FRAMEWORK USING INTERPLANETARY FILE SYSTEM AND BLOCKCHAIN

The implementation of our proposed working model for data storage is carried out using blockchain and IPFS-distributed file-sharing system, where each transaction recorded by their hash value is returned by the IPFS-distributed peer-to-peer storage system. The experimental setup consists of python anaconda and python flask. The setup is performed on Intel(R) Xeon(R) W-2175 CPU @ 2.50 GHZ running Window x64-based processor with 128 GB of RAM and 1 TB of local storage.

2.6.1 UPLOADING TRANSACTIONS TO THE INTERPLANETARY FILE SYSTEM AND BLOCKCHAIN NETWORK

As shown in Figs. 2.13 and 2.14, the student information gets uploaded into the IPFS and blockchain. There are two different types of transactions being shown, that is, text and image. To store the image as a transaction on blockchain, we have used IPFS-distributed peer-to-peer file system that returns the hash of image which is much smaller than the original size of an image.

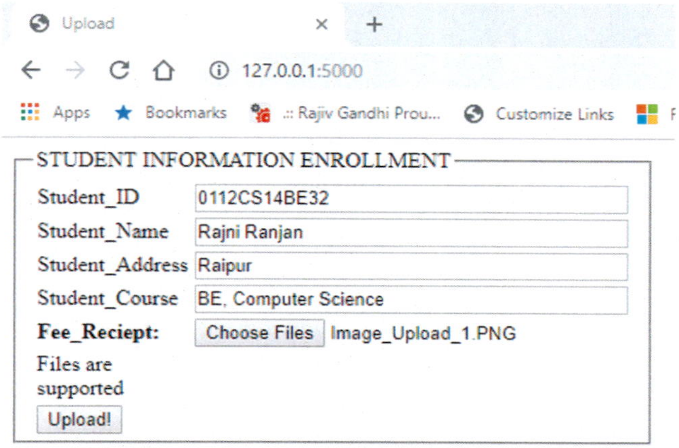

FIGURE 2.13

Uploading student information to the blockchain network.

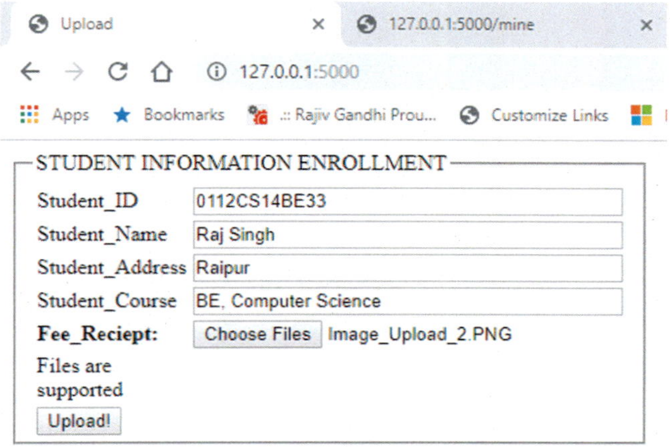

FIGURE 2.14

Uploading student information to the blockchain network.

In this experiment we have used jpg and png formats of image to generate report of the student. The same format gets uploaded to the IPFS network, and the returned hash gets added to the blockchain availability.

2.6.2 VALIDATING THE TRANSACTION USING MINING PROCESS

Mining is a process that is used to validate and add the transaction to the existing ledger of blockchain which is distributed among all the peers of the blockchain network. The mining process creates the hash of a block of transactions to ensure integrity of the blockchain without involvement of the central system. The proof of work gets generated during the mining process which ensures integrity of the block.

As shown in Figs. 2.15 and 2.16, we have applied mining activity to validate the transactions. The uploaded transactions are verified using mining process (which is discussed in Figs. 2.5 and 2.6). In our proposed scheme of mining, we have created the IPFS hash for the uploaded image (fee receipt of student) rather than storing original file into the blockchain network. The created proof of work maintains integrity among the peers of blockchain network. The previous hash maintains the chain of storage and ensures the immutability of records.

2.6.3 ADDING TRANSACTION INTO BLOCKCHAIN NETWORK

As shown in Fig. 2.17, the list of transactions are recorded into the blockchain storage after the mining process. The first block of blockchain is genesis block that stores the transaction which is

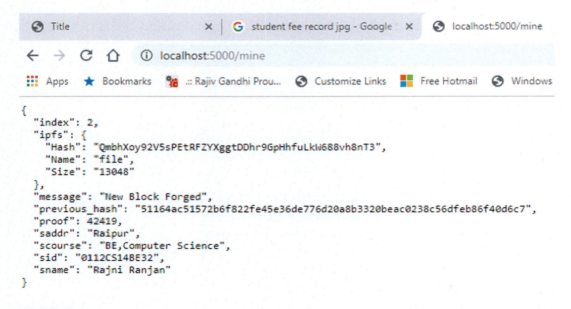

FIGURE 2.15

Process of mining to transaction validate and interplanetary file system hash generation.

2.6 IMPLEMENTATION OF FRAMEWORK USING IFPS AND BLOCKCHAIN 51

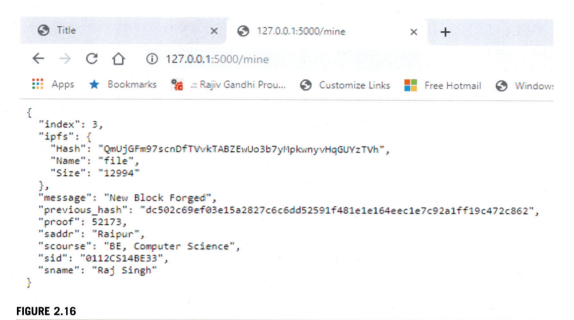

FIGURE 2.16

Process of mining to transaction validate and interplanetary file system hash generation.

initialized not by specific and meaningful values. The rest of the block consists of information of the student according to the transaction uploaded by the peers in the blockchain network. The timestamp of the blockchain provides the history of transaction such as when this transaction is created [29]. The blockchain ensures trust among the peers owing to the provenance of record list. The index number specifies the block number in our blockchain network. The block of information ensures the availability, the list of previous hash maintains immutability and fingerprint of the block structure, the proof of work ensures integrity, and the IPFS hash ensures efficiency owing to the reduced size of an original transaction in the blockchain network. Moreover, the hash generated by the IPFS can be used as a content-based access of the transaction. Furthermore, the transactions that are added to the blockchain network cannot be altered by the peers of the blockchain network, and this feature of the blockchain structure ensures the tempered evidence distributed to the ledger. Each block contains cryptographic hash of previous block which is secured by the design. This cryptographic hash of the block gets changed whenever the transactions of block are being modified by any source in the blockchain network, either by peers or by regulator.

2.6.4 ACCESS OF TRANSACTION USING CONTENT-ADDRESSED TECHNIQUE

As shown in Figs. 2.18 and 2.19, the transaction that is recorded into the blockchain network by their IPFS hash value gets accessed by the http://127.0.0.1:8080/ipfs/hashofthefile. In this experiment, we are accessing the fee receipt of the students by their hash value. The access scheme which

52 CHAPTER 2 BLOCKCHAIN-BASED FRAMEWORK FOR DATA STORAGE

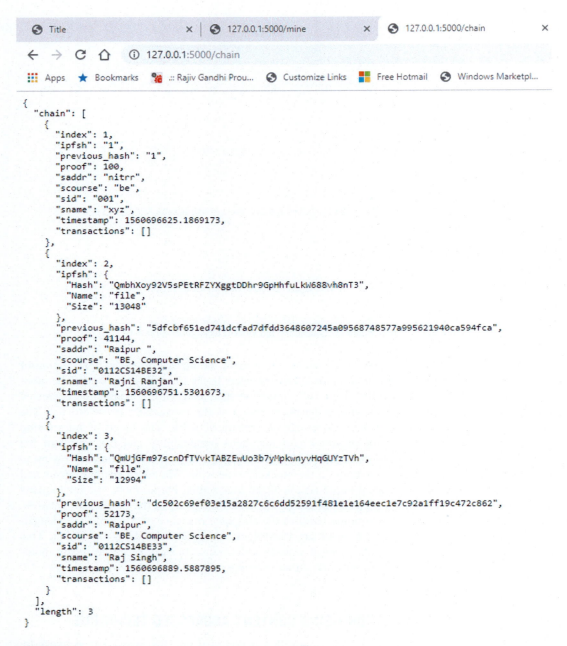

FIGURE 2.17

List of student record on blockchain network.

2.6 IMPLEMENTATION OF FRAMEWORK USING IFPS AND BLOCKCHAIN

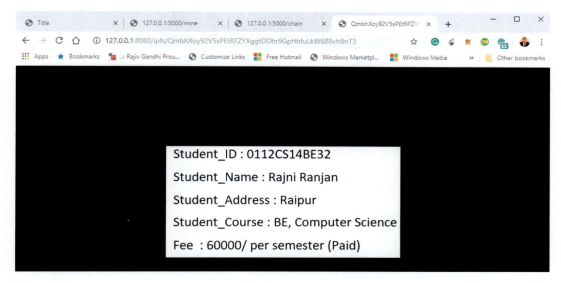

FIGURE 2.18

Access of student record from interplanetary file system using its hash.

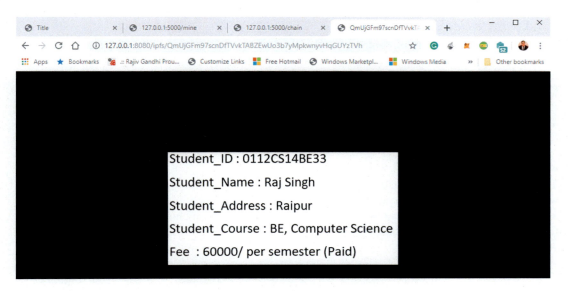

FIGURE 2.19

Access of student record from interplanetary file system using its hash.

54 CHAPTER 2 BLOCKCHAIN-BASED FRAMEWORK FOR DATA STORAGE

we are using in our working model is known as content-addressed scheme. Moreover, this scheme accesses the record by the hash of the file rather than its location.

The proposed scheme of content-addressed includes various advantages.

2.6.4.1 Advantage of content addressed over location addressed

The content-addressed scheme facilitates to identify the content by its fingerprint (IPFS Hash) rather than by identifying its location. The access of content with the fingerprint ensures the access of the copy of content hold by the any peer in the network rather than access the content from the specific location. The location-based access of content can have a problem during access once the location is changed, whereas content-addressed always accesses the copy of content from the peers in the present network. To access and identify the content, cryptographic hash-SHA (IPFS hash) is utilized which never gets changed in the distributed IPFS network. Moreover, the content addressed guarantees that the hash will always return the same content regardless of where the content is retrieved from and when the content was added.

2.6.4.2 Maintains integrity

The process of decentralization increases the integrity of data owing to the links that are content addressed. Moreover, we can validate the data or content by its fingerprint (unique IPFS hash) against the links of file, name of file, and location of files. The proposed techniques and working model provide efficiency with large-scale data sets that are accessed by millions of peers at same time. The term location addressed in all the connections is brittle, whereas content addressed the connections is more reliable and resilient.

2.6.4.3 Increases durability of data

The decentralized structure of content addressed radically increases durability of data. The content-addressed approach ensures that data cannot become endangered as long as any peer in the network is holding the copy of the data. If any peer in the network holds the data set, then it cannot be lost [30]. The peers do not have to worry about single points of failure. The content-addressed approach verifies the peer ID before access of the transactions from IPFS-distributed network. This scheme also verifies the list of peers accessing the same transactions from the different peers ID or not. Hence, this approach also ensures security in order to access of transactions.

2.6.4.4 Permanent storage of files

The content-addressed approach is permanent which exactly points to that content. The content-addressed protocol has significance in the field of computer science in terms of persistent data storage. In this working model, we are only using few implications including sharing and storing the transactions using a content-addressed protocol.

2.6.5 ADDING PEERS INTO BLOCKCHAIN NETWORK

As shown in Fig. 2.20, we have added the peer on the blockchain network using different port number 5005 with the URL http:/127.0.0.1:5005/nodes/register. The peer is added on the main chain URL http://127.0.0.1:5005 of blockchain network. The peer with port 5005 can see all the details of port 5000 which is added on the blockchain. The blockchain consists of a set of peer nodes that

2.6 IMPLEMENTATION OF FRAMEWORK USING IFPS AND BLOCKCHAIN

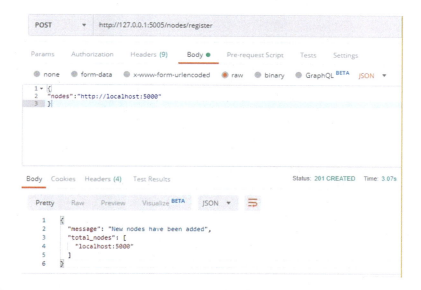

FIGURE 2.20

Adding peers to the blockchain network.

hold the chain. In this experiment, we have shown the interaction between peers and main chain. To register the peer into the blockchain network, we have used postman browser where json structure has been used. The node 5005 gets registered on peer 5000 to interact with the main chain of blockchain network. The postman browser makes easy to interact and register the node to the blockchain network. To register the node, we have used POST method to transfer the registration data of nodes in the network. The peers in the network hold the instance of chain for blockchain network which ensures prevention from the single point of failure.

As shown in Fig. 2.20, all the added peers must have to follow the consensus of the blockchain network to maintain the consistency in the network. Moreover, the peers are also allowed to verify the transaction after the mining process is complete. Once the verification of transactions completed by the pees, then the transaction gets added to the block. To access the same chain by the peers, the consensus must be resolved (which is discussed in next subsection of use of consensus in blockchain network).

2.6.6 USE OF CONSENSUS IN BLOCKCHAIN NETWORK

In contrast to the centralized system, the blockchain network follows decentralized structure that works on large scale without any single authority [6]. The blockchain involves thousands of peers (participants) who verify the transactions occurring in the blockchain network. The dynamic changing behavior of the blockchain needs to be efficient, real time, reliable, functional, secure, and fair to ensure the transactions happening in the network are real and genuine, and each peer in the network agrees on the consensus and the status of the chain (ledger). As shown in Fig. 2.21, we have

56 CHAPTER 2 BLOCKCHAIN-BASED FRAMEWORK FOR DATA STORAGE

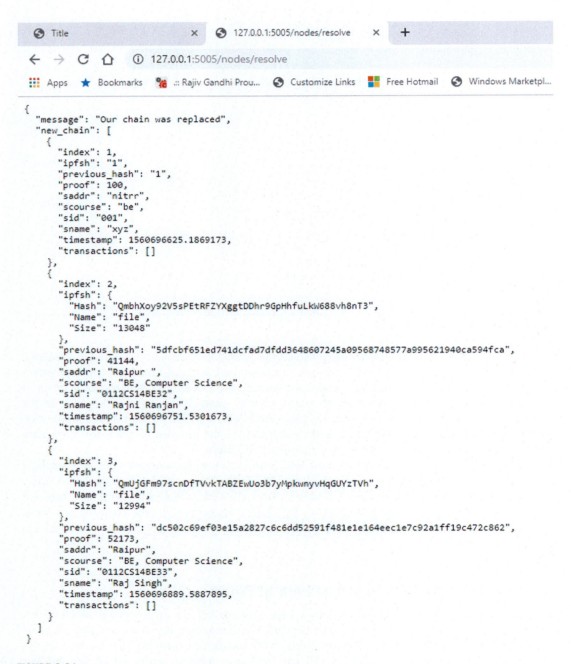

FIGURE 2.21

Access of similar chain (Ledger) of student record by the peer (5005).

used proof of work as a consensus algorithm to maintain the genuine status of the chain (ledger). The peer follows the same consensus to get the details of chain in the blockchain network. To access the consensus by the peers, we have used the following URL http:/127.0.0.1:5005/nodes/resolve. As we can see, the peers with port 5005 obtain same chain (ledger) of the blockchain network. Similarly, the set of peers can be added to the blockchain network using postman browser and get resolved by the proof-of-work consensus algorithm to get the similar chain of the blockchain network. This experiment shows that each peer can access the same chain once the peers get resolved. The characteristics of consensus make redundant chain (ledger) by providing access to set of peers in the network which ensures prevention against single point of failure (centralized structure).

2.7 CONCLUSION

In the proposed working model, we have implemented the student record storage framework using the blockchain and IPFS, which ensure the immutability, integrity, and availability of the record. The IPFS has been used for the purpose of generating hash of the fee receipt (jpg format image) which is much smaller than the original image in terms of storage. To reduce the size of blockchain network, we have used the IPFS-distributed peer-to-peer file system storage. Moreover, we have also applied the access control rules, where the peers are only allowed to validate and read the transactions of chain (ledger). The existing centralized storage scheme has various limitations, such as single point of failure, information can be compromised, not reliable, and nonavailability of information. To deal with the drawbacks of the existing storage system, we have used our working model that is efficient to handle all these limitations. We have also discussed the utilization of content-based addressed scheme in contrast to location-based addressed scheme to access the transaction. The proposed working model can also be applied to store other formats like video and audio. Moreover, the working model scheme can also be applied on larger set of data owing to the hash creation feature of IPFS and distributed storage scheme of blockchain network.

REFERENCES

[1] Q.K. Nguyen, Q.V. Dang, Blockchain technology for the advancement of the future, 2018 4th International Conference on Green Technology and Sustainable Development (GTSD), IEEE, 2018, pp. 483−486.

[2] Q. Liu, Q. Guan, X. Yang, H. Zhu, G. Green, S. Yin, Education-industry cooperative system based on blockchain, 2018 1st IEEE International Conference on Hot Information-Centric Networking (HotICN), IEEE, 2018, pp. 207−211.

[3] X. Gong, X. Liu, S. Jing, G. Xiong, J. Zhou, Parallel-education-blockchain driven smart education: Challenges and issues, Chinese Automation Congress (CAC), IEEE, 2018, pp. 2390−2395.

[4] G. Sahonero-Alvarez, Blockchain and peace engineering and its relationship to engineering education, 2018 World Engineering Education Forum-Global Engineering Deans Council (WEEF-GEDC), IEEE, 2018, pp. 1−6.

[5] J. Benet, Ipfs-content addressed, versioned, p2p file system, arXiv preprint arXiv:1407.3561.

[6] S. Nakamoto, et al., Bitcoin: A peer-to-peer electronic cash system, 2008.

[7] K. Al Harthy, F. Al Shuhaimi, K.K.J. Al Ismaily, The upcoming blockchain adoption in higher-education: requirements and process, 2019 4th MEC International Conference on Big Data and Smart City (ICBDSC), IEEE, 2019, pp. 1−5.

[8] A. Mikroyannidis, J. Domingue, M. Bachler, K. Quick, Smart blockchain badges for data science education, 2018 IEEE Frontiers in Education Conference (FIE), IEEE, 2018, pp. 1−5.

[9] J. Zhong, H. Xie, D. Zou, D.K. Chui, A blockchain model for word-learning systems, 2018 5th International Conference on Behavioral, Economic, and Socio-Cultural Computing (BESC), IEEE, 2018, pp. 130−131.

[10] M. Vukolić, The quest for scalable blockchain fabric: Proof-of-work vs. bft replication, International workshop on open problems in network security, Springer, 2015, pp. 112−125.

[11] M. Fukumitsu, S. Hasegawa, J. Iwazaki, M. Sakai, D. Takahashi, A proposal of a secure p2p-type storage scheme by using the secret sharing and the blockchain, 2017 IEEE 31st International Conference on Advanced Information Networking and Applications (AINA), IEEE, 2017, pp. 803−810.

[12] J. Sidhu, Syscoin: A peer-to-peer electronic cash system with blockchain-based services for e-business, 2017 26th international conference on computer communication and networks (ICCCN), IEEE, 2017, pp. 1−6.

[13] G. Chen, B. Xu, M. Lu, N.-S. Chen, Exploring blockchain technology and its potential applications for education, Smart Learn. Environ. 5 (1) (2018) 1.

[14] B. Wu, Y. Li, Design of evaluation system for digital education operational skill competition based on blockchain, 2018 IEEE 15th International Conference on e-Business Engineering (ICEBE), IEEE, 2018, pp. 102−109.

[15] C. Udokwu, A. Kormiltsyn, K. Thangalimodzi, A. Norta, The state of the art for blockchain-enabled smart-contract applications in the organization, 2018 Ivannikov Ispras Open Conference (ISPRAS), IEEE, 2018, pp. 137−144.

[16] H. Shen, Y. Xiao, Research on online quiz scheme based on double-layer consortium blockchain, 2018 9th International Conference on Information Technology in Medicine and Education (ITME), IEEE, 2018, pp. 956−960.

[17] A. Alammary, S. Alhazmi, M. Almasri, S. Gillani, Blockchain-based applications in education: A systematic review, Appl. Sci. 9 (12) (2019) 2400.

[18] M. Turkanović, M. Hölbl, K. Košič, M. Heričko, A. Kamišalić, Eductx: A blockchain-based higher education credit platform, IEEE Access 6 (2018) 5112−5127.

[19] X. Liu, A small java application for learning blockchain, 2018 IEEE 9th Annual Information Technology, Electronics and Mobile Communication Conference (IEMCON), IEEE, 2018, pp. 1271−1275.

[20] Y. Chen, H. Li, K. Li, J. Zhang, An improved p2p file system scheme based on ipfs and blockchain, 2017 IEEE International Conference on Big Data (Big Data), IEEE, 2017, pp. 2652−2657.

[21] J. Pearce, Blockchain in academia: A literature review, 2019.

[22] A. Grech, A. F. Camilleri, Blockchain in education, 2017.

[23] W. Gräther, S. Kolvenbach, R. Ruland, J. Schütte, C. Torres, F. Wendland, Blockchain for education: lifelong learning passport, in: Proceedings of 1st ERCIM Blockchain Workshop 2018, European Society for Socially Embedded Technologies (EUSSET), 2018.

[24] D. Puthal, N. Malik, S.P. Mohanty, E. Kougianos, C. Yang, The blockchain as a decentralized security framework [future directions], IEEE Cons. Electr. Mag. 7 (2) (2018) 18−21.

[25] T. Aste, P. Tasca, T. Di Matteo, Blockchain technologies: the foreseeable impact on society and industry, Computer 50 (9) (2017) 18−28.

[26] S. Kolvenbach, R. Ruland, W. Gräther, W. Prinz, Blockchain 4 education, in: Proceedings of 16th European Conference on Computer-Supported Cooperative Work-Panels, Posters and Demos, European Society for Socially Embedded Technologies (EUSSET), 2018.

REFERENCES

[27] M. Fenwick, W.A. Kaal, E.P. Vermeulen, Legal education in the blockchain revolution, Vand. J. Ent. Tech. L 20 (2017) 351.

[28] X. Yang, X. Li, H. Wu, K. Zhao, The application model and challenges of blockchain technology in education, Mod. Dist. Educ. Res. 2 (2017) 34–45.

[29] R. Neisse, G. Steri, I. Nai-Fovino, A blockchain-based approach for data accountability and provenance tracking, Proceedings of the 12th International Conference on Availability, Reliability and Security, ACM, 2017, p. 14.

[30] S. Kim, S. Park, Y. Park, J. Kim, Y.-H. Chc, J.-Y. Choi, et al., A feature based content analysis of blockchain platforms, 2018 Tenth International Conference on Ubiquitous and Future Networks (ICUFN), IEEE, 2018, pp. 791–793.

CHAPTER

INTEGRATION OF BLOCKCHAIN AND INTERNET OF THINGS

3

S. Porkodi and D. Kesavaraja

Department of Computer Science and Engineering, Dr. Sivanthi Aditanar College of Engineering, Tiruchendur, India

3.1 INTRODUCTION

As there is lots of Internet of Things (IoT) adaption in the present and in the future, to preserve the integrity of the data in a huge scale there is a need of some protection other than the traditional techniques such as cryptography [1]. IoT is mainly based on the internet is basically insecure as lot of hackers threaten to steal the data, data loss, continuous data transfer, manual handling of devices, etc. so security to the data is very much important [2]. Also both the IoT and the internet have different architecture. Higher computing capability between the objects and extended network connectivity also with much lesser computing power can be achieved in the IoT by using sensors and temporary devices to collect, exchange, and use data minimum human works [3]. Implementing costly security to internet and extending the computational demand is not a solution and also it is not a practical approach [4]. As the internet has huge IoT data to store and process, cloud storage is used above the internet [5]. In some cases, as there are lots of data generated in the real time, they are stored in cloud servers which are accessed and processed via distributed cloud computing method [6,7]. Cloud service is also insecure due to the connection of the internet and they are vulnerable to cybercrime attacks such as data tampering [8], structured query language (SQL) injection [9], and node failure [10]. Data integrity and data availability [9] cannot be ensured by cloud service to the IoT applications as expected.

Blockchain can be used as it is a distributed ledger, tamper resistant, not susceptible to corruption. It also has the capacity to find the critical security problems of IoT mainly reliability and data integrity [11]. Blockchain gives access to software to send and receive data transactions or events in a distributed also trustworthy manner. Nowadays, the popularity of the blockchain is increasing due to the usage of the digital asserts [12], smart contracts [13], and distributive storage [14]. The major application of blockchain in the IoT includes events recording such as moisture, pressure, temperature change, location changes, also it has ledgers which are tamper resistant that can only read by the authenticated users. The requirements of the security feature in the IoT can be satisfied by the blockchain technology [15]. The features of the blockchain integrated with IoT applications to increase the security of IoT.

Handbook of Research on Blockchain Technology. DOI: https://doi.org/10.1016/B978-0-12-819816-2.00003-4
© 2020 Elsevier Inc. All rights reserved.

62 CHAPTER 3 INTEGRATION OF BLOCKCHAIN AND INTERNET OF THINGS

3.1.1 INTEGRITY

In blockchains, the transactions can be stored permanently with the verifiable techniques. The sender is made to register their signature in the transaction of data to give guarantee to the nonrepudiation and integrity of data transaction. The blockchains are with a hash chain structure to make sure that every transaction is recorded which cannot be updated. The blockchain protocol can be guaranteed with consistent and valid record, also can tolerate attacks and failures. Some of the attacks include, at Proof of Work (PoW) attackers attack with lesser than ½ hash power, minimum of 1/3 node present in the Practical Byzantine Fault Tolerance (PBFT) protocol [16]. These are the most critical application belongs to IoT in which the data can be created, processed with heterogeneous devices or network.

3.1.2 DECENTRALIZATION

Blockchain has peer-to-peer network architecture which is more suitable for IoT network, blockchain is also a distributed network. Some examples include blockchain in Mobile Ad Hoc Network (MANET) or Vehicular Ad Hoc Network (VANET) [17,18] and blockchain in Bitcoin system. Blockchains are used to record traction of data between different parties without a central authority. So this technique can ensure to minimize the risk in single point failure, provide flexible configuration of network.

3.1.3 ANONYMITY

To ensure privacy and anonymity, blockchain uses public key (PK) which is changeable at any time is used as user's identity [19]. This can be highly useful in many applications of IoT mainly when a user wants privacy and confidentiality [20].

The integration of blockchain with IoT has already aroused in the industries and research area for the purpose of providing security [21−27]. By using this way, in IoT applications cloud is used to give a distributive access also the integrity of storage is secured by blockchain and tampering of data is prevented. The cloud can be integrated with blockchain as distributed blockchain-based cloud [28].

Already existing blockchain technologies are not enough for IoT applications, due to the previously mentioned development in the IoT devices and technologies such as massive amount of data in sensor, nonhomogeneous structure of the network with very strong partitioning, user demands for larger capacity in the blockchain, speed of block generation, or transaction speed [29]. While the data are stored in different locations in a distributed way, the inconsistency among records can be caused by physical characteristics and networks in IoT devices such as network topology, connectivity, link delays, and bandwidth. Also, the speed of record generation is to be in control by the speed of the block propagation these are the blockchain's data units.

In the current blockchain system, there are some unexpected operations in the application layer that neglects the physical aspects which are of devices and networks which in turn reduces the speed of the block generation much slower than block propagation which results in the inefficient usage of blockchain. For the survey, we can see about the benefits and key challenges faced in the blockchain and the IoT applications. The data structures and protocols are analyzed to find the

3.2 RESEARCH SURVEY 63

current state-of-art in blockchain technologies. In the current blockchain technologies, the limitations also the future potential direction of the research are listed.

There are lots of recent surveys on the general blockchain technologies and blockchain specialized in the area of IoT applications they are worth to point out [30−33]. These surveys are mostly based on the design and also in application areas. But this survey, to the contrast, this chapter is about the background of the blockchain in a theoretical manner. In this chapter, the gaps and limitations in the existing theories are identified; the impacts on scalability of the design of the blockchain in IoT applications are understood.

3.2 RESEARCH SURVEY

This survey consists of the summarization of the research on blockchain and blockchain used in IoT applications in recent times. The research references are mostly found from the IEEE Xplorer, Google Scholar, Elsevier, and Web of science also from the online sources such as developer communities and webpages to give an up-to-date summary of the technologies in the blockchain. The following are the important five aspects of the blockchain in IoT technologies.

The security issues of the IoT are aimed first, that is the IoT characteristics, performance of security analysis in IoT. The unique feature in the network of IoT and applications have been specially seen such as large number of devices, decentralized network, high throughput, low cost, mobility, large data generation in the IoT devices, and instability in the connection. On referencing the recent researches on the IoT security, this work describes the issues in the security of the IoT applications such as communication channel attacks, denial of service (DoS) attacks, sensor data attacks, software attacks, and network protocol attacks.

The first application of the blockchain this can explain is about bitcoin, the preliminaries of blockchain along with data structure chain, general problems of Byzantine at last consensus protocol has been discussed. On analyzing the attacks in existing blockchain technologies, limitations in the security of the Internet of Things are analysed in this survey. The attacks are distributed DoS attack, consensus protocol attack, double spending attack, and eclipse attack. The blockchain is also disturbed by leakage of private key, smart contract vulnerability, and programming fraud. The survey on real-time blockchains based on the IoT projects and applications in industries is discussed in the section further in the work, which has two structures of blockchain-based on the IoT projects and applications, they are IoT integration with blockchain and also blockchain acting as service of IoT. The requirements of the IoT and performance of the blockchain are studied to find the critical challenges of integration of blockchain in the IoT applications. The suitable technologies and designs of the blockchain that can be integrated with IoT applications are researched which is followed on discussing about the access control, identity, and privacy in the blockchain integrated with IoT applications.

Then, the latest blockchain technique which suits the IoT in the major three categories of the blockchain is elaborated in the further survey. The categories are public blockchain technology, private blockchain technology, and hybrid blockchain technology. The most popular validation mechanism of blocks listing Byzantine Fault Tolerance, PoW, and proof of X is elaborated in detail. In addition to the structure of chain, the data structure which brings improvement to the performance

64 **CHAPTER 3** INTEGRATION OF BLOCKCHAIN AND INTERNET OF THINGS

of the blockchain and benefits to the blockchain of IoT applications such as GHOST and Direct Acyclic Graph (DAG) are listed. Some technologies and projects that create impacts are compared and analyzed on the basis of scalability, capacity, and some unique or specific features.

Finally, the survey is taken on the direction of the future research, opportunities to build the gap which resides in between the IoT requirements and limits of the blockchain technologies in the current research works. The possible direction of the research includes consensus algorithm in specialization with IoT, editable blockchain, and simplified verification of payment.

3.3 **LIMITATIONS IN SECURITY OF INTERNET OF THINGS**

The superiority of the IoT network is the ability to have interconnection within a large number of devices with different sensing and with computing ability which works with less human interaction [34]. Various applications are developed in the IoT network to sense and operate devices in a heterogeneous way. Such applications are smart transport, smart home, smart grid [35], and eHealth. The IoT architecture consists of these layers, interface layer, service layer, network layer, and perception layer from top to bottom.

The first layer on top is interface layer, which is used for interaction between data and object in the applications [36]. Service layer is to provide and manage the necessary service for the given request to satisfy the IoT applications. Network layer connects all the network devices, sever, and other small products in the IoT application. Perception layer is mainly for data collection, it is also called as sensor layer [37]. They have sensors and devices to collect the data from the environment and process the data with necessary functions to extract needful information such as moisture, pressure, acceleration, temperature change, location changes, and motion. This layer is absolutely necessary in all the IoT applications. But the sensors and devices can be changes according to necessary of the application which connects the gap between the digital world and physical world. Some of the devices at the end to collect data include wireless sensors, actuators, mobile phones, Radio-Frequency Identification (RFID) tags, and Near-Field Communication. RFID is a microchip which is been attached along with an antenna. RFID tags can be attached with objects to identify, monitor, and track the needed data in the necessary areas. The architecture of IoT is mentioned in Fig. 3.1.

3.3.1 **INTERNET OF THINGS CHARACTERISTICS**

The IoT application has the ability to interfere in all the aspects in routine life of a human. The different domains could be classified as healthcare, transportation, smart environment such as smart house, logistics, and social or personal applications [38]. To meet the demands and targets of specific applications, different sensors, and actuators are used to gather data and communicate with other devices or send data to other devices with the help of networking. The main aspects which changes between applications are as follows (Fig. 3.2).

3.3.1.1 *High capacity and low-cost performance*

The IoT devices generally are heterogeneous which consist of different types of abilities and hardware. Sensor is one among the IoT devices which occupies a smaller space and it requires only

3.3 LIMITATIONS IN SECURITY OF INTERNET OF THINGS

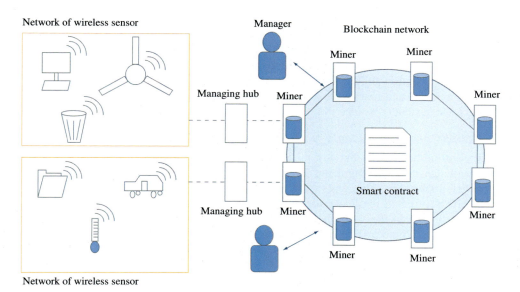

FIGURE 3.1

General architecture of blockchain in IoT.

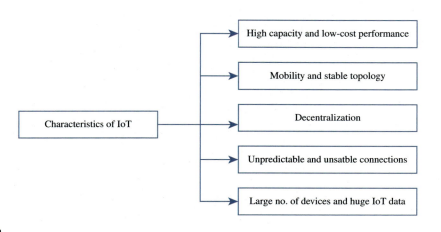

FIGURE 3.2

Characteristics of IoT.

minimum resource to do the processing, storing, or even communicating with other device. These sensors are usually low in cost and can be installed in larger area to measure as moisture, pressure, acceleration, temperature change, location changes, and motion chemical changes [39]; also medical parameters can also be measured in a human. The communication takes place with the help of

66 **CHAPTER 3** INTEGRATION OF BLOCKCHAIN AND INTERNET OF THINGS

the wireless mesh or even ad hoc network like Zigbee [40]. These sensors only need limited battery to create limited energy, which is the major aspect needed. Nowadays, new communication technique like narrowband internet of things (NB-IoT) has be used to make sure that the lifetime span of the sensor has been extended [41]. But those sensors still are much minimized in processing, storage, and communication ability. Some other IoT devices are highly expensive but also powerful which includes vehicles and mobile phones, with larger battery and high capability to perform computing process and sufficient storage. So, these devices can give maximum capacity.

3.3.1.2 Mobility and stable topology

The mobility within the sensors and devices in the end leads to unpredictable network connectivity and brings challenges in managing them. The difference in the speed can lead to topological difference in the IoT application. The application which is with mobile topology is Vehicular Ad Hoc Network (VANET) used in the applications of transportation and stable topology is smart home. The sensors or devices in the smart home system are stable also it consists of network which is with stable topology. Whereas vehicle's transportation speed is high and also different vehicles travels at different speed; thus it has time varying topology [41,42].

The sensors or devices in the end are heterogeneously implemented with different protocols, the common characteristics in IoT network are as follows.

3.3.1.3 Decentralization

The major characteristics of IoT are heterogeneity and decentralization. When there are a large number of IoT devices, the concept of decentralization is much important as in case of smart city as there is a very large amount of data to be processed at the same time [43,44]. The IoT sensors and devices are used together, process and also store the data in decentralized way. Clustering algorithm in the area of wireless sensor network, decentralization computing are some of the algorithms in the decentralization where scalability and capacity of the IoT can be maintained [45].

3.3.1.4 Unpredictable and unstable connections

The state of sleep/ideal condition or mobility of the IoT device is the major cause for the unpredictable connections and unstable conditions. The unreliable links in the wireless IoT devices are also a reason for this condition [46]. Due to these circumferences, the network of the IoT devices is disconnected into partitions and the number of partitions could vary as time goes on.

3.3.1.5 Large number of devices and huge Internet of Things data

The IoT devices will highly increase with time. By 2020, it will be expected to reach above 20 billion [47]. The problem in IoT is not only to handle a large number of devices but also the demand of handling the capacity is growing high because of the huge amount of data been generated from the sensors and devices from the data collection end.

3.3.2 ANALYSIS OF SECURITY ON INTERNET OF THINGS APPLICATIONS

The data security becomes one of the severe problems due to some characteristics in IoT [48]. The first problem is that the IoT sensors and devices are mostly installed in a remote or unattended area and also they are unfriendly to handle. Also there are enormous number of devices all those are

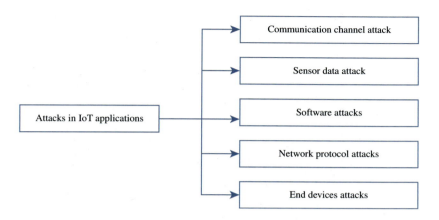

FIGURE 3.3

Attacks in IoT applications.

impossible to be seen or protected by a human in all times. This leads to vulnerable harms from many different dimensions [49]. Some advertising company tries to control the end sensor or devices to enter into the IoT network [50]. Asymmetric encryption, which are traditional mechanism for security demands computational ability of IoT devices in a limited manner [51]. The data gathered from the sensors and devices could be store, transferred over network, processed to extract information with the help of intermediate system, this gives the risk of forgery and tampering. The open channels that are wireless and unreliable gives extra risk to the security of data. The vulnerability increases with the complex nature of the IoT systems [52]. Some of the attacks caused in IoT network are listed (Fig. 3.3) [53].

3.3.2.1 Communication channel attack

In the transmitting channel, opponent parties can eavesdrop to find profit in their plans, broadcast nature of the radio can be exploiting. So to avoid third party or opponent party stealing data, encryption can be used. Even though the signals are encrypted, the opponents or third party can extract some kind of private information like source address or destination address [54]. They can also send some noisy signals to jam the network of wireless channel [55].

3.3.2.2 Denial of service attack

The congest situation of services and the exhausted resources are some attacks that represents DoS attack. If a sleep routine is programmed, to break it sleep deprivation attack is used and it keeps the device to stay awake all times as far as the battery run out of power supply [56]. DoS can cause sudden great damage even with limited communication resource and network channels on the IoT devices. This kind of attacks causes exhaustion of the limited energy present in the sensors or devices decreases the connectivity of the network, entire network can be paralyzed and network can be reduced even for the lifetime.

68 **CHAPTER 3** INTEGRATION OF BLOCKCHAIN AND INTERNET OF THINGS

3.3.2.3 Sensor data attack

Ad hoc protocols are used to communicate between IoT devices where the messages are been transmitted in hop-by-hop fashion until the message reaches the destination. This creates an opportunity for the hackers or opponents to inject the false data or to tamper the data. The hacker or opponent tampers the data and forward the data to next node this process is called as tampering data. To prevent the tampering of data, authentication algorithms have been used [57]. Injecting false data to the original message and passing it to the targeted network is called as false data injection attack, it is done by using fake identities so the culprit cannot be found easily. As the false information reaches the receiver end it will be a just an error message which leads to providing wrong service, damaging the reliability of the IoT network and application. If vehicles started accepting the false or error road direction or assisting messages, and then it will lead to a traffic collapse or leading to a state of congested situation. This type of situations can be prevented with the help of authentication algorithm.

3.3.2.4 Software attacks

Backdoor attacks are most common in the software attacks, which refers to series of attacks using backdoors of software by which the software and control operations are been modified [58]. The software attacks also include virus, malicious scripts, and worms [59]. Internet security techniques such as intrusion deduction system are used handle the attacks in software [60]. Security is the most important aspect of the IoT applications especially the integrity of IoT objects and data produces by the objects [61]. The integrity, confidentiality, and authentication of the IoT communication are to be protected with effective mechanisms such as encryption and authentication algorithm [62]. The data integrity is to be maintained from the origin of the data which then transferred over trusted third parties such as identity provider [63]. The security of data should also be maintained in the service of data storage [64].

3.3.2.5 Network protocol attacks

The vulnerability of the network protocol is used by the hackers to do man in middle attack, reply attack, wormhole attack, Sybil attack, etc. The Sybil device pretends to be authenticated identity in the IoT system applications. These attacks will lead to reduction of accuracy and efficiency in multipath routing system or voting mechanism.

3.3.2.6 End devices attacks

Node capture attack is used to capture the devices or sensors physically and controlling them. The hidden secret data which are stored in those captured sensor or devices like certificates or private keys are exposed to the hackers so that they can use the data to act as the authenticated user and perform attacks on the devices or system [65].

3.4 EXISTING TECHNOLOGIES OF BLOCKCHAIN

Blockchain is a distributed tamper resistant ledger, it consist of data blocks called records; each transaction block is recorded and maintained with a timestamp. The centralized body which needs to store data can use the blockchain technology. The blockchain has the capacity to maintain the data security in the network of IoT, since blockchain is a decentralized network with the properties of anonymity and integrity [66]. In the year 2008, the very first blockchain was found and proposed by Satoshi Nakamoto in the field of cryptocurrency—Bitcoin [67].

3.4 EXISTING TECHNOLOGIES OF BLOCKCHAIN

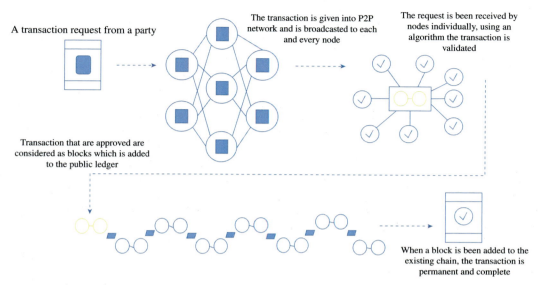

FIGURE 3.4

Working method of block chain.

The data of the blockchain are recorded in a distributed and secure way. The traction is said as basic unit of record of the blockchain as seen in Fig. 3.4. Whenever generation of new transaction takes place, it is been broadcasted to entire network of blockchain. When a node receives a transaction, it can verify by evaluating the signature and finding it to be valid which is been attached with the transaction. Miners do the verification of the transaction helps to maintain cryptographic secure blocks. These nodes are termed to be block miners or miner. In the distributed system, the consensus problem is solved to give way for a miner to generate a block. The solved problem of consensus by the miner [68], broadcast the new block all throughout network.

When a new block is received the miners should be ready to solve consensus problem which is been appended to the block of their own chain block which is locally maintained by miners. When all the transactions present in single block is verified then that block is proven that to provide necessary correct answer for consensus problem. The previous block is linked with the new block in chains with the implementation of cryptography algorithms. All the miners can do synchronization to their chains with a regular interval of time and the ledger is shared among distributed network is ensured. The bitcoin's blockchain is the only blockchain which has the longest chain also it has discrepancy in the chains.

The key components in the blockchain such as data structure, consensus protocol, security, and smart contracts in Fig. 3.5 are been analyzed in blockchain as follows.

3.4.1 DATA STRUCTURE

The transaction is said to be the basic measuring unit in the blockchain. Transactions are just records of data or events which are been observer in the network by miners. To sign transaction,

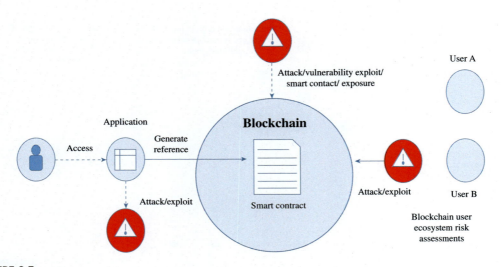

FIGURE 3.5

Smart contracts in blockchain.

private key is used in cryptography. The signature of the sender is attached in the transaction which is very important, since it gives a mathematical proof to the transaction which confirms to come from owner of the transaction with their private key. The miner verifies the genuineness of transaction with the PK which is corresponding to that particular private key. The PK is used as source address present in transaction, PK is preloaded to all the miners or the PK and its digital certificate is attached to signature for the transmission. Transaction binds event with its initiator by the power of cryptography. The transaction was used first in the bitcoin blockchain to catch financial interaction among two different financial members. To know about programmable events and ownership rights, transaction can be used [69,70].

Each and every miner in the network has a local transaction record, which is maintained as backward, ordered linked list. If a transaction enters the ledger, each block is encapsulated with batch of the verified transaction. All the blocks consist of header with link of previous parent block; it is hashed value belonging to parent block in blockchain of bitcoin and a response to reply to consensus problem. The header block also consists of other fields like timestamp according to the demands needed. In the block header cryptography algorithm is used to generate hash value, by which the blocks can be identified uniquely.

Each block is linked to its parent block with the sequence of the hash function operations. Thus creating tamper-resistant chains in which back trace can be done to reach the first block created. By this all the blocks are been chained together which acts as ledger in each separate nodes, which is shown in Fig. 3.6. The current block's hash value is affected since the link of parent block lies inside the header of the block. If any one block is modified in the available chain, then its child block and grandchild blocks are also needed to be recalculating in order to meet consensus problem relevant to it, but all those recalculations as prohibited. In order to secure the data in intractability

3.4 EXISTING TECHNOLOGIES OF BLOCKCHAIN

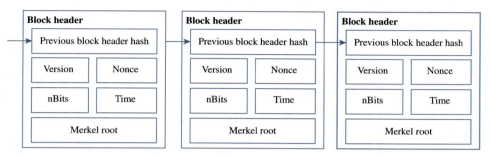

FIGURE 3.6

Data structure of the blocks.

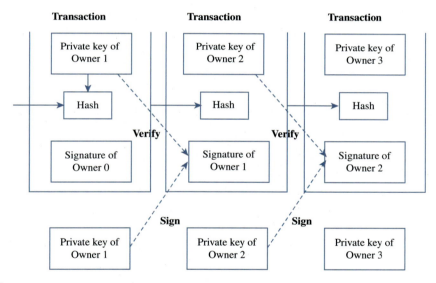

FIGURE 3.7

Data structure of transaction.

in tampering, the blocks are maintained in a long chain and ledger with tamper-resistant ability is constructed. The local blockchains are maintained regularly and updated throughout the network. The entire ledger consists of only one longest chain of the bitcoin blockchain, which is been accepted publicly that is locally maintained and updated correctly.

The header blocks consist of fields with the information on all transactions which is present in the current block [71]. For example, Merkle root in the blockchain of bitcoin [72] shown in Fig. 3.7. The transactions are the leaves built in Merkle tree in order to improve the efficiency of the storage in every block. The tree structure of the Merkle tree has transaction as leaf node and all

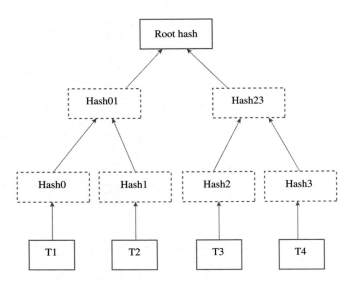

FIGURE 3.8

In Merkle tree the transaction is hashed.

other nonleaf nodes are just hash of their child node as given in Fig. 3.8. Merkle root is the name of the root node in the tree. On transporting into the Merkle tree of blockchain network, if a transaction is mined in block with the verification of hash to respective branch can be confirmed. With this structure the network capacity, memory, and storage could be much reduced.

Each block and transaction that enter the blockchain network are transferred and verified in the distributed consensus. An example for this scenario is, Bitcoin. First a valid transaction is generated by the node; it sends the inv (inventory) message which consists of the transaction's hash identification (TID) instead of original transaction data, which is sent to all the nearby neighbor nodes. The nearby node who does not have the transaction asks it to sender. Then, the transaction can be send to the nearby nodes. Then that transaction is verified and further sent to their neighbor nodes. This process continues till the transaction reaches the entire network.

The node where the transaction is generated is solely responsible for the circulation of transaction. If a transaction is not reached by any node, then the responsible not can rebroadcast if required. Also miners who generate new blocks, solving consensus problem also takes responsibility to circulate the transaction across the entire network.

3.4.2 CONSENSUS PROTOCOL AND BYZANTINE GENERAL PROBLEM

The consensus problem is locally generated and developed on the basis of last public accepted block on blockchain, this is mainly for miners. The miner tries to mine the block or transaction, and then the complexity of problem is specified in last block accepted into the blockchain. The miners are capable of verifying each other's transaction or block based on predefined criterion. The consensus protocol is an open access network that allows untrustworthy and unverified miners also

3.4 EXISTING TECHNOLOGIES OF BLOCKCHAIN 73

to mine the transactions or blocks even without the need of verifying the identity of the miners, they are basically public blockchains. The main consensus protocols in the public blockchain are: the network is open access, proof of shake (PoS) is adopted in peer coin, and PoW is adopted in bitcoin. There are independent miners who can mine different blocks in same time but these cause disruptions known as fork in blockchain growth where the local chain of block that are maintained turns out to be inconsistent among different nodes. There are many miners who expand their own resources to mine in same block which leads to delay and energy waste.

There is a fundamental theory in which blockchain extensively faces the Byzantine general problem [73]. It is basically generalized as agreement problem. When peers are trying to reach consensus where the hackers hidden among peers betray the peers and preventing them on reaching consensus, this is the problem [74]. The hackers may attack in some strategies they are forging the messages, fake messaging, and ignore messages also two-face behavior this can be found when, different nodes may send conflicting opinions of messages. These problems lead into Byzantine failure of network where consensus is required [75]. The worst mode of failure in the distributed system is Byzantine failure. It consists of detectable authentication Byzantine failures where Byzantine fault server forges the detectable by using authentication mechanism. When there is an early or late correct result delivery by the server then it is performance failure. When a request to the service gets a delayed service response or does not get any response at all then it is omission failure. The server state where it is having a crash failure then other servers which functions correctly can deduct the failures, then it is called as fail stop failures. These attacks are important in blockchain; they can also create inconsistency to the blockchain. The peers can join the bitcoin network or leave it freely whenever needed. If the peers are leaving then they can be termed as fail stop failure or crash failure. Omission failure can stop the mined blocks in broadcasting on to the entire remaining network. The hackers betray the original peers in tow faced attack.

In the area of replication, there are lots of research techniques which are used to handle the Byzantine failures, many researches in the area of replication are to focus on fault tolerant systems. In early times, the protocol of Byzantine agreement was employed to signal expensive also recursive confirmation in order to gain full whole system picture before trying to solve Byzantine general problem. The overhead of the communication protocol is high which is exponential with the number of peer nodes. Even without assuming the behavior of the fault process, the Byzantine Fault Tolerance can give solution to the blockchain system. A popular method for fault tolerant mechanism is state machine replication; it replicates the servers to fulfill client interaction [76].

The key factor on maintaining consistent and distributed ledger also with decentralized coordination that provides Byzantine general problem a solution is consensus protocol in blockchain. The law of block selection and generation is defined by the consensus protocol. The consensus problem is solved to mine blocks from blockchain by miners. It prevents from hacking into the network or hijacking the process of block generation. The service provider of the blockchain can announce the consensus problem can also be generated in distributed way by the following global agreement of criteria.

There are also private and permitted blockchains [77]. They need authentication to allow miners to participate in mining which notifies all the blocks in peer-to-peer manner to observer transaction. When a miner synthesizes the observations made by them they are intended to use Byzantine Fault Tolerance algorithm, which produce consistent transaction or block in distributed system.

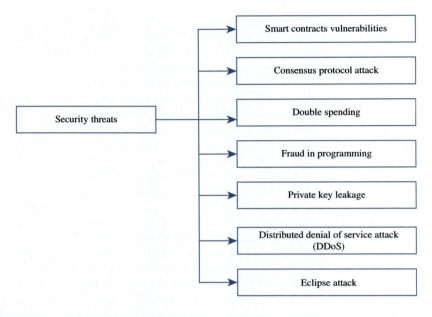

FIGURE 3.9

Security threats blockchain security.

3.4.3 ANALYSIS OF BLOCKCHAIN SECURITY

As there is a huge tamper-resistant property in the blockchain which is also a decentralized network, it gets lot of attentions. Also there is no need for trusting peers with each other. Even in this situation vulnerability is still there in the blockchain. Some of the security threats are as follows (Fig. 3.9).

3.4.3.1 Smart contracts vulnerabilities

The openness nature and irreversibility in blockchain make the smart contracts more liable to be influenced. Even the frauds and bugs are visible transparent to opponents or hackers or public. Since there is irreversibility in the blockchain, it is more challenging in order to create a bug and fix it in the smart contracts. For example, in 2016 there was an attack in Decentralization Autonomous Organization (DAO) well known as DAO attack, this leads to a result of forked Ethereum in blockchain [78].

3.4.3.2 Consensus protocol attack

The security of consensus protocol can be broken by attackers by having a huge amount of chunk file partition in the computing power in entire network. These attackers may also control or even reconstruct the chain. For example, there were around 51% of attacks mainly in PoW of bitcoin blockchain. If the hackers own more than half of hash power, then illegitimate blocks are produced in the blockchain after solving consensus problems even much faster than any other peers. Currently, 33% of hash power is proved to be sufficient in order to overpower PoW [79].

3.4.3.3 Double spending

The opponent tries to mislead the block or transaction receivers with a conflict transaction. For example, the same coin is spent in bitcoin. The possible attacks are to send conflict transaction; one or more blocks are premined to generate conflict transactions that are accepted by blockchain [80].

3.4.3.4 Fraud in programming

The frauds are implanted by attackers into the programming codes to get the blockchain properties like in 2018 the privacy attack are reported [81,82].

3.4.3.5 Private key leakage

The private key of the particular account can be stolen to take over that account without the knowledge of the real user. This is done by capturing the physical nodes or using traditional network attacks [83].

3.4.3.6 Distributed denial of service attack

The opponents try to fully exhaust the resources of blockchain (exhausting whole processing capability of the network). This is done by a collaborative attack launch. The opponents took hold with underprice instruction of ethereum virtual machine (EVM) which leads to slow down of block processing speed in 2016. There were large amount of accounts consisting of low balance which are been produced by the opponents, they lead to that DDoS attack [84].

3.4.3.7 Eclipse attack

The attack caused in peer-to-peer network in which the opponents monopolize every connection with its legitimate nodes, preventing those nodes from getting connected to any true peers, this is eclipse attack. This type of attack was first used in bitcoin system by using randomized protocol in that a certain nodes of bitcoin is connected to certain selected nearby neighbor odes to have a constant peer-to-peer network communication and related functions of blockchain. The most recent eclipse attack was exposed in Ethereum by adopting Kademlia p2p network in Ethereum [85].

3.5 APPLICATIONS OF BLOCKCHAIN IN INTERNET OF THINGS

The network of IoT is said to be data centric, where a large number of sensors or end devices upload large data. This raises the vulnerability to both the device and data, making these as the target for any potential attacks in IoT. The data from the sensors of the IoT system are sensitive or even personal data. For example, it could be the medical data of a patient (medical IoT) [86] or it could be the data of any national application which should be kept under privacy (smart grid based IoT or nuclear power plant) [87]. The privacy and integrity of data are much important. The blockchain only holds key in order to settle down data integrity, reliability, and security of the IoT

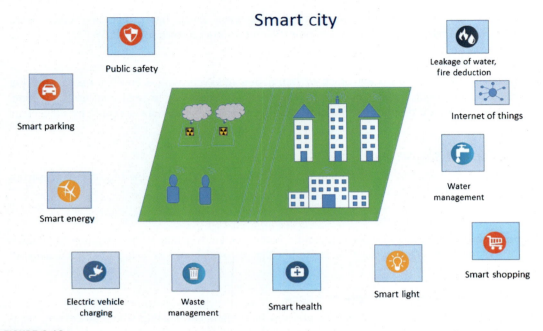

FIGURE 3.10

Applications in smart city.

network. Currently, there are lots of applications developed on the basis of IoT such as smart city [88] in Fig. 3.10, supply chain management [89], and bitcoin management. These technologies should handle the security issues of both end device and sensor data.

3.5.1 MALICIOUS BEHAVIORS IN INTERNET OF THINGS DEVICES

There are three ways to describe the malicious behaviors of IoT devices in blockchain. They are (1) transaction is sent with the false signature, this can be detected or rejected or even punished by blockchain; or (2) transaction is sent with correct signature but has false data in it, this can be found and eliminated by an algorithm called false data detection and even punish the source node responsible for transaction; or (3) exhausting the resources completely, for example, DoS attack. Transaction free mechanism can prevent DoS attack [90].

3.5.2 SENSOR DATA CORRECTNESS

The blockchain in the IoT network consists of data which is divided into relevant data to the blockchain such as account name or number, current balance, transaction fees, and relevant data to the IoT such as sensor data. The data relevant to the blockchain is verified on the basis of previous transaction such as expense list which must be less than that of the current balance; this can be done in any applications of blockchain. The data relevant to the IoT are been protected with the help of signature present in the transaction this gives an assurance that messages are sent only by authenticated users and authorized IoT devices. The correctness of data relevant to the IoT can only be given by Oracle service by providing the authenticated data. The trust worth of the sensor data can be ensured by backward linked and hashed structure recorded in the ledger of IoT blockchain [91,92].

3.5.3 INTEGRATION OF BLOCKCHAIN AND INTERNET OF THINGS PROJECTS AND APPLICATIONS

Mostly the application layer is focused on the existing blockchain-based IoT technologies. Here, the peer-to-peer network is used where there is no physical restriction for device or network or bandwidth. For example, the recently developed blockchain network with peer-to-peer decentralized network called enigma which is used to manage the personal data.

IOTA provides the solution for blockchain-based IoT network. In 2016, based on the technology "Tangle," IOTA was built; it has no block, no fees, and no chain. The properties such as distributed ledger, tamper-resistant qualities of blockchain are inherited to tangle. Where tangle uses DAG structure but bitcoin transaction uses chain structure to store data in IOTA. Every transaction should confirm previously published two transactions. Work and energy are saved by the flexible structure while the transaction is been mined. Parallel the transactions are verified and accepted instantly by tangle, it also gives high capacity transaction rate in IOTA. IOTA has no transaction fee since if the transaction fee could be more than the recorded value so that the participants can get discouraged.

There are four varieties of nodes that IOTA supports; they are Android wallet, light wallet [93], full node, and headless node. In 2017, beta version is released by IOTA to give support for light wallet. The current status of network, transaction can be obtained by connecting the light wallet to server in which IRI (IOTA reference implementation) is running. In order to produce the legal transaction, computations are to be performed by light wallet which is then required for DAG structure [94]. Some limited ability IoT devices are not allowed to use light wallet in IOTA. An example of limited ability IoT device is battery power nodes.

For specific targets, there are many other IoT platforms based on the blockchains. The International Business Machines Corporation (IBM) proposes a project based on blockchain after the partnership along Samsung electronics. The project is autonomous decentralized peer-to-peer technology along with advocate device democracy to bring the future to IoT. IBM also launched data sharing service based on the blockchain for industries and businesses, so that the IoT data could be shared privately through ledgers of blockchain preventing problems within business opponents or partners.

3.5.4 STRUCTURE OF BLOCKCHAIN INTEGRATED WITH INTERNET OF THINGS APPLICATIONS

Depending on the ability of the IoT devices, there are two structures to be applied in blockchain-based IoT applications.

3.5.4.1 Blockchain involved with Internet of Things

In blockchain network, IoT devices are joined to become a part of core function in the blockchain. The core functions would be transaction creation for raw sensor data, transaction verification, and block mining. The important roles which need support in network of blockchain and IoT are, full node, light node, and miner [95]. The sufficient application to run on this network structure is Vehicular Ad Hoc Network (VANET) in Fig. 3.11. All the blocks are stored in full block which includes header of the block, body of the block, but full block does not involve in the mining of the block. It needs a little computation and a very huge storage. The light nodes are the IoT sensors or end devices. The private keys are either generated independently or registered to certificate authority (CA) by the IoT devices for authentication and access control. Light node only stores block header and creates transactions but do not mine the blocks. Light nodes are supported with simplified payment verification technique. Light node only needs less computation power and less storage compared with other two. A special kind of light node is wallet, which requires minimum computational power and minimum storage. It does the most basic transaction functions; it also acts as full node when retrieving mined data into blocks. The miner's work is to mine the transactions to form blocks which are stored. There is a high demand for computation and storage. For example, hyper ledger where a new client (new IoT sensor or device) is to be registered and enrolled in CA and get a private key which is maintained. The private key is light node used for signature generation in transaction to denote the valid owner of the particular transaction. The client's work is to generate transaction and broadcast it.

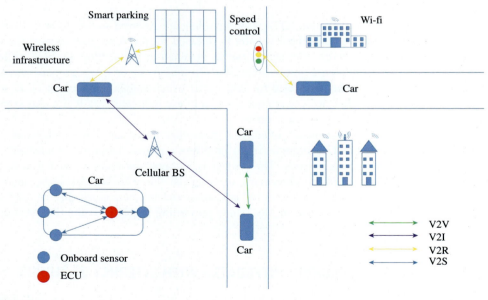

FIGURE 3.11

Vehicular Ad Hoc Network Architecture (VANET).

3.6 SECURITY IN BLOCKCHAIN WITH CIPHERTEXT POLICY AND HYBRIDIZED WEIGHTED ATTRIBUTE-BASED ENCRYPTION

User privacy and data security are main aspects when transferring data via internet. Ciphertext policy and hybridized weighted attribute-based encryption (CP–HWABE) is used to protect the data and privacy in the blockchain network. The transaction in blockchain is done anonymously, so that if an illegal action is involved the identity of the user cannot be obtained to find the culprit also to keep the system decentralized CP–HWABE encryption algorithm can be used. Here, the identities of the users are encrypted with access rights policy and transaction is done. Wallet key pair is generated based on the identities of the user which can be used later to decrypt and fine who the user causing illegal transactions in the network. Only the authorized regulation nodes can reveal the identities of the suspected account. Even the account can be blacklisted in order to avoid future problems. This system is secure from chosen plaintext attack under Bilinear Diffie–Hellman Exponent in standard mode. The system is capable of revealing the criminal identity who is involved in illegal activities (Fig. 3.12).

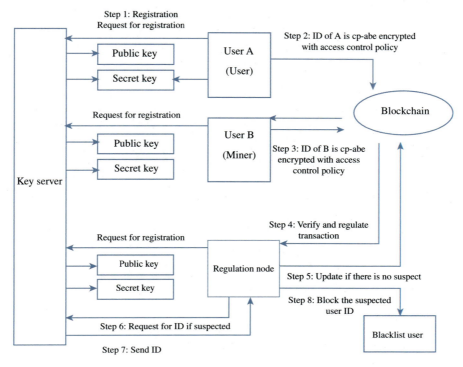

FIGURE 3.12

Bitcoin regulation system.

3.6.1 EXPERIMENTAL SETUP

The admin calls setup function to get, public parameters. Data sender chooses the security parameters and then requests admin to yield the private key SK components and sends the encrypted attributes to CA (Central Authority). Whereas, the CA verifies user attributes and if the attributes are valid then CA utilizes PK and MK of the system to create PK and SK to a newly registered user. If a user is identified with fake identity then no key pair is generated. The weight attribute authority allocates weight priority to the attributes according to their importance (Fig. 3.13).

3.6.2 WORKING OF CIPHERTEXT POLICY AND HYBRIDIZED WEIGHTED ATTRIBUTE-BASED ENCRYPTION ALGORITHM

Data security and user privacy are the major aspects while transferring data through internet. The main objectives can be done with CP–HWABE. In many attribute based encryption (ABE) algorithms single authority is used to handle both PK and secret key. But most of the time, user holds attribute set in multiple authorities while the holder of the data shares the content of message file with associated user, managed by the relevant authority. Lots of multiple authority based attribute encryption is developed. In the access policy (AP) tree, to update the ciphertext the user should be present in online all times, the attributes are also given similar characters. In this system, the weights for the attributes are assigned to provide data security and privacy of the user.

Encryption and decryption advanced standard encryption are used. The weight allocation authority (WAA) generates priority based on the attributes of user. Secure hash algorithm-512 is used to digest the larger data to the smaller data. Then with access key policy and key values the messages are encrypted and send to the receiver. The proposed is reliable and has high security. In real time the system is applicable. The proposed system's encryption ensures fine grained access control, collusion resistant and supports multiple authorities. In the systemic level of the proposed work, higher-level operations are explained, such as setup the system, registration of user, collection of user details, file creation, accessing file, and file transfer.

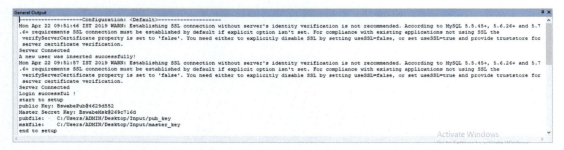

FIGURE 3.13

Experimental setup of CP–HWABE algorithm.

- Step 1: Collect user's attributes form the user who registers to create an account.
- Step 2: Take the user's attributes as input generate hash values of the key pair—PK and Master Secret Key (MK) with secure message digest algorithm SHA-1. WAA allocates weight for the attributes based on their importance.
- Step 3: Take the subset of user's attributes and MK as the input to generate Private Key (SK) as output.
- Step 4: In the sender's side, the public parameters (PK_{abe}), message to be sent (m), access structure (M, ρ), and weighed AP are taken as the input for encryption where AES algorithm is used to give ciphertext (CT) as the output.
- Step 5: In the receiver's side, when the user's attributes set matches the weighted AP of the ciphertext, then the user is the authorized user. Find the minimum set S_u, which satisfies the AP to decrypt the CT to get the original plaintext.

3.6.3 EXPERIMENTAL RESULTS

3.6.3.1 User registration

The necessary attributes alone with a unique attribute is gathered from the users who are then registered into the application. As a user is registered, a key pair along with the account address is distributed to the user (Figs. 3.14 and 3.15).

Then, the user can normally login with the phone number and the password whereas all other attributes of the user are manage safely in the database which are later used in the access control policy.

3.6.3.2 Key pair generation

While a user login for the first time into server, a unique user ID is been generated for user by CA. The attribute set of user is send to WAA; the attributes are authenticated by attribute authority.

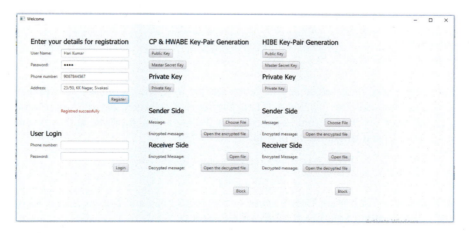

FIGURE 3.14

User registration.

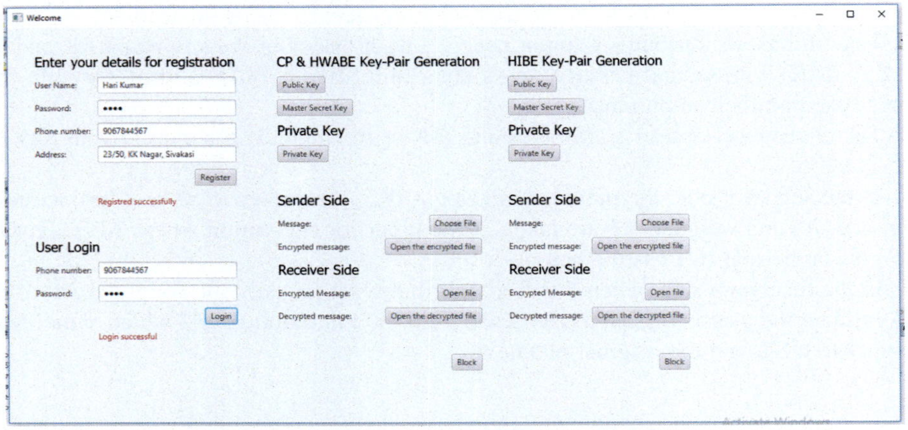

82   CHAPTER 3 INTEGRATION OF BLOCKCHAIN AND INTERNET OF THINGS

FIGURE 3.15

User login.

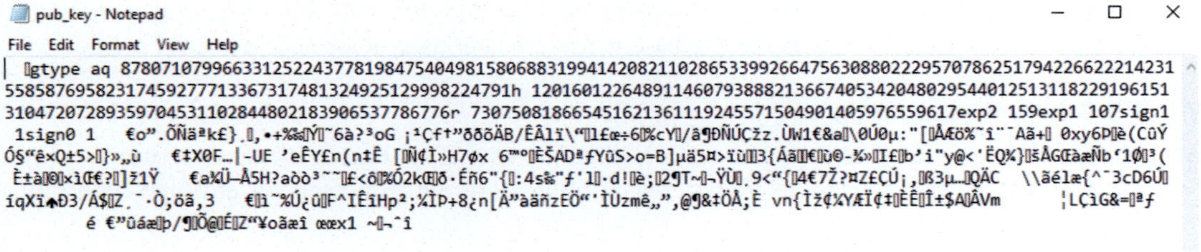

FIGURE 3.16

Public key generation.

If the attributes are valid, then secret key is generated for user who login for first time, based on their importance of the attributes by WAA. Then, the receiver's secret key of system and attribute secret key of the user are transferred separately by WAA and CA. Both authority and central authority setup are initiated to get the keys (Figs. 3.16—3.18).

$$PK_{abe} = (g, g^a, Y = e_2(g,g)^\alpha, u'_1 \ldots u'_n, U_1 \ldots U_n) \tag{3.1}$$

$$MK_{abe} = g^\alpha \tag{3.2}$$

The inputs of this algorithm are User's set of attributes ($S_u \subseteq S$) and Master Secret Key (MK_{abe}). The output of this algorithm would be the Private Key (SK_u).

$$SK_u = (D = g^{\alpha+ar}, D' = g^r, \{D_w\}_{\forall w \in S_u}) \tag{3.3}$$

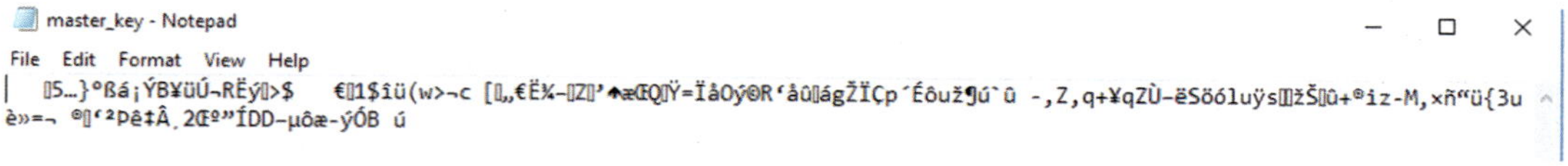

3.6 SECURITY IN BLOCKCHAIN WITH CP−HWABE

FIGURE 3.17

Master secret key generation.

FIGURE 3.18

Private key generation.

3.6.3.3 CP - HWABE Encryption

Before the plaintext message file is uploaded to internet, data sender first have to login through uniquely generated ID for them. Then advanced standard encryption is to encrypt the file to be sent. Data sender also defines a threshold weighted access structure (W) for receivers of the data and then the message data file is encrypted with W and sent over internet (Figs. 3.19−3.21).

- Step 1: Gather all the necessary attributes from the user to register into the application by creating a new account.
- Step 2: From the gathered attributes, PK and MK are generated and distributed to the user.
- Step 3: Take a selected list of attribute along with PK and MK to generate the private key.
- Step 4: In the sender side, the user is logged into the registered account; the plaintext is encrypted with CP−HWABE algorithm, where each attribute is assigned with a weight according to their priority by the weight attribute authority.
- Step 5: The weight and the secret key is sent to the central authority whereas the encrypted message is sent to the receiver via internet.

$$CT = (\check{C}, C, \{C_{\sigma 1}, C_{\sigma 2}\}_{\forall \sigma \varepsilon [1,d]}) \tag{3.4}$$

84 CHAPTER 3 INTEGRATION OF BLOCKCHAIN AND INTERNET OF THINGS

FIGURE 3.19

Select message to be encrypted.

FIGURE 3.20

Original text or plaintext.

where

$$C_{\sigma 1} = g^{as\sigma}\left(u'_j \Pi_{\delta=1}^{k} u_{j\delta}^{w1/j\delta}\right)^{-r\sigma} \tag{3.5}$$

$$C_{\sigma 2} = g^{r\sigma} \tag{3.6}$$

$$\check{C} = mY^s \tag{3.7}$$

$$C = g^{-s} \tag{3.8}$$

3.6 SECURITY IN BLOCKCHAIN WITH CP−HWABE

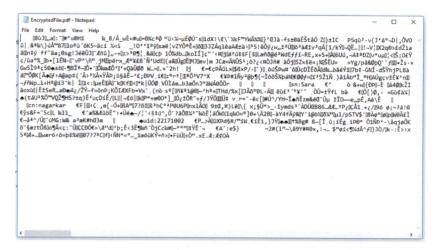

FIGURE 3.21

Encrypted text or ciphertext.

3.6.3.4 CP - HWABE Decryption

The received files are been downloaded by the data receiver and the message file is decrypted. Only if the attribute secret key of the receiver is authenticated, system generates priority values to its attributes based on the level of attributes. Then with respect to W the receiver decrypts the file. If a user is found to be invalid, user is not allowed to decrypt, the hackers who try to hack the message will not have the real attribute of that particular user and thus they cannot get the original message and it is hard to hack. The invalid user is reported. If they are found to be the attacker, then they are added to the blacklist. As one unique ID is used as an attribute the different accounts using the same identity can also be blacklisted (Fig. 3.22).

- Step 1: Receiver logs into their registered account. Before decrypting the message, the user is verified a valid user or not.
- Step 2: Only if the attribute secret key of the receiver is authenticated, system generates priority values to its attributes based on the level of attributes. Then with respect to W the receiver is allowed to decrypt the file.
- Step 3: If a user is found to be invalid, user is not allowed to decrypt.
- Step 4: If the user is found valid, then the message is decrypted with CP−HWABE algorithm to get the original plaintext or message.

When a user attribute set matches the AP of the ciphertext (CT) then the user is the authorized user. If the authorized user with the secret key $SK_u = (D, D', \{D_w\}_{\forall w \varepsilon Su})$ wants to decrypt the ciphertext $(CT) = (\check{C}, C, \{C_{\sigma 1}, C_{\sigma 2}\}_{\forall \sigma \varepsilon [1,d]})$, then the user should match their attributes to find the minimum set S_u, which satisfies the AP from S_u (Fig. 3.23).

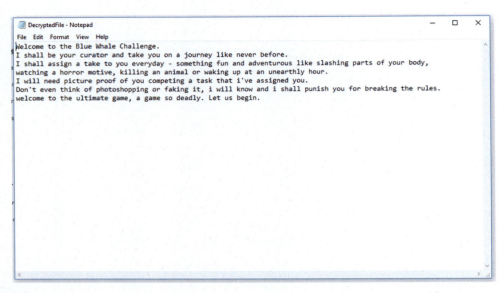

FIGURE 3.22

Decrypted text or plaintext.

FIGURE 3.23

Successfully blocked the user illegally using the application.

3.6.4 RESULT ANALYSIS

Here CP−HWABE requires minimum amount of time to perform encryption and produce the encrypted message or ciphertext, due to the AP with the hierarchical structure and weighted attributes (Fig. 3.24).

Here CP–HWABE requires minimum amount of time to perform decryption and produce the decrypted message or plaintext, due to the AP with the hierarchical structure and weighted attributes (Fig. 3.25).

Ratio between file size and data encrypted time is defined as throughput. It can also be said as dividing the total plaintext in the megabytes encrypted on the total encryption time [96] for the algorithm. Throughput can be calculated as (Fig. 3.26):

$$\text{Throughput} = \frac{\text{File size(mb)}}{\text{Encryption time}} \quad (3.9)$$

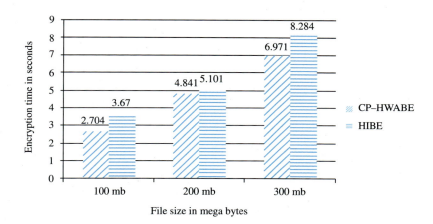

FIGURE 3.24

Encryption time comparisons of files between CP–HWABE and HIBE.

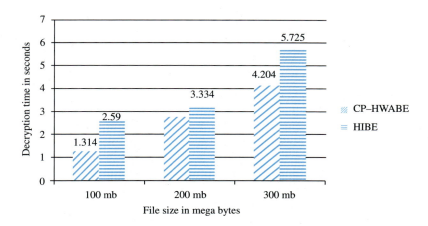

FIGURE 3.25

Decryption time comparisons of files between CP–HWABE and HIBE.

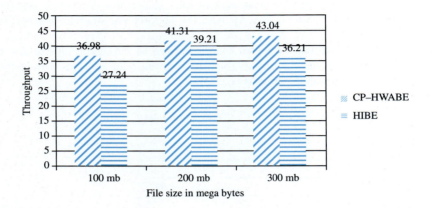

FIGURE 3.26

Throughput comparisons in encryption between CP–HWABE and HIBE.

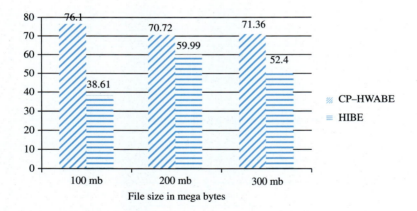

FIGURE 3.27

Throughput comparisons in decryption between CP–HWABE and HIBE.

Here, the throughput is expressed in megabytes per second is increased in the proposed CP–HWABE algorithm. The performance analysis shows that CP–HWABE transmits more number of bits over a certain period of time compared to hierarchical identity based encryption (HIBE), since the time taken for the encryption in CP–HWABE is much smaller (Fig. 3.27).

Here, the throughput expressed in megabytes per second is increased in the proposed CP–HWABE algorithm. The performance analysis shows that CP–HWABE transmits more number of bits over a certain period of time compared to HIBE, since the time taken for the encryption in CP–HWABE is much smaller.

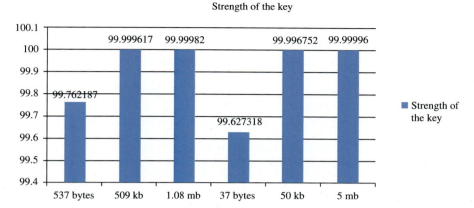

FIGURE 3.28

Strength of the key in the proposed system.

3.6.4.1 Avalanche effect

In cryptographic algorithm, one of the most desirable properties is avalanche effect. In the hash functions in cryptography and block ciphers, within which if there is slight input change (e.g., flip/change a single bit), then the output should change significantly that is in the output more than half bits are flipped. So if a tiny change in key or plaintext results in a huge drastic ciphertext change, then it is the block cipher with high quality. Avalanche effect [97] can be measured by dividing the number of switched bits by the number of total bits in the cipher.

$$\text{Avalanche effect} = \frac{\text{Number of flipped bits in the ciphertext}}{\text{Number of total bits in the ciphertext}} \tag{3.10}$$

The strength of the key increases as the number of bits or characters in the plaintext gets increased. Since a single bit or character change in the plaintext leads to a vast change in the ciphertext or encrypted text generated. When calculating the strength of the key according to the avalanche effect, it is found to be more than 99% (Fig. 3.28).

3.7 CONCLUSION

In this chapter, a survey is made to integrate the blockchain to the IoT devices. The causes of huge number of devices, low bandwidth for communication, and less computing power on the blockchain performance were studied. The current state-of-art technologies of the blockchain are been analyzed, by comparing the technologies with IoT environment. For the security and privacy of the data CP−HWABE is been added to the blockchain and is been analyzed and found it consume less computation power, higher throughput, and the strength of the key is found to be around 99%. If this algorithm is used in the blockchain, then it would be both decentralized and secure and if there is found any illegal transaction involved then necessary action can be taken.

REFERENCES

[1] M. Katagi, S. Moriai, Lightweight Cryptography for the Internet of Things, Sony Corporation, 2012, 1–4.

[2] B. Fabian, O. Günther, Security challenges of the EPCglobal network, Commun. ACM 52 (7) (2009) 121–125.

[3] K. Rose, S. Eldridge, L. Chapin, The Internet of Things: An Overview, The Internet Society, 2015, pp. 1–50.

[4] R.H. Weber, Internet of Things—new security and privacy challenges, Comput. Law Secur. (2010) 23–30.

[5] Y. Liu, B. Dong, B. Guo, J. Yang, W. Peng, Combination of cloud computing and internet of things (IoT) in medical monitoring systems, Int. J. Hybrid Inf. Technol. 8 (12) (2015) 367–376.

[6] H.F. Atlam, A. Alenezi, A. Alharthi, R.J. Walters, G.B. Wills, Integration of cloud computing with internet of things: challenges and open issues, in: Proceedings IEEE International Conference in Internet of Things (iThings) and IEEE Green Compututing Communications (GreenCom) and IEEE Cyber, Physical Social Computing (CPSCom) and IEEE Smart Data (SmartData), 2017, pp. 670–675.

[7] X. Lyu, W. Ni, H. Tian, R.P. Liu, X. Wang, G.B. Giannakis, et al., Optimal schedule of mobile edge computing for internet of things using partial information, IEEE J. Sel. Areas Commun. 35 (11) (2017) 2606–2615.

[8] N. Chidambaram, P. Raj, K. Thenmozhi, R. Amirtharajan, Enhancing the security of customer data in cloud environments using a novel digital fingerprinting technique, Int. J. Digit. Multimedia Broadcast (2016).

[9] G. Booth, A. Soknacki, A. Somayaji, Cloud security: attacks and current defenses, in: Proceedings 8th Annual Symposium Information Assurance (ASIA13), 2013, pp. 4–5; N. Kshetri, Can blockchain strengthen the internet of things? IT Professional 19 (4) (2017) 68–72.

[10] N. Kshetri, Can blockchain strengthen the internet of things? IT Professional 19 (4) (2017) 68–72.

[11] Z. Zheng, S. Xie, H.-N. Dai, H. Wang, Blockchain challenges and opportunities: a survey, Int. J. Web Grid Serv 14 (4) (2018) 352–375.

[12] B. Betts, Blockchain and the Promise of Cooperative Cloud Storage, 2017, ComputerWeekly.com. Retrieved from: <https://www.computerweekly.com/feature/Blockchain-and-the-promise-of-cooperative-cloud-storage>.

[13] A. Dorri, S.S. Kanhere, R. Jurdak, Towards an optimized blockchain for iot, in: Proc. 2rd Int. Conf. Internet-of-Things Design Implementation, ACM, 2017, 173–178.

[14] K. Christidis, M. Devetsikiotis, Blockchains and smart contracts for the internet of things, IEEE Access 4 (2016) 2292–2303.

[15] X. Zha, X. Wang, W. Ni, R.P. Liu, Y.J. Guo, X. Niu, et al., Blockchain for IoT: the tradeoff between consistency and capacity, Chin. J. Internet Things 1 (1) (2017).

[16] B. Leiding, P. Memarmoshrefi, D. Hogrefe, Self-managed and blockchain-based vehicular ad-hoc networks, in: Proc. ACM Int. Joint Conf. Pervasive Ubiquitous Comput.: Adjunct, New York, 2016, pp. 137–140.

[17] M. Castro, B. Liskov, Practical Byzantine fault tolerance, in: Proc. 3rd Symp. Operating Syst. Des. Implementation (OSDI'99), New Orleans, LA, 1999.

[18] P.K. Sharma, S.Y. Moon, J.H. Park, Block-VN: a distributed blockchain based vehicular network architecture in smart city, J. Inf. Process. Syst. 13 (1) (2017) 184–195.

[19] M.C.K. Khalilov, A. Levi, A survey on anonymity and privacy in bitcoin-like digital cash systems, IEEE Commun. Surveys Tut (2018).

REFERENCES 91

[20] X. Zha, K. Zheng, D. Zhang, Anti-pollution source location privacy preserving scheme in wireless sensor networks, in: 13th Annual IEEE International Conference on Sensing, Communication, and Networking (SECON), 2016, pp. 1−8.

[21] G. Zyskind, O. Nathan, A.S. Pentland, Decentralizing privacy: using blockchain to protect personal data, in: Proc. IEEE Secur. and Privacy Workshops (SPW'15), IEEE, 2015, pp. 180−184.

[22] B. Panikkar, S. Nair, P. Brody, V. Pureswaran, ADEPT: An IoT Practitioner Perspective, 2014, IBM Inst. Bus. Value, New York, NY, USA, White Paper, 2015, pp. 1−18.

[23] G. Zyskind, O. Nathan, A. Pentland, Enigma: Decentralized Computation Platform With Guaranteed Privacy, arXiv preprint arXiv:1506.03471.

[24] IOTA, IOTA, March 2017. Retrieved from: <http://iotatoken.io/>.

[25] C. O'Connor, What Blockchain Means for You, and the Internet of Things, February 2017. Retrieved from: <https://www.ibm.com/blogs/internet-of-things/watson-iot-blockchain/>.

[26] P. Brody, V. Pureswaran, Device Democracy: Saving the Future of the Internet of Things, IBM, 2014. Retrieved from: <https://www.ibm.com/downloads/cas/Y5ONA8EV>.

[27] IBM, Watson Internet of Things, March 2017. Retrieved from: <https://www.ibm.com/internet-of-things/>.

[28] J. Fedak, How Can Blockchain Improve Cloud Computing, 2016. Retrieved from: <https://medium.com/iex-ec/how-blockchain-can-improve-cloud-computing-1ca24c270f4f>.

[29] M. Chen, S. Mao, Y. Liu, Big data: a survey, Mobile Netw. Appl. 19 (2) (2014) 171−209.

[30] F. Tschorsch, B. Scheuermann, Bitcoin and beyond: a technical survey on decentralized digital currencies, IEEE Commun. Surveys Tut. 18 (3) (2016) 2084−2123.

[31] M. Swan, Blockchain, O'Reilly Media, 2015.

[32] M. Crosby, P. Pattanayak, et al., Blockchain technology: beyond Bitcoin, Appl. Innov. 2 (2016) 6−10.

[33] A. Lewis, A Gentle Introduction to Blockchain Technology, 2015. Retrieved from: <https://j2-capital.com/gentle-introduction-blockchain-technology>.

[34] M. Pticek, V. Podobnik, G. Jezic, Beyond the internet of things: the social networking of machines, Int. J. Distrib. Sens. Netw. 12 (6) (2016).

[35] C. Perera, A. Zaslavsky, P. Christen, D. Georgakopoulos, Context aware computing for the internet of things: a survey, IEEE Commun. Surv. Tut. 16 (1) (2014) 414−454.

[36] L.D. Xu, W. He, S. Li, Internet of things in industries: a survey, IEEE Trans. Ind. Inform. 10 (4) (2014) 2233−2243.

[37] A. Al-Fuqaha, M. Guizani, M. Mohammadi, M. Aledhari, M. Ayyash, Internet of things: a survey on enabling technologies, protocols, and applications, IEEE Commun. Surveys Tut. 17 (4) (2015) 2347−2376.

[38] L. Atzori, A. Iera, G. Morabito, The internet of things: a survey, Comput. Netw. 54 (15) (2010) 2787−2805.

[39] P. Sethi, S.R. Sarangi, Internet of things: architectures, protocols, and applications, J. Electr. Comput. Eng. 2017 (2017) 1−25.

[40] J.S. Lee, Y.W. Su, C.C. Shen, A comparative study of wireless protocols: Bluetooth, UWB, ZigBee, and Wi-Fi, in: Proc. 33rd Annu. Conf. the IEEE Ind. Elect. Soc. (IECON'07), 2007, pp. 46−51.

[41] R. Ratasuk, B. Vejlgaard, N. Mangalvedhe, A. Ghosh, NB-IoT system for M2M communication, in: 2016 Proc. IEEE Wireless Commun. and Netw. Conf. Workshops (WCNCW' 16), 2016, pp. 1−5.

[42] X. Zha, W. Ni, X. Wang, R.P. Liu, Y.J. Guo, X. Niu, et al., The impact of link duration on the integrity of distributed mobile networks, IEEE Trans. Inf. Forensics Secur. 13 (9) (2018) 2240−2255.

[43] O. Vermesan, P. Friess, P. Guillemin, et al., Internet of things strategic research roadmap, Internet of Things-Global Technological and Societal Trends (2011) 9−52.

[44] C. Ren, X. Lyu, W. Ni, H. Tian, R.P. Liu, Distributed onlinelearning of fog computing under non-uniform device cardinality, IEEE Internet Things J. (2018).

[45] C.-W. Tsai, C.-F. Lai, M.-C. Chiang, L.T. Yang, et al., Data mining for internet of things: a survey, IEEE Commun. Surveys Tut. 16 (1) (2014) 77−97.

[46] X. Zha, W. Ni, K. Zheng, R.P. Liu, X. Niu, Collaborative authentication in decentralized dense mobile networks with key predistribution, IEEE Trans. Inf. Forensics Secur. 99 (2017).

[47] Internet of Things (IoT) Connected Devices Installed Base Worldwide From 2015 to 2025 (in Billions), 2016.

[48] I. Makhdoom, M. Abolhasan, J. Lipman, R.P. Liu, W. Ni, Anatomy of threats to the internet of things, IEEE Commun. Surv. Tut., 2018.

[49] G. Gan, Z. Lu, J. Jiang, Internet of things security analysis, in: 2011 Int. Conf. Internet Technol. and Appl., 2011, pp. 1−4.

[50] E. Alsaadi, A. Tubaishat, Internet of things: features, challenges, and vulnerabilities, Int. J. Advanced Comput. Sci. Inf. Technol. 4 (1) (2015) 1−13.

[51] X. Liu, M. Zhao, S. Li, F. Zhang, W. Trappe, A security framework for the internet of things in the future internet architecture, Future Internet, 9 (3).

[52] R. Roman, P. Najera, J. Lopez, Securing the internet of things, Comput. 44 (9) (2011) 51−58.

[53] J. Lin, W. Yu, N. Zhang, X. Yang, H. Zhang, W. Zhao, A survey on internet of things: Architecture, enabling technologies, security and privacy, and applications, IEEE Internet Things J. 4 (5) (2017) 1125−1142.

[54] K. Mehta, D. Liu, M. Wright, Protecting location privacy in sensor networks against a global eavesdropper, IEEE Trans. Mobile Comput. 11 (2) (2012) 320−336.

[55] N. Namvar, W. Saad, N. Bahadori, B. Kelley, Jamming in the internet of things: a game-theoretic perspective, in: 2016 IEEE Global Commun. Conf. (GLOBECOM), 2016, pp. 1−6.

[56] A. Mosenia, N.K. Jha, A comprehensive study of security of internet-of-things, IEEE Trans. Emerg. Top. Comput. 5 (4) (2017) 586−602.

[57] R. Xu, R. Wang, Z. Guan, L. Wu, J. Wu, X. Du, Achieving efficient detection against false data injection attacks in smart grid, IEEE Access 5 (2017) 13787−13798.

[58] J.S. Perry, Anatomy of an IoT Malware Attack, 2017. Retrieved from: <https://developer.ibm.com/articles/iot-anatomy-iot-malware-attack/>.

[59] X. Wang, W. Ni, K. Zheng, R.P. Liu, X. Niu, Virus propagation modeling and convergence analysis in large-scale networks, IEEE Trans. Inf. Forensics Secur. 11 (10) (2016) 2241−2254.

[60] X. Wang, K. Zheng, X. Niu, B. Wu, C. Wu, Detection of command and control in advanced persistent threat based on independent access, in: Proc. 2016 IEEE Int. Conf. Commun. (ICC'16), 2016, pp. 1−6.

[61] J.E. Boritz, Is practitioners' views on core concepts of information integrity, Int. J. Account. Inf. Syst. 6 (4) (2005) 260−279.

[62] A. Fongen, Identity management and integrity protection in the internet of things, in: Proc. 3rd Int. Conf. Emerg. Secur. Technol. (EST'12), 2012, pp. 111−114.

[63] H.C. Pöhls, JSON sensor signatures (JSS): end-to-end integrity protection from constrained device to iot application, in: Proc. 9th Int. Conf. Innovative Mobile Internet Serv. in Ubiquitous Comput. (IMIS'15), 2015, pp. 306−312.

[64] C. Wang, Q. Wang, K. Ren, W. Lou, Privacy-preserving public auditing for data storage security in cloud computing, in: Proc. 29th Annu. IEEE Int. Conf. Comput. Commun. (INFOCOM'10), 2010, pp. 1−9.

[65] K. Zhang, X. Liang, R. Lu, X. Shen, Sybil attacks and their defenses in the internet of things, IEEE Internet Things J. 1 (5) (2014) 372−383.

[66] S. Davidson, P. De Filippi, J. Potts, Economics of Blockchain. Retrieved from: <https://papers.ssrn.com/sol3/papers.cfm?abstract_id=2744751>.

REFERENCES 93

[67] S. Nakamoto, Bitcoin: A Peer-to-Peer Electronic Cash System, 2008. Retrieved from: <https://bitcoin.org/bitcoin.pdf>.

[68] A. Miller, J. Litton, A. Pachulski, et al., Discovering Bitcoin's Public Topology and Influential Nodes. Retrieved from: <https://www.cs.umd.edu/projects/coinscope/coinscope.pdf>.

[69] How do bitcoin transactions work? Available from: <http://www.coindesk.com/information/how-dobitcoin-transactions-work/>, 2015.

[70] O. Wyman, Blockchain in Capital Markets, 2016. Retrieved from: <https://www.oliverwyman.com/content/dam/oliver-wyman/global/en/2016/feb/BlockChain-In-Capital-Markets.pdf>.

[71] Bitcoin developer guide. Available from: <https://bitcoin.org/en/developer-guide>, 2017.

[72] R.C. Merkle, Protocols for public key cryptosystems, in: Proc. 1st IEEE Symp. Secur. Privacy (SP'80), 1980, pp. 122−122.

[73] D. Ghosh, How the Byzantine General Sacked the Castle: A Look Into Blockchain, 2016. Retrieved from: <https://medium.com/@DebrajG/how-the-byzantine-general-sacked-the-castle-a-look-into-blockchain-370fe637502c>.

[74] L. Lamport, R. Shostak, M. Pease, The Byzantine generals problem, ACM Trans. Program. Lang. Syst. 4 (3) (1982) 382−401.

[75] S. Poledna, Fault-Tolerant Real-Time Systems: The Problem of Replica Determinism, vol. 345, Springer Sci. & Business Media, 2007.

[76] P.J. Marandi, M. Primi, F. Pedone, High performance statemachine replication, in: Proc. 41st Int. Conf. Depend. Syst. Netw. (DSN' 11), IEEE, 2011, pp. 454−465.

[77] V. Buterin, On Public and Private Blockchains, Ethereum Blog. Available from: <https://blog.ethereum.org/2015/08/07/on-publicand-private-blockchains/>.

[78] N. Atzei, M. Bartoletti, T. Cimoli, A survey of attacks on ethereum smart contracts (SoK), in: Proc. 6th Int. Conf. Principles Secur. Trust, 2017, pp. 164−186.

[79] I. Eyal, E.G. Sirer, Majority is not enough: Bitcoin mining is vulnerable, in: Proc. Int. Conf. Financial Cryptography Data Secur. (FC'14), Springer, 2014, pp. 436−454.

[80] H. Finney, Best Practice for Fast Transaction Acceptance—How High Is the Risk, 2011. Retrieved from: <https://bitcointalk.org/index.php?topic=3441.0>.

[81] S.S. Team, Billions of Tokens Theft Case Cause by ETH Ecological Defects, March 2018. Retrieved from: <http://cxytiandi.com/blog/detail/18431>.

[82] X. Wang, X. Zha, G. Yu, W. Ni, R.P. Liu, Y.J. Guo, et al., Attack and defence of ethereum remote APIs, in: 2018 Proc. IEEE Globecom Workshops (GC Wkshps'18), 2018.

[83] M. Smache, N.E. Mrabet, J.J. Gilquijano, A. Tria, E. Riou, C. Gregory, Modeling a node capture attack in a secure wireless sensor networks, in: 2016 Proc IEEE 3rd World Forum on Internet of Things (WF-IoT'16), 2016, pp. 188−193.

[84] M. Vasek, M. Thornton, T. Moore, Empirical analysis of denial-of-service attacks in the bitcoin ecosystem, in: Proc. Int. Conf. Financial Cryptography Data Secur. (FC'14), Springer, 2014, pp. 57−71.

[85] E.H. Yuval Marcus, S. Goldberg, Low-Resource Eclipse Attacks on Ethereum's Peer-to-Peer Network, 2018. Retrieved from: <https://www.cs.bu.edu/~goldbe/projects/eclipseEth.pdf>.

[86] Y. Yang, X. Liu, R.H. Deng, Lightweight break-glass access control system for healthcare internet-of-things, IEEE Trans. Ind. Inform. PP (99) (2017).

[87] W.L. Chin, W. Li, H.H. Chen, Energy big data security threats in IoT-based smart grid communications, IEEE Commun. Mag. 55 (10) (2017) 70−75.

[88] K. Biswas, V. Muthukkumarasamy, Securing smart cities using blockchain technology, in: Proc. 18th IEEE Int. Conf. High Performance Comput. Commun.; 14th IEEE Int. Conf. Smart City; 2nd IEEE Int. Conf. Data Sci. Syst. (HPCC/SmartCity/DSS'16), 2016, pp. 1392−1393.

94 **CHAPTER 3** INTEGRATION OF BLOCKCHAIN AND INTERNET OF THINGS

[89] K. Korpela, J. Hallikas, T. Dahlberg, Digital supply chain transformation toward blockchain integration, in: Proc. 50th Hawaii Int. Conf. Syst. Sci., 2017.

[90] G. Wood, Ethereum: A Secure Decentralised Generalised Transaction Ledger, Ethereum Project Yellow Work. Retrieved from: <https://gavwood.com/paper.pdf>.

[91] F. Zhang, E. Cecchetti, K. Croman, A. Juels, E. Shi, Town crier: an authenticated data feed for smart contracts, in: Proc. 23rd ACM Conf. Comput. Commun. Secur. (CCS' 16), ACM, New York, 2016, pp. 270−282.

[92] Oraclize. Available from: <http://www.oraclize.it>, 2018.

[93] Wallet knowledge base. Available from: <https://iotasupport.com/walletknowledgebase.shtml>, 2018.

[94] Why chain, not iota, is ideal for iot apps. Available from: <https://medium.com/0chain/why-0chain-not-iota-isideal-for-iot-web-enterprise-apps-bdd1154d148f>, 2017.

[95] J. McKinney, Light Client Protocol, 2017. Retrieved from: <https://github.com/ethereum/wiki/wiki>.

[96] A. Lemma, Gebremedhn, M. Bravo, Performance analysis on the implementation of data encryption algorithm used in network security, Int. J. Comput. Sci. Technol. 4 (4) (2015) 711−717.

[97] C. Echeverri, Visualization of the Avalanche Effect in CT2, Bachelor's Thesis at University of Mannheim, 2016, p. 10.

FURTHER READING

Q. Cui, Y. Wang, K. Chen, W. Ni, I. Lin, X. Tao, et al., Big data analytics and network calculus enabling intelligent management of autonomous vehicles in a smart city, IEEE Internet Things J. 6 (2) (2018).

CHAPTER

A DEEP DIVE INTO SECURITY AND PRIVACY ISSUES OF BLOCKCHAIN TECHNOLOGIES

4

Neha Gupta
MRIIRS, Faridabad, India

4.1 INTRODUCTION

Blockchain technology is a combination of various algorithms and techniques like cryptography, mathematics, peer-to-peer networking, and so on that are used to solve the synchronization problems of distributed databases. Blockchain can be defined as integrated infrastructure of multifield applications like financial market, IOT, medical field, and so on. Bitcoin is one example of blockchain technology that is gaining popularity these days. Other application areas of blockchain are protection of intellectual property, international payments, prediction market, hyper ledger, ethereum, and so on.

Financial industry (FinTech) is using blockchain as the core technology for its operation, but security of blockchain has always remain an area of concern for users. Security is often measured in terms of integrity, confidentiality, and availability. Most of the public blockchain systems are working on distributed systems and are low on confidentiality, although these systems promise integrity and availability. Availability of data is always on a higher end in these blockchain systems. Availability of readable data is higher as data are replicated on distributed systems but is low for write availability. Although the core architecture of any blockchain system is very secure, the implementations of innovative technologies have exploited the security aspects of blockchain. Blockchain system is also vulnerable to leakage of transactional privacy because of the visibility of all public keys of network to everyone. In the recent past, various security vulnerabilities have been reported related to ethereum and smart contracts. For example, in June 2016, criminals have used recursive calling vulnerability to attack smart contracts and have stolen approximately $60 million.

Various studies have been conducted on security and privacy issues of blockchain, but none of them have given the solutions related to security enhancements. In this chapter, we will try to elaborate more comprehensive perspective of blockchain technologies and their related security issues, and will also discuss various security risks related to popular blockchain systems, giving example of real attacks and will analyze the vulnerabilities exploited. This chapter also discusses various encryption techniques used by blockchain systems to provide data safety and protect vulnerability.

Handbook of Research on Blockchain Technology. DOI: https://doi.org/10.1016/B978-0-12-819816-2.00004-6
© 2020 Elsevier Inc. All rights reserved.

95

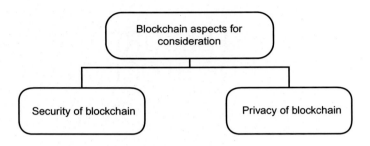

FIGURE 4.1

Aspects of blockchain [1].

4.2 BLOCKCHAIN ASPECTS FOR CONSIDERATION

The key aspects to be considered for a blockchain system are security issues and countermeasures of a blockchain system and the various privacy issued involved (Fig. 4.1).

4.2.1 SECURITY OF BLOCKCHAIN

Security of blockchain is a key concept as it involves the protection of information and the data used in cryptocurrency transactions and in blocks against various malicious and nonmalicious attacks. Protection involves implementation of various security policies, tools, and IT services to detect and protect the threats.

1. *Defense in penetration.* In this strategy, various corrective measures are enforced to protect the data. It works on the principle of protecting data in multiple layers instead of implementing single layer of security.
2. *Minimum privilege.* This method reduces the accessibility of data to the lowest possible level to strengthen and enhance the security level.
3. *Manage vulnerabilities.* Vulnerabilities are checked and manage by modifying, identifying, authenticating the user, and patching the gap.
4. *Manage risks.* Risks of an environment are processed by identifying the risk, assessing the level of risk, and by controlling the possibility of risks.
5. *Manage patches.* In this strategy, we patch the administered part like code, application, operating system by acquiring, testing, and installing patches.

Blockchain technology uses several techniques to ensure the security required in transaction data or block data, irrespective of the usage or data in the block. Several applications like Bitcoin use cryptographic techniques for data safety.

The other most secure idea of blockchain is that the longest chain is the legitimate one. This dispenses with the security chances because of 51% dominant attack and fork issues. As the longest chain is the most genuine, alternate assaults end up invalid and void as they end up being stranded forks.

4.3 SECURITY ISSUES OF BLOCKCHAIN TECHNOLOGY 97

4.2.2 PRIVACY OF BLOCKCHAINS

Privacy is the capacity of a solitary individual or a gathering to disconnect themselves or information along these lines conveying everything that needs to be conveyed discerningly. Security in blockchain implies having the capacity to perform exchanges without spilling recognizable proof data. In the meantime, privacy enables a client to stay agreeable by discerningly unveiling themselves without exhibiting their action to the whole system.

The objective of enhancing privacy in blockchain is to make it incredibly difficult for different clients to duplicate or utilize other clients' crypto profile. An immeasurable volume of variations can be perceived when applying blockchain technology [2].

1. *Stored data sorting.* Blockchain gives the edibility to store all types of data. The privacy viewpoint in blockchain changes for individual and organizational data. Despite the fact that privacy rules are applicable on individual data, increasingly stringent privacy rules apply to sensitive and organizational data.
2. *Storage distribution.* Full nodes in the network are the nodes that stores the entire copy of the blockchain. Full nodes when combines with the append-only characteristics of blockchain system often leads to redundancy in data. This redundancy in data in blockchain system adds two new features: transparency and variability. Transparency and variability levels in the network are decided by the compatibility level of application with its data minimization.
3. *Append-only.* It is not possible to change the information of previous blocks within the blockchain undiscovered. In some cases, the append-only feature of blockchain does not curtail to correction of users, particularly, if information is recorded incorrectly. Special attention has to be provided while distributing rights to data subjects in blockchain technology.
4. *Private versus public blockchain.* Blockchain accessibility is remarkable from a privacy point of view. At an advanced level, the restricted data on a block can be encrypted by authorized users for conditional access, as each node in the blockchain holds a copy of the entire blockchain.
5. *Nonpermissioned versus permissioned types of blockchain.* With public or unauthorized blockchain applications, all users are allowed to add data in principle. Distribution of network control can be restored by allowing trusted mediators.

4.3 SECURITY ISSUES OF BLOCKCHAIN TECHNOLOGY

Blockchain has captured a lot of attention in recent times. Although various features of blockchain technologies have given us convenient and reliable services, the security and privacy issues are still an area of concern that needs attention. Various authors have conducted studies on security issues and privacy issues of blockchain technology, but a detailed and systematic study is needed which will try to cover all the important aspects. Various security issues are as follows.

1. 51% Vulnerability
2. Double spending
3. Mining pool attacks
4. Client side security threats

98 CHAPTER 4 A DEEP DIVE INTO SECURITY AND PRIVACY ISSUES

5. Forking
6. Criminal activity
7. Private key security
8. Transaction privacy leakage

4.3.1 51% VULNERABILITY

To develop mutual trust, blockchain works by integrating distributed consensus mechanism. In this mechanism, the computing power is distributed among all the available data miners. Work of these data miners is to check the hashes generated by CPU cycles. If these miners join together, they can become a big mining pool having maximum computing power. If the mining pool has 51% or more computing power, they can take a control of the blockchain and can cause serious security risks. For example, in proof of work (POW)-based blockchains, if the hashing power of a single miner is 50% more than the total hashing power of a complete blockchain system, then the miner can easily launch 51% attack and can cause vulnerabilities like:

- reverse transaction attacks
- double spending
- exclude transactions
- modify transactions
- disturbing operations of other miners
- termination of verification process

In other examples, a single miner working on a proof of stake (POS)-based blockchain can have 50% of the total coins, can launch a 51% attack, and can modify and exploit the information of blockchain system.

4.3.2 DOUBLE SPENDING

If a consumer is using same cryptocurrency for multiple transactions, then it is double spending. An attacker can execute race attacks to initiate double spending. In POW-based blockchain, these types of attacks are comparatively easy to implement because attacker can easily exploit the time between initiation of two transactions as well as confirmation of two transactions. Before the attackers second transaction got invalid, he got the output of first transaction which may lead to double spending.

Model to depict double spending behavior of an attacker:

Assumption:

- Vendor address is known to the attackers before initiation of the attack.
- Let us have two transactions: T1 and T2.
- Same Bitcoin address as input for both the transactions.

Working:

- Set T1 recipient address as targeted vendor address.
- Set T2 recipient address as colluding address that is controlled by attacker.

4.3 SECURITY ISSUES OF BLOCKCHAIN TECHNOLOGY 99

- Initiate T1 so that it will be added to the wallet of the vendor.
- Before confirmation of T1, initiate T2 as well.
- While T2 is in process, attacker will get the confirmation of T1 and has successfully completed the transaction.
- When T2 completes, T1 will be mined as invalid, while the attacker has already used the same cryptocurrency twice.
- Because of the colluding address of T2 which is owned by attacker, he still owns the BTC and is enjoying the service without spending BTC.

4.3.3 MINING POOL ATTACKS

To increase the computing power or the hash power of a block, mining pools are created. These pools directly affect the time required to verify a block. These mining pools also increase the chances of winning the mining reward.

Mining pools are evolving and the vulnerability to exploit these pools are also increasing. Mining pool attacks are of two types:

1. internal attacks
2. external attacks

Internal attacks are the attacks in which the miner maliciously collect more than the required rewards and disrupt the normal functionality causing pool to disregard successful mining attempts.

External attacks are caused when the miner uses higher hash power to attack the pool causing double spending. Mining pool attacks includes selfish mining, hopping attacks, block withholding, bribery attack, and so on.

4.3.4 CLIENT SIDE SECURITY THREATS

Popularity of various cryptocurrencies increased a number of users to join blockchain networks. Every user on blockchain network has a set of private public keys to access its cryptocurrency wallets. Therefore it is necessary to manage these keys securely. Important aspect of client side security is, if the client loose or compromise the keys, then he/she will not be able to access its wallet and will lead to irrevocable monetary loss. The client security is compromised using various mechanisms like hacking, using buggy software, or by incorrect usage of the wallet.

4.3.5 FORKING

Forking refers to the agreement that takes place between decentralized nodes, when the software upgrades. It is an important issue as it involves many blockchains at once. In forking, whenever a new version of the software related to blockchain is published, a new agreement in consensus rule also changed between all the decentralized nodes. Because of the above process, the nodes in the blockchain are divided into two types, that is, old nodes and new nodes. The new nodes thus formed may or may not agree with the transaction blocks sent by the old nodes. Similarly, old

100 CHAPTER 4 A DEEP DIVE INTO SECURITY AND PRIVACY ISSUES

nodes may or may not agree with the transaction blocks sent by the new nodes. Because of this, fork problem arises. Fork problem is divided into two types:

1. soft fork
2. hard fork

Hard fork happens when system upgrades to a new version and is not compatible with the old version. New version nodes did not agree with the mining of old version nodes and hence both form their own blockchain. When hard fork occurs, all the old nodes in the network are requested to upgrade to new agreement. If the old nodes do not upgrade themselves, then they can continue to work as a different chain and hence the ordinary chain of nodes will fork into two chains: old and new.

When the new version or the agreement is not compatible with the old version, and the new nodes do not agree with the transaction mining of old nodes, then soft fork happens. In the soft fork, nodes in the network do not have to upgrade themselves to the new agreement immediately; instead, this process happens gradually and will not affect the stability and effectiveness of the system. Also soft fork has only one chain. A soft fork can also be the result of a temporary divergence. When a particular miner is using nonupgraded software, then the clients on their nodes violate a new consensus rule that their nodes do not understand.

4.3.6 CRIMINAL ACTIVITY

Multiple Bitcoin addresses can be attached to one single user, and the address is not related to their true identity in life. Bitcoin was therefore used in illegal activities. Bitcoin is supported by various third-party platforms and the users can buy or sell the products using these platforms. It is quite difficult to track the behavior of the users as the process of selling and buying using third-party platforms are anonymous. Criminal activities that are frequently carried out using Bitcoin are the following.

1. *Ransomware.* It is frequently used by criminals to extort money using Bitcoin as a trade currency. In July 2014, a ransomware called CTB-Locker spread throughout the world as a mail attachment. If the user clicks the attachment, the ransomware runs in the system background and encrypts approximately 114 types of each. The victim must pay a certain amount of Bitcoin Wi to the attacker within 96 hours. Otherwise the encrypted_les will not be restored.
2. *Underground market.* Bitcoin is often used in the underground market as a currency. Silk Road, for example, is an anonymous, international online marketplace which operates as a hidden Tor service and uses Bitcoin as its currency. The top 10 item categories on the Silk Road are listed in Table 4.1. Most of the products sold on the Silk Road are drugs or some other controlled products in the real world.

   Since international transactions account for a large proportion on the Silk Road, Bitcoin makes the transaction more convenient on the underground market, which is harmful to social security.
3. *Money laundering.* Because Bitcoin has features such as anonymity and virtual network payment and has been adopted by many countries, Bitcoin carries the lowest risk of money laundering compared with other currencies. Oh, Cody et coll. Propose Dark Wallet, a Bitcoin

Table 4.1 Top 10 Categories of Items Available in Silk Road

Number	Category	Items	Percentage
1	Weed	3338	13.7
2	Drugs	2194	9.0
3	Prescription	1784	7.3
4	Benzos	1193	4.9
5	Books	955	3.9
6	Cannabis	877	3.6
7	Hash	820	3.4
8	Cocaine	630	2.6
9	Pills	473	1.9
10	Blotter (LSD)	440	1.8

application that can completely stealthy and private the Bitcoin transaction. Dark Wallet can encrypt and mix the user's transaction information with valid coins making money laundering much easier.

4.3.7 PRIVATE KEY SECURITY

In blockchain systems, the user's private key is recognized as a security and authentication credential created by the user and no third party is involved in this process. Whenever a user creates a wallet for a cryptocurrency, he/she must import the private key into the wallet as well. This private key is imported into the wallet to guarantee the security and authentication of the cryptocurrencies. If private key is lost or stolen, it cannot be recovered, which means that the user cannot access the wallet with any other alternative means and that all his cryptocurrencies in the wallet are unavailable.

Blockchain systems are not controlled by third-party institutions, so lost or stolen private key scenarios lead to the risk of data being altered by untraceable attackers.

4.3.8 TRANSACTION PRIVACY LEAKAGE

Since the behaviors of users in the blockchain are traceable, the blockchain systems take measures to protect the privacy of users' transactions. They use one-time accounts in Bitcoin and Zcash to store the received cryptocurrency. In addition, the user must assign a private key for each transaction. Similarly, the attacker cannot determine whether the cryptocurrency is recovered in different transactions received by the same user. In Monero, users can include certain chaff coins (called "mixins") when initiating a transaction, so that the attacker cannot determine the linkage between the actual coins spent by the transaction.

The data protection measures in the blockchain are unfortunately not very robust. Miller et al. [3] empirically assess two weaknesses in the Mixin sampling strategy of Monero and found that 66.09% of all transactions do not contain mixins. The 0-mixin transaction will lead to the sender's privacy leakage. Since users can use 0-mixin transaction outputs as mixins, these mixtures can be

102 CHAPTER 4 A DEEP DIVE INTO SECURITY AND PRIVACY ISSUES

Table 4.2 Linkability Analysis of Monera Transaction Inputs With Mixins			
	Not Deducible (%)	Deducible (%)	In Total (%)
Using newest TXO	15.07	4.60	19.67
Not using newest TXO	22.61	57.72	80.33
In total	37.68	62.32	100

deducted. In addition, they study the mixin sampling method and find that the mixin selection is not really random. More frequently, new TXO (transaction outputs) are used. They also find that 62.32% of mixin transaction inputs are deductible, as shown in Table 4.2 [3]. By taking advantage of these weaknesses in Monero, the actual transaction inputs can be determined with 80% accuracy.

4.4 PRIVACY ISSUES OF BLOCKCHAIN TECHNOLOGY

Nowadays, discussions about blockchain technology seem to be everywhere, with potential applications covering industries as diverse as banking, healthcare, property, law enforcement, entertainment, and even wine and jewelry sales. Different blockchain applications present different and unique data security and privacy challenges and opportunities, but three general categories currently concern legal privacy experts. The first involves the necessary bridge between the physical and cyberspace limits; the second involves sensitive information that is actually stored on the blockchain; and the third involves the very existence of blockchains. Each of these options offers trade-offs between security, privacy, speed, and functionality, and different applications require different blockchain networks to work according to each application's specific requirements.

4.4.1 PHYSICAL—CYBERSPACE BOUNDARY

The "physical—cyberspace boundary" refers to the concept that when a flesh-and-blood person interacts in cyberspace, they do so through an "online identifier." For example, if you want to interact with users on Facebook, you need to create a username and log on; so Facebook's network knows who you are. The same applies to any online interaction, whether it is banking, purchasing concert tickets, or downloading music—a connection between you and your online identifier needs to be established to participate in a transaction. The ID is pseudo-anonymous (e.g., a bank account or email address that has no real name attached to it), but at some point the physical—cyberspace boundary has to be a bridge of the physical—cyberspace boundary. At present, this scaffold is cultivated fundamentally through username and secret key blends, sometimes with the expansion of multifaceted verification techniques. In the near future, biometric identifiers will supplant usernames and passwords as the methods for intersecting the physical—cyberspace boundary. One problem with this system is that in order for a physical person to log into a network, the network must have a copy of the login credentials of that person coupled with the online identifier of that person. These credentials must only be stored in one place in a centralized system (e.g., On Facebook or your bank's central severs). These credentials would be stored in a blockchain network on all the nodes containing the blockchains with which you want to interact, some of which can be more easily compromised than a secure central server.

This is particularly important when it comes to biometric identifiers, which are not easily changed once they are compromised by identity thieves. Compounding this issue is the fact that, as will be mentioned underneath, the character of a blockchain approach that all facts stored on a blockchain remains stored as additional blocks are added to the chain, which means sensitive private data can be saved in cyberspace for all time.

Some other issue with blockchain are the absence of a strong central authority; it may be difficult to save you from hackers having access to sensitive statistics once a person's login credentials are compromised.

For example, if someone hacks your bank account or steals your credit card information, you can call your bank and update your login information or cancel your old credit card. In a blockchain network with no strong central authority, it can be difficult to update your login credentials, and even possible for a hacker to lock you out by updating the credential once they have access.

Not only is the potential for hacking of this sensitive information problematic from a security standpoint but it also creates uncertainty concerning who, if anyone, is responsible for notifying individuals if their login credentials are compromised. Most states have passed data breach notification laws requiring personal information custodians ("PII") to notify PII owners if their PII is compromised. At this stage, it is unclear how these laws are applied to a distributed network, such as blockchain, or whether they are even written applicable.

4.4.2 INFORMATION STORAGE AND INFERENCE

Some of the data which will be stored on blockchains will be particularly sensitive—blockchain networks are currently being explored as means of recording and updating healthcare records, genomic sequences, and biometric credentials (as discussed above). While any sensitive information stored on the blockchain will (as a best practice) be encrypted, because of the distributed nature of the blockchain, hackers may target those specific nodes that, for one technical reason or another, can be more easily compromised to access the encrypted information, or where the laws are inadequate to prevent such hacking. This concern is compounded when it comes to government-employed hackers who can benefit from the physical location of nodes in countries where information is more easily hacked or where the laws are insufficient to prevent such hacking. While privacy risks can be mitigated by operating in closed networks, there are benefits to open networks that require at least some blockchains containing sensitive information to operate in networks that are less than completely closed. Another concern about open networks is that, although the information itself is encrypted, sensitive information can be gathered from the fact that transactions take place. For example, if two large banks engage in a high volume of transactions within a short period of time, information can be extrapolated from this information by other banks or private individuals who can see the transactions happening, even if they cannot see the transaction details themselves. On a more personal level, if a doctor accesses the patient's health records to make changes, a hacker can see the transaction if he knows the doctor's and the patient's online identifiers. Although the hacker cannot see the health records or what has changed without accessing and decrypting the records, he or she can at least conclude that the patient has seen a particular doctor on a specific date, information that a patient might want to keep privately.

104 CHAPTER 4 A DEEP DIVE INTO SECURITY AND PRIVACY ISSUES

Equally problematic is the fact that, at this point, it is unclear who, if anyone might be legally liable in the event, this information is accessed and harm results to the owner of the information, or to a third party as a result of unauthorized use of the information.

4.4.3 NATURE OF THE BLOCKCHAIN—ETERNAL RECORDS

One of the great challenges faced by data protection in the 21st century is the combined advances in data retention, data cataloging, and data search capabilities. As we create more and more data on our lives and as these data are cataloged and easily searched, the data becomes eternal and visible to the general public in a way that has never been before. The technology of blockchain is likely to accelerate this trend. One of the as-advertised advantages of blockchain is that it records all transactions back to the genesis block to keep records almost perfectly. As the types of transactions stored in blockchains increase, eternal records of each transaction will increase. In the future, it will be possible for every transaction you undertake to be stored on a blockchain and you will have no control over where this information is stored or how it is used and no way to delete it. There are numerous concerns about privacy (and laws) involved in these eternal records. To begin with, the simple fact that these records exist could pose problems for anyone who does not want to have a complete record of all their transactions for all time. In addition, there is currently no clear agreement on who "owns" the information in these records as a legal matter. It is possible that blockchain networks could sell the information contained in these records without any input from the persons involved in the transactions, and these persons would not have recourse. In the absence of clear ownership rules, public bodies and private citizens may also have access to these data without the consent of the persons involved in the transaction. In blockchains with a weak or no central authority, it can be impossible to correct bad data that enters the chain.

4.5 TYPES OF ATTACKS

In this section, we have tried to analyze the real attacks on blockchain systems, and have discussed the vulnerabilities associated with these attacks.

4.5.1 SELFISH MINING ATTACK

The attack on selfish mining is carried out by attackers (i.e., selfish miners) to obtain undue rewards or to lose the computer power of honest miners [4]. The attacker privately holds discovered blocks and tries to forge a private chain. Later, selfish miners mine in this private chain and try to keep a private branch longer than the public branch because they hold more newly discovered blocks in private. Meanwhile, honest miners are continuing to exploit the public chain. New blocks mined by the attacker would be revealed when the public branch approaches the length of the private branch, so that honest miners end up with a loss of computing power and no reward, as selfish miners publish their new blocks just before honest miners. As a result, selfish miners gain a competitive advantage and honest miners are encouraged to join the selfish miners' branch. This attack undermines the decentralization nature of the blockchain by further consolidating the mining power in favor of the attacker. A Selfish-Mine attack strategy was proposed such that it can force honest miners to perform wasteful computations on the stale public sector. The length of the public chain and the private chain are the same in the initial circumstances of Selfish-Mine. The Selfish-Mine involves the following three scenarios:

1. The public chain is longer than the private one. Since the computing power of selfish miners may be smaller than that of the honest miners, selfish miners update the private chain according to the public chain, and in this scenario, selfish miners cannot reward themselves.
2. The first new block is found almost simultaneously by selfish miners and honest miners. In this scenario, selfish miners publish the newly discovered block and two forks of the same length will be present at the same time. Honest miners in either branch, while selfish miners in the private chain, continue to mine. If selfish miners first find a new block, they will immediately publish that block. At this stage, selfish miners receive two blocks of rewards simultaneously. Since the private chain is longer than the public chain, the private chain is the ultimate branch of industry. If honest miners first find the second new block and this block is written into the private chain, selfish miners receive the first new block and honest miners receive the second new block. Otherwise, if this block is written in the public block, honest miners will receive the rewards of these two new blocks and selfish miners would not receive any rewards.
3. They also find the second new block after the selfish miners find the first new block. In this scenario, these two new blocks are held privately by selfish miners and continue to mine new blocks in the private chain. When honest miners find the first new block, selfish miners release their own new block. When honest miners find the second new block, selfish miners publish their own new block immediately. This response will then be followed by selfish miners until the public chain length is only 1 larger than the private chain, after which selfish miners will publish their last new block before honest miners find this block. The private chain is considered valid at this point and, consequently, selfish miners gain the rewards of all new blocks.

4.5.2 DAO ATTACK

Some other attacks that exploit smart contracts' vulnerabilities are:

Attack Case	Related Vulnerabilities
King of the Ether throne	Out-of-gas send, exception disorder
Multiplayer games	Field disclosure
Rubixi attack	Immutable bug
GovernMental attack	Immutable bug, stack overflow, unpredictable state, timestamp dependence
Dynamic libraries attack	Unpredictable state

The Decentralized Autonomous Organization (DAO) is an intelligent contract deployed in Ethereum on May 28, 2016 that implements a platform for crowd financing. The DAO contract was only attacked after 20 days and had been deployed. Before the attack, DAO had already raised $150 million, the largest crowdfund ever. The attacker stole approximately US$60 million. In this case, the attacker exploited the reentrancy vulnerability. First, the attacker publishes a malicious

106 **CHAPTER 4** A DEEP DIVE INTO SECURITY AND PRIVACY ISSUES

intelligent contract that includes a withdrawal () call to the DAO function in its callback. The withdrawal () sends Ether to the street, which also has a call form. It will therefore invoke the callback function of the malicious intelligent contract again. The attacker can thus steal the entire Ether from the DAO. There are 15 more cases that exploit smart contracts' vulnerabilities.

4.5.3 **BGP HIJACKING ATTACK**

BGP is a de facto routing protocol that regulates how IP packets are sent to their destination. To intercept blockchain network traffic, attackers can either use or manipulate BGP routing. BGP hijacking typically requires network operators to be controlled, which could be used to delay network messages. Many authors have thoroughly analyzed the impact of routing attacks on Bitcoin, including both node and network attacks, and have shown that the number of Internet prefixes successfully hijacked depends on the distribution of mining power. Due to the high centralization of some Bitcoin mining pools, it will have a significant effect if they are attacked by BGP hijacking. The attackers can divide the Bitcoin network effectively or delay block propagation speed. Attackers hijack BGP to intercept the connections of Bitcoin miners to a mining pool server analyzed by Dell SecureWorks in 2014. By redirecting traffic to an attacker-controlled mining pool, the victim's cryptocurrency could be stolen. This attack raised an estimated cryptocurrency of US $ 83,000 over a period of 2 months. Since BGP security extensions are not widely used, network operators must rely on surveillance systems that report rogue announcements, such as BGP-Mon. However, even if an attack is detected, it still costs hours to solve a hijacking, because it is a human-driven process that changes the configuration or disconnects the attacker. For example, YouTube ever took about 3 hours to resolve a hijacking of its prefixes by a Pakistani Internet Service Provider (ISP).

4.5.4 **ECLIPSE ATTACK**

Some other attacks caused by the eclipse attack are as follows.

- Engineering block races that lead to the loss of mining power on orphan blocks.
- Splitting mining power that can cause a 51% vulnerability to be triggered.
- Selfish mining attackers can earn more than normal mining rewards.
- 0-Double spending confirmation: the vendor would not receive rewards for its service.
- Double confirmation.

The eclipse attack allows an attacker to monopolize all incoming and outgoing links between the victim and the other network peers. The attacker can then filter the victim's view of the blockchain or allow the victim to use obsolete views of the blockchain for unnecessary computing power. In addition, the attacker can leverage the computer power of the victim to perform his own malicious acts. Authors considered two types of eclipse attacks on the peer-to-peer network of Bitcoin, namely botnet attacks and infrastructure attacks. The botnet attack is started with bots with a variety of IP addresses. The attack on infrastructure models the threat from an ISP, company, or nation state with adjacent IP addresses. The Bitcoin network may be disrupted, and the view of a victim of the blockchain is filtered due to the eclipse attack. An eclipse attack is also a useful base for other attacks.

4.5.5 **LIVENESS ATTACK**

Liveness attack is an attack that can delay the confirmation time of a target transaction as much as possible. They also show two instances of this attack against Bitcoin and Ethereum. The attack consists of three phases, namely the attack preparation phase, the denial phase of the transaction, and the blockchain retardation phase.

- *Phase of attack preparation.* Like a selfish mining attack, an attacker gains advantage over honest miners somehow before TX is transmitted to the public chain. The assailant builds the private chain longer than the public chain.
- *Transaction denial phase.* The attacker holds privately the block containing TX to prevent TX from being placed in the public chain.
- *Blockchain retarder phase.* In the growth process of the public chain, TX cannot be held in private for a certain period of time. In this case, the attacker will publish the block containing TX. When the depth of the block containing TX is higher than a constant, TX is considered valid in some blockchain systems. The attacker will therefore continue to build a private chain to build an advantage over the public chain.

The attacker will then publish its blocks in private into the public chain in good time to slow the growth rate of the public chain. The liveliness attack ends when TX is checked in the public chain as valid.

4.6 **SECURITY ENHANCEMENT TO BLOCKCHAIN SYSTEMS**

In this section, we summarize security enhancements to blockchain systems, which can be used in the development of blockchain systems.

4.6.1 **SMARTPOOL**

As mentioned above, there is already a mining pool with over 40% of the total blockchain computing power. This poses a serious threat to the nature of decentralization, making blockchain vulnerable to attacks of various kinds. Fig. 4.2 shows the execution process of SmartPool:

SmartPool receives Ethereum node client transactions (i.e., parity or geth) that contain information about mining tasks. The miner then performs a task-based hacking calculation and returns the completed shares to the SmartPool client. When the number of the shares completed reaches a certain amount, they are committed to the Ethereum SmartPool contract. The SmartPool contract verifies shares and provides the customer with rewards. SmartPool system has following advantages in comparison with traditional peer-to-peer (P2P) pool.

1. *Decentralized.* The core of the SmartPool is implemented in the form of an intelligent contract in a blockchain. Miners first need to connect to Ethereum to me via the customer. The mining pool can rely on the consensus mechanism of Ethereum. Similarly, the nature of pool miners is decentralized. The state of the mining pool is maintained by Ethereum, and the pool operator is no longer required.

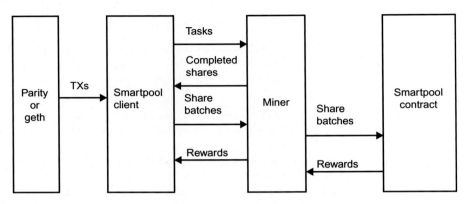

FIGURE 4.2

Execution process of SmartPool.

2. *Efficiency*. Miners can send the completed shares in batches to the SmartPool contract. In addition, miners must send only part of the shares to be verified but not all of it. SmartPool is therefore more efficient than the P2P pool.
3. *Secure*. SmartPool uses a new data structure that can prevent the attacker from re-sending shares in various batches. In addition, the SmartPool verification method can guarantee that honest miners will receive expected rewards even if there are malicious miners in the pool.

4.6.2 QUANTITATIVE FRAMEWORK

There exist trade-offs between blockchain's performance and security. Fig. 4.3 shows the framework.

There are two components to the quantitative framework: blockchain stimulator and safety model. The stimulator imitates the execution of the blockchain, the inputs of which are consent protocol and network parameters. It can gain performance statistics of the target blockchain by analyzing the simulator, including block propagation times, block sizes, network delays, block rate, throughput, and so on. The stale block refers to a block mined in the public chain, but not written. The transaction throughput is the number of transactions that the blockchain can handle. Stale block rate is transferred as a parameter to the component of the security model, which is based on Markov Decision Processes to defeat double spending and selfish mining. The framework ultimately produces an optimal adversarial strategy against attacks and facilitates the establishment of security measures.

4.6.3 OYENTE

Oyente has been proposed to detect bugs in intelligent contracts with Ethereum. Oyente uses symbolic execution to analyze the bytecode of intelligent contracts and complies with the EVM execution model. Since Ethereum stores smart contract bytecodes in its blockchain, Oyente can be used

4.6 SECURITY ENHANCEMENT TO BLOCKCHAIN SYSTEMS

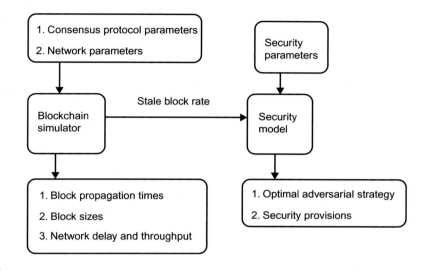

FIGURE 4.3

Components of quantitative framework.

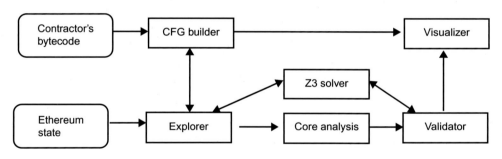

FIGURE 4.4

Oyente architecture and execution process.

to detect bugs in contracts deployed. Fig. 4.4 shows the architecture and execution process of Oyente.

It takes the bytecode and the Ethereum global state of the smart contract as inputs. First, based on bytecode, Control Flow Graph (CFG) BUILDER will statically build smart contract CFG. According to Ethereum and CFG information, EXPLORER simulates the execution of intelligent contracts by leveraging static symbolic performance. In this process, CFG will be further enriched and improved because some jump targets are not constants; they should instead be calculated during symbolic performance. The CORE ANALYSIS module detects four different vulnerabilities using the related analysis algorithms. The VALIDATOR module confirms the vulnerabilities and vulnerabilities detected. Confirmed vulnerability and CFG information are finally displayed in the

110 **CHAPTER 4** A DEEP DIVE INTO SECURITY AND PRIVACY ISSUES

VISUALIZER module, which users can use to debug and analyze the program. Oyente is currently open source for public use.

4.6.4 **HAWK**

Leakage of privacy is a serious threat to blockchain. In the era of blockchain 2.0, not only transactions but also contractual information are public, such as bytecodes of contracts, parameters of invoking, and so on. Hawk is a new framework proposed for the development of intelligent contracts to preserve privacy. Using Hawk, developers can write smart private contracts and they do not have to use any encryption or obstruction code. Furthermore, the financial transaction's information will not be explicitly stored in blockchain. When programmers develop Hawk contract, the contract can be divided into two parts: private portion and public portion. The private data and financial function-related codes can be written into the private portion, and codes that do not involve private information can be written into the public portion. The Hawk contract is compiled into three pieces.

1. A program to be executed on all virtual node machines, just like Ethereum's smart contracts.
2. The program executed by smart contract users only.
3. The program to be carried out by the manager who is a special trustworthy Hawk party. The Hawk Manager is executed in the Intel SGX enclave and you can see the contract privacy information, but it would not.

Hawk can protect not only privacy from the public, but also privacy between various Hawk contracts. If the manager aborts the Hawk protocol, it is automatically penalized financially and users receive compensation. In general, Hawk can protect the privacy of users when using blockchain in large part.

4.7 **FUTURE DIRECTIONS**

On the basis of the above systematic examination of the safety of current blockchain systems, we list some future directions for encouraging research in this field. First, the most common consensus mechanism in blockchain today is POW. But the waste of computer resources is a major disadvantage of POW. Ethereum tries to develop a hybrid consensus mechanism between POW and POS to solve this problem. Research and the development of more efficient consensus mechanisms will play an important role in the development of blockchain. Second, with the increase in the number of feature-rich dAPPs, the risk of blockchain leakage in privacy will be more serious. Both a dAPP itself and the communication process between the dAPP and the Internet face risks to privacy. Some interesting techniques can be used in this problem: code obfuscation, application hardening and computing with trust (e.g., Intel SGX), and so on. Third, the blockchain generates a lot of data, including block data, transaction data, bytecode, and so on. However, all the data stored in blockchain are not valid. For example, an intelligent contract may delete its code by SUICIDE or SELFDESTRUCT, but the contract address is not deleted. In addition, there are many intelligent contracts that do not contain a code or completely the same code in Ethereum and many intelligent contracts are never executed after their deployment. To improve the performance efficiency of blockchain systems, an efficient data cleanup and detection mechanism is required.

4.8 CONCLUSION

Although a lot of studies have been carried out on the security and privacy issues of blockchain, but a systematic examination on the security of blockchain systems is still missing. In this chapter, we have tried to demonstrate a systematic illustration on the security threats to blockchain and presented a detailed description related the corresponding real attacks by examining popular blockchain systems. This chapter has discussed the security and the privacy of blockchain along with their impact with regards to different trends and applications. The chapter has focused on the key security attacks and the enhancements that will help develop better blockchain systems.

REFERENCES

[1] Z. Cai, Z. He, X. Guan, Y. Li, Collective data-sanitization for preventing sensitive information inference attacks in social networks, IEEE Trans. Depend. Secure Comput. 15 (4) (2016) 577−590. Available from: https://doi.org/10.1109/TDSC.2016.2613521.

[2] J.A. Garay, A. Kiayias, N. Leonardos, The bitcoin backbone protocol: analysis and applications, EUROCRYPT 9057 (2) (2015) 281−310. Available from: https://doi.org/10.1007/978-3-662-46803-6_10.

[3] A. Miller, M.M. Oser, K. Lee, A. Narayanan, An Empirical Analysis of Linkability in the Monero Blockchain, arXiv preprint:1704.04299, 2017.

[4] X. Li, P. Jiang, et al., A survey on the security of blockchain systems, Future Gen. Comput. Syst. (2017). https://doi.org/10.1016/j.future.2017.08.020. <http://www.sciencedirect.com/science/article/pii/S0167739X17318332>.

FURTHER READING

N. Capurso, T. Song, W. Cheng, J. Yu, X. Cheng, An android-based mechanism for energy efficient localization depending on indoor/outdoor context, IEEE Inter. Things J. 4 (2017) 299−307. Available from: https://doi.org/10.1109/JIOT.2016.2553100.

F. Chen, P. Deng, J. Wan, D. Zhang, A.V. Vasilakos, X. Rong, Data mining for the internet of things: literature review and challenges, Int. J. Distr. Sens. Netw. 11 (2015) 431047. Available from: https://doi.org/10.1155/2015/431047.

A. Dorri, S.S. Kanhere, R. Jurdak, Blockchain in Internet of Things: Challenges and Solutions, (2016), arXiv preprint, arXiv:1608.05187.

A. Dorri, M. Steger, S.S. Kanhere, R. Jurdak, Blockchain: a distributed solution to automotive security and privacy, IEEE Commun. Mag. 55 (2017) 119−125. Available from: https://doi.org/10.1109/MCOM.2017.1700879.

Z. Duan, M. Yan, Z. Cai, X. Wang, M. Han, Y. Li, Truthful incentive mechanisms for social cost minimization in mobile crowdsourcing systems, Sensors 16 (2016) 481. Available from: https://doi.org/10.3390/s16040481.

112 CHAPTER 4 A DEEP DIVE INTO SECURITY AND PRIVACY ISSUES

A.S. Elmaghraby, M.M. Losavio, Cyber security challenges in smart cities: safety, security and privacy, J. Adv. Res. 5 (2014) 491−497. Available from: https://doi.org/10.1016/j.jare.2014.02.006.

N. Gupta, R. Agrawal, NoSQL security, Adv. Comput. 109 (2018) 101−132.

A. Joshi, M. Han, Y. Wang, A survey on security and privacy issues of blockchain technology, Math. Found. Comput. 1 (2) (2018) 121−147. Available from: https://doi.org/10.3934/mfc.2018007.

I.-C. Lin, T.-C. Liao, A survey of blockchain security issues and challenges, Int. J. Netw. Sec. 19 (5) (2017) 653−659. Available from: https://doi.org/10.6633/IJNS.201709.19(5).01.

Working of blockchain, <http://aimsciences.org/article/doi/10.3934/mfc.2018007>.

CHAPTER

BLOCKCHAIN IMPLEMENTATION FOR INTERNET OF THINGS APPLICATIONS

5

Vibha Nehra[1], Ajay K. Sharma[2] and Rajiv K. Tripathi[3]

[1]*Department of Computer Science and Engineering, National Institute of Technology Delhi, New Delhi, India*
[2]*Department of Computer Science and Engineering, National Institute of Technology Jalandhar, Jalandhar, India*
[3]*Department of Electronics and Communication Engineering, National Institute of Technology Delhi, New Delhi, India*

5.1 INTRODUCTION

Banks, email service providers, and social networking websites ensure security and privacy of our digital assets. But, the fact is that these unbiased observers can always be hacked, compromised, or manipulated. This brings in the notion of distributed consensus and anonymity, but arises a requirement of someone who validates, safeguards, and preserves the possible data.

The first decentralized virtual currency, bitcoin, was introduced by Satoshi Nakamoto [1] in 2008. This digital currency was based on cryptographic proof instead of authentication from trusted third party. The most popular cryptocurrency, bitcoin (in service since early 2009), drives the multimillion dollar market of anonymous transactions globally without entanglement of trusted third party. The fluctuating value, cost, use, and validity of cryptocurrency endorse to its interest in both research and market community.

5.1.1 BITCOIN

For a bitcoin transaction taking place between two communicating entities, all the transaction data are broadcasted anonymously to each participant in the network, whose responsibility is to validate the transactions. This validation includes a race among network participants for solving a cryptographic puzzle [named as Proof of Work (PoW)] known as mining. The winner of this race is rewarded few bitcoins or may charge a fees for solving the puzzle. The solution to mining process is duly verified by all the participants before its addition as transaction for a block in the blockchain.

Numerically, in January 2009, 50 bitcoins were awarded/generated as computation reward for solving PoW of 1 transaction block. With a limit of total 21 million bitcoins to be generated by 2140, the rate of generation of coins is controlled by increasing the level of difficulty of PoW such that in addition to halving the reward every 2016 blocks, one new block generation takes approximately 10 minutes for the increasing amount of computational power [2]. This implies that the

Handbook of Research on Blockchain Technology. DOI: https://doi.org/10.1016/B978-0-12-819816-2.00005-8
© 2020 Elsevier Inc. All rights reserved.

114 CHAPTER 5 BLOCKCHAIN IMPLEMENTATION FOR IOT APPLICATIONS

difficulty level is changed approximately every 2 weeks (computation time for 2016 blocks for 10 minutes per block). Furthermore, this is to be noted that after 2140, no new bitcoins can be generated, that is, the reward for computing PoW shall be the transaction fees only.

5.1.2 BITCOIN ENERGY CONSUMPTION INDEX

The bitcoin's energy footprint ranges from a small to medium sized country. Bitmain, the largest fabricator of bitcoin mining machines, has already warned us to not to produce new coins due to its enormous carbon footprint [3].

Bitmain floated S15 and T15 antminers as the new generation of mining machines in November 2018 with publicized energy consumption of 57 and 96 Joule/TeraHash, respectively, against extant S9, T9, and comparable machines. Although introduction of this new generation in the cluster reduces the profit proportion of existing ones due to their higher power consumption. Though it was a tough year for bitcoin economy even then S15 and T15 appears to be lucrative pact. In the prevailing market conditions (January 2019), a vested passion in antminer S15 shall return as much as 34% annually over a period of 2 years [3]. In view of monthly network hash rate of 11% over past 1 year, then 5% monthly increase in network hash rate shall return mere 4% annually.

Contrarily, a new advanced generation of machines advocates that earlier archaic ones are doomed to be discarded. The S9 and T9 machines (accessible since mid-2016 till first half of 2018) stand to become e-waste with arrival of S15 and T15 machines, creating as high as 19,000 metric tonnes of e-waste, where a single machine weighs around 4.5 kg. Therefore, a single generation accounts to massive 28,000 metric tonnes of e-waste. This implies that unreal bitcoin has realistic environmental consequences.

The bitcoin sustainability report issued in January 2019 [3] presents that bitcoin has already employed energy comparable to a country such as Singapore in just first month of the year. It further pointed toward an annual increase of 12% in energy consumption than 2018, while mining revenues amounted lower at the beginning of the year than the previous year. The sustainability report published for February 2019 provided data as compared to previous month, showed a drop of 8% in mining revenues and hike of 23% in transaction fees for average fees per transaction as $0.30. In addition to this, marked 395 kWh per unique transaction, which is equivalent to power 1 US household for more than days.

5.1.3 BLOCKCHAIN

The bitcoin is built on blockchain technology. Blockchain is a distributed, permanent database of all transactions that have ever happened. The term distributed, here implies to being shared, and every block (constituent unit of blockchain) is duly verified by concord of the majority of participants. In other words, blockchain devise a system of distributed concord in the online digital world, analogous to banks as a credible arbitrator in the physical world.

The constituents of blockchain, the blocks are tethered in the sense that each block consists of the hash of the previous block, where the foundation block named as *genesis block* is customarily hardcoded. Thus, a block usually comprises a set of transactions, the hash of the previous block, and a nonce that is in accordance to the current target T. The target T is calculated periodically so as to control the degree of difficulty [2]. The degree of difficulty figures out the volume of work to

5.1 INTRODUCTION

be done for calculation of PoW. This implies that for increasing the level of difficulty, more amount of time is required to calculate the PoW (depicted in Fig. 5.1). The miners want to calculate nonce such that $H(B*N*PreH) < T$, where B is the set of transactions to be comprehended in the current block, N is the nonce value, $PreH$ is the hash of previous block, H is the cryptocurrency hash function [e.g., SHA256(SHA256(B)) is the bitcoin hash function], and * is the concatenation function.

The high acceptance of blockchain is credited to its sturdy approach to a possible castrate. This accounts to the fact that the blockchain copy is preserved at each participant of the system, unlike a single centralized copy at an unbiased observer say bank. Therefore, an evesdropper willing to corrupt the data stored in a blockchain needs to recalculate the PoW for all the blocks in the blockchain for every participant that too within the time frame of addition of new block in the authenticated blockchain, which amounts to whooping computation task which is usually very high as compared to the value of information being stored on blockchain.

Contribution: Considering the high-carbon footprint of standard blockchain implementation and resource-constrained nature of Internet of Things (IoT) applications, the work in this chapter outlines the challenges and research opportunities in the domain. In addition to this, the work proposes a simplified blockchain that is more applicable to IoT. The simplified blockchain is simple in the sense that it has reduced the amount of computation effort for calculation of PoW to an extent. The system taken into consideration is an ambient living or ambient assisted living application, specialized for elderly care in a big hall to collect physical parameters such as temperature, pressure, humidity, sound, and light on a single-board computer (Raspberry Pi). The physical parameters namely temperature, pressure, and humidity have been choosen for the application taking into consideration healthy indoor air quality, sound sleep, and bed sores (for bed-ridden elderly folks); similarly, light and sound parameters have been taken into consideration to ensure safety against a possible intrusion or to detect an abnormal elderly behavior (such as no movement for a long

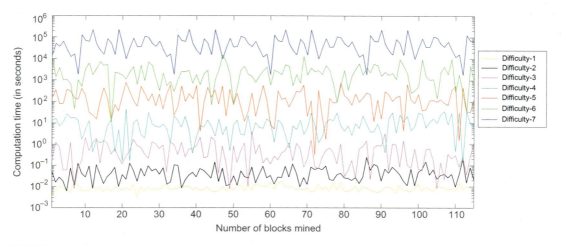

FIGURE 5.1

Computation time for increasing level of difficulty.

116 **CHAPTER 5** BLOCKCHAIN IMPLEMENTATION FOR IOT APPLICATIONS

period of time, no light or sound even after sunset), so that family or nearest healthcare center may be informed well in time. In addition to making self-sufficient blockchain-based IoT (BIoT) system, where all computations are carried out at IoT devices instead of a setting up a separate machine for PoW computation, the system also assigns the PoW computation task to a few compute-efficient machines in order to reduce the computation time. The standard blockchain implementation for above application could compute PoW only till difficulty level 7; after this, the application for computation got killed (even on repeated trials). Therefore, the current work compares the standard blockchain with simplified blockchain only till difficulty level 7.

Chapter Organization: The work in this section has been organized as follows: Section 5.2 describes the work done in the domain of BIoT. The same has been summarized in Table 5.1. This follows the identification of literature gap. Section 5.3 introduces the domain of blockchain implementation in IoT along with challenges in doing so and possible solutions in literature. In addition

Table 5.1 Literature Domain.	
Reference	**Marco Conoscenti et al. [4]**
Advantages Lag Remarks	Design identified 18 use cases for blockchain utility, 4 specifically for IoT. Issues on integrity, anonymity, and adaptability. Identified blockchain as pseudoanonymous.
Reference	**Oscar Novo [5]**
Advantages Remarks	Used proof of concept (PoC) as access management technology. Deviced a fully distributed access control architecture for IoT.
Reference	**Khan et al. [6]**
Advantages Remarks	Investigated security threats for IoT, their achievable solutions, layerwise IoT, security issues, and their counter measures. Discussed, analyzed the efficacy of blockchain in IoT.
Reference	**Chan Hyeok et al. [7]**
Advantages Remarks	Enhanced the anonymity of blockchain using zero-knowledge proof to avoid personal information disclosure. Worked on process of device authentication.
Reference	**Mandrita Banerjee et al. [8]**
Advantages Remarks	Hypothesized the possibility of using blockchain to ensure the integrity. Posed following research questions: • Usage of blockchain as a collective security system for IoT and related systems? • Investigation on optimization of blockchain and its related schema? • Vulnerability against possible compromise in hardware or software of an IoT device in case of physical accessibility? • Cost-effective approach to device a mature blockchain-based security solution?
Reference	**Yash Gupta et al. [9]**
Advantages Remarks	Presented a blockchain consensus model for secure data communication in IoT. Addressed feasibility of Blockchain in IoT.

Table 5.1 Literature Domain. *Continued*	
Reference	**Fernandez-Carames et al. [10]**
Advantages	Presented possible changes required to implement blockchain to IoT.
	Talked about key and hash function optimization. Pointed for non-applicability of every consensus algorithm on IoT.
Remarks	Addressed specific challenges such as privacy, security, energy efficiency, bandwidth, infrastructure, adoption rate, usability, multichain management, versioning, mining boycott, smart contract enforcement, and autonomy.
Reference	**Sriram et al. [11]**
Advantages	Profiled energy consumption in blockchain implementation.
Remarks	Worked on real-time workload.
Reference	**Jie Wan et al. [12]**
Advantages	Discussed wide adoption of blockchain in IoT for effective and reliable healthcare.
Remarks	Worked on ambient living/ambient assisted living application.
Reference	**Daniel Minoli et al. [13]**
Advantages	Proposed a novel IoT protocol architecture for critical e-health and assisted living application.
Remarks	Concerned security requirement for IoT.
Reference	**Mashail et al. [14]**
Advantages	Identified need for situation awareness in effective domain analysis.
	Future prediction using decision trees and association analysis algorithms.
Remarks	Multimodal data analysis.

to this, the research opportunities available too have been deliberated. The next section proposes an energy-efficient solution to compute PoW. This follows the details of experimental setup used on various machines for energy analysis. Section 5.5 presents the results of implementation of proposed energy-efficient solution against standard blockchain implementation. Sections 5.6 and 5.7 conclude and torch the way forward.

5.2 LITERATURE WORK

The work done in the literature domain has been detailed and tabulated in Table 5.1.

Marco Conoscenti et al. [4] started with whether blockchain and its peer-to-peer approaches are applicable to decentralized and private-by-design IoT. Among 18 use cases identified for blockchain, four use cases were peculiarly designed for IoT, with issues on integrity, anonymity, and adaptability. Blockchains were identified as pseudoanonymous, integrity was dependent on the level of difficulty of PoW and number of trustworthy miners, while the difficulty of PoW obstructed adaptability.

Oscar Novo [5] used blockchain technology to device a fully distributed access control architecture for IoT. It used proof of concept (PoC) as access management technology in real-time scalable IoT scenarios.

Khan et al. [6] presented a detailed investigation of security threats and their achievable solutions for IoT, layerwise IoT security issues, and their counter measures, discussed, and analyzed the efficacy of blockchain to secure IoT systems.

Chan Hyeok et al. [7] worked on enhancing the anonymity of blockchain using zero-knowledge proof in order to avoid personal information disclosure during the process of device authentication.

Mandrita Banerjee et al. [8] presented a survey on IoT-based security solutions since January 2016. Considering the lack of availability and potential risks in sharing available IoT data sets, they hypothesized the possibility of using blockchain technology to ensure the integrity of such data sets. The paper also delineates possible research questions regarding utility of blockchain as collective security system, optimization of blockchain, vulnerability against potential hardware or software attack in case of physical accessibility, and cost-effective mature blockchain-based system for IoT and similar systems.

Yash Gupta et al. [9] addressed the feasibility of applicability of blockchain technology in IoT by presenting a blockchain consensus model to implement secure data communication in resource-constrained IoT network.

Fernandez-Carames et al. [10] presented possible changes required to implement blockchain to specific challenges of IoT applications so as to develop BIoT applications in various domains such as government, democracy and law enforcement, telecommunication and information systems, defense and public safety, farming, energy, transportation, financial transactions, collaborative and crowd sensing, fleet monitoring and management, industry, personal sensing, smart cities, healthcare, logistics, and smart living. The paper also suggested a flow diagram to decide whether blockchain is actually required for the current application domain. Furthermore, the paper also talks about optimization of blockchain for IoT applications that include consideration of different possible ways to achieve consensus as it is usually not possible to implement blockchain on resource-constrained IoT devices. The key exchange and hash functions too need optimization so as to meet the complexity of IoT applications. The paper concludes with possible challenges for BIoT.

Considering resource-constrained nature of embedded devices in IoT applications, which incur an accountable amount of time in processing and verifying transactions in blockchain implementation, Sriram et al. [11] profile the energy consumption and analyze the different energy-performance tradeoffs. The paper also quantifies the amount of energy consumed in different operations on the ethereum platform for real-time workloads.

Regarding the current work that focuses on ambient living or ambient assisted living as IoT application, Jie Wan et al. [12] discussed wide adoption of IoT in order to provide reliable and effective healthcare for independent living of elderly folks. They identified applications, challenges, and research opportunities in IoT for ambient assisted living.

Daniel Minoli et al. [13] proposed a novel IoT protocol architecture for e-health and assisted living application concerning security requirement for IoT. They further pointed toward more efforts in the security field in view of criticality to human life.

Mashail et al. [14] identified the need for situation awareness as environment smartness for effective domain analysis in providing ambient assisted living. The data obtained from different sensors including cameras and microphones are worked upon to attain useful information from them. The work briefs existing research work for multimodal data analysis so as to improve the surrounds of the elderly. This work also considers multimodal sensing in an elderly ambient assisted living environment for getting current information about application domain and future prediction using decision trees and association analysis algorithms.

Literature Gap. Sufficient amount of work is being carried out in implementation of blockchain for IoT applications and also to the domain of ambient living and ambient assisted living applications by proposing new architectures or using other consensus strategies for mining processes such as proof of stake, proof of authority, proof of concept, proof of burn, proof of time, etc. However, efforts should be done on standard problem considering the requirement and criticality of the application domain, instead of eliminating standard PoW. The most energy-consuming process in bitcoin mining is the mining process, that is, calculating the PoW. So, the aim should be to simplify the problem of PoW computation still satisfying the security demands of the application domain.

5.3 BLOCKCHAIN FOR INTERNET OF THINGS

IoT implies to anything (any device, anybody, any service, any business, any place, any context, any time) that is connected to the Internet or may be controlled by the Internet. IoT applications range from wearable devices to consumer products for automation, healthcare, that is ambient living and ambient assisted living, smart farming to industrial applications, smart parking to the smart city, etc. Considering the robust attitude of blockchain against any form of compromise, the current work stores the data on blockchain so as to protect the system against any security threat. Moreover, instead of storing data on the connected edge device, considering the lifetime of the data set, the historical data may be backed up on cloud. The data stored on the cloud may be verified for potential security lapse from blockchain data, that is, blockchain also facilitates self-healing of compromised devices.

5.3.1 CHALLENGES

There exist a few challenges for implementation of blockchain in IoT applications. The important ones are detailed in the following sections.

5.3.1.1 Privacy

Blockchains are pseudoanonymous as different blockchain system participants can be pinpointed on the basis of their public key or its corresponding hash. Third-party agencies can obtain exact user identities by analyzing the transactions in the system [15,16]. User authentication in IoT can be crucial: an administrator answerable to user authorization can even block a particular user. Permissioned blockchain [17] may be used in such a situation for secure management of multiple IoT devices in a pool. The recommended solution strengthens the security by using certificate-based authentication along with hash function substitution. On the other hand, centralized identity management system [18] focused on automatic authentication system for IoT. The solution to this problem has been given as a blockchain-based system for IoT smart homes, which authenticate user and appliance by automatically obtaining appliance signatures. Another critical domain is access management, which includes exact specifications regarding capabilities, access list, and rights of particular user. Blockchain-based multilevel mechanism [19] solved this problem.

For a public blockchain, using different addresses for every new transaction make the data analysis difficult, while usage of unique address for each different communicating entity shall be more practical but less anonymous solution. Contrarily, in a private blockchain, where accesses are controlled by neutral access controller, a potential solution could be maintaining an autonomous

120 **CHAPTER 5** BLOCKCHAIN IMPLEMENTATION FOR IOT APPLICATIONS

blockchain for each different entity being communicated with. This solution increases complexity but secludes each user from unwanted monitoring [20].

Another technique to boost privacy is to collect transaction data from different IoT devices and events along with different addresses with whom communication is being carried out, but this too is vulnerable to statistical disclosure attacks [21], where malicious users may even steal money in case of financial transactions.

Zero-knowledge proving methods can also contribute to enhancing privacy [22–24]. The method includes proving that a particular user has certain information regarding a counterparty without letting them know about the information [25]. The zero-knowledge proofs authenticate without revealing identity of a user or a device.

Cryptonote [26] based cryptocurrencies such as Bytecoin [27] and Monero [28] are based on ring signatures where tracking blockchain does not reveal identity of communicating entities. An entity with either of the users' private key or the communicating entities themselves can have the transaction information. The concept of ring signatures specifies the possible set of signers, but not the exact signers.

Homomorphic encryption [29,30] too upholds privacy by encryption followed by data processing through third-party agencies without revealing the plain text being communicated. Furthermore, this is to be noted that the cryptographic techniques being implemented to enhance privacy should be feasible for resource-constrained IoT devices.

5.3.1.2 Security

The three pillars of security, that is, confidentiality, integrity, and availability, ensure a secure application. The confidentiality of data associated with information being communicated is related to privacy (already discussed in Section 5.3.1.1). In secure cloud-based or centralized setup for information storage, the stored information is protected against any possible threats and internal leaks [31,32]. While the blockchain-based system is decentralized and consensus based, thus protected even if one of the participating machines is compromised.

An evesdropper needs the private key of a user in order to masquerade as authentic user. Coniks [33] is a key management system to ease the users from responsibility of encryption key management. Blockchain defends IP spoofing and forgery attacks [34] for IoT devices.

Certificate-based security does fail occasionally [35]. Google's certificate transparency system [36] monitors and audits SSL certificates at almost real time by using Merkle hash trees in distributed environment.

The basic blockchain architecture facilitates that data stored on blockchain cannot be modified. But, for a few very exceptional instances (e.g., 2014 Vericoin case where a hacker stole almost 30% of the total coins), in order to ensure data integrity against a serious threat, hard forks have been done for permanent divergence from previous version of blockchain. IoT depend on third-party agencies for integrity services. Blockchain technology liberates IoT devices against such agencies by providing a framework for cloud-based IoT applications [37].

Distributed nature of blockchain platform ensures availability even if a few participating entities have been compromised. Still, the availability of blockchain can be breached by a few attacks. The most popular attack, the majority attack (or the 51% attack), keeps the data available, but the transactions being carried out may be controlled by single miner that has the ability to control whole blockchain's consensus.

5.3 BLOCKCHAIN FOR INTERNET OF THINGS **121**

5.3.1.3 Adoption rate

Pseudoanonymity of blockchain hinders to its acceptability by government bodies. Government bodies appeal direct link between real-world and online entities, so that culprit may be traced in case of emergency.

The number of participants in blockchain-based system directly impacts the value and security of information being stored. This implies that a higher number of participants make the application more robust against the most formidable 51% attack.

The participating entities in BIoT application also demand that the participants are competent enough to handle computation requirement of the system.

5.3.1.4 Forks and multichain management

Forks do occur in blockchain for administrative and versioning purposes, which are difficult enough to be handled by IoT applications where resources are already constrained.

Generation of new blocks in the system sometimes leads to instance where multiple blockchains need to be handled. If such instance occurs in IoT application, then system should be robust enough to handle the same.

5.3.1.5 Smart contract administration

Smart contract duly in place as designed by a governing body needs to be administered to resolve a dispute. Moreover, the issue of binding real-world contracts to smart contracts [38] needs to be addressed.

5.3.1.6 Throughput

Large number of transactions can be processed per unit time by increasing the device computation power, by processing large blocks, etc. [39]. While bitcoin can process maximum seven transactions per second [40], but this is very slow as compared to up to 24,000 transactions per second in VISA [41].

On the other hand, a given IoT application may need to handle large number of transactions per unit time. This high computation power requirement may be a hurdle for blockchain implementation in IoT.

Blockchain transaction processing is a time-consuming process. For example, bitcoin takes an average of 10 minutes to process a block, still users are suggested to wait for approximately an hour for a transaction to get confirmed, while VISA (VisaNet) needs only a few seconds for the similar task [41].

For minimizing the time taken in completing the consensus mechanism, a variation in blockchain, which is comparatively faster than standard SHA256, can be a possible solution. For example, Litecoin [42] uses scrypt [43].

5.3.1.7 Energy efficiency

The resource-constrained, battery-powered IoT devices always expect the application to be energy efficient, while the blockchains are usually portrayed as power hungry attributed to the mining process and P2P communication. Ball et al. [44] suggest a few outcomes where the energy consumed in computation of PoW can be used parally for some other jobs. Alternative mining mechanisms,

122   **CHAPTER 5** BLOCKCHAIN IMPLEMENTATION FOR IOT APPLICATIONS

which can any how simplify the mining process, could be a possible solution (e.g., Gridcoin [45], Primecoin [46]). One such energy-efficient solution working on PoW has been proposed in Section 5.4. Proof of Capacity is a greener solution to PoW [47]. Burstcoin [48] uses PoC where user has to show justifiable interest in a particular service by assigning certain memory space.

The participating entities in a blockchain communicate with peers in order to distribute blocks and send updates. Though the more the updates the better the blockchain, but these updates consume the fixed battery power. Mini-blockchain [49] could be a solution for IoT devices to directly reach out with the blockchain as they maintain record of only the latest transactions.

The popularity of SHA256 is contributed to the fact that it is used by bitcoin, but algorithms such as scrypt [43], X11 [50], Blake-256 [51], and Myriad [52] are another option that promise less energy consumption.

5.3.1.8 Infrastructure

Blockchain implementation in resource-constrained IoT domain needs to be proportioned according to limitations of IoT application. For instance, small transaction data may consume large amount of energy in communication or large transactions involving huge amount of data, which is not capable of resource-constrained IoT system, etc.

The matter of fact is that most of the IoT applications are not competent enough to handle even small fraction of standard traditional blockchain.

Therefore, in order to suit IoT applications, lightweight participating entities can be used, which do not store data but just perform transactions on blockchain. This architecture needs certain powerful machines that can store data. Another approach uses the concept of mini-blockchain [49,53], where account tree stores only current state of every participant of the blockchain and the blockchain only in case of new participant joins the system.

5.3.2 RESEARCH OPPORTUNITIES

Even though blockchain technology as a helping hand to IoT appears to be quite lucrative and a dazzling future can be foreseen, there still exist compelling challenges that need to be addressed.

- Scalability, security, cryptographic developments, and cohesion prerequisites of BIoT application are still a challenge. In addition to this, tendency of centralized approach problem needs to be worked upon.
- *Interoperability and standardization*: An amalgam of two different technologies, that is, resource-constrained IoT and energy-hungry blockchain, needs adjustment on all the collaborating participants of the system. The adjustment scales from different tradeoffs along with legal issues to international standards of trust, access control, authorization, etc. For example, at international level, authentications are provided on the basis of Level of Assurance (LoA) where according ISO/IEC 29115:2013 standard LOA are defined on scale ranging from LoA1 till LoA4. The higher the LoA, better the system. This standard defines the risk, aftereffects of an error in authentication, exploitation of credentials, etc.
- *Government regulatory aspects*: Design of a regulatory framework is a significant aspect to be worked upon in BIoT. This framework shall bring in interest of different capitalists to invest and popularize the domain.

5.4 ENERGY-EFFICIENT BLOCKCHAIN FOR INTERNET OF THINGS

- *Field testing*: The BIoT applications need to be tested in various real-time domains, so that different loopholes may be identified and worked upon in order to improve social acceptability. The testing process standardizes the system on the basis of numerous aspects as listed in Section 5.3.1.

5.4 ENERGY-EFFICIENT BLOCKCHAIN FOR INTERNET OF THINGS

The standard blockchain implementation using PoW as consensus technique is not likely on resource-constrained IoT applications. This is due to the fact that:

For block 100 (in 2009), the signature requirement was 8 consecutive zeroes, which has increased to 18 consecutive zeroes for block 568512 as on March 25, 2019, for an average hash rate of 45.66 EH/s [54], illustrates that PoW computation is the most time consuming thus energy-consuming component in blockchain system. Furthermore, this level of security is not required for IoT applications, nor is this high hashing capacity available at normal machines. Therefore, in order to suit IoT applications, available hardware, and considering the value and age of information, the PoW puzzle needs to be simplified.

Assume that hash output H obtained from cryptographic hash function be $x_0x_1x_3x_4\ldots x_{63}$, where x_i is a hexadecimal digit representing four bits. The standard blockchain system specifies the standard difficulty in terms of the number of initial hexadecimal digits, say k, to be 0(s), that is,

$$x_0x_1x_2x_3x_kx_{k+1}x_{k+2}\ldots x_{63}$$

where $x_i = 0$, $i < k$.

Table 5.2 illustrates the standard blockchain implementation that uses first "k" x_i to be zero for different levels of difficulty "k."

The current work proposes a more flexible solution, which is more applicable to IoT systems due to its simplicity. The solution allows first k hexadecimal digits to be a value from set X = {0,1,2,3,4,5,6,7,8,9,a,b,c,d,e,f}, that is, hexadecimal (0−F). That is, we modify the difficulty as follows:

$$x_0x_1x_2x_3x_kx_{k+1}x_{k+2}\ldots x_{63}$$

where $x_i = p$, $i < k$ ($0 \leq p \leq F$).

Table 5.2 Standard Blockchain Hashes for Different Difficulty Levels.

Difficulty Level	Eligible Hash Output
1	0abcdefghijkl...yz
2	00bcdefghijkl...yz
3	000cdefghijkl...yz
4	0000defghijkl...yz
k	00(k-zeros)kl...yz

124 CHAPTER 5 BLOCKCHAIN IMPLEMENTATION FOR IOT APPLICATIONS

Table 5.3 Modified Blockchain Hashes for Different Difficulty Levels.

Difficulty Level	Eligible Hash Output
2	33bcdefghijkl...yz
3	555cdefghijkl...yz
4	2222defghijkl...yz
5	99999efghijkl...yz
k	k-consecutive kl...yz

In other words, the modified blockchain for IoT applications considers first "k" x_i to be a value from set "X" for different levels of difficulties "k," where X = {0,1,2,3,4,5,6,7,8,9,a,b,c,d,e,f}, such that all x_i have same value from X for a given hash output (illustrated in Table 5.3). Here this too is to be noted that for modified blockchain, the difficulty level has been considered from $k = 2$ because, for $k = 1$, all the possible hashes shall be the eligible hashes, which will lead to the irrelevance of value of target T.

Furthermore, another consideration is that the standard blockchain implementation allowed the following number of hashes from all possible hashes for a given difficulty level:

$$\textit{Difficulty level } 1 \quad = \frac{1}{16} \times 100 = 6.25\%^1$$

$$\textit{Difficulty level } 2 \quad = \frac{1}{16} \times \frac{1}{16} \times 100 = 0.3906\%$$

$$\textit{Difficulty level } 3 \quad = \left(\tfrac{1}{16}\right)^3 \times 100 = 0.000244\%$$

$$..$$
$$..$$

$$\textit{Difficulty level } n \quad = \left(\tfrac{1}{16}\right)^n \times 100\%$$

While the modified blockchain allowed the following number of hashes for a given difficulty level:

$$\textit{Difficulty level } 1 = \frac{16}{16} \times 100 = 100\%$$

(This implies allows every possible hash output as an eligible hash, which defeats the purpose of target T. Therefore, ignoring level 1 for comparison purpose.)

$$\textit{Difficulty level } 2 \quad = \left(\frac{16}{16 \times 16}\right) \times 100 = 6.25\%$$

$$\textit{Difficulty level } 3 \quad = \left(\frac{16}{16 \times 16 \times 16}\right) \times 100 = 0.3906\%$$

$$..$$
$$..$$

$$\textit{Difficulty level } n \quad = \left(\tfrac{1}{16}\right)^{n-1} \times 100$$

This implies that the modified solution allows 16 times more hashes as eligible hashes at a particular level of difficulty, which implies that less computation time will be required to find eligible hash, thus saving energy as compared to standard blockchain implementation.

5.4.1 EXPERIMENTAL SETUP

The current work assumes an IoT application as an intrusion detection system in a big hall where physical parameters such as temperature, pressure, humidity, light, and sound are to be recorded for an ambient living application say elderly care. The system collects data using three digital sensors [BME280 (*temperature*, *pressure*, *humidity*), proximity sensor (*light*), and KY-038 (*sound*)] connected to a single-board computer (Raspberry Pi 2 model B) as shown in Fig. 5.2. The collected data are stored on the Raspberry Pi from where it has been used for different analysis tasks.

The system collects data values from the sensors at an interval of 60 seconds, making it a transaction. Sixty transaction data are combined to form a block. This implies the system forms one transaction per minute, one block per hour, which counts to total 24 blocks per day.

With ever-increasing computation power, in order to ensure security in the vulnerable communication world, the difficulty level is increased periodically. The difficulty level for a particular PoW computation task is determined in terms of time and hashing power required to find eligible hash (signature). For the given experimental setup, the PoW computation task was carried out on attached Raspberry Pi, so that data storage, along with the corresponding blockchain, could ensure security and privacy of the available system. Raspberry Pi could compute PoW up to difficulty level 7, but after this, it was not able to compute further and the process got killed by the system (even on repeated trials) as shown in Fig. 5.1. Therefore, the current work takes into consideration only to difficulty level 7 for resource-constrained IoT devices.

Later, the PoW computation task has been assigned to comparatively more powerful machines in order to cut down the computation time as shown in Fig. 5.3. The different machines taken into consideration are detailed in Table 5.4.

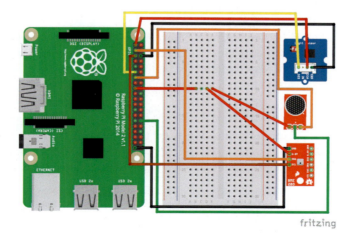

FIGURE 5.2

Experimental setup.

126 CHAPTER 5 BLOCKCHAIN IMPLEMENTATION FOR IOT APPLICATIONS

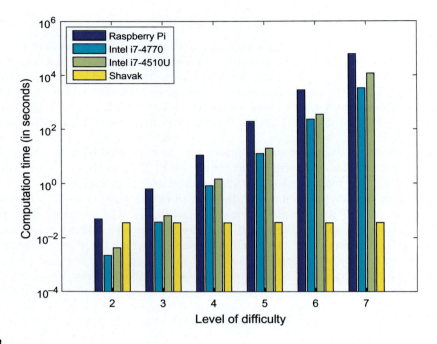

FIGURE 5.3

PoW computation task on different machines.

5.4.2 ENERGY CONSUMPTION IN ATTACHED MACHINES

- *Raspberry Pi*: On connecting to 5 V voltage (recommended voltage for Raspberry Pi) and an ammeter in series with the power supply to the Raspberry Pi, 0.29 A current (averaged over 3000 values) was recorded for any mining process. While the time for computation increased with an increase in the level of difficulty. This implies that energy consumed in Raspberry Pi is a factor of time for a fixed power consumed to run the hardware.
- *Intel i7-4770*: Energy consumed by an Intel device is computed as follows:

$$\text{Computation energy} = EPI \times T_{cpu} \times X$$

where EPI is energy consumed per instruction, T_{cpu} is the total time the CPU is active, and X is the processor clock frequency. Here this is to be noted that EPI value and processor clock frequency are fixed for a particular processor, which means computation energy is directly proportional to CPU active time.

- *Intel i7-4510U*: The above technique is applicable in this system too being an Intel device. But there exists one more Linux utility to compute power consumption in battery-powered devices known as "Powertop." Powertop takes power estimates of 197 measurements on battery power, where each measurement is taken at time duration of 20 seconds. It gives average power consumption for running the standard blockchain implementation as 376 J (or W/s). Therefore, energy, which is the product of power and time, can be estimated from the time factor that is increasing for an increase in the level of difficulty.

5.4 ENERGY-EFFICIENT BLOCKCHAIN FOR INTERNET OF THINGS **127**

Table 5.4 Technical Specifications of Different Machines Used.

Device	Raspberry Pi 2
Type	(single-board computer)
Architecture	armv71
Byte order	Little endian
CPU(s)	4
Threads per core	1
Core(s) per socket	4
Model	5
Model name	ARM v7 Processor rev 5 (v7l)
Device	**Genuine Intel**
Type	(wired computer)
Architecture	$\times 86\_64$
Byte order	Little endian
CPU(s)	8
Threads per core	2
Core(s) per socket	4
Model	60
Model name	Intel(R) Core(TM) i7-4770
	CPU @ 3.40 GHz
Device	**Genuine Intel**
Type	(battery-powered computer)
Architecture	$\times 86\_64$
Byte order	Little endian
CPU(s)	4
Threads per core	2
Core(s) per socket	2
Model	69
Model name	Intel(R) Core(TM) i7-4510U
	CPU @ 2.00 GHz
Device	**Param Shavak**
Type	(the super computer)
Architecture	$\times 86\_64$
Byte order	Little endian
CPU(s)	24
Threads per core	1
Core(s) per socket	12
Model	63
Model name	Intel(R) Xeon(R) CPU
	E5-2670 v3 @ 2.30 GHz

- *Param Shavak*: Similarly, the power consumption for Param Shavak can be estimated from the same as in wired computer (Intel i7-4770), again being an Intel device.

Therefore, for the sake of simplicity, the energy consumption has been depicted in terms of computation time.

5.5 RESULTS

Fig. 5.4 shows the results for implementation of modified solution for different machines under consideration as in Table 5.4 for difficulty levels from 2 to 7 in terms of computation time (in seconds) for the mining process.

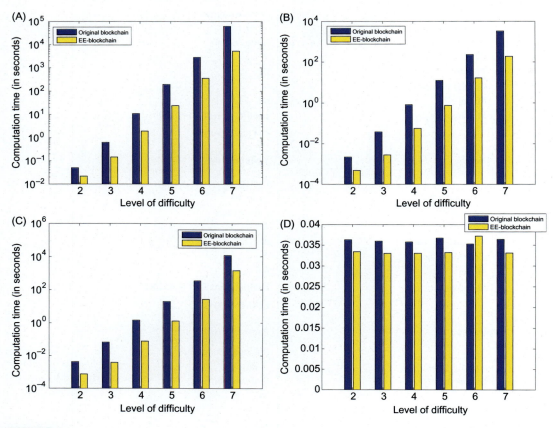

FIGURE 5.4

Energy-efficient algorithm implementation on different machines. (A) for Raspberry Pi (Single board computer), (B) for i7-4770 (wired computer), (C) for i7-4510U (battery powered computer), (D) for Param shavak (the super computer).

Table 5.5 Computation Time (in Seconds) for Mining Process at Different Machines and Increasing Difficulty Level.

Difficulty Level	Raspberry Pi	Intel i7-4770	Intel i7-4510U	Param Shavak
Diff-2	0.049692	0.002229	0.00426636	0.03631
EE-Diff-2	0.021689	0.00048	0.000754	0.033428
Diff-3	0.633516	0.038282	0.066523	0.035977
EE-Diff-3	0.145693	0.002841	0.003871	0.033043
Diff-4	10.90047	0.0831858	1.458152	0.035789
EE-Diff-4	1.94733	0.056478	0.078535	0.033052
Diff-5	192.3865	12.39542	19.48517	0.036703
EE-Diff-5	24.32805	0.761665	1.309791	0.033228
Diff-6	2819.878	229.4927	346.5656	0.035291
EE-Diff-6	353.9819	16.4279	26.8063	0.037208
Diff-7	61403.99	3326.107	11831.3	0.036403
EE-Diff-7	5268.56	187.5868	1450.932	0.033136

Diff, Level of difficulty, *EE-Diff*, energy-efficient implementation of a difficulty level.

Fig. 5.4A depicts the implementation of modified blockchain for IoT application where mining is carried out on a single-board computer (Raspberry Pi), which collects all the data from all digital sensors. The energy-efficient implementation shows less computation time due to the higher number of possible hashes accepted. Similarly, Fig. 5.4B and C too shows comparatively less amount of time for computation task due to higher acceptance ratio. While figure does not show any significant change in computation time even with an increase in the level of difficulty because the computation resources available with the super computer is incredibly high as compared to computation task at hand. Moreover, this too is to be noted that some resources are mandatorily consuming energy to make such large system work. In energy-efficient implementation, the results also show similar behavior but comparatively less computation time for less computation task at hand. The results for computation time (in seconds) at given difficulty level for standard and modified energy-efficient blockchain implementation for IoT applications are given in Table 5.5. Here this is to be noted that each value in Table 5.5 is averaged over 113 blocks' computation time. In addition to this, it is to be reported that, for over 113 blocks, the data of over 4 days have been collected to calculate computation time for PoW.

5.6 CONCLUSION

Considering computation resources available with IoT devices, along with value and validity of information in IoT applications, suggests that standard blockchain implementation is neither possible nor required for IoT application. Further taking into account, the criticality in terms of security and privacy for IoT applications, which can directly impact human life (as in case of ambient living or ambient assisted living applications), blockchain is a good solution for such applications, but with some simplicity in the computation of PoW. Therefore, the current work proposes modified energy-efficient blockchain implementation for IoT applications, which consume less energy in terms of computation time. In addition to this, the current work answers the issues raised in literature work (section 5.2) as follows:

- Blockchain can be used as collective security to secure application in IoT and related systems.
- Blockchains and blockchain-based platforms can be optimized by applying some simplification in solving PoW puzzle.
- Blockchain can be used to reduce the possibility of hardware and software vulnerabilities in a physically approachable IoT device, by matching the hash at a particular block level, which cannot be changed easily by an intruder.
- The proposed solution in the current work may be a cost-effective approach to device a mature blockchain-based security solution.

5.7 FUTURE WORK

The future work in this domain can be on implementation of blockchain on image data obtained from a camera, industrial application of blockchain, and biomedical application of blockchain such as storing biomedical data, for example, reports of patients, date of birth of newborns, etc.

ACKNOWLEDGMENTS

The authors sincerely acknowledge the help and support for lab facility provided by Indian Institute of Space Science and Technology, Thiruvananthapuram, Kerala, India.

REFERENCES

[1] S. Nakamoto, Bitcoin: a peer-to-peer electronic cash system, <www.bitcoin.org>.
[2] J.K. O'Dwyer, D. Malone. Bitcoin mining and its energy footprint. 2014, 280−285.
[3] https://digiconomist.net.
[4] M. Conoscenti, A. Vetro, J.C. De Martin, Blockchain for the Internet of Things: a systematic literature review, 2016 IEEE/ACS 13th International Conference of Computer Systems and Applications (AICCSA), IEEE, 2016, pp. 1−6.
[5] O. Novo, Blockchain meets iot: an architecture for scalable access management in iot, IEEE Internet Things J. 5 (2) (2018) 1184−1195.
[6] M.A. Khan, K. Salah, Iot security: review, blockchain solutions, and open challenges, Future Gener. Comp. Syst. 82 (2018) 395−411.
[7] C.H. Lee, K.-H. Kim, Implementation of iot system using block chain with authentication and data protection, 2018 International Conference on Information Networking (ICOIN), IEEE, 2018, pp. 936−940.
[8] M. Banerjee, J. Lee, K.-K.R. Choo, A blockchain future for Internet of Things security: a position paper, Digital Commun. Netw. 4 (3) (2018) 149−160.

REFERENCES 131

[9] Y. Gupta, R. Shorey, D. Kulkarni, J. Tew, The applicability of blockchain in the Internet of Things, 2018 10th International Conference on Communication Systems & Networks (COMSNETS), IEEE, 2018, pp. 561–564.

[10] T.M. Fernández-Caramés, P. Fraga-Lamas, A review on the use of blockchain for the Internet of Things, IEEE Access 6 (2018) 32979–33001.

[11] S. Sankaran, S. Sanju, K. Achuthan, Towards realistic energy profiling of blockchains for securing Internet of Things, 2018 IEEE 38th International Conference on Distributed Computing Systems (ICDCS), IEEE, 2018, pp. 1454–1459.

[12] J. Wan, X. Gu, L. Chen, J. Wang, Internet of Things for ambient assisted living: challenges and future opportunities, 2017 International Conference on Cyber-Enabled Distributed Computing and Knowledge Discovery (CyberC), IEEE, 2017, pp. 354–357.

[13] D. Minoli, K. Sohraby, B. Occhiogrosso, Iot security (iotsec) mechanisms for e-health and ambient assisted living applications, Proceedings of the Second IEEE/ACM International Conference on Connected Health: Applications, Systems and Engineering Technologies, IEEE Press, 2017, pp. 13–18.

[14] M.N. Alkhomsan, M.A. Hossain, S.M.M. Rahman, M. Masud, Situation awareness in ambient assisted living for smart healthcare, IEEE Access 5 (2017) 20716–20725.

[15] S. Meiklejohn, M. Pomarole, G. Jordan, K. Levchenko, D. McCoy, G.M. Voelker, et al., A fistful of bitcoins: characterizing payments among men with no names, Proceedings of the 2013 Conference on Internet Measurement Conference, ACM, 2013, pp. 127–140.

[16] M. Möser, R. Böhme, D. Breuker, An inquiry into money laundering tools in the bitcoin ecosystem, 2013 APWG eCrime Researchers Summit, IEEE, 2013, pp. 1–14.

[17] D.W. Kravitz, J. Cooper, Securing user identity and transactions symbiotically: iot meets blockchain, 2017 Global Internet of Things Summit (GIoTS), IEEE, 2017, pp. 1–6.

[18] A. Dorri, S.S. Kanhere, R. Jurdak, Towards an optimized blockchain for iot, Proceedings of the Second International Conference on Internet-of-Things Design and Implementation, ACM, 2017, pp. 173–178.

[19] S.H. Hashemi, F. Faghri, P. Rausch, R.H. Campbell, World of empowered iot users, 2016 IEEE First International Conference on Internet-of-Things Design and Implementation (IoTDI), IEEE, 2016, pp. 13–24.

[20] G. Greenspan, Multichain private blockchain, White paper, <http://www.multichain.com/download/MultiChain-White-Paper.pdf>.

[21] G. Danezis, A. Serjantov, Statistical disclosure or intersection attacks on anonymity systems, International Workshop on Information Hiding, Springer, 2004, pp. 293–308.

[22] https://www.zerocoin.org.

[23] https://www.zerocash-project.org.

[24] https://www.z.cash.

[25] M. Schukat, P. Flood. Zero-knowledge proofs in M2M communication. 2014, 269–273.

[26] https://www.cryptonote.org.

[27] https://www.bytecoin.org.

[28] https://www.getmonero.org.

[29] C. Moore, M. O'Neill, E. O'Sullivan, Y. Doröz, B. Sunar, Practical homomorphic encryption: a survey, 2014 IEEE International Symposium on Circuits and Systems (ISCAS), IEEE, 2014, pp. 2792–2795.

[30] H. Hayouni, M. Hamdi, Secure data aggregation with homomorphic primitives in wireless sensor networks: a critical survey and open research issues, 2016 IEEE 13th International Conference on Networking, Sensing, and Control (ICNSC), IEEE, 2016, pp. 1–6.

[31] R.M. Jabir, S.I.R. Khanji, L.A. Ahmad, O. Alfandi, H. Said, Analysis of cloud computing attacks and countermeasures, 2016 18th International Conference on Advanced Communication Technology (ICACT), IEEE, 2016, pp. 117–123.

[32] A.O.F. Atya, Z. Qian, S.V. Krishnamurthy, T. La Porta, P. McDaniel, L. Marvel, Malicious co-residency on the cloud: attacks and defense, IEEE INFOCOM 2017-IEEE Conference on Computer Communications, IEEE, 2017, pp. 1−9.
[33] https://coniks.cs.princeton.edu.
[34] N. Kshetri, Can blockchain strengthen the Internet of Things? IT Prof. 19 (4) (2017) 68−72. Available from: https://doi.org/10.1109/MITP.2017.3051335.
[35] T. Chen, S. Abu-Nimeh, Lessons from stuxnet, Computer 44 (4) (2011) 91−93.
[36] https://www.certificate-transparency.org.
[37] B. Liu, X.L. Yu, S. Chen, X. Xu, L. Zhu, Blockchain based data integrity service framework for iot data, 2017 IEEE International Conference on Web Services (ICWS), IEEE, 2017, pp. 468−475.
[38] N. Fabiano, The Internet of Things ecosystem: the blockchain and privacy issues. the challenge for a global privacy standard, 2017 International Conference on Internet of Things for the Global Community (IoTGC), IEEE, 2017, pp. 1−7.
[39] N.T. Courtois, P. Emirdag, D.A. Nagy, Could bitcoin transactions be $100\times$ faster? 2014 11th International Conference on Security and Cryptography (SECRYPT), IEEE, 2014, pp. 1−6.
[40] M. Vukolić, The quest for scalable blockchain fabric: proof-of-work vs. bft replication, International Workshop on Open Problems in Network Security, Springer, 2015, pp. 112−125.
[41] https://usa.visa.com/run-your-business/small-business-tools/retail.html.
[42] https://www.litecoin.com.
[43] http://www.tarsnap.com/scrypt.html.
[44] M. Ball, A. Rosen, M. Sabin, P.N. Vasudevan, Proofs of useful work, IACR Cryptology ePrint Archive 2017 (2017) 203.
[45] http://gridcoin.us.
[46] http://primecoin.org.
[47] S. Dziembowski, S. Faust, V. Kolmogorov, K. Pietrzak, Proofs of space, Annual Cryptology Conference, Springer, 2015, pp. 585−605.
[48] https://www.burst-coin.org.
[49] J.D. Bruce, The mini-blockchain scheme, White paper. 2014.
[50] https://dashpay.atlassian.net/wiki/spaces/doc/pages/1146918/x11.
[51] J.-P. Aumasson, L. Henzen, W. Meier, R.C.-W. Phan, Sha-3 proposal blake, Submission to NIST 229 (2008) 230.
[52] http://myriadcoin.org.
[53] BF. França, Homomorphic mini-blockchain scheme. 2015.
[54] https://www.btc.com.

CHAPTER 6

BLOCKCHAIN IMPLEMENTATION USING SMART GRID-BASED SMART CITY

M. Afshar Alam and Sapna Jain
Department of Computer Science and Engineering, Jamia Hamdard University, New Delhi, India

6.1 INTRODUCTION

Blockchain generation can contribute to a city's capability to reach each of the six important dimensions that is, smart economic gadget, smart environment, smart government, smart dwelling, smart mobility, and smart human beings. Sustainability and smart cities circles have constantly been the way to permit cities to be extra self-sustaining with distributed renewable strength structures. The method rapid changes to current cities, the growing volumes of embedded renewable generation, which includes wind and solar photovoltaic (PV). Blockchain has pulled in the thought of the power business with its capacity to discharge an essentialness change in which the two utilities and clients will convey and sell control. The sharp home machines related with an imperativeness trading stage could always look for the best offer and thus substitute to another essentialness provider through a sharp contract. This experience will empower the customers to interface from their home or office truly to imperativeness vendors.

Blockchain could offer a strong, ease course for financial or operational trades to be recorded and affirmed over an appropriated framework with no basic issue of intensity. Basically, the "prosumers" can pitch their surplus essentialness to other customer in the framework clearly through contracts set up and affirmed through blockchain. Blockchain is an essential advancement that can be used to make new arrangements of activity and bolster business, money related, and social structure. While various blockchain use cases have been proposed for the essentialness business, the one getting the most balance at present is peer-to-peer (P2P) control trading, where owners of little scale age can offer excess age honestly to various buyers. Today, concentrated control of circled imperativeness distributed energy resources (DERs) cutoff points to whom and when DER owners can offer their essentialness back to the network. A blockchain empowered P2P model allows significantly progressively essential flexibility and could be an astounding engaging impact for a customer-driven transactive essentialness schedule. To help the headway of blockchain-based responses for the imperativeness portion, a lot of affiliations are setting up Blockchain Labs with the purpose of stimulating new blockchain applications, for instance, passed on record courses of action and its usage cases. In case the new applications are productive for mass appointment, it would impactly influence the strategies of the entire imperativeness part regard chain. Europe is the most unique district for the blockchain pilots, with utilities wearing down electric vehicle (EV)

134 CHAPTER 6 BLOCKCHAIN IMPLEMENTATION USING SMART GRID

charging, related home, rebate settlement, and lab creation tries. At the present time, in Germany various blockchain pilots are in various periods of progression.

The world's first P2P imperativeness trade occurred in New York City in 2016. In the principle quarter of 2017, the DoE issued a fragile for Blockchain adventures with a consideration on security. In South Africa, Sun Exchange interfaces money-related masters to associations moreover, systems who need access to sensible influence. In Japan, Marubeni is taking off bitcoin portions for customers and in China, Wanxiang is expecting to place $30billion in a blockchain maintained Smart City Project. Power Ledger, a start-up in Perth, Australia is wearing down various endeavors over the region. In a pilot in Netherlands, Vandebron will work with customers who guarantee an EV to make the point of confinement of their vehicle batteries available to help the cross section director balance the structure while guaranteeing the battery life. Sustainable energy assets renewable energy source (RES) have experienced enormous improvement as of late, empowered through privatization, unbundling of the power zone and helped with the asset of financial impetuses and power inclusion ventures. In 2016, 24.6% of the United Kingdom gross power utilization moved toward becoming produced by the method for the utilization of RES, particularly from coastal and seaward wind homesteads and PV sun vegetation, representing 44.9% and 12.5% of the whole 35.7 GW setup RES limit, individually. RES are variable, hard to foresee and depend on atmosphere conditions, along these lines improve new difficulties in charge and task of vitality frameworks, as additional adaptability measures are required to make certain safe activity and soundness. Adaptability estimates typify the blend of fast showing up convey, call for reaction, and power stockpiling contributions. Adding to the transformational exchange because of disseminated quality sources (DERs) and renewables, power structures are on the purpose of getting into the virtual period as exhibited with the guide of the huge sending of keen meters in a few endeavors. In the assembled kingdom without anyone else, 53 million power and gas sharp meters are intentional to be mounted by method for 2020, one for each home and little modern organization. To aggregate imposing outflow decrease objectives, vitality frameworks will require enormous financing. It is foreseen that inside independent from anyone else, the progress inside the course of an additional supportable and comfortable vitality machine may require a venture of €200 billion unfaltering with 365 days for age, network, and power execution improvement. $2 trillion in power arrange upgrades might be required through 2030 inside the United States. To direct required subsidizing smart control and control should be pursued, obligations which may be progressively hard as quality frameworks are developing to turn out to be increasingly dynamic, decentralized, complex, and "multispecialist," with an ever increasing number of entertainers and feasible activities. progressed verbal trade and insights trades among one of a kind added substances of the power network are to a developing amount required, causing basic to control and task progressively hard. Neighborhood appropriated control and control procedures are required to suit those decentralization and digitalization patterns. Blockchains or administered record period distributed ledger technology (DLT) has been regularly intended to encourage apportioned exchanges by getting rid of crucial control. Accordingly, blockchains might need to help tending to the requesting circumstances went up against by means of decentralized vitality structures. A keen city utilizes data innovation to coordinate and oversee physical, social, and business frameworks to supply higher administrations to its occupants while making certain practical and best use of reachable assets. With the expansion of advances like snare of Things (IoT), distributed computing, and interconnected systems, savvy urban areas will convey inventive arrangements and extra direct cooperation and coordinated effort among voters and in this manner the specialists. In spite of an assortment of potential focal points, computerized interruption represents a few difficulties related with information security and protection. In late decades, the planet has more seasoned unexampled urban development because of populace increment, worldwide environmental change, and deficiency of assets. Ongoing examination demonstrates that extra people abide in urban communities (54%) than country zones (46%) and this range can increment to 66 by 2050. To manage these emergencies, urban areas territory unit represents considerable authority in vogue advancements further as going to downsize costs, use assets ideally, and make extra decent urban setting. The various progressions in IoTs and remote correspondences have made it easy to interconnect an assortment of gadgets and help them to transmit learning pervasively even from remote areas.

6.2 BACKGROUND

The blockchain innovation is the thing that is behind the digital money known as Bitcoin. It was made by Satoshi Nakamoto and first distributed in the original white paper Bitcoin: A Peer-to-Peer Electronic Cash System [1]. The enthusiasm for blockchain has developed since its initiation in 2008 [2]. Yli-Huumo [2] characterizes the blockchain as a "decentralized exchange and information the executives innovation," a definition that is steady with the meaning of different creators. He distinguishes seven specialized difficulties and restrictions that blockchain innovation has. Those difficulties are throughput, idleness, size and data transmission, security, squandered assets, ease of use, forming hard forks, and various chains. There is an agreement that the blockchain is a momentous innovation that can be as significant as the rise of the web itself, yet shockingly for an innovation with so much potential Yli-Hummo (2016) discovers that the greater part of the examination is focused on the digital money subject instead of the blockchain innovation. Additionally, as per Yli-Hummo.

Vitalik Buterin, prime supporter of Ethereum and Bitcoin magazine, was additionally an underlying supporter of the Bitcoin codebase, yet ended up baffled around 2013 with its programming restrictions and pushed for a flexible blockchain. Met with opposition from the Bitcoin people group, Buterin set out to construct the second open blockchain called Ethereum (2016), most distribution types originate from meetings and workshops while a minority originate from book sections, diaries, and symposiums. In spite of the fact that blockchain is discussion about like there is just one sort of it, three sorts of blockchain exist, open blockchain, private blockchain, and half and half blockchain [3]. Siba [3] portrays the open blockchain as an "advanced record [that] is totally decentralized and can be open to any web client" [3]. A private blockchain contrasts in the sense that there is a focal expert giving authorizations of who can compose on the blockchain, that being the principle distinction with the open blockchain. The crossover blockchain is a blend of the private furthermore, open blockchain. No more subtleties are given about that blockchain type. Each kind of blockchain can connected with the four highlights that as indicated by Siba [3] the blockchain innovation has, those are permissioned organize, resources, exchanges, and agreement. Carlozo [4] says that as per a 2017 overview 60% huge organization administrators were learned somewhat about what blockchain is. Organizations like IBM and Maersk have been looking into it to make genuine executions. Benton [5] says that in 2018 IBM and Maersk found out during a multimonth consider that a shipment from Kenya to Port Rotterdam containing just blooms, had more than 200 paper transactions. Those transactions sum practically 20% of the expense of the

136 CHAPTER 6 BLOCKCHAIN IMPLEMENTATION USING SMART GRID

transportation process. Well-known reasoning accepts that the innovation will be misused uniquely by business yet Manski [6] makes the case that blockchain innovation could be of incredible potential to the helpful development. The helpful development is antagonistic to both corporate free enterprise and state communism [6] since the two frameworks depend on a focal control. He introduces six key propensities of the blockchain innovation that lead toward a mechanical province also, other six that will avoid a mechanical province. The six positive key inclinations are disintermediation, thrustless trade, expanded client control of data, strong, secure decentralized systems, straightforwardness and unchanging nature, upkeep of high caliber, precise information. The six negative key propensities are uncertain specialized difficulties, agitated administrative condition, cybersecurity also, security concerns, difficulties to far reaching appropriation, work misfortune due to computerization, and diminished corporate responsibility. Woodside [7] on the other hand completes a PEST examination (Political, Monetary, Social, and Technical) of the condition of the blockchain innovation. In the political perspective, Woodside [7] says that at present in the United States, the internal revenue service (IRS) considers every single virtual money property, and consequently should pay with the appropriate laws to it. (That piece of the PEST investigation is worried about the digital currency side of the point as opposed to the blockchain itself.) In the monetary part of the investigation, there are a few figures that says that 30% of the retail banking occupations will be lost due the usage of blockchain, given the robotization of a few errands [7]. The social part of the blockchain concurring the PEST investigation made by Woodside [7] says that client will have more control and straightforwardness of their data. The cost to pay is that once data are transferred, there could be security issues with it. R3 builds up a protection focused blockchain called Codra. Their site portrays it as pursues, "Not at all like conventional blockchain stages, Corda limits data spillage by just sharing exchange information with members that require it" (R3, 2018). The specialized side of the investigation by Woodside [7] says that the blockchain is a standout among the most progressive systems ever made.

The fundamental element of the blockchain is its thrustless correspondence since, each hub on the system do not depend on a focal server to get what is considered the "truth." In addition, the blockchain get increasingly secure each time another square is added to chain, since it is more earnestly to change it. Nakamoto [1] in his original white paper says, "To change a past block, an aggressor would need to retry the proof-of-work of the block and all block after it and afterward get up to speed with and outperform the work of the legitimate hubs." As Halaburda [8] puts it, the blockchain use cryptography components. The motivation behind why is practically difficult to hack the blockchain is on the grounds that is too exorbitant to even consider rewriting history, since that would require a great deal of processing control. All that is a piece of its cryptographic component. Halaburda [8] explains that a appropriated record, a keen contract and encryption are isolated things. Individuals frequently think each one of those things are one and the same. They structure some portion of the blockchain, however, they need not bother with one another to work. A smart contract is a sort of agreement that can make certain move under a few criteria.

6.3 BLOCKCHAIN CONCEPT

Blockchain is a decentralized and distributed ledger, where, blocks containing a set of transactions are chained together by cryptographic hash. Transactions originating from a node are validated by participating nodes and a set of transactions are added into a block by a "mining" node. Any

6.3 BLOCKCHAIN CONCEPT **137**

mining node with sufficient compute power that solves a cryptographic puzzle can generate and broadcast a new block containing the set of validated transactions.

The term "blockchain" depicts a technique for putting away information in a manner that is compositionally not quite the same as the customary value-based database the executives frameworks (records) that went before it. By using blockchain clients may just attach records, never erase or alter. Second, information is circulated over a network of associated distributed gadgets, every one of which stores an indistinguishable full or fractional duplicate of the database. Features of blockchain are as follows:

1. New made records are assembled in sequential request and gathered into a particular estimated block.
2. Each new data block is attached to the centralised database.
3. Each data block contains an information component that is an encoded reference to the previous blocks along these lines making a "chain."
4. When a square is attached, every server among the distributed system other than the one that made the data block must affirm that it contains a substantial arrangement of exchanges and a legitimate cryptographic hash. Different servers will dismiss a data block that comes up short the legitimacy test. Each blockchain network has an accord procedure to guarantee enough taking an interest servers endorse a square before it very well may be seen as a perpetual piece of the chain.
5. Since each square alludes iteratively to the square promptly earlier, the chain has ordered respectability.
6. Since information is just affixed and not adjusted, and has demonstrated sequential respectability, it cannot be changed retroactively and is said to be "permanent." However, a superior term might be "alter apparent." While it is workable for an awful on-screen character to change a chronicled record, the altering is quickly clear.
7. Any endeavor to include false exchanges will be defeated by the way that all hubs screen the ongoing exchange stream and approve that new squares are precise. Simply after enough hubs have approved, a square can be viewed as perpetual.

Blockchains are probabilistic frameworks, by plan. Hubs, or the power costs (PCs) in the system, freely choose and agree whereupon "chain of blocks" is the longest and generally legitimate. As a square is made and set around the system, every hub forms the blocks and chooses where it fits into the current general blockchain ledger.

Most blocks essentially broaden the present primary blockchain as in Fig. 6.1. These are designated "primary branch blocks." Some blocks reference a parent obstruct that is not at the current blockchain tip. These blocks are classified "side branch blocks." Some blocks reference a parent hinder that is not known to the hub preparing the square. These are classified "vagrant blocks." Side branch blocks are especially fascinating. They may not right now exist in the primary branch, yet on the off chance that more work is done on them which means different blocks are mined that reference them as a parent, there is the likelihood that a specific side branch will be redesigned into the fundamental branch. This rearrangement happens on the grounds that the "principle" part of the blockchain is the one that has had the most work done on it. As new blocks are attached to the blockchain, it turns out to be progressively hard to "overwrite" existing blocks in light of the fact that the most substantial chain is the one that has had the most work done on it. The shared idea of a blockchain network naturally implies it is dispersed, yet a blockchain is not really decentralized.

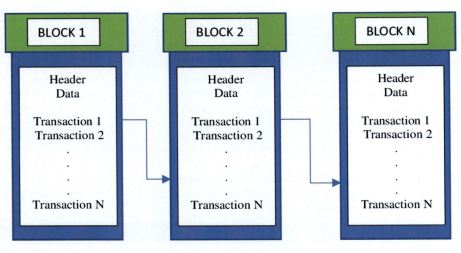

FIGURE 6.1

Blockchain architecture.

Regardless of whether a blockchain is unified or decentralized is characterized by the plan, nature, and sort of framework, which is dictated by the members of that blockchain network. The idea of blockchain however not simply the term was developed by Satoshi Nakamoto, a pen name an obscure individual or gathering in 2008 as a technique to record exchanges of the cryptographic money, Bitcoin. The innovation has advanced essentially since 2008, albeit all blockchains still look to some extent like Nakamoto's plan.

6.4 BLOCKCHAIN TYPES

Beginning with Bitcoin in 2008, a wide range of structures have been created to meet fluctuating specialized, business and administration plan goals. The term "blockchain" does not have an unmistakable reasonable definition and is utilized for a huge number of various improvements. There are wide contrasts in properties of different blockchain types, for example, level of decentralization, exchange capacities, accord instruments, availability, unchanging nature, adaptability, or straightforwardness. Subsequently, blockchain authorities and the media often claim summed up articulations about blockchain as if they apply to all blockchains, when in all actuality the declarations simply relate to only one blockchain or a gathering of blockchain propels (Fig. 6.2).

The following types of blockchain exists:

1. Public/permissionless blockchain

    Public blockchains, for instance, Bitcoin, Ethereum, and various others, offer absolutely open access and can be scrutinized or formed by anyone, without preauthorization. Game-speculative motivating forces are used to propel trust between darken center points. Open blockchains are seen as totally decentralized. Protectors ensure that since creators of usages

6.4 BLOCKCHAIN TYPES

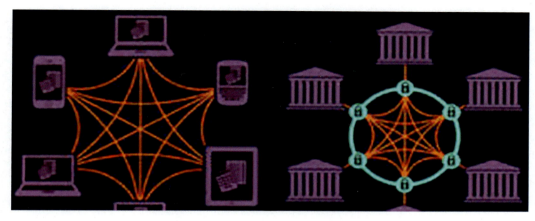

FIGURE 6.2

Nodes in public and private blockchains.

have no pro to modify or change organizes once being utilized, self-governance, and control check are ensured.

In perspective on this trademark, advertisers of open blockchains consider the advancement as an enabling specialist for responsiveness, straightforwardness, and lack of bias. In any case, without specific or legitimate parts to approve consistence, open blockchains may be in threat of slipping by into strife. Open blockchains require strong security and create organization to prompt the fundamental trust and sureness among those that would build apparently interminable business or sensitive applications on it. An open blockchain designing infers that the data and access to the system are available to any person who is glad to partake (for instance, Bitcoin, Ethereum, and Litecoin blockchain structures are open).

2. Private/permissioned and consortium

Private/permissioned blockchains have ascended as a choice as opposed to open blockchains in order to use the development among a great deal of portrayed, known individuals. In a private (permissioned) blockchain designing, create approvals (the ability to add a record to a database) are surrendered unmistakably to embraced centers. Complete or obliged read assent may be made available to all center points or to the overall public or by and large limited.

Private blockchains can be appealing for some business use circumstances where a particular dimension of security, auditability, and organization is required. All individuals inside a private blockchain can be recognized, yet do not generally need to trust in each other. Confined information may be obvious to general society, or not. As opposed to open blockchains, any described master can change the standard set for the blockchain. The understanding arrangement of private blockchains can be a lot more straightforward, with a solitary hub or gathering of hubs having expert to approve new squares. A unique kind of private blockchain regularly thought about a particular sort is the consortium blockchain, which can be comprehended as a blockchain where agreement is inferred by an approved arrangement of hubs, for example, hubs having a place with a gathering of money-related establishments. This kind of blockchain can be portrayed as mostly decentralized, in that nobody hub has full control, yet nor is any hub permitted to join and take an interest freely.

In any case, in light of the sort of blockchain structure and its particular circumstance, the system can be logically united or decentralized. This fair suggests the blockchain designing structure and who controls the record. A private blockchain is seen as logically bound together since it is compelled by a particular social affair with extended security. Notwithstanding what may be normal, an open blockchain is open-completed and thusly decentralized. In an open blockchain, all records are unquestionable to the overall public and anyone could take an interest in the getting method. Of course, this is less gainful since it requires a great deal of venture to recognize each new record into the blockchain building. As to, the perfect open door for each trade in an open blockchain is less eco-obliging, since it requires a huge proportion of count power appeared differently in relation to private blockchain designing.

Property	Public Blockchain	Consortium Blockchain	Private Blockchain
Consensus determination	All miners	Selected set of nodes	Within one organization
Read permission	Public	Public or restricted	Public or restricted
Immutability level	Almost impossible to tamper	Could be tampered	Could be tampered
Efficiency (use of resources)	Low	High	High
Centralization	No	Partial	Yes
Consensus process	Permissionless	Needs permission	Needs permission

Fig. 6.3 provides a detailed comparison among these three blockchain systems.

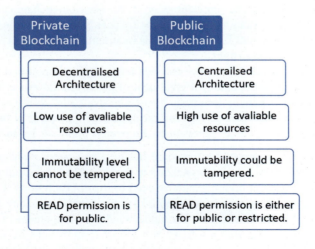

FIGURE 6.3

Blockchain types comparison.

6.5 CENTER COMPONENTS OF BLOCKCHAIN ARCHITECTURE: HOW DOES IT WORK

These are the center blockchain design parts:

1. Node—client or PC inside the blockchain design in which every node an autonomous duplicate of the entire blockchain record.
2. Transaction—littlest structure block of a blockchain framework which stores records, data that fill in as the motivation behind blockchain.
3. Block—an information structure utilized for keeping a lot of exchanges which is disseminated to all hubs in the system.
4. Chain—a succession of blocks in a particular request.
5. Miners—explicit hubs which play out the block check process before adding anything to the blockchain structure.
6. Consensus (agreement convention)—a lot of guidelines and courses of action to complete blockchain activities.

Any new record or exchange inside the blockchain suggests the structure of another block. Each record is then demonstrated and carefully marked to guarantee its validity. Before this square is added to the system, it ought to be confirmed by most of hubs in the framework. The following is a blockchain architecture diagram that shows how this actually works in the form of a digital wallet.

Each blockchain block comprises of certain information, the hash of the block, the hash from the past block as shown in Fig. 6.4. The information put away inside each block relies upon the sort of blockchain. For example, in the Bitcoin blockchain structure, the block keeps up information about the beneficiary, sender, and the measure of coins.

A hash resembles a unique finger impression long record comprising of certain digits and letters. Each block hash is created with the assistance of a cryptographic hash calculation (SHA 256). Thusly this distinguishes each square in a blockchain structure effectively. The minute a square is made, it naturally appends a hash, while any progressions made in a square influence the difference in a hash as well. Basically expressed, hashes help to identify any adjustments in squares. The last component inside the block is the hash from a past block. This makes a chain of block and is the principle component behind blockchain engineering's security. For instance, square 45 points to square 46. The absolute first block in a chain is somewhat exceptional that is affirmed and approved blocks which are gotten from the beginning of the blocks. Any degenerate endeavors incite the blocks to change. All the accompanying blocks at that point convey wrong data and render the entire blockchain framework invalid. Then again, in principle, it could be conceivable to change every one of the block with the assistance of solid PC processors. Be that as it may, there is an answer that disposes of this plausibility called evidence-of-work. This enables a client to hinder the procedure of formation of new squares. In Bitcoin blockchain engineering, it takes around 10 minutes to decide the important evidence-of-work and add another block to the chain. This work is finished by diggers extraordinary hubs inside the Bitcoin blockchain structure. Excavators get the chance to keep the exchange expenses from the block that they confirmed as a reward. Each new client (hub) joining the distributed system of blockchain gets a full duplicate of the framework. When another block is made, it is sent to every hub inside the blockchain framework. At that point,

142   CHAPTER 6 BLOCKCHAIN IMPLEMENTATION USING SMART GRID

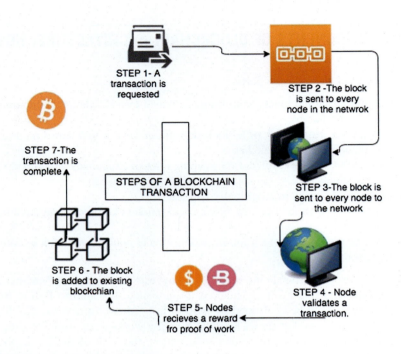

FIGURE 6.4

How blockchain works.

every hub confirms the block and checks whether the data expressed there is right. In the case of everything is okay, the block is added to the neighborhood blockchain in every hub. Every one of the hubs inside a blockchain design make an agreement convention. An agreement framework is a lot of system rules, and if everybody keeps them, they become self-upheld inside the blockchain.

For instance, the Bitcoin blockchain has an accord guideline expressing that an exchange sum must be sliced down the middle after each 200,000 blocks. This implies if a square delivers a check reward of 10 BTC, this esteem must be divided after each 200,000 blocks. Also, there must be 4 million BTC left to be mined, since there is a limit of 21 million BTC set down in the Bitcoin blockchain framework by the convention. When the diggers open this many, the supply of Bitcoins closes except if the convention is changed.

6.6 CARBON CREDIT

A carbon credit is a grant or testament enabling the holder to emanate carbon dioxide or other ozone harming substances. As far as possible the emanation to a mass equivalent to one ton of carbon dioxide. The issuance of carbon credits plans to diminish the discharge of ozone depleting substances into the environment.

The clean development mechanism (CDM) official board issues the carbon credits. Carbon credits under the Kyoto Protocol are a market component that take into consideration an advantage to creating nations to keep growing in any case, in a progressively "greener" way than most existing created nations. Also, while it is ordinarily comprehended that sustainable power sources are "free—the sun and wind" the innovation is more expensive than "nothing new" by and large coal-terminated power plants. Carbon credits are a market instrument made by the United Nations under the Kyoto Protocol and will before long be tradable under Article 6 of the Paris Agreement. As a manner by which nations and partnerships can attempt to achieve focuses on their carbon outflows by purchasing a credit produced somewhere else on the planet to limit or counterbalance their very own carbon impression. Carbon credits work as pay frameworks allowing balance between new Greenhouse Gas (GHG) emanations and relating amounts of guaranteed alleviations. At the end of the day, nations and companies required to alleviate their outflows can balance their carbon liabilities by buying alleviations (as affirmed carbon credits), from enlisted venture exercises. Such a framework has been recognized, together with carbon charge, as the most financially savvy relief methodology to be received around the world. Truth be told, it permits organizations that transmit carbon that cannot bear the cost of direct alleviations to repay their discharges through credit acquisitions, while remunerating net alleviations makers by enabling them to sell their affirmed relief activities. Normally known as "carbon credits." There are a few sorts of carbon credits. They are mostly characterized upon their motivation. These reasons for existing are verified emission reduction (VERs) for intentional counterbalancing and certified emissions reductions (CERs) for consistence balancing. The CDM encourages CERs, which are the best balancing instrument at present being used today and will be moved as far as the Paris Agreement with included reasonable improvement objectives presented by the UNFCCC (dependent on the Millennium Development Goals) and host nations endorsement.

Our cutting edge world is being looked with an expansive scope of biological dangers. One of these dangers, which has exponentially become in the course of recent years, is the expansion in carbon outflows. As nature ends up unfit to adjust the difficulties related with environmental change and a dangerous atmospheric deviation, governments around the globe have started issuing laws inside their nations to constrain carbon outflows. In addition, with the presentation of square chain innovation, carbon emanations have been appeared to increment through the high-vitality request they intrinsically require.

6.7 THE PARIS AGREEMENT

The Paris Agreement flag a move toward a base up methodology where atmosphere exercises, as reflected in their Nationally Determined Contributions (NDCs), are driven by national reality, financial development, and political needs. While empowering wide cooperation, the World Bank perceives that undeniably additionally financing is required to enable nations to actualize their NDCs and eventually come to the worldwide 1.5−2C objective. The World Bank Group (WBG) accepts that the utilization of business sectors will have a significant task to carry out in the fruitful, practical usage of the Paris Agreement by diminishing expenses and encouraging more noteworthy asset assembly.

WBG additionally perceives that the Paris Agreement gives atmosphere advertises a genuinely necessary, reestablished reason for help by empowering Parties to willfully collaborate in accomplishing their NDCs through Internationally Transferred Mitigation Outcomes regularly known as "carbon credits" under Article 6. As far back as the marking of the Kyoto convention in 1997, nations have begun restricting the measure of ozone depleting substances they produce every year. One way they have done that has experienced the foundation of Emissions Trading Schemes around the globe where businesses in specific districts are required to partake.

This sets a cost for each metric ton that it expenses to diminish ozone harming substances (GHG) past a foreordained top, as members are compelled to purchase balances. It has additionally prompted a wilfull market where people and organizations can purchase carbon balances to constrain their very own carbon impression.

With the execution of the Paris Agreement, the Carbon Offsetting and Reduction Scheme for International Aviation is presently likewise an objective market with which Carbon Chain will likewise try to play a noteworthy job. The Paris Accord was set up by a mind a lot of the United Nations part states as a framework developing the motivation behind the Kyoto Protocol to conform to and moderate overall ecological change. The Paris Accord fuses a great deal of courses of action that are proposed to respond to overall biological changes in order to shield and to amass human advancement so it could thrive in a predominant offset with nature.

Carbon credits are a framework, which licenses signatory get-togethers to the Paris Accord to adjust and trade rights to exude carbon outperforming the amounts stipulated in the Paris Accord and its endorsement on the national measurement.

The United Nations has developed a show, which allows part states to set up top and trade procedures, carbon charge and a carbon credit instrument on a national measurement to control the productive economy under the Paris Accord.

Associations willing to partake in the framework can do all things considered purposefully by procuring carbon credits. Carbon credits license individuals, associations and governments to outperform their offer in carbon emissions, which for the most part include nonsustainable power sources and related outflows and to exchange these credits for spread rights or use them to help a selected endeavor that reduces carbon releases. Such assignment can for instance be a reasonable power source adventure. A carbon recognize is portrayed as a unit of exchange issued by The Clean Development Mechanism— board at the United Nations as a similarity one ton in CO_2 in surges rights.

Contemporary instruments overseeing the usage are stacked with challenges identifying with the organization of issuance of carbon credits, auditability and following the introduction of endeavors qualified inside national carbon credits plans. The troubles by and large contain administrational issues, originating from the endeavors being alluded to being regulated by administrators and authoritative associations. On a very basic level, this sort of organization model offers rise to issues in affirmation and check of the show of the enrolled errands, much also as issues in exercises in worldwide progression have administrational issues. Blockchain development can assuage administrational challenges. Blockchain grants constant record of trades, and a system for affirmation basically invulnerable to creation and deception. All relevant datasets and records may be constantly affirmed, and shared in an open record available to self-sufficient commentators, associates and regulating associations. With the help of blockchain development in actuality, incites related to association, the board and corruption can be enough directed, as everyone in the natural framework have a phase each on-screen character can rely upon for validity.

Carbon Chain is a phase that offers willing individuals a passageway to trade carbon credits, and apply them in their own assignments or an endeavor encouraged by another part in the segments stipulated by the Paris Accord and the United Nations or one of the signatories to the Paris understanding. At the point when a money-related expert has made a theory into Carbon Chain—tokens, the tokens may be exchanged for power carbon credits with the United Nations. The carbon credits may then be placed assets into qualified endeavors.

Past blockchain, what makes Carbon Chain tale, is that the stage offers individuals the opportunity to trade carbon credits. While ensure, high all-out resources money-related authorities may in all likelihood share in this instrument through a concentrated stage, Carbon Chain makes the segment open to the more vital open. The exercises partner with carbon credits and along these lines similarly by means of Carbon Chain—tokens, must be selected with the United Nations in order to get carbon credits. Usually, these exercises contain enhancements that help in easing carbon releases, for instance, practical power source adventures, carbon-fair structure, carbon sinks, cleantech adventures, reaches out in repetitive economy, and so on.

The stage continues running on a ton of sharp contracts on the Ethereum arrange. At first, the tokens are appropriated through a Discounted Private Sale related with a Token Generation Event.

Carbon Chain is constrained by an association, Carbon Chain International. The association pledges 30% of its benefits in procuring carbon credits from United Nations Framework Convention on Climate Change, UNFCCC, and in selecting new endeavors that fit the bill for carbon credits the world over. The rest of the tokens are placed assets into existing enrolled adventures that fit the bill for carbon credits. The endeavor behind Carbon Chain itself is to be enlisted with the UNFCCC CDM board.

A holder of Carbon Chain tokens may trade their Carbon Chain tokens on open exchanges for anticipated buyers at the spot cost, or offer the tokens back to Carbon Chain International to adjust their carbon charge commitment or to display their adherence to the Paris Accord. The exchange to carbon credits happens on a Carbon Credit Exchange.

6.8 CARBON CREDIT EXCHANGE

Carbon Credit Exchange is an open exchange kept up by means of Carbon Credit International that gives an instrument to trade computerized cash for Carbon Credit tokens and Carbon Chain tokens for carbon credits. The full scale number of carbon credits reflects the proportion of carbon credits the enrolled assignments on it meet all prerequisites for. The carbon credit buyer goes into a radiations decline purchase simultaneousness with Carbon Chain International. The association, Carbon Credits International thoughts up to 30% as pay share for token holders through a pivoting store each time another buyer enters the market. A Carbon Credit Trade happens when a client holding the Carbon Chain Tokens goes into a concurrence with Carbon Chain International. If a client uses their tokens to purchase carbon credits under Paris Accord, the genuine carbon credits are real for a period of 1 year for each trade. At the completion of the trade, the tokens are returned to course.

A Members-Only Exchange is open for those clients who have looked into the Token Generation Event by methods for Private Sale. This exchange is proposed to contain tokens that have been purchased by means of carbon credit buyers by methods for open exchanges. These

146 **CHAPTER 6** BLOCKCHAIN IMPLEMENTATION USING SMART GRID

tokens are offered toward the completing of each Carbon Credit Trade to Token Generation Event—individuals yearly, and each part has 7 days to recognize the airdrop. The tokens left over from the airdrop will be appropriated to the individuals, who have recognized them during the magnificence time span. These endeavor tokens can be exchanged with Carbon Chain International on a yearly explanation behind tradeable Carbon Chain tokens moderately to trading volume the earlier year. The tokens would then have the option to be returned back to stream in open exchanges.

The token is held by the clients in an item wallet, and traded the exchange. The exchange is blockchain controlled, so it might be believed to be a decentralized exchange.

There are two sorts of tokens made by the Carbon Chain:

1. Carbon Chain Token (CCT), which is a token that can be traded direct for carbon credits, picture CCT. These tokens are exchanged direct for carbon credits.
2. Carbon Chain Project Token, which is a token that can be traded for Carbon Chain Tokens, consistently scattered to private arrangement individuals. This token is a reward instrument, whose total depends on yearly arrangements.

6.8.1 CARBON CREDITS AND CARBON MARKETS

There is a complexity between Carbon Credits and Carbon Markets. Carbon credits are seen compensation revelations for GHGs surges, and can be used to offset lightening tries, from now on being possibly exchanged. On the other hand, Carbon markets which are a respectable yet insufficient response for make an overall impact—address the most relevant, yet not exceptional, approach to manage carbon credits exchange.

Carbon credits, generally called carbon counterbalancing, were considered as a triumph win procedure in empowering control of GHGs, thusly transforming into a key gadget in doing combating natural change. Immediately exhibited as segments inside the Kyoto Protocol, carbon credits have filled in as "compensation systems" yielding harmony between new GHGs releases likewise, looking at measures of guaranteed mitigations. By the day's end, performers required to mitigate their radiations can adjust their commitments by getting mitigations from various on-screen characters. Such a structure has been perceived, together with carbon charge, as the most common-sense balance method to be grasped far and wide. In reality it licenses performers that cannot deal with the expense of direct mitigations to compensate their spreads every through credit's acquisitions, while compensating net mitigations creators by allowing them to sell their guaranteed mitigations. Such a structure has incited the progression of proper carbon markets for the credits exchange. All spread declines made through parity exercises and carbon credits delivered from balance endeavors must be certifiable, additional, specific, constant, and enforceable. Any carbon credit that is made and checked identifies with mitigations that have starting at now happened, therefore permitting its practical impact on GHGs transmissions. There are a couple of sorts of carbon credits, basically portrayed upon their inspiration.

- VERs

   These are carbon credits routed to stamp where clients willfully counterbalance their GHGs alleviations to build their positive ecological and social effect.
- CERs

   CERs speak to the last result of the Clean Advancement Mechanism (CDM). The CDM is the best counterbalancing instrument presented by the Kyoto Protocol. A CER, otherwise called

CER, is an endorsement issued by the United Nations to part countries for counteracting one ton of carbon dioxide discharges. These are typically issued to part states for tasks accomplishing ozone depleting substance decreases using CDMs. CDMs make it workable for these undertakings to happen and set a standard for future emanation targets.

Nations with created or conventional economies under the Kyoto Protocol use CERs to enable them to achieve their discharge targets. Those countries can accomplish their targets and can set future objectives as it tries of lessening ozone depleting substances increasingly practical numerous nations.

- Renewable Energy Certificates

    The tradable, nonunmistakable vitality items in the United States that speak to verification that 1 MWh of power was created from a qualified sustainable vitality asset (inexhaustible power) and was nourished into the mutual arrangement of electrical cables which transport vitality.

- Reducing Emissions from Deforestation and forest Degradation

    This component created by Parties to the UNFCCC. It makes a money related an incentive for the carbon put away in woodlands by offering motivating forces for creating nations to diminish emanations from forested grounds and put resources into low-carbon ways to manageable improvement.

6.9 CARBON EMISSION EFFECT

Financial development and urbanization move pair, as financial development and ozone harming substance discharges have for in any event the most recent 100 years. Since most financial movement is packed in urban regions, urban areas have a key job in environmental change. Wealth and way of life decisions decide ozone depleting substance outflows, furthermore, truly created nations have had more prominent ozone harming substance outflows than creating nations. The world is urbanizing rapidly and under the same old thing situation, ozone harming substance discharges will likewise increment significantly.

Urban areas are real supporters of ozone depleting substance discharges. Half of the total populace lives in urban communities, an offer that is probably going to achieve 70% in 2050 (Fig. 6.5). Urban communities expend as much as 80% of vitality creation worldwide and represent a generally equivalent offer of worldwide ozone harming substance discharges. As improvement continues, ozone harming substance discharges are driven less by modern exercises and more by the vitality administrations required for lighting, warming, and cooling. The International Energy Agency (IEA) gauges that urban territories as of now represent more than 67% of vitality related worldwide ozone depleting substances, which is expected to ascend to 74% by 2030. It is evaluated that 89% of the expansion in CO_2 from vitality use will be from creating nations (IEA, 2008). Urban populace is required to twofold by 2030; anyway the worldwide developed territory is relied upon to triple during a similar period.

This structure out as opposed to working up will significantly expand vitality necessities and expenses of new framework. Inadequately overseen urban areas fuel gigantic new requests for vitality and framework venture (Fig. 6.6).

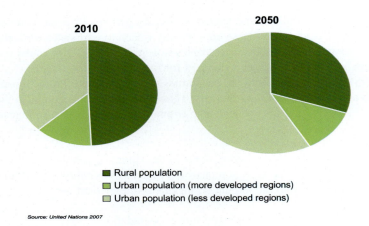

FIGURE 6.5

Share of urban and rural population in 2010 and 2050.

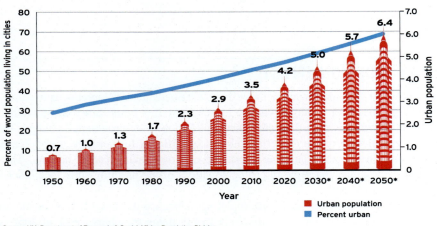

FIGURE 6.6

People living in cities percentage of world population and total.

Fig. 6.7 presents a thorough rundown of as of now surveyed urban ozone depleting substance baselines for around 70 urban areas, detailed as qualities per capita, with a for each capita stock an incentive for the relating nation. The association in charge of setting up each stock is shown. While the strategy and information accessible for every city may fluctuate, the information is a significant beginning stage for future consistency in urban stock revealing. The table is presently accessible on UNEP, UN-HABITAT, the Global City Indicators Facility and World Bank sites. In taking a gander at the inventories exhibited in Fig. 6.7, some significant patterns rise: creating nations will in general have lower per capita outflows than created nations; thick urban areas will in general have

6.9 CARBON EMISSION EFFECT

Country/city	GHG emissions (tCO$_2$e/capita) and year		Country/city	GHG emissions (tCO$_2$e/capita) and year	
ARGENTINA	**7.64**	2000	**NORWAY**	**11.69**	2007
Buenos Aires	3.83	1	Oslo	**3.5**	2005, 3
AUSTRALIA	**25.75**	2007	**PORTUGAL**	**7.71**	2007
Sydney	20.3	2006, 2	Porto	**7.3**	2005, 3
BANGLADESH	**0.37**	1994	**REPUBLIC OF KOREA**	**11.46**	2001
Dhaka	0.63	1	Seoul	**4.1**	2006, 3
BELGIUM	**12.36**	2007	**SINGAPORE**	**7.86**	1994
Brussels	**7.5**	2005, 3	**SLOVENIA**	**10.27**	2007
BRAZIL	**4.16**	1994	Ljubljana	**9.5**	2005, 3
Rio de Janeiro	2.1	1998, 3, i	**SOUTH AFRICA**	**9.92**	1994
São Paulo	1.4	2000, 3, i	Cape Town	**7.6**	2005, 5, i
CANADA	**22.65**	2007	**SPAIN**	**9.86**	2007
Calgary	17.7	2003, 3	Barcelona	**4.2**	2006, 5, i
Toronto (City of Toronto)	9.5	2004, 4	Madrid	**6.9**	2005, 3
Toronto (Metropolitan Area)	**11.6**	2005, 5, i	**SRI LANKA**	**1.61**	1995
Vancouver	4.9	2006, 6	Colombo	1.54	1
CHINA	**3.4**	1994	Kurunegala	9.63	1
Beijing	10.1	2006, 3, i	**SWEDEN**	**7.15**	2007
Shanghai	11.7	2006, 3, i	Stockholm	**3.6**	2005, 3
Tianjin	11.1	2006, 3, i	**SWITZERLAND**	**6.79**	2007
Chongqing	3.7	2006, 7	Geneva	**7.8**	2005, 5, i
CZECH REPUBLIC	**14.59**	2007	**THE NETHERLANDS**	**12.67**	2007
Prague	**9.4**	2005, 5, i	Rotterdam	**29.8**	2005, 3
FINLAND	**14.81**	2007	**THAILAND**	**3.76**	1994
Helsinki	**7**	2005, 3	Bangkok	**10.7**	2005, 5, i
FRANCE	**8.68**	2007	**UK**	**10.5**	2007
Île-de-France (Region incl Paris)	**5.2**	2005, 3	London (City of London)	6.2	2006, 11
GERMANY	**11.62**	2007	**London (Greater London Area)**	**9.6**	2003, 5, i
Frankfurt	**13.7**	2005, 3	Glasgow	**8.8**	2004, 3
Hamburg	**9.7**	2005, 3	**USA**	**23.59**	2007
Stuttgart	**16**	2005, 3	**Austin**	**15.57**	2005, 3
GREECE	**11.78**	2007	Baltimore	14.4	2007, 12
Athens	**10.4**	2005, 3	Boston	13.3	13
INDIA	**1.33**	1994	Chicago	12	2000, 14
Ahmedabad	1.2	1	Dallas	15.2	13
Delhi	1.5	2000, 8	**Denver**	**21.5**	2005, 5, i, †
Kolkata	1.1	2000, 8	Houston	14.1	13
ITALY	**9.31**	2007	Philadelphia	11.1	13
Bologna (Province)	**11.1**	2005, 3	Juneau	14.37	2007, 15
Naples (Province)	**4**	2005, 3	**Los Angeles**	**13**	2000, 5, i
Turin	**9.7**	2005, 3	Menlo Park	16.37	2005, 16
Veneto (Province)	**10**	2005, 3	Miami	11.9	13
JAPAN	**10.76**	2007	**Minneapolis**	**18.34**	2005, 3
Tokyo	**4.89**	2006, 3, i	**New York City**	**10.5**	2005, 5, i
JORDAN	**4.04**	2000	**Portland, OR**	**12.41**	2005, 3
Amman	**3.25**	2008, 9, i	San Diego	11.4	13
MEXICO	**5.53**	2002	San Francisco	10.1	13
Mexico City (City)	4.25	2007, 10	**Seattle**	**13.68**	2005, 3
Mexico City (Metropolitan Area)	2.84	2007, 10	Washington, DC	19.7	2005, 17
NEPAL	**1.48**	1994			
Kathmandu	0.12	1			

FIGURE 6.7

Per capita greenhouse gas emissions by country and city.

*Values in bold are peer-reviewed and considered comparable. Inventory year, source, and content are indicated in Annex B. All per capita national emissions are calculated from national inventories submitted under the UNFCCC and exclude LULUCF; national population figures are from the World Development Indicators, World Bank data, and correspond to the inventory year.

moderately lower per capita emanations especially those with great transportation frameworks, urban areas will in general have higher discharges, if in a cool atmosphere zone. The most significant perception is that there is no single factor that can clarify varieties in per capita discharges crosswise over urban areas; the varieties are because of an assortment of physical, monetary, and social variables explicit to the remarkable urban existence of every city. The subtleties of each stock and its capacity to experience friend survey are basic to creating and checking a viable relief technique. gives instances of the distinctions in carbon emanations of three people living in various worldwide urban communities. The three nations in the precedents Colombia, Canada, and Tanzania—have various dimensions of business and mechanical action, which accommodate changing ways of life and utilization, while advising the lifecycle carbon discharges related with those exercises. The national discharges for the three nations spoke to underneath are as per the following: Canada has the most elevated GHG per capita at 22.65 tCO_2e; Colombia is 3.84 tCO_2e per capita; and Tanzania is 1.35 tCO_2e per capita. In the precedents that pursue, the people have ozone harming substance discharges that vary essentially from the national per capita qualities. This features the significance of figuring emanations at different scales counting national, local, and city to catch separation.

Urbanization and expanded thriving have occurred with urban spread and expanded interest for land. In spite of the fact that the urban populace has multiplied, involved urban land has significantly increased. In created nations, this development has been especially broad in rural territories as interest for space increments with pay, and land costs are regularly lower in rural zones. Expanding thickness could altogether diminish vitality utilization in urban regions as depicted in Figs. 6.8 and 6.9. Urban areas represent an interesting test to engineers in that they require concentrated vitality supplies. Most urban areas are provided with power from huge scale power plants,

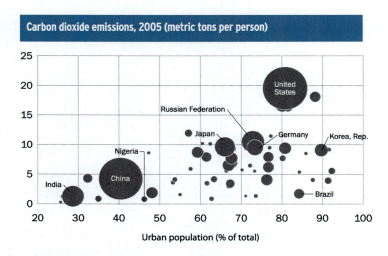

FIGURE 6.8

Development and CO_2 emissions.

6.9 CARBON EMISSION EFFECT

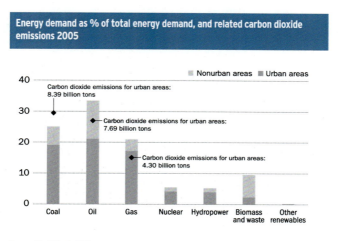

FIGURE 6.9

Emissions from urban and nonurban sources.

transmitted over a separation as short as conceivable to decrease transmission misfortunes. Also, trucks, autos, and flying machine require fuel with high-vitality content. Changing to electric vehicles will probably just strengthen the requirement for concentrated wellsprings of vitality and again requires an unpredictable fuel appropriation arrange. As water accessibility diminishes, urban areas may likewise require extra vitality hotspots for desalination. Sustainable power sources, for example, wind and sunlight based, will be a significant and developing wellspring of vitality for urban areas, yet as at present imagined, they will probably not have the option to supplant the more thought hydroelectric, carbon-based, and atomic vitality sources. Significant changes in vitality supply to lessen GHG emanations will likewise expect changes to the vitality use for instance, less car use and more vitality productive structures.

6.9.1 BLOCKCHAIN APPLICATION FOR DISTRIBUTED ENERGY RESOURCES MANAGEMENT IN SMART CITY

Blockchain innovation can possibly be most promptly helpful in segments where there is no physical trade, for example, in the money-related segment. In such segments, blockchains can give solid records of exchanges without the requirement for confirmation of physical trade. Of the segments with physical trade, be that as it may, the power segment is maybe progressively powerless than others to the incorporation of blockchain innovation. Power deals and buys are cleared total on brought together exchanging stages like stock trades and other money-related advertise stages.

In recent years, development of blockchain ventures that try to improve power area markets and tasks. Today, there are more than 120 associations included in such activities and around 40 sent pilot projects. Today, like never before, it is urban areas that are choosing what's to come. Over portion of the total populace as of now lives in urban communities, and the urbanizing pattern is

152 **CHAPTER 6** BLOCKCHAIN IMPLEMENTATION USING SMART GRID

not probably going to back off; by 2030, 60% of mankind will be city-inhabitants. Since the finish of the First World War, the worldwide urban populace has grown nine-overlap. As urban communities keep on developing, it is winding up progressively indispensable to discover better methods for dealing with these populaces and the administrations they require.

Populace thick urban communities are clearly enormous wellsprings of intensity request, expending 66% of the world's vitality and delivering a comparable extent of worldwide carbon discharges. This spots urban communities at the core of the environmental change discourse, and an immense inquiry for present day city experts, organizers and utilities is: by what method can our electrical framework be created such that supports financial development of life additionally incorporating more sustainable power sources than any other time in recent memory, and profoundly decreasing our urban communities' effect on nature These activities want to discover application in discount and retail power markets, shared vitality commercial centers, the arrangement of "adaptability" or adjusting administrations, electric vehicle charging and coordination, organize security which are essential tools for development of Smart City.

The distributed energy resources reduce variable costs of retail payment processing and accounting. They improve greater transparency into billing and greater customer choice of energy supply. Blockchain could improve retail power advertises by utilizing digital currencies for bill repayment and other "meter-to-money" forms. By empowering the immediate settlement of exchanges, blockchain could lessen the variable expenses of installment handling and bookkeeping to that of executing a keen contract. Some imagine blockchain-based meter-to-money computerization expelling the requirement for discount to-retail mediators through and through. Blockchain could further improve retail clients by empowering more prominent straightforwardness into vitality charges and bill segments, the capacity to enter and leave vitality contracts all the more smoothly, and more prominent decision and straightforwardness into vitality supply.

6.10 **BLOCKCHAIN IMPLEMENTATION IN POWER GENERATION, TRANSMISSION AND DISTRIBUTION**

As indicated by the World Economic Forum, in 2016 solar and wind energy ended up less expensive than nonrenewable energy sources, demonstrating that the fight against an Earth-wide temperature boost could turn into a rewarding business. A similar article shows that by late 2016, 47 creating nations had refreshed their vitality designs by raising their reusable vitality utilization focuses to 100%. In the meantime, Bill Gates, Jeff Bezos, Mark Zuckerberg, Jack Ma, and others put $1 billion USD in Breakthrough Energy Ventures, a reserve for developing vitality source explore. The World Economic Forum article brings up that numerous new interests in vitality framework today go to sustainable power source. The greatest speculations originate from Asia, where India and China have submitted tremendous activities in sun-based vitality, which turned into the least expensive sustainable power source a year ago, well in front of the first figure. The article additionally says that most worldwide organizations have 100% environmentally friendly power vitality reception focuses for their tasks. Substituting the transportation frameworks in enormous urban areas, which used to devour around 50% of the all-out creation of petroleum derivatives, with electrical, savvy, clean frameworks can likewise prompt an extreme diminishing in the interest of nonrenewable energy sources. Uttar Pradesh has plentiful sustainable power source assets. Land-based breeze, the most promptly accessible for advancement. Uttar Pradesh has a sun oriented vitality capability of 22,300 MW which the Yogi government has chosen to bridle to meet the objective of 10,700 MW by 2022. Consequently, the state government is going full scale to advance sunlight-based power plants by presenting different sponsorships for those ready to do it. Power created by these sun oriented power plants is moved to the lattice, an interconnected system for moving power from makers to buyers. This is a piece of the express government's choice to advance private support in age of power through sun oriented power.

6.10.1 SMART ENERGY GRIDS

Smart Energy Grids are portrayed by a two-route stream of power, data and are equipped for observing everything from power plants to client inclinations to person apparatuses. This innovation consolidates into the lattice the advantages of dispersed processing and interchanges to convey constant data and empower the near instantaneous equalization of free market activity at the gadget's are created on the accompanying primary components:

- Active Network Management: They are a blend of programming, computerization and control frameworks that screen the network continuously to guarantee it stays inside its working limits.
- Dynamic Line Rating: They empower transmission proprietors to determinate limit and apply line rating progressively.
- Automatic Voltage Control: They guarantee that voltage and power factor of the particular transports are inside the preset qualities and decrease the power loss of the lattice because of superfluous responsive power flow.
- Phasor Measurement Unit: Electronic gadgets that measure alternating current (AC) phasors and synchronize these estimations under the control of global positioning system (GPS) reference source.
- Reactive Power Compensation: Electronic gadgets for the pay of responsive power.
- Advanced Metering Infrastructure (AMI): The framework incorporates savvy meters, correspondence organizes in various dimensions of the framework chain of importance, Meter Data Management Frameworks (MDMSs), and intends to incorporate the gathered information into programming application stages. The proposed improved arrangement permits the formation of a decentralized vitality advertise that can essentially move the equalization of consumption toward vitality speculations of appropriated assets, while making a potential redistribution of power to new vitality showcase partners, uniquely in contrast to the manner in which the power is right now circulated and controlled as depicted in Fig. 6.10.

6.11 SMARTGRID BLOCKCHAIN IMPLEMENTATION IN INDIA—A CASE STUDY OF PUDUCHERRY PROJECT

To assess the genuine advantages and to recognize appropriate advancements/models of the Smart Grid, Ministry of Power, Govt. of India proposed 14 pilot activities the nation over with various functionalities of Smart Grid. At present all these pilot undertakings are under starting phase of

154 CHAPTER 6 BLOCKCHAIN IMPLEMENTATION USING SMART GRID

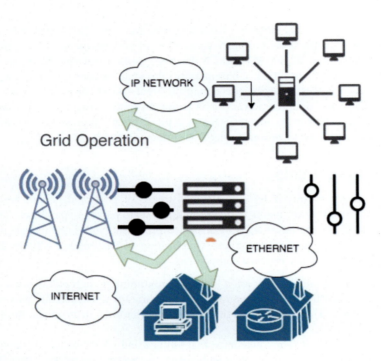

FIGURE 6.10

Smart grid blockchain architecture.

execution. The principle destinations of these pilots are indigenization of innovation, advancement of adaptable and replicable models, raising of appropriate benchmarks and guidelines dependent on these pilot undertaking encounters. Puducherry Smart Grid undertaking is one of the proposed pilots which is being grown mutually by Power Grid Corporation of India. The features of the project are as follows:

1. Puducherry Electricity Distribution System: Puducherry is one of the Union Territories in India which has roughly a million of populace with education rate of 96%. Puducherry electricity department (PED) is not yet unbundled and dissemination shrewd it is isolated into ten divisions for the viable task and support. Keen network pilot undertaking covers one Division viz. Division-I which covers every one of the highlights of brilliant matrix like AMI, Peak Load Management (PLM), Outage Management System (OMS), Power Quality Management (PQM), Renewable Energy Integration and vitality stockpiling. After fruitful execution of the above in to the Smart Grid, the pilot is required to be stretched out to a keen city that highlights water the board, gas the executives, e-medicinal, e-instruction and e-transportation, e-administration, and so forth.
2. Smart Grid Pilot Project (Division-I) profile: Division-I of Puducherry has 100% charge. Approximately 87,035 users, 79% use household purpose and 21% for different purposes such as business, horticulture, road lighting and others. The whole territory is provided by one number of 110/22/11 kV substation, which feeds to 7 numbers of 22 kV overhead feeders,

6.11 SMARTGRID BLOCKCHAIN IMPLEMENTATION IN INDIA 155

Key parameters of the feeder		
Description	Unit	Qty
Length of line OH	Km	5.11
Length of line UG	Km	3.83
Number of DTs	Nos	49
DT capacity	KVA	16425
HT consumers	Nos.	5
HT consumers CMD	KVA	2910
AB switches	Nos	25

Description	Unit	Qty
Peak loading	Amps	125
Power loss	KW	146.46
	%	3.41
Energy loss	MU	0.658
	%	3.25
Minimum voltage	KV	21.38
Voltage drop	%	2.84

FIGURE 6.11

22 kV feeder mapping.

5 numbers of 11 kV underground link feeders and 325 numbers of dissemination transformers taking care of an absolute heap of 127.8 MVA. The single line chart of one 22 kV feeder is appeared in Fig. 6.11.

3. Interim Smart Grid Pilot in activity: Among the 12 feeders of Division-I, under open coordinated effort, a between time pilot covering around 1400 buyers in 22 kV Town feeder with nine Distribution Transformers (DTs) is finished by Power Grid which exhibits all functionalities of shrewd framework, displaying different correspondence advances of keen meters and a condition-of-workmanship brilliant network control focus furnished with head closures, MDMS, and Demo frameworks.

6.11.1 SYSTEM FUNCTIONALITY

The framework is created for encouraging savvy metering, charging activity among clients, control advertise, age organizations, retailers, and little autonomous family control makers in future brilliant matrices. The fundamental engineering of proposed framework appeared in Fig. 6.12 which comprises of brilliant meters, correspondence systems, database and the executives framework,

156 CHAPTER 6 BLOCKCHAIN IMPLEMENTATION USING SMART GRID

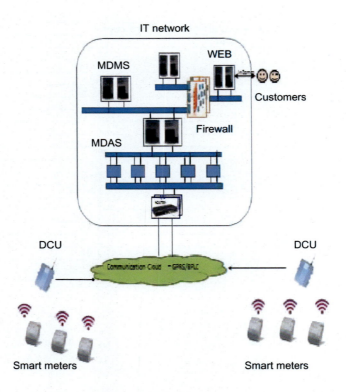

FIGURE 6.12

System functionality diagram.

show and control units. General task of the proposed framework incorporates a few stages as in Fig. 6.12. The vitality utilization information is estimated by shrewd meters. The Data Concentrator Unit (DCU) gathers and moves information from the keen meters to the control focus. Radio Frequency (RF), Power Line Carrier Communication (PLCC), and General Packet Radio Service (GPRS) systems are utilized to move information from meters to nearby servers. Local and neighborhood databases store all the utilization information and client data. This information base will enable customers to settle on increasingly educated choices about their vitality utilization, modifying both the planning and amount of their power use.

The Smart Grid components are used in the system are as follows:

1. **Smart Meters:** Smart meter with AMI is the fundamental segment in keen matrix with two-way correspondence entryway for customer and utility collaboration. Brilliant metering can possibly decrease both purchaser and utility expenses. A noteworthy advantage of the AMI is that it bolsters customer by achieving mindfulness their prompt kWh power utilization and estimating that guides the utilities likewise in burden decrease needs. As a rule, considers have appeared on the off chance that the shoppers are made mindful of degree of their vitality utilization, at that point they decrease their utilization by around 7%.

Today, the expanding infiltration of keen meters enable homes to interface with pervasive information systems and savvy framework that provides for customer just as utilities, perceivability of ongoing free market activity adjusting. Show of utilization information and vitality protection projects urge customers to give back a portion of their vitality use as a by-product of setting aside cash other than spare the expenses toward structure extra age limit required to fulfill future basic pinnacle need rise. Fig. 6.13 demonstrates a portion of the GPS area of introduced brilliant meters, DCU, and DT at between time pilot venture. Burden limitation is one of the striking highlights of the savvy meter exhibited in the meantime pilot as appeared in Fig. 6.14.

FIGURE 6.13

GPS location of smart meters, DCU, and Distribution Transformer (DT).

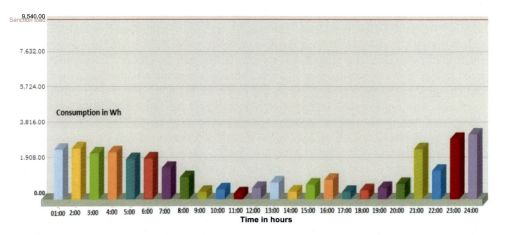

FIGURE 6.14

Day wise load profile of consumer.

158 CHAPTER 6 BLOCKCHAIN IMPLEMENTATION USING SMART GRID

2. Data Collector Units (DCUs): DCUs gather ongoing utilization information from the smart meters and transmit to local information base server framework. Vitality utilizations are put away in the DCU in schedule opening of 30 minutes to execute time-of-utilization include and join the constant costs from the power advertise. One of the significant worries for the shrewd network framework is the unlawful tapping of power. To maintain a strategic distance from such pilferage, edge recognition is included the DCU. The normal power utilization throughout the previous a while, which can be gotten from the utilization history, is utilized to distinguish power robbery through examination between the use and the limit. Typically, the normal utilization is steady. In the event that the use for the earlier month is much lower than the edge, the proposed framework will remind the utility to organize staff to check wrong doing. The framework has adaptability for adding more capacities as indicated by different requests and necessities. However, the examination says that one DCU can take into account the need of around 100 m as a rule, in the meantime pilot, it was discovered that one DCU can take into account not in excess of 50 m by and large particularly when the correspondence is of RF.

3. Communication Network: Communication systems utilized for the metering framework incorporate RF correspondence, control line bearer correspondence (PLC) and GPRS systems. Low Power Radio (LPR) utilizes the nonauthorized industrial, scientific and medical (ISM) (modern, logical, and therapeutic) band for RF correspondence frequently around 865−867 MHz and ZigBee based 2.5 GHz. In LPR case, each meter is outfitted with a RF transmitter that enables correspondence to the information concentrator legitimately, or to different meters with RF transmitters which go about as repeaters or forward the information for example in a coincided system setup. The ZigBee is for the most part utilized for the applications which require a low information rate. The PLC is an innovation to move information flag through the current power transmission and conveyance organize. Contrasted and other correspondence innovations, the PLC requires no additional charge for structure organize. Work arrange development for information move from one meter to other meter till it achieves the DCU is appeared in Fig. 6.15.

4. Database and Management System: Customer data are put away in focal database. The information from the DCUs is gathered through Meter Data Acquisition System (MDAS) and dependent on which the Meter Data Management System (MDMS) gives the fundamental information in the endorsed arrangements to both Utility and purchaser. A common MDMS is appeared in Fig. 6.16, which likewise gives the charging subtleties to purchasers according to booked charging cycles for various classes and to the utility subtleties of alters, if any adjacent to the events of any power deficiencies/disappointments in the framework for remedial activities.

5. Online energy review is one of the lone highlights of this venture which finds the immediate specialized and business misfortunes. Month-to-month vitality review has been completed for the shoppers covering of one circulation transformer as depicted in in Fig. 6.17 and it is discovered that unaccounted vitality is contained inside 7−10 rate utilizing shrewd meters in brilliant matrix. Consequently, the online vitality review helps in checking the unaccounted vitality that can be decreased further, if preventive moves are made. Significant troubles for the vitality review are reconciliation of various correspondence and producing advances and mix of which is a genuine test ahead for the smart matrix administrators.

6.11 SMARTGRID BLOCKCHAIN IMPLEMENTATION IN INDIA

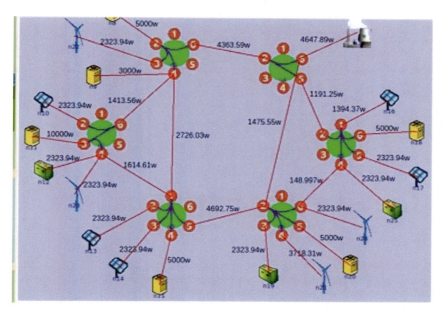

FIGURE 6.15

The mesh network formation of PLC-based smart meters.

FIGURE 6.16

Load monitoring from MDMS solution.

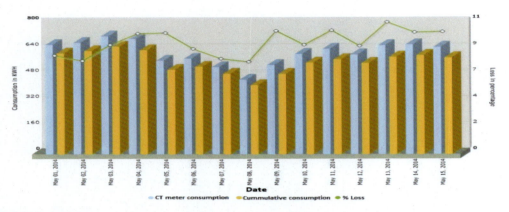

FIGURE 6.17

Distribution Transformer wise daily energy audit.

    Approval Estimation and Editing (VEE) is a basic capacity of MDMS. As purchaser subtleties alongside endorsed burden and meter information being stacked into the database, information syntactic and semantic checks are performed. VEE is then led to make an interpretation of the crude information into a lot of steady and usable information for charging. Current VEE work frequently considers chronicled information insights just as climate, occasions, account comparability, and so on. Principles are characterized to recognize "opening" and conceivable "alter" or "robbery." The VEE arrangement gives gauges that are illustrative of authentic utilization. It is performed dependent on precharacterized reference esteems and additionally principles identified with individual client without ongoing feeder and transformer stacking reflections. Both the low power radio frequency (LPRF) and PLC correspondences are utilized to move information from meters to the organizer (DCU). The principle favorable circumstances of LPRF/ZigBee and PLC are simple access and minimal effort of establishment. Likewise, double correspondence systems can diminish the information mistake rate and increment the unwavering quality of information transmission. Information is moved from meters to the DCUs which are additionally the hubs of correspondence organize. Contingent upon the viewable pathway and nature of correspondence, switches and repeaters are additionally conveyed as organizers to gather the information from several meters and send the information to the nearby database through the DCUs utilizing GPRS correspondence. Those meters with GPRS modems legitimately speak with the focal information base control focus. The principle advantage with the GPRS correspondence meters is that they can be filled in the holes where the PLC or LPR meters cannot be conveyed as found in the meantime pilot.

6. Peak Load Management: Peak load the board arrangement makes the electric lattice substantially more effective and adjusted by helping the buyers to lessen their general electric interest, and additionally moves the time span when they utilize their power staying away from high pinnacles and related high duties.
7. Power Quality Management: PQM arrangements are expected to address occasions like voltage flash, unequal stage voltages and symphonious mutilated sullied supply, and so forth.

8. Outage Management System: OMS comprises of fault passage indicator (FPI), feeder remote terminal unit (FRTU), Sectionalizers, and distribution transformer monitor (DTMS) and so forth and enables utility to oversee booked and unscheduled blackouts of dispersion framework viz. DTs and, high tension (HT) and low tension (LT) feeders.
9. Supervisory Control and Data Acquisition Distribution Management System: The Supervisory Control and Data Acquisition framework will give constant observing and control elements of dispersion arrange from a local area.

6.12 BLOCKCHAIN IMPLEMENTED PROJECTS WORLDWIDE IN ENERGY CONSERVATION

LO3 Energy banded together with Transactive Grid to build up a network microgrid in New York. The organization accomplished the first P2P blockchain exchange between a private PV maker and his neighbor. From that point forward the organization has declared more pilot projects. In association with vitality supplier Energie Sudwest and Karlsruhe Institute of Technology, they are building up a local energy advertise in the Lazarettgarten microgrid in Landau, Germany. Solar boards and vitality stockpiling gadgets are a piece of the microgrid. The project centers around market component structure and administrative changes required for further take off of comparative nearby marketplaces. With Allgauer Uberlandwerk they will test blockchain advances in the Allgau microgrid. They mean to explore the enthusiasm of purchasers in such markets and how neighborhood microgrids and commercial centers can be integrated into existing vitality systems. In Australia, in partnership with Yates Energy Service, they intend to research transactive energy market models that advance age from sustainable sources and energy conservation. Also, LO3 Energy is wanting to start a venture in Texas with Direct Energy concentrating on business and industrial purchasers of energy. In the UK, LO3 Energy, and CentraCare intending to build up a neighborhood vitality advertise in the system constressed territory of Cornwall, that means to diminish high sustainable.

Power Ledger is an Australian start-up associated with an assortment of blockchain applications for vitality frameworks. The most develop application is the advancement of a private P2P power exchanging deface keyplate among prosumers and nearby shoppers, the first of its sort in Australia. The start-up ran a preliminary and showed critical potential for vitality charges investment funds and extra incomes for PV makers. For example, PV prosumers are normally payed 7 c/kWh when exporting excess power back to the principle matrix, while customers are charged 25 c/kWh. Power Ledger's P2P pilot task concurred a valuing plan of 20 c/kWh of vitality bought through the stage. 75% of electricity charges went to prosumers and 25% to the service organization. In future deployments, a little cut will be taken by the blockchain stage developers. Moreover, Power Ledger joined forces with Vector Energy, New Zealand's biggest vitality circulation organization, and implemented the first P2P blockchain-empowered vitality exchanging platform across a directed conveyance arrange in Auckland. The organization recently raised $34 million AUS dollars through offers of their POWR cryptocurrency (power ledger), which is tradable on the open Ethereum blockchain. POWR can be changed over to Ecochain stablecoin token (SPARKZ), the commercial center's local currency, which can be exchanged for power on the organization's private blockchain. Power Ledger is likewise dynamic in the fields of discount power trading, electric e-portability, IoT smart

162 CHAPTER 6 BLOCKCHAIN IMPLEMENTATION USING SMART GRID

gadgets and computerization, framework oversee, and green endorsements carbon exchanging, anyway most exercises in this space are still in the early advancement stage with beta discharge arranged toward the finish of 2018. Power Ledger is engaged with more pilot extends in a few nations including Tasmania, India, Thailand, and Lichtenstein. Power Ledger has likewise banded together with Kepco, the Kansai Electric Power Corporation. They intend to look at blockchain feasibility for exchanging overabundance vitality of prosumers, who may possess generating or capacity gadget resources. The underlying preliminary will occur in Osaka, Japan and will include exchanging between 10 households.

Vector, the biggest power and gas wholesaler in New Zealand, are testing a P2P nearby vitality commercial center in New Zealand, created by Power Ledger. Members in this progressing task incorporate 500 residential PV prosumers, schools, and network groups. Vattenfall also run a blockchain preliminary of Power peers, a stage that enables energy exchanging through a P2P organize, where people determine from whom to purchase or offer self-created power, however they selected out for a customary stage arrangement. In Japan, the Eneres venture involves in excess of 1000 family units, vitality prosumers that possess PV, small wind age or combined heat power (CHP) that can exchange their vitality surplus with other families through blockchains.

In Amsterdam, The Netherlands, Alliander is teaming up with Spectral Energy to build up a P2P vitality sharing stage, called Jouliette at De Ceuvel in Amsterdam [189] on the premise of a private and permissioned blockchain arrangement that can accomplish quicker transactions and improve execution. Ghostly Energy has propelled an energy token called "Jouliette" that can be utilized to encourage P2P vitality transactivities. De Ceuvel is a private behind-the-meter keen lattice which consists of 16 places of business, a nursery, a café, a little homeland a few PV boards. Vitality is traded inside the smart network on a P2P design. The Jouliette stage can show constant power streams of the network and uses artificial intelligence (AI) calculations to predict energy generation and consumption. Multichain, the blockchain platform utilized, permits open permissionless and private and permissioned blockchains. In the last case, the agreement system sent uses round-robin component to distribute square age among a set of known validators.

In France, Bouygues Immobilier is working together with the city of Lyon to build up a blockchain demonstrator venture for direct energy exchanges between sun oriented makers and vitality shoppers. Blockchain technology is utilized to confirm and check the vitality delivered and consumed at various areas in the framework, as vitality is exchanged between various pads in a structure. Smart contracts are utilized to derive geolocation of hubs in the framework that empower precise figuring's of power misfortunes as vitality is being transmitted. Blockchain infrastructure is created by Stratumn. The agreement calculation is called Proof of Process. Confirmation of information is decoupled from the source of data making a zerolearning evidence for example framework individuals can verify that an agreement has been respected without knowing the careful terms of the contracts itself. Evidence of process utilizes run of the mill know your customer (KYC) strategies thus represents a progressively incorporated blockchain approach where network members pursue a various leveled request of trust. Condole centers on P2P vitality exchanging neighborhood networks, where prosumers can sell their vitality surplus to nearby families or associations. They are involved in two pilot extends in Kettwig and Mulheim, in Germany. The German start-up, upheld by Innogy and Tokyo electric power company (TEPCO), is likewise developing a disseminated stage where vitality makers, purchasers, energy capacity and adaptability suppliers can execute without central management.

6.12 BLOCKCHAIN IMPLEMENTED PROJECTS WORLDWIDE   **163**

Divvi is an Australian new business that expects to build up a distributed commercial center for sustainable power source with spotlight on community energy frameworks and new plans of action for sustainable power source possession. The conveyed stage depends on Ethereum keen contracts. In a similar application zone, Energo Labs is a Chinese start-up company that uses blockchain answers for network vitality undertakings including prosumers, buyers, vitality stockpiling, and smart grid gadgets to empower P2P vitality sharing. Energo Labs envisions achieving decentralized and self-sufficient vitality frameworks, where members of networks or microgrids can trade vitality, information and installments continuously. Energo Labs utilizes two tokens: the WATT token is identical to 1 kWh of vitality put away in the microgrid or storage asset, and TSL a digital currency that empowers access to the microgrid energy stockpiling frameworks. TSL are premined and 80% are appropriated to storage proprietors. Vitality utilization at a neighborhood level is prioritized. Quantum blockchain moves from PoW arrangements and empowers a decentralized application improvement stage, coordinated with brilliant meters or EV charging stations. P2P vitality exchanging valuing instrument is similar to the task of securities exchanges and request book tables. Automated exchanging is practiced by the utilization of wise operators and use of user versatile applications. Energo Labs has effectively exhibited their solution can effectively gather information from generation and consumption ends. In 2018, they intend to control vitality use by AI and a savvy homeapp. They have undertakings made arrangements for Philippines, Australia and Southeast Asia in the size of hundreds MWs. Energy Bazaar centers around nearby vitality advertises in rising counter attempts and especially in India. They plan to encourage vitality exchanges for family units, business customers, microgrid administrators, utilities, and transmission administrators. The organization intends to build up a suite of technologies, for example, keen programming operators, AI for improved forecasting of vitality generation and utilization, and game-hypothetical market design that will give motivators to adaptability administrations and matching of free market activity. Blockchain will shape the trust layer that enables transactions between various stakeholders. In Denmark, BLOC or Blockchain Labs for Open Collaboration is a digital arrangements organization engaged with two tasks in Copenhagen and Samso.

Energy Block is a network pilot venture in Copenhagen that investigates neighborhood and network vitality markets, and sensor retrofitting for use with blockchains. They are additionally engaged with a project in Samso, Community Power that expects to examine how blockchain advancements can empower coresponsibility for assets, retrofitting existing RES resources for associations into blockchain systems and information check for cooperation in vitality and carbon trading markets. Greeneum is centering in consolidating fake intelligence and AI procedures with blockchains to develop a decentralized vitality showcase that permits installment settlement and constant vitality exchanges. Greeneum has been engaged with pilot projects in Israel and Cyprus. Our solar grid centers around network energy systems and P2P vitality exchanging among prosumers and purchasers at a local level, controlled by Ethereum. They accept that buyers might be willing to pay a premium for privately delivered inexhaustible energy, which can demonstrate to be an extra motivator for RES investment.

Power-ID is a P2P vitality exchanging pilot venture in Switzerland between 20 prosumers and buyers. The venture will research the potential of DLT for more attractive system expenses and lower framework task costs. According to an as of late distributed report, venture engineers are still investigating both open and private blockchain arrangements. StromDAO, a German start-up, offers a stage where purchasers can contribute in community and inexhaustible undertakings and decrease their

164 CHAPTER 6 BLOCKCHAIN IMPLEMENTATION USING SMART GRID

vitality bills. They aim to give blockchain answers for vitality framework partners incompliance with regular vitality showcase structures. StromDAO depends on the Fury Network blockchain and uses Proof of Authority. In the Netherlands, To Blockchain has built up a P2P digital platform for vitality trade, called PowerToShare. Vitality transactions are overseen through a token mechanism.

PowerToShare is currently being tried at the Green Village venture in Netherlands. In Switzerland, Hive Power utilizes blockchain-empowered smart meters to check amounts of vitality delivered. Empowered by an Ethereum solution and brilliant contracts, prosumers can participate in decentralized energy exchanging.

In Thailand, BCPG has built up a blockchain-based application that takes out all delegates, for example, vitality suppliers and utilities and empowers P2P vitality exchanging between consumers. Too much energy is a blockchain start-up that plans to make a P2P platform for customers and vitality prosumers. OneUp is delicate product designers represent considerable authority in information investigation and blockchain technologies, with aptitude in the Ethereum platform. OneUp was awarded for building up a decentralized vitality exchanging stage at start-up rivalry in 2017 and has been supporting a few companies interested in blockchain advances, most eminently PWC.

6.13 CONCLUSION AND RECOMMENDATIONS

This chapter has dealt with the proposal of an innovative high technology-based architecting the Smart Environment Pillar of the Smart City evolutionary process. In particular, the improvement of the Quality of Life and the enhancement of the Quality of Services for the citizens of a Smarter City has been addressed by this paper, proposing a disruptive synergy between the so-called Smart Energy Grid and the emerging Blockchain technology. Indeed, it has been proved that it possible to make cities smarter promoting innovative solutions by use of Information and Communication Technology for collecting and analyzing large amounts of data generated by several sources, such as sensor networks, wearable devices, and IoT devices spread among the city. A Smart Grid framework in the current dissemination arrangement of Puducherry has been effectively created. Starter studies show improvement in productivity and unwavering quality of the dissemination organize. Circulated processing and interchanges which convey ongoing data and empower close immediate parity of free market activity at the gadget level has been joined. By giving mindfulness and preparing on brilliant lattice highlight to purchasers viz. load confinement and information utilization show, the shopper additionally determines benefits by checking altering loads other than sparing the vitality through interest reaction. The utility likewise receive the savvy matrix rewards viz. request side administration fitting the Time of Use (ToU) taxes, streamlined blackout the executives frameworks, reconciliation of renewables by net metering. The chapter focuses on the blockchain technology utility and implementation areas. Blockchain could help make a decentralized vitality commercial center. In what might be the most troublesome situation for the power showcase, we trust the mix of blockchain and interchanges innovation could encourage secure exchanges and installment between a large number of gatherings, by towns that do not have power get to, this application could have more an incentive for India than anyplace else on the planet. For example, a start-up called Grid Singularity is utilizing blockchain to investigate "pay-as-you-go" sun oriented in creating nations where lattice foundation is less refined and administrative obstacles might be lower blockchain drives increasingly conveyed network foundation. The capacity to execute in the vitality advertises as a limited generator would probably drive a greater move toward advancements that empower a circulated network. The upset would incorporate shrewd blockchain systems and gadgets, yet in addition Internet of Things (IoTs) machines and electric vehicles, just as power assets like housetop sun oriented, vitality stockpiling, and even energy components. Blockchain could end the requirement for net metering. We trust the selection rate of dispersed sun oriented has to a great extent profited by arrangements, for example, net and gross metering, which bolster the financial matters of going sunlight based versus paying for lattice control. In any case, the more extended term viewpoint for net metering is not sure attributable to developing restriction from utilities. The appropriated vitality makers would grasp an option in contrast to offering back to the matrix, for example, selling into a restricted vendor advertise, for which blockchain could give the conveyed and secure value-based spine to empower a decentralized commercial center. Consolidating blockchain with the IoT could empower the arrangement of disseminated control exchanges. By utilizing conveyed remote or wireline information interfaces in a work arrange appropriated makers could naturally communicate data on abundance control accessibility alongside applicable span data. People organizations will work as small scale power exchanging organizations. On a fundamental level, purchasers could consequently react with their capacity needs. Utilizing a blockchain-based record, machine intermediaries of makers and shoppers can arrange valuing and go into a power deal exchange dependent on a smart contracts.

GLOSSARY

Blockchain Shared, trusted, open record of exchanges, that everybody can investigate however which no single client controls. It is a cryptographed, secure, alter safe appropriated database. It takes care of a complex numerical issue to exist. A blockchain is an ideal spot to store esteem, characters, understandings, property rights, certifications, and so forth.

Private blockchain It is a completely private blockchain is a blockchain where compose consents are held unified to one association. Peruse authorizations might be open or limited to a discretionary degree. Likely applications incorporate database the board, evaluating, and so on inside to a solitary organization, thus open meaningfulness may not be essential much of the time by any means, however in different cases open auditability is wanted.

Private key Each time a client runs a digital currency wallet out of the blue an open private key pair gets produced.

Public blockchain It is an open blockchain is a blockchain that anybody on the planet can peruse, anybody on the planet can send exchanges to and hope to see them included on the off chance that they are substantial, and anybody on the planet can take an interest in the agreement procedure.

Smart contracts These are PC conventions that encourage, check, or authorize the arrangement or execution of an agreement, or that hinder the requirement for an authoritative condition. Keen contracts as a rule likewise have a user interface (UI) and frequently imitate the rationale of authoritative provisos.

Token A token is a computerized character for something that can be possessed. Verifiably, tokens began as meta data encoded in straightforward Bitcoin exchanges, subsequently exploiting the Bitcoin blockchain's solid unchanging nature.

166 CHAPTER 6 BLOCKCHAIN IMPLEMENTATION USING SMART GRID

Smart grid Smart Grid is an idea with respect to computerized innovation application and electric power organize. It offers a great deal of important innovations that can be utilized inside the not so distant future or are as of now being used today. Savvy Grid incorporates electric system, computerized control apparatus, and shrewd observing framework.

Smart city Smart city is an idea that has been the subject of expanding consideration in urban arranging and administration during ongoing years.

Hash A hash capacity is a scientific capacity that takes a dataset as information and returns a fixed esteem, a hash esteem.

Full node Any PC that downloads the whole history of the blockchain and can check new squares, for example Bitcoin.

Light node Downloads just header data and cannot confirm all data. Light hubs cannot mine for example Cell phones. Customer Software that gives cryptographic money wallets to clients.

Consortium A gathering that together consolidates assets to achieve a shared objective. Regularly pools of excavators.

DISCUSSION QUESTIONS

1. What is blockchain technology?
2. Is it workable for a client to leave a blockchain when he knows about certain private information?
3. If it is conceivable to join or leave a blockchain organize, what will be the system to pursue?
4. What are the advantages of blockchain technology?
5. What are carbon emissions?
6. How are they produced and what is their effect?
7. What are smart grids?
8. Given that sustainable sources give just a little level of our vitality and that atomic power is so costly, what can we reasonably do to get off petroleum products as quickly as time permits?
9. What is greenhouse effect?
10. Why worried about the greenhouse effect now?
11. What is carbon credit?
12. How does carbon dioxide contribute to climate change?
13. How can blockchain reduce carbon emission?
14. Is carbon trading the answer to the greenhouse problem?
15. What is the benefit of turning emission reductions into financial products?

REFERENCES

[1] S. Nakamoto, Bitcoin: A Peer-to-Peer Electronic Cash System, 2008, <https://bitcoin.org/bitcoin.pdf, www.bitcoin.org>.

[2] J. Yli-Huumo, D. Ko, S. Choi, S. Park, K. Smolander, Where is current research on blockchain technology?-a systematic review, PLoS One. (2016). <https://doi.org/10.1371/journal.pone.0163477> eCollection 2016. <https://www.ncbi.nlm.nih.gov/pubmed/27695049>.

[3] K. Siba, Tarun, A. Prakash, Block-chain: an evolving technology 8 (4) (2016). <https://doi.org/10.18311/gjeis/2016/15770>, <http://www.informaticsjournals.com/index.php/gjeis/article/view/15770>.

FURTHER READING 167

[4] L. Carlozo, Why CPAs need to get a grip on blockchain, 2017, <https://www.journalofaccountancy.com/news/2017/jun/blockchain-decentralized-ledger-system-201716738.html>.

[5] M.C. Benton, N.M. Radziwill, Quality and Innovation with Blockchain, Technology 20 (1) (2017). <www.asq.org>.

[6] S. Manski, Building the blockchain world: technological commonwealth or just more of the same? Wiley Online Library, 2017. <https://doi.org/10.1002/jsc.2151>.

[7] J.M. Woodside, F.K. Augustine Jr, W. Giberson, Blockchain technology adoption status and strategies, J. Int. Technol. Inform. Manag 26 (2) (2017). <http://scholarworks.lib.csusb.edu/cgi/viewcontent.cgi?article = 1300&context = jitim>.

[8] H. Halaburda, Blockchain revolution without the blockchain, Commun. ACM 61 (7) (2018) 27–29. Available at SSRN: <https://ssrn.com/abstract = 3133313> or <https://doi.org/10.2139/ssrn.3133313>.

FURTHER READING

U. Ahsan, A. Bais, Distributed big data management in smart grid, Wirel. Opt. Commun. Conference (WOCC) 2017, IEEE, 2017, pp. 1–6.

M. Andoni, V. Robu, D. Flynn, S. Abram, D. Geach, D. Jenkins, et al., Use of Block Chain technology in providing quality reliable uninterrupted power by mini grids having different generating technologies (Solar PV/Wind/ Bio Mass etc.), in: Conference on Creation of Eco-system Using for Renewable Energy, Distributed Energy Generation & Supply.

M. Andoni, V. Robu, D. Flynn, S. Abram, D. Geach, D. Jenkins, et al., Blockchain Technology in the Energy Sector: A Systematic Review of Challenges and Opportunities, Contents Lists Available at ScienceDirect Renewable and Sustainable Energy Reviews, 2019. Available from: <https://reader.elsevier.com/reader/sd/pii/S1364032118307184?token=F1756D45991448B7E4C49238162C8EBE22C427694D97A453F3793AD73FF808EC04F1B221AE38B00AC326CC8A4BC208BB>.

Bitcoin: A Peer-to-Peer Electronic Cash System. Available from: <https://bitcoin.org/bitcoin.pdf>.

Blockchain in Electricity: A Critical Review of Progress to Date Blockchain References, 2018.

P. Bronski, J. Creyts, M. Crowdis, S. Doig, J. Glassmire, L. Guccione, The Economics of Load Defection: How Grid-Connected Solar-Plus-Battery Systems Will Compete With Traditional Electric Service—Why It Matters, and Possible Paths Forward, 2015. Available from: <https://www.rmi.org/wp-content/uploads/2017/04/2015-05_RMI-TheEconomicsOfLoadDefection-FullReport-1.pdf>.

S. Bruyn, Blockchain an Introduction, 2017. Available from: <https://beta.vu.nl/nl/Images/werkstuk-bruyn_tcm235-862258.pdf>.

Carbonchain.org

J. Chen, F.N. Lee, A.M. Breipohl, R. Adapa, Scheduling direct load control to minimize system operational cost, Systems 24 (3) (1995) 1199–1207.

COMPUTABLE, Tennet Test Blockchain Voor Energienet, May 4, 2017. Available from: <https://www.computable.nl/artikel/nieuws/security/6014258/250449/tennet-test-blockchain-voor-energienet.html?utm_source=nieuwsbrief&utm_medium=email&utm_campaign=Dagelijks_04_05_2017&utm_content=topartikelen>.

CONNECT, Blockchain Is Hype, April 18, 2017. Available from: <https://agconnect.nl/blog/blockchain-hype?utm_source=nb_agc_20170418&utm_medium = email&utm_term=&utm_content=&utm_campaign=18-04-2017>.

Consensys, Grid + : Welcome to the Future of Energy (White Paper), n.d.

Conference on Creation of Eco-system Using for Renewable Energy, Distributed Energy Generation & Supply 2018-10-17_blockchain_white_paper_final.pdf.

G. Deconinck, An evaluation of two-way communication means for advanced metering in Flanders (Belgium), in: IEEE International Instrumentation and Measurement Technology Conference Victoria, 2008, pp. 900–905.

Department for Business. Energy & Industrial Strategy (BEIS), DIGEST of United Kingdom Energy Statistics 2017 Chapter 6: Renewable sources of energy. Available from: <https://assets.publishing.service.gov.uk/government/uploads/system/uploads/attachment_data/file/643414/DUKES_2017.pdf>.

Drift, Available from: <https://www.joindrift.com/>.

C. Eid, P. Codani, Y. Perez, J. Reneses, R. Hakvoort, Managing electric flexibility from distributed energy resources: a review of incentives for market design, Renew. Sust. Energy Rev. 64 (2016) 237–247.

eMotorWerks, 2018. Available from: <https://emotorwerks.com/>.

Energy Union Package. A Framework Strategy for a Resilient Energy Union With a Forward-Looking Climate Change policy.

J.P. Green, S.A. Smith, G. Strbac, Evaluation of electricity distribution system design strategies, IEEE Proc. Gener., Transm. Distrib. 146 (1) (1999) 53–60.

Grid + , Available from: <https://gridplus.io/>.

http://siteresources.worldbank.org/INTUWM/Resources/340232-1205330656272/4768406-1291309208465/PartIII.pdf

http://www.energie-nachrichten.info/file/01%20Energie-Nachrichten%20News/2018-05/80503_Eurelectric_1_blockchain_eurelectric-h-DE808259.pdf

http://www.indiasmartgrid.org/reports/White%20Paper%20on%20Blockchain.pdf

https://blockchainhub.net/blockchain-glossary/

https://eur-lex.europa.eu/legal-content/en/TXT/?Uri = COM%3A2015%3A80%3AFIN), (2015).

https://medium.com/@tuomassantakallio/carbon-chain-trading-carbon-credits-on-the-blockchain-50566bce6093

https://mlsdev.com/blog/156-how-to-build-your-own-blockchain-architecture

https://static1.squarespace.com/static/58ec123cb3db2bd94e057628/t/5b4e59751ae6cf086c4450a5/1531861368631/EFI_Blockchain_July2018_FINAL + .pdf

https://www.buschsystems.com/resource-center/knowledgeBase/glossary/what-is-the-certified-emissions-reduction-cer

https://www.forbes.com/sites/bernardmarr/2018/02/16/a-very-brief-history-of-blockchain-technology-everyone-should-read/#6e8cd1e7bc47

https://www.pluralsight.com/guides/blockchain-architecture

Ideo CoLab, Smart Solar. Available from: <https://www.ideocolab.com/prototypes/smartsolar> (last accessed 25.04.18).

International Energy Agency (IEA): Technology Roadmap Smart Grid, OECD/IEA, Paris, France. IEA Smart Grid, 2011.

A. Ipakchi, Implementing the smart grid: enterprise information integration. Grid Interop Forum. National Institute of Standards and Technology (NIST)—a report on Interoperability standards road map-07, Systems. Proceeding of IEEE Power Engineering Society(PES'09), 2007, pp. 1–8.

S.M. Jomar, Current State of Blockchain Technology A Literature Review, 2018. Available from: <https://www.researchgate.net/publication/329622321>.

B. Kamanashis, Vallipuram Muthukkumarasamy (2017), Securing Smart Cities Using Blockchain Technology. Available from: <https://www.researchgate.net/publication/311716550>.

R. Kappagantu, S. Senn, M. Mahesh, S. Arul Daniel, Smart grid implementation in India—a case study of Puducherry pilot project, Int. J. Eng. Sci. Technol. 7 (3) (2015) 94–101. Available from: https://doi.org/10.4314/ijest.v7i3.11S.

M.V. Krishna Rao, J.V. Pandurangam, R. Peri, K.N. Clinard, Indian power utility, in: IEEE Power & Energy Society General Meeting (PES'09), Development and Evaluation of a Distribution Automation, 1995.

FURTHER READING 169

R.N. Lahiri, A. Sinha, S. Chowdhury, S.P. Chowdhury, C.T. Gaunt, Importance of Distribution Automation System for, 2009.

X. Luo, J. Wang, M. Dooner, J. Clarke, Overview of current development in electrical energy storage technologies and the application potential in power system operation, Appl. Energy 137 (2015) 511−536.

P. McCallum, A. Peacock, Blockchain technology in the energy sector: a systematic review of challenges and opportunities, Renew. Sustain. Energy Rev. 100 (2019) 143−174. Available from: https://doi.org/10.1016/j.rser.2018.10.014.

S. Nakamoto, The Creator(s) of Bitcoin. Available from: <https://en.wikipedia.org/wiki/Satoshi_Nakamoto>.

Nikhil lohade, dubai aims to be a city build on blockchain, 2017. Available from: <https://www.wsj.com/articles/dubai-aims-to-be-a-city-built-on-blockchain-1493086080?__prclt = dyd2krrg>.

T. Parikshit Jain, Blockchain in Power Generation, Transmission and Distribution.

M. Pedrasa, T. Spooner, I. MacGill, Coordinated scheduling of residential distributed energy resources to optimize, applications on the appliances of the home environment, IEEE. Netw. 23 (6) (2010) 8−16.

Policy Paper, Promising Blockchain Applications for Energy: Separating the Signal From the Noise, July 2018. Available from: <file:///Users/sapna/Desktop/dstproject/blockchainchapter/architecturepaper1.pdf>.

Share & Charge, Available from: <http://shareandcharge.com/en/>.

SolarCoin, Available from: <https://solarcoin.org/en/node/6>.

I. Steklac, H. Tram, Advanced AMR Benefits—Utility Case Studies. Electric Light & Power, July/August 2005. Available from: <http://www.iea.org/papers/2011/smartgrids_roadmap.pdf>.

C. Su, D. Kirschen, Quantifying the effect of demand response on electricity markets, IEEE Trans. Power Syst. Indian Utility. IEEE Trans. Power Deliv. 10 (1) (2009) 452−458.

S.L. Tompros, N.P. Mouratidis, M. Draaijer, A. Foglar, H. Hrasnica, Enabling applicability of energy saving, IEEE Trans. Power Syst. 10 (4) (2009) 1994−2001.

H. Tram, AMI Benefits in T&D − What Is for Real? Distribu TECH Conference, 2007.

T. Santakallio, Carbon Chain—Trading Carbon Credits on the Blockchain, 2018.

P.P. Varaiya, F.F. Wu, J.W. Bialek, Smart operation of smart grid: risk-limiting dispatch, Proc. IEEE 99 (1) (2011) 40−57.

Whitepaper Revision, July 2018. Available from: <https://climatecoin.io/uploads/WHITEPAPER-OFFICIAL-V5.3-1.pdf>.

Q. Wu, P. Wang, L. Goel, Direct load control considering interrupted energy assessment rate in restructured power, smart home energy services, IEEE Trans. Smart Grid 1 (2)) (2009) 134−143.

S. Zhou, M.A. Brown, Smart meter deployment in Europe: a comparative case study on the impacts of national policy schemes, J. Clean Prod. 144 (2017) 22−32.

Office of Gas and Electricity Markets (Ofgem). Transition to Smart Meters. Available from: <https://www.ofgem.gov.uk/gas/retailmarket/metering/transition-smart-meters>.

CHAPTER 7

CLOUD-BASED BLOCKCHAINING FOR ENHANCED SECURITY

D. Jeyabharathi[1], D. Kesavaraja[2] and D. Sasireka[3]

[1]*Department of IT, Sri Krishna College of Technology, Coimbatore, India* [2]*Department of Computer Science and Engineering, Dr. Sivanthi Aditanar College of Engineering, Tiruchendur, India* [3]*VV College of Engineering, Tisaiyanvillai, India*

The pictorial representation of cloud-based security mechanism is shown in Fig. 7.1.

Cloud computing offers numerous services to users on demand and gains it significance in the current era. Securing the information stored in the cloud data center is one of the prominent issues. Sensitive information such as medical images attains much concentration regarding its security, when it is uploaded to data centers for outsourcing. Cryptographic algorithms are generally used for the encryption process, in which block cipher algorithms are frequently applied for image encryption methods. An efficient Image Encryption Method based on blockchaining in Cloud Computing as a Security Service. The proposed work provides enhanced security using blockchaining in cloud environment. While including blockchaining concept in cloud that leads to two main challenges such as:

1. Poor access control

Virtual machine (VM) measurement data in IaaS cloud are owned by users. They may contain sensitive information. The access to such data should be restricted. Nevertheless, the original blockchain for Bitcoin is a permissive network with a public distributed ledger. Anyone can download, look through, and verify a replica of the data. The inconsistency with the access control demand on VM measurement data makes one noticeable challenge.

2. Time-intensive PoW task

The confidence of data integrity on blockchain relies on the time-intensive PoW tasks from the mining process. But the mining causes a main defect in inefficiency of data storage: high confirmation latency and low throughout. The latency is a time interval between data submission and storage confirmation. In Bitcoin, it is 10 minutes, approximately equal to the computation time of a block. The throughput is about seven transactions per second [1]. Comparing with classical data storage, the blockchain is rather poor in efficiency. This forms another challenge of delivering the same integrity.

Especially, to enhance the integrity, a three-layer blockchain network is introduced. In the first layer, access policy is checked. Based on overall VM measurement, proposed blockchain-based security registry enhances the security policy. This is done in the second layer. Finally in the third layer, security checking is done on corresponding VM's data packages.

Handbook of Research on Blockchain Technology. DOI: https://doi.org/10.1016/B978-0-12-819816-2.00007-1
© 2020 Elsevier Inc. All rights reserved.

172 CHAPTER 7 CLOUD-BASED BLOCKCHAINING FOR ENHANCED SECURITY

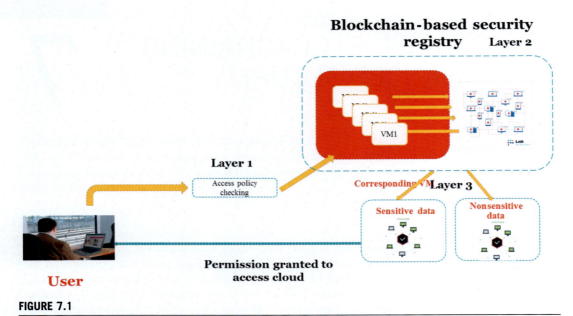

FIGURE 7.1

Cloud-based security mechanism using blockchaining.

In the third layer, the main work is the separation of sensitive and nonsensitive data. The three types of blockchains are public blockchain, consortium blockchain, and private blockchain. Everyone can check the transaction and verify it, and can also participate the process of getting consensus is called public blockchain.

Consortium blockchain means the node that had authority can be choose in advance, usually has partnerships like business-to-business, the data in blockchain can be open or private, can be seen as Partly Decentralized. Like Hyperledger and R3CEV are both consortium blockchains. In the private blockchain, node will be restricted, not every node can participate this blockchain, has strict authority management on data access. In the proposed work, public blockchain is used for nonsensitive data.

7.1 INTRODUCTION

7.1.1 BLOCKCHAINING

Technically, blockchain is a chain of blocks each containing particular data about specific value: money, a share of the company, property ownership certificate, kilowatt of energy, and literally any value, depending on blockchain type.

Along with worth, every block stores encrypted details concerning blockchain peers who exchanged that value. For instance, buyer and seller who exchanged a product for money; sender and receiver who participated in a transaction; and so on.

7.1 INTRODUCTION

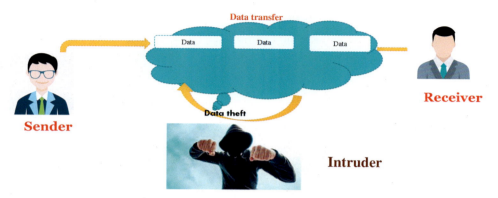

FIGURE 7.2

Intruder attack during transaction.

Blockchain has disrupted businesses in various industries, from banking, prediction markets, law enforcement, healthcare, and energy management, to retail, education, Internet of things, charity, and many others. The typical intruder detection is given in Fig. 7.2.

Blockchain technologies are not just only single one technique, but also contain cryptography, mathematics, algorithm, and economic model, combining peer-to-peer networks and using distributed consensus algorithm to solve traditional distributed database synchronize problem, it is an integrated multifield infrastructure construction [2–4].

The blockchain technologies composed of six key elements. The basic feature of blockchain means that blockchain does not have to rely on centralized node anymore, the data can be recorded, stored, and updated distributedly.

7.1.2 TRANSPARENT

The data' record by blockchain system is transparent to each node, it also transparent on update the data, that is why blockchain can be trusted. Most blockchain system is open to everyone, record can be checked publicly and people can also use blockchain technologies to create any application they want.

7.1.3 AUTONOMY

Because of the base of consensus, every node on the blockchain system can transfer or update data safely, the idea is to trust form single person to the whole system, and no one can intervene it.

7.1.4 IMMUTABLE

Any records will be reserved forever, and cannot be changed unless someone can take control more than 51% node in the same time.

174 CHAPTER 7 CLOUD-BASED BLOCKCHAINING FOR ENHANCED SECURITY

7.1.5 ANONYMITY

Blockchain technologies solved the trust problem between node to node, so data transfer or even transaction can be anonymous, only need to know the person's blockchain address.

7.2 BLOCK CHAIN STRUCTURE

In general in the block, it contains main data, hash of previous block, hash of current block, timestamp, and other information.

7.2.1 MAIN DATA

Depending on what service is this blockchain applicate, for example: transaction records, bank clearing records, contract records, or Internet of Things (IOT) data record. When a transaction executed, it had been hash to a code and then broadcast to each node. Because it could be contained thousands of transaction records in each node's block, blockchain used Merkle tree function to generate a final hash value, which is also Merkle tree root. This final hash value will be recorded in block header (hash of current block), by using Merkle tree function, data transmission, and computing resources can be drastically reduced.

7.2.2 TIMESTAMP

Time of block generated.

7.2.3 TYPES OF BLOCKCHAIN

Blockchain technologies can be roughly divided into three types.

1. Public blockchain

   Everyone can check the transaction and verify it, and can also participate the process of getting consensus. Like Bitcoin and Ethereum are both public blockchains. Fig. 7.2 shows public blockchain.
2. Consortium blockchains

   It means the node that had authority can be choose in advance, usually has partnerships like business-to-business, the data in blockchain can be open or private, can be seen as Partly Decentralized. Like Hyperledger and R3CEV are both consortium blockchains.
3. Private blockchain

   Node will be restricted, not every node can participate this blockchain, has strict authority management on data access.

7.2.4 **BLOCKCHAIN SECURITY PROCESS**

Blockchain stages are organized environments where exchange information and parameters (esteem and state) are near business rationale. Blockchain exchanges are the most part dependent on cryptographic and other numerical models executed for exchanging accomplices. The most mainstream cryptographic procedure is utilized for blockchain exchanges and information is unbalanced key cryptography (likewise alluded to as open/private key cryptography). In this model, there is a couple of keys—an open key and a private key—utilized for marking the exchanges and confirming the marks in the accompanying way (Bozic et al., 2016):

Private key is utilized to create exchange advanced marks. Open key is utilized to confirm the mark created with private key. The open key might be made known to numerous clients without influencing the security of the entire exchange. The private key, then again, must be made known just to the key proprietor. Topsy-turvy cryptography ensures that the private key cannot be resolved dependent on the learning of the open key.

Cloud figuring (Wang et al.) collects huge systems of virtualized administrations: equipment assets (CPU, stockpiling, furthermore, system) and programming assets (databases, messagequeuing frameworks, checking frameworks, and load balancers). In the industry, these administrations are alluded to as "Foundation as a Service" (IaaS), "Stage as a Service (PaaS)", and "Programming as a Service" (SaaS). Cloud figuring administrations are facilitated in enormous server farms, regularly alluded to as "information ranches." In view of asset and information the executives and the related security and protection issue, we can recognize three principle kinds of cloud stages: (1) open cloud, (2) private cloud, furthermore, and (3) half breed cloud. Open clouds offer boundless access to shared information and assets for a wide gathering of clients, yet there is no certification that clients' information will be secured. Access to assets and information in private clouds is confined furthermore, every client must be approved through solid authorization what is more, confirmation methodology. Private cloud bunches are generally claimed by endeavors and work under explicit cloud models. Crossover clouds appear to be a perfect model of coordination of the numerous private clouds into a joint worldwide foundation. Such coordination is done through the upper-level open layer. The fundamental issue with that model is to achieve an understanding among private cloud suppliers to work under a brought together open cloud standard. In this manner, the "many cloud model," where the appropriated private cloud groups are associated by utilizing the standard P2P arrange is a considerably more practical situation. There are two principle strategies for combination of the cloud with blockchain stages:

1. Utilizing the cloud for the improvement of blockchain applications and supporting the joining with endeavor systems (private clouds) to encourage capacity, replication, and access to value-based information.
2. Utilizing the blockchain techniques to improve the security of undertaking, client and information the executives in the clouds.

Difficulties, uncommon conditions and primary issues related to information and clients' security, alongside the ongoing thoughts and improvements are quickly talked about in the area underneath. Cloud support for blockchain exchanges and information difficulties and prerequisites. The quantity of exchanges in blockchain systems can be tremendous. The enormous volumes of produced information need versatile information preparing administrations. Flexibility and adaptability

176 CHAPTER 7 CLOUD-BASED BLOCKCHAINING FOR ENHANCED SECURITY

are probably the most significant functionalities of the cloud frameworks to give on-request cloud assets to progressively evolving remaining task at hand.

Open clouds can offer an enormous scale system of assets accessible for the clients who pay just for the used ones. Private clouds more often than not should be upgraded for dealing with huge informational collections. From the security viewpoint, cloud frameworks can adequately conceal the physical area of information. Tuning exercises can be done persistently with insignificant effect on sent applications, which is significant for a proficient execution of the majority of the blockchain calculations. Any blockchain framework must consider information sway principles and store and procedure information just in the areas allowed by the guidelines. It implies that the cloud administration supplier enables their clients to have authority over the areas in which their information is put away and prepared.

Another significant issue in regard to blockchain systems is framework flexibility and adaptation to noncritical failure. It implies that a disappointment of any single hub in the blockchain system ought to not influence crafted by the entire framework. Cloud administrations help in such cases through the replication of information put away in information focuses and the utilization of different programming applications.

At long last, the execution of blockchain calculations in clouds may improve the security of the blockchain framework itself. Programming can be midway kept up in a dispersed cloud condition with information put away on a neighborhood information server. The ongoing instances of such effective combination of the blockchain with cloud stages are Oracle Blockchain Cloud Service venture (Prophet, 2017) and iEx.ec venture (iEx.ec, 2018). Blockchain support for the cloud clients, task, and information the board—new thoughts.

The latest mechanical advancements and anonymization of the client's data and information in the cloud environments are roused by blockchain innovations (Bozic et al., 2016). Blockchain is by all accounts a promising technique for guaranteeing namelessness in huge scale clouds, an electronic wallet for client secrecy (Park and Park, 2017). Such an electronic wallet is introduced when utilizing the blockchain innovation, and after that it must be safely erased from the framework to maintain a strategic distance from the private client's data being gotten to by outsiders.

Another new thought is to utilize blockchain confirmation of idea calculations for secure information and undertaking booking in the cloud. For instance, the "numerous clouds" model is utilized for representing the disseminated P2P cloud group engineering. Every hub in that P2P system relates to the cloud Specialist co-op (SP). The SP hub may have a complex inside engineering: one SP hub might be the Alliance Chicago Exchange (ACE) hub for the neighborhood information and computational servers (slave hubs).

The execution of the produced ideal calendar can be moreover checked by the blockchain framework all together to produce the suggestion rundown of the information stockpiling servers, cloud administrations, and cloud asset suppliers.

This is absolutely new idea, which is presently one of our research assignments in our cloud advancement work.

7.3 RELATED WORKS

A new blockchain innovation has pulled in enormous interests recently. It was initially utilized in a computerized money Bitcoin with entrancing alter obstruction and nonrepudiation properties for record stockpiling [5]. Exchanges are recorded and approved whenever without a concentrated controller. Specialists have exhibited the blockchain could be connected to different settings. For example, an open source permissioned organize Multichain has been created to guarantee high exchange throughput on a private cloud [6]. A middleware Tierion was given to transfer and distribute information records into Blockchain organize [7]. A blockchain put together naming and capacity framework with respect to top of Namecoin was proposed by Blockstack Labs in Princeton University [8]. The blockchain innovation likewise could be connected to the fields of information stockpiling for good respectability and auditability. For instance, Gyskind et al. executed a computerized access control supervisor without requiring trust in an outsider [9]. They utilized blockchain exchanges in their framework to convey get to directions and secure individual information. A design ProvChain was introduced to gather and confirm cloud information provenance by installing them in blockchain exchanges [10]. A blockchain-based database in cloud environments was proposed to give the ideal certifications on information trustworthiness, execution, and solidness [11]. These arrangements are structured explicitly for their very own application situations. Be that as it may, when applying the blockchain innovation to the safe stockpiling of VM estimations in IaaS cloud, new difficulties will emerge.

7.4 INCORPORATION OF BLOCKCHAINING WITH CLOUD ENVIRONMENT

Cloud computing offers numerous services to users on demand, and gains it significance in the current era. Securing the information stored in the cloud data center is one of the prominent issues. Sensitive information such as medical images attains much concentration regarding its security, when it is uploaded to data centers for outsourcing. Cryptographic algorithms are generally used for the encryption process, in which block cipher algorithms are frequently applied for image encryption methods. An efficient Image Encryption Method based on blockchaining in Cloud Computing as a Security Service.

To improve security in Cloud environment blockchain-based security registry is proposed.

This is done in this system with the help of three layers.

1. Layer 1: Access policy checking.
2. Layer 2: VM measurement security checking.
3. Layer 3: Sensitive and nonsensitive data separation and data package security checking process.

7.4.1 LAYER 1: ACCESS POLICY CHECKING

Access policy rules are configured to manage the type of device used to access the portal and resources. Devices that you can specify include iOS and Android mobile devices, computers that run either Windows 10 or macOS operating systems, Web Browser, Workspace ONE App, and All Device Types.

The policy rule with device type Workspace ONE App defines the access policy for launching applications from the Workspace ONE app after signing in from a device. When this rule is the first rule in the policy list, after users are authenticated, they can stay signed in to the Workspace ONE app and access their resources for up to 90 days according to the default setting.

178 CHAPTER 7 CLOUD-BASED BLOCKCHAINING FOR ENHANCED SECURITY

The policy rule with device type Web Browser defines an access policy using any kinds of web browser, regardless of device hardware types and operations' systems. Access policy checking is done in the first layer.

7.4.2 LAYER 2: VIRTUAL MACHINE MEASUREMENT SECURITY CHECKING PROCESS

Layer 2 maintains overall VM measurement. After policy checking, to enhanced security proposed work move on to blockchain-based security registry. Blockchain is a novel technology that, besides its application to cryptocurrency, features fascinating properties concerning integrity, distribution, and control of data. More specifically, we utilize so-called smart-contracts, that is programs deployed and executed autonomously on blockchain.

The blockchain-based registry offers a set of core functionalities upon which our governance proposal is built. The chain of VM measurement is checked for security purpose.

VMs are fundamental working units in IaaS cloud [12,13]. They play out the appointed errands. Along these lines the reliability of VMs is significant. It intently identifies with the security of an IaaS cloud. VM estimations are one kind of delegate confirmations. They are basic, and getting to be a somewhat engaging focus for digital assaults. The most effective method to guarantee the security over such information has dependably been the focal point of studies. The VM estimations have various structures: hash esteems, properties, logs, or test reports [2,14]. Heaps of work have been proposed to verify the capacity of the estimations. Trusted equipment units (TPM [3], SGX [15], and TrustZone [16]) and immense server farm are two well-known procedures. As the previous, the equipment helped techniques have normal alter safe capacities, so the respectability and secrecy of information can be ensured great. Regardless, the information access needs explicit directions or solicitations which cannot be prepared in parallel. Other than the inside extra rooms are constrained. These imperfections confine the utilization of confided in equipment units in VM estimations secure capacity. They cannot fulfill the necessities of adaptable information get to, parallel handling and monstrous information stockpiling. As the second system, a gigantic information focus is constrained by one single expert regularly. It is simple to concentrate on security ensures. Cryptographic devices and access control arrangements are traditional and compelling. While, they may come up short when there are inner vulnerabilities. The bargain of keys and arrangements will enable intangible infringement to information, driving unusual outcomes (obscure control). Henceforth, it is important to look for new ways to deal with secure VM estimations in IaaS cloud.

In the proposed work, VM measurement policy checking is done. The chain of VM measurement history is in blockchain-based security registry.

7.4.3 LAYER 3: SENSITIVE AND NONSENSITIVE DATA SEPARATION AND DATA PACKAGE SECURITY CHECKING PROCESS

Each of the nodes in the blockchain may receive candidate transactions submitted by end-users. These transactions are then propagated to other nodes in the working group network.

This operation, however, does not actually save the transaction in the blockchain. Subsequent to this process, mining nodes need to add the aforementioned transactions to the blockchain. Until then the committed transactions wait in the "transaction pool" (a queue). As mentioned before, the mining nodes are responsible for keeping the blockchain up-to-date by publishing freshly committed blocks. This process performs the actual operation of adding transactions to the blockchain. Thus a "block" is composed of validated transactions. To this end, the providers of transactions, who are shown in the input values of each transaction, must cryptographically sign the transaction to ensure its "legitimacy," meaning that each of them had access to the appropriate private key.

No blocks containing invalid transactions will be accepted in the blockchain. To this aim, the rest of the mining nodes in the network check the validity of each and every transaction in the published block. Once a block is created, it must be hashed. To this purpose, a 518 digest, which represents the block, will be created. The immutability of data is ensured by this method since even a change in a single bit of the block would drastically change the generated hash. In addition, a copy of the hash of every block is shared among all the nodes in order to improve security. This system prevents any change since every node can check if the hash matches.

Each block typically consists of the following components:

- The block number also known as "block height"
- The current block hash value
- The previous block hash value
- The Merkle tree root hash
- A timestamp
- The size of the block
- A list of transactions within the block

The generated hash is stored in a data structure called "Merkle tree" instead of the header of the block. The hash values of the gathered data are combined by the Merkle tree until there is a singular root called "Merkle tree root hash." The presence of transactions within blocks and their summary can be efficiently verified by means of the aforementioned root. In addition, this data structure enables the system to detect any changes to the underlying data, therefore assuring that the data sent through the network is valid.

7.5 CONCLUSION

In this chapter, we discuss about blockchain mechanism and its usage in cloud environment. To enhance security in cloud environment, we present a blockchain-based security registry. To enhance the integrity, a three-layer blockchain-based network is introduced. In the first layer, access policy is checked. In the second layer, security was provided based on overall VM measurement. Finally in the third layer, one-to-one VM-user-node and package-policy relation was done.

The future plan of improving our Mchain design is extending the current design to a general secure storage approach for more data types with a better and flexible access control. The PoW function used in current work depends on finding an eligible nonce, which is the same as Bitcoin system. It only generates tamper-resistant metadata.

REFERENCES

[1] F. Reid, M. Harrigan, An analysis of anonymity in the bitcoin system, in: Proc. Privacy Secur. Risk Trust, 2011, pp. 1318–1326.

[2] L. Chen, R. Landfermann, H. Löhr, M. Rohe, A.-R. Sadeghi, C. Stüble, A protocol for property-based attestation, in: Proc. ACM Workshop Scalable Trusted Comput. 2006, pp. 7–16.

[3] P. Fan, et al., An improved vTPM-VM live migration protocol, Wuhan Univ. J. Nat. Sci. 20 (6) (2015) 512–520.

[4] J. Chen, H. Wee, Semi-adaptive attribute-based encryption and improved delegation for Boolean formula, in: Proc. Int. Conf. Secur. Cryptogr. Netw. 2014, pp. 277–297.

[5] Bitcoin: A Peer-to-Peer Electronic Cash System. [Online]. Available from: <https://cryptohuge.company/en/bitcoin-paper.html> (accessed 25.03.18).

[6] Multichain Private Blockchain White Paper. [Online]. Available from: <http://www.multichain.com/download/MultiChainWhite-Paper.pdf> (accessed 25.03.18).

[7] Tierion Api. [Online]. Available from: <https://tierion.com/app/api> (accessed 25.03.18).

[8] M. Ali, J. Nelson, R. Shea, M.J. Freedman, Blockstack: a global naming and storage system secured by blockchains, in: Proc. USENIX Annu. Tech. Conf. (USENIX ATC), 2016, pp. 181–194.

[9] G. Zyskind, O. Nathan, A. Pentland, Decentralizing privacy: using blockchain to protect personal data, in: Proc. IEEE Secur. Privacy Workshops, May 2015, pp. 180–184.

[10] X. Liang, S. Shetty, D. Tosh, C. Kamhoua, K. Kwiat, L. Njilla, ProvChain: a blockchain-based data provenance architecture in cloud environment with enhanced privacy and availability, in: Proc. IEEE/ACM Int. Symp. Cluster, Cloud Grid Comput. 2017, pp. 468–477.

[11] E. Gaetani, L. Aniello, R. Baldoni, F. Lombardi, A. Margheri, V. Sassone, Blockchain-based database to ensure data integrity in cloud computing environments, in: Proc. Italian Conf. Cybersecur. 2017, pp. 1–10.

[12] Q. Duan, Y. Yan, A.V. Vasilakos, A survey on service-oriented network virtualization toward convergence of networking and cloud computing, IEEE Trans. Netw. Serv. Manag. 9 (4) (2012) 373–392.

[13] B. Guan, J. Wu, Y. Wang, S.U. Khan, CIVSched: a communicationaware inter-VM scheduling technique for decreased network latency between co-located VMs, IEEE Trans. Cloud Comput. 2 (3) (2014) 320–332.

[14] T. Zhang, R.B. Lee, CloudMonatt: an architecture for security health monitoring and attestation of virtual machines in cloud computing, in: Proc. ACM/IEEE 42nd Annu. Int. Symp. Comput. Archit. (ISCA), vol. 43, no. 3, 2015, pp. 362–374.

[15] A. Baumann, M. Peinado, G. Hunt, K. Zmudzinski, C.V. Rozas, M. Hoekstra, Secure execution of unmodified applications on an untrusted host, in: Proc. Symp. Oper. Syst. Princ. 2013, p. 1.

[16] S. Zhao, Q. Zhang, G. Hu, Y. Qin, D. Feng, Providing root of trust for ARM trustzone using on-chip SRAM, in: Proc. 4th Int. Workshop Trustworthy Embedded Devices, 2014, pp. 25–36.

FURTHER READING

J. Becker, D. Breuker, T. Heide, J. Holler, H.P. Rauer, R. Böhme, Can we afford integrity by proof-of-work? Scenarios inspired by the bitcoin currency, The Economics of Information Security and Privacy, Social Science Electronic, Rochester, NY, 2013, pp. 135–156.

D. Boneh, C. Gentry, A Fully Homomorphic Encryption Scheme, Ph.D. dissertation, Dept. Comput. Sci. Stanford Univ. Stanford, CA, 2009.

FURTHER READING **181**

L. Chen, M. Ryan, Attack, solution and verification for shared authorisation data in TCG TPM, in: Proc. Int. Workshop Formal Aspects Secur. Trust, 2009, pp. 201–206.

C. Chen, H. Raj, S. Saroiu, A. Wolman, cTPM: a cloud TPM for cross-device trusted applications, in: Proc. 11th USENIX Symp. Netw. Syst. Design Implement. 2014, pp. 1–16.

H. Dang, Y.L. Chong, F. Brun, E.-C. Chang, Fine-grained sharing of encrypted sensor data over cloud storage with key aggregation, IACR Cryptol. ePrint Arch. 2015.

P. Dinadayalan, S. Jegadeeswari, D. Gnanambigai, Data security issues in cloud environment and solutions, in: Proc. IEEE World Congr. Comput. Commun. Technol. February/March 2014, pp. 88–91.

I. Eyal, A.E. Gencer, E.G. Sirer, R. van Renesse, Bitcoin-NG: a scalable blockchain protocol, in: Proc. USENIX Conf. Netw. Syst. Design Implement. 2016, pp. 45–59.

J. Garay, A. Kiayias, N. Leonardos, The bitcoin backbone protocol: analysis and applications, in: Proc. Annu. Int. Conf. Theory Appl. Cryptograph. Techn. 2015, pp. 281–310.

V. Goyal, O. Pandey, A. Sahai, B. Waters, Attribute-based encryption for fine-grained access control of encrypted data, in: Proc. 13th ACM Conf. Comput. Commun. Secur. 2006, pp. 89–98.

R.M. Jogdand, R.H. Goudar, G.B. SayedPratik, B. Dhamanekar, Enabling public verifiability and availability for secure data storage in cloud computing, Evolving Syst. 6 (1) (2015) 55–65.

OpenSSL. [Online]. Available from: <https://www.openssl.org/> (accessed 25.03.18).

Python. [Online]. Available from: <https://www.python.org/> (accessed 25.03.18).

S. Rajak, A. Verma, Secure data storage in the cloud using digital signature mechanism, Int. J. Adv. Res. Comput. Eng. Technol. 1 (4) (2012) 489.

R. Saikeerthana, A. Umamakeswari, Secure data storage and data retrieval in cloud storage using cipher policy attribute based encryption, Indian J. Sci. Technol. 8 (9) (2015) 318–325.

F. Schuster et al., VC3: trustworthy data analytics in the cloud using SGX, in: Proc. IEEE Symp. Secur. Privacy, May 2015, pp. 38–54.

D.-A. Suciu, R. Sion, Droidsentry: efficient code integrity and control flow verification on trustzone devices, in: Proc. Int. Conf. Control Syst. Comput. Sci. May 2017, pp. 156–158.

Tripwire. [Online]. Available from: <https://www.ibm.com/developerworks/cn/aix/library/au-usingtripwire> (accessed 25.03.18).

L. Zhou, V. Varadharajan, M. Hitchens, Trust enhanced cryptographic role-based access control for secure cloud data storage, IEEE Trans. Inf. Forensics Secur. 10 (11) (2015) 2381–2395.

CHAPTER 8

BLOCKCHAIN INTEGRATION WITH LOW-POWER INTERNET OF THINGS DEVICES

Sandeep B. Kadam[1] and Shajimon K. John[2]

[1]*Department of Electronics & Communication Engineering, Saintgits College of Engineering, Kottukulam Hills, Pathamuttam, Kerala* [2]*Saintgits College of Engineering, Kottukulam Hills, Pathamuttam, Kerala*

8.1 INTRODUCTION

Progression in electronics and wireless communication has encouraged phenomenal development of our society. The decrease in the expense of a generation of electronic devices has prompted expansion in a number of devices, and the manner by which we collaborate with the world has changed. Internet of Things (IoT) has turned out as a lot of technologies that can sense ecological changes, work, and communicate over the Internet. IoT devices can be any smart devices, for example, door lock, wearables, or any hardware that communicates through Internet, and IoT has cleared the way to smart grids (SGs), smart homes, smart cities, astute transport frameworks, and SGs, and this is only a start of a completely digital and associated world. It is assessed that, by the year 2020, more than 20 billion IoT devices will be associated on the Internet [1].

IoT is a novel standard that is forming another time in the digital world and is changing the reason for current wireless communication. The essential idea of this thought is available on this day and age in which sensors, actuators, cell phones, and so on are communicating with one another to achieve a shared objective [2]. Every one of these devices communicates with one another by means of the Internet and consequently the name, IoT. Be as it may, the meaning of IoT isn't straight, and the definition can change depending on the setting of utilization. IoT applications create an enormous measure of data and expect power to associate all the time. Difficulties in associated devices are restricted power, arrange multifaceted nature, constrained memory, and security of data from low-power IoT devices. Heterogeneity among the devices makes it hard to incorporate and communicate with one another. Present IoT architecture is a centralized architecture, and the security of the data produced depends up on the trust of the focal hub. In the existing cloud-based IoT architecture, cloud goes about as storehouses for putting away the data. There are no current techniques to know whether the data in the cloud are manipulated/falsified/tampered by focal hubs. This issue can be wiped out by receiving decentralized architecture for interfacing IoT devices. Blockchain is a case of decentralized architecture which can guarantee security and protection to IoT data by actualizing IoT architecture depending on it [3]. This venture considers and assesses different blockchain-based IoT architecture and attempts to think of an ideal IoT-blockchain architecture that can defeat current difficulties looked by joining of both (Fig. 8.1).

184 CHAPTER 8 BLOCKCHAIN INTEGRATION WITH LOW-POWER IOT DEVICES

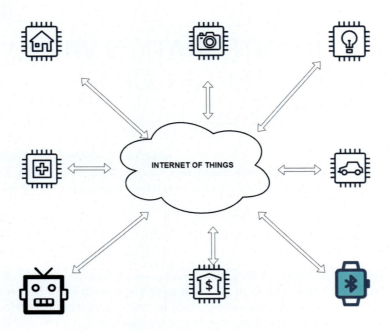

FIGURE 8.1

Internet of Things landscape [4].

The principal nature of the IoT thought is the high impact it will have on a couple of parts of normal day-to-day presence and lead to potential clients. From the point of view of a private client, the clearest effects of the IoT introduction will be discernible in both working and household fields. In this unique circumstance, domotics, e-health, and improved learning are only two or three cases of possible application circumstances, where the new perspective will accept the primary employment soon. Correspondingly, from the perspective of business clients, the most evident results will be also be observable in the fields, for instance, computerization and present-day assembling, logistics, business/process the board, and canny transportation of goods. IoT devices produce and process private delicate data, and along these lines, it is inclined to different sorts of cyberattacks. IoT devices are low-power devices that are constrained by memory and handling power. Along these lines, executing asset comprehensive security techniques are troublesome. Additionally, conventional security techniques are centralized structures that are not appropriate for IoT devices.

8.1.1 THE CONCEPTS OF INTERNET OF THINGS AND BLOCKCHAIN

These days, an IoT device can be an electronic device from a wearable to an equipment in advanced stage, and the scope of utilization where it tends to be utilized envelops numerous regions of the general public. The IoT assumes a focal job in transforming flow urban communities into smart urban areas, electrical grids into SGs, and houses into smart homes, and this is just the start. Per different research reports, the quantity of associated devices is anticipated to reach somewhere in the range of 20—50 billion by 2020 [5], primarily because of the tremendous number of devices that the IoT can put on the scene. The IoT imagines a completely associated world, where things can impart estimated information and connect with one another. This makes conceivable a computerized portrayal of this present reality through which many smart applications in an assortment of businesses can be created. These include smart homes, weareables, smart urban communities, health care, automotive, environment, smart water, smart framework, and so on. IoT arrangements are being conveyed in numerous zones, streamlining the creation and digitizing businesses. IoT applications have very specific attributes; they create huge volumes of information and require network and power for extensive stretches. This, together with the constraints in memory, PC limit, systems, and restricted power supply, represents a high number of difficulties. In the literature review, we discuss more IoT security threats, basics of blockchain, ethereum, and smart contracts.

8.1.1.1 Internet of Things security and threats

IoT security is unique in relation to classic Internet security because of the lossy low-power nature of the IoT devices, and henceforth high asset-expending encryption technique cannot be actualized. Sorts of security measures to be taken are the privacy of the data, authorization, authentication, and encryption. Kinds of dangers looked by current IoT architecture are talked about in Alaba et al. [4]. IoT nodes are powerless against listening in, hacking, and altering. IoT depends on low-power and lossy network (LLN) and has dynamic requirements, for example, handling power and memory influences the dimension of security execution. These perspectives are not considered for the standard Internet. LLNs experience mind-boggling data mishaps on account of node impersonation. For instance, during the time spent data transmission, if an attacker can connect with the system using any character, the aggressor can be acknowledged as a genuine node [6].

The security features and necessities of both the IoT and standard system devices are one of a kind. In the IoT recognition layer, sensor nodes have confined computational power and low stockpiling point of confinement, which make the repeat bouncing correspondence application and open key encryption to confirm the IoT devices incomprehensible [7]. Lightweight encryption, which incorporates lightweight cryptographic calculation, is used for the IoT devices. The IoT system has security issues, for instance, man-in-the-center and phony attacks, in the system layer. The two ambushes can get from and send fake data to imparting centers in the system [8]. Character authentication and data privacy instrument are utilized to avert unapproved nodes. At the application layer, data sharing is the main element. Data sharing makes security issues in data privacy, get to control, and divulgence of information [8]. The security necessities for the application layer include authentication, key understanding, and insurance of client privacy crosswise over heterogeneous systems.

The regular security architecture is planned depending on the viewpoint of clients and not appropriate for communication among machines. The security issues in the two systems might be comparative; however, various methodologies and techniques are utilized in handling each system security issue [9]. The IoT communication includes data exchange/sharing among the IoT devices or between different IoT layers. With the colossal capability of the IoT in various domains, the entire IoT communication framework is inconsistent from the security viewpoint and powerless against insurance disaster from the point of perspective on end clients [10]. The IoT communication

186 CHAPTER 8 BLOCKCHAIN INTEGRATION WITH LOW-POWER IOT DEVICES

medium fills in as a decision point for attackers. The potential attacks in the channel are depicted as follows.

Man-in-the-middle attacks

In cryptography and PC security, a man-in-the-middle attack (MitM) is an attack where the attacker furtively transfers and perhaps modifies the correspondences between two gatherings who trust they are legitimately communicating with one another. Attacks like MitM must be deflected to keep up information uprightness in the midst of a discussion. In MitM, an aggressor unobtrusively transmits, and no doubt changes the correspondence between the two IoT devices that clearly talk with each other. Reliable data, for instance, the health status of a patient, billing data of SGs, or even secret keys of smart doors, can be manufactured and balanced by an attacker with MitM, along these lines causing genuine security issues [11]. MitM attacks speak to a genuine risk to the IoT security, since they outfit an attacker with the capacity to seize and control a communication channel. Thus, aggressors can get to sensitive data, logically communication among centers and secure command over the channel. The attacker by then cases a relationship with the genuine center point and goes about as a go between to scrutinize, occupy, install, and change the traffic between the client and the authentic node. For instance, an assailant may need to fake the temperature information from an observing gadget inside the IoT to encourage the gadget to overheat, which can keep the gadget from working. This action can make load the gadget and can in like manner make brief physical mischief and cash-related disasters [12].

Eavesdropping

Eavesdropping is a block endeavor of information between two communicating nodes. Eavesdropping occurs on the system layer in the IoT and shows up as data sniffing. A particular program is utilized for sniffing and recording bundles from the system layer, which are thusly fixed on or perused using cryptographic instruments for examination and decoding. Privacy is used as a strategy for giving viable access control and security against eavesdropping in the midst of data correspondence [13]. Eavesdropping in like manner presents unique troubles to the IoT architecture, particularly when an attacker centers on the correspondence channels to expel data from the stream information. This assault type is performed by listening clearly to the message or data sniffing [14]. Therefore, MitM and eavesdropping attacks in the IoT occur among dynamic sensor nodes that don't require agave concentrated server, not in any manner like the standard system where a submitted server is used for traffic control and observing [15].

Data privacy

Data privacy is the capacity to control the information one uncovers about oneself over the Internet, and who can get to that information has turned into a growing concern. These worries include whether e-mail can be put away or perused by outsiders without assent, or whether outsiders can continue to follow the sites that somebody has visited. Another worry is if the sites that are visited can gather, store, and perhaps share by and by recognizable information about clients. The clients' privacy and trust must be guaranteed for the IoT to be totally sent and completely acknowledged. Data privacy and characterization for business frameworks are up 'til now essential issues, and finding down to business courses of action stay testing [16]. Client data privacy must be guaranteed in light of the way that clients require the most extraordinary affirmation for

8.1 INTRODUCTION **187**

their very own information. Trust includes safeguarding of client privacy, which incorporates individual client data, by the methodology and prospect of clients in a versatile manner. Transmitting and figuring trust among different nodes in a heterogeneous IoT is a troublesome issue in light of the fact that various system nodes have distinctive trust criteria [17]. The security administrations given by IEEE 802.15.4 are data legitimacy, data order, and replay affirmation. The principle risks to this show are encrypted ACK frames, NO arranged edge counters, and NULL security level. Exactly when the ACK casing is unencrypted, an interloper can block an MAC casing and style an ACK outline with an arrangement number that brings about casing mishap with no retransmission [18].

8.1.2 THE CONCEPT OF BLOCKCHAIN

The issue of trust in data frameworks is very perplexing when no verification nor review systems are given, particularly when they need to manage delicate data, for example, monetary exchanges with virtual monetary forms. Blockchain is an instrument that enables exchanges to be verified by a gathering of problematic on-screen characters. It gives a dispersed, changeless, straightforward, secure, and auditable record. The blockchain can be counseled straightforwardly and completely, enabling access to all exchanges that have happened since the first exchange of the framework and can be verified and gathered by any substance whenever. The blockchain protocol structures data in a chain of blocks, where each block stores a lot of Bitcoin exchanges performed at a given time. Blocks are connected together by a reference to the past block, shaping a chain. Blockchain is a digitized ledger, which can store data that are cryptographically hashed. Blockchain is appropriated to all of the nodes in the system which means a comparative copy of data is secured in all nodes [8]. Data are secured in blocks that are cryptographically hashed with past block. These two properties of blockchain make it permanent and invulnerable to single point disappointment.

The routing capacity is important to take part in the peer-to-peer network, and this includes transaction and block spread. The capacity is in charge of keeping a duplicate of the chain in the node (the whole chain for full nodes, and just a piece of it for light nodes). Wallet administrations give security keys that enable clients to arrange transactions, that is, to work with their Bitcoins. Finally, the mining capacity is in charge of creating new blocks by solving the proof of work (POW). The nodes that play out the POW (or mining) are known as miners, and they get recently produced bitcoins, and expenses, as a reward. The idea of POW is one of the keys to empowering trustless consensus in the blockchain network. The POW comprises of a computationally intensive errand that is essential for the age of blocks. This work must be intricate to understand and in the meantime effectively verifiable once finished.

When a miner finishes the POW, it distributes the new block in the network, and the remainder of the network verifies its legitimacy before adding it to the chain. Since the generation of blocks is completed simultaneously in the network, the blockchain may briefly fork in various branches (created by various miners). This inconsistency is unraveled by considering that the longest part of blocks is the one that will be considered as substantial. This, together with the intensive idea of the blockage procedure, gives a novel, circulated trustless-consensus instrument. It is in all respects computationally costly for a noxious attacker to adjust a block and degenerate the blockchain, since the remainder of the believed miners would beat the attacker in the blockage process, and therefore the believed part of blocks will invalidate the one created by the attacker. In specialized terms, in a request for a manipulated block to be effectively added to the chain, it is important to fathom the POW quicker than the remainder of the network, which is computationally too costly, and it requires having control of at any rate 51% of the computing assets in the network. Because of the enormous computational limit expected to change the blockchain, the defilement of its blocks is essentially inconceivable. This implies, regardless of whether the members are not totally legit about the utilization of Bitcoin, a consensus has constantly come to in the network as long as the majority of the network is framed by honest members. The arrangement proposed by Nakamoto was an incredible insurgency in the unwavering quality of inconsistent entertainers in decentralized frameworks.

More insights regarding blockchain engineering can be found in Zheng et al. [19]. Blockchain on a theoretical measurement is an associated linked rundown, where each rundown is associated with the past rundown. In the blockchain, the linked rundown is called as a block, since it contains different information. Blockchain is realized either in structure data type or in JSON format. Blockchain was first used in bitcoin by Satoshi Nakamoto in 2008 [20]. Blockchain innovation has gone mainstream in an assortment of ventures, for example, finance, insurance, logistics, and farming. With its ability to digitize trades effectively and proficiently, this innovation is promising an important change in standpoint in making a couple of methodologies progressively thin, faster, and increasingly direct. From the abnormal state point of view, blockchains have an overwhelming usage of cryptography to give "trustless" systems without united specialists so that data executing nodes can accomplish faster trade-off. Since the trademark features of blockchains set out the foundations of serverless record keeping, a couple of researchers are endeavoring tries to utilize blockchains to decentralize IoT correspondences and to discard the necessity for bound together trusted in specialists [2].

The likelihood of a blockchain-based IoT has gathered extensive research eagerness, since decentralizing the IoT through blockchains has the accompanying potential points of interest:

- The move from centralized to blockchain-based IoT improves adjustment to internal disappointment and ousts singular points of disappointments. It moreover neutralizes the bottleneck that was inborn in a developing IoT subject to centralized servers [21]. A decentralized texture for taking consideration of IoT data also dodges untouchable substances to control the individual data of IoT clients.
- A decentralized shared system architecture empowers IoT gadget self-administration, and start to finish correspondences don't have to encounter a centralized server for performing robotization administrations. Individuals in blockchain systems can affirm the trustworthiness of the data they are sent, similarly as the character of the sending part. The safe, painstakingly structured limit in blockchains moreover empower sending secure software updates to IoT contraptions.
- Since no single component controls the substance of a blockchain, IoT data and occasion logs set away on the blockchain are permanent; thus, there is guaranteed obligation and detectability. Unwavering quality and trustless IoT participations are an essential responsibility of blockchains to the IoT.
- Blockchains offer the convenience of programmable rationale through smart contracts [22] and can treat IoT communications as exchanges. They can help perform security functions such as access control, arrangement, and affirmation to improve the security in a blockchain-based IoT.

- Blockchains open up doors for an IoT biological system, where administrations can be adjusted in an extremely just manner. The trustless system state of blockchains licenses secures little scale trades for IoT administrations and data.

Blockchain has additionally given an innovation, where the idea of a smart contract can emerge. When all is said in done terms, a smart contract alludes to the PC protocols or on the other hand programs that enable an agreement to be consequently executed/implemented, taking into the record a lot of predefined conditions. For instance, smart contracts define the application rationale that will be executed at whatever point a transaction happens in the trading of cryptographic money. In smart contracts, capacities and conditions can be defined past the trading of digital forms of money, for example, the approval of resources in a certain scope of transactions with nonmoney-related components which make it an ideal part to extend blockchain innovation to other regions. Ethereum [23] was one of the pioneer blockchains to include smart contracts. Today, smart contracts have been included in most of existing blockchain usage, for example, Hyperledger [24], a blockchain intended for organizations that enables parts to be conveyed according to the requirements of clients (smart contracts, administrations, or conferences among others) with the help of huge organizations, for example, IBM, JP Morgan, and Intel.

8.1.3 FEATURES OF BLOCKCHAIN

The most noteworthy features that change the blockchain innovation into something with the capability of definitely reshape a couple of undertakings are:

- *Decentralization:* In centralized system frameworks, data exchanges (i.e., the trades) are endorsed and affirmed by trusted in central outcast substances. This gain costs the extent that centralized server support, similar to execution cost bottlenecks. In blockchain-based establishments, two nodes can participate in trades with each other without the need to place trust upon a central component to keep up records or perform endorsement.
- *Immutability:* Since each and every new area made in the blockchain is settled upon by partners by methods for decentralized consensus, the blockchain is without oversight and is about hard to modify. Basically, all as of late-held records in the blockchain are in like manner permanent, and, in order to change any past records, an attacker would need to bargain a larger piece of the nodes related to the blockchain organize. Something else, any modifications in the blockchain substance are successfully distinguished.
- *Auditability:* All peers hold a copy of the blockchain and would in this way have the option to get to all timestamped exchange records. This straightforwardness empowers peers to look upward and affirm exchanges, including unequivocal blockchain addresses. Blockchain addresses are not related to characters, everything considered, so the blockchain gives a method for pseudo-namelessness. While records of a blockchain address can't be followed back to the owner, unequivocal blockchain addresses can definitely be viewed as mindful, and surmising can be made on the exchanges a specific blockchain address participates in.
- *Fault tolerance:* All blockchain peers contain indistinguishable copies of the records. Any faults or data spillages that occur in the blockchain system can be distinguished through decentralized consensus, and data spillages can be mitigated utilizing the multiplications set away in blockchain peers.

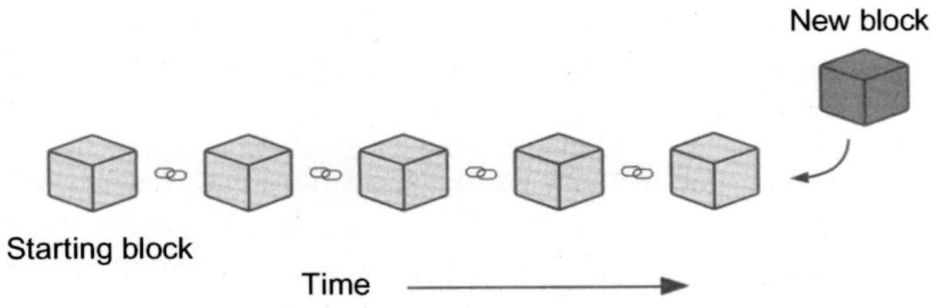

190 **CHAPTER 8** BLOCKCHAIN INTEGRATION WITH LOW-POWER IOT DEVICES

FIGURE 8.2

Illustration of blockchain.

8.1.4 STRUCTURE OF BLOCKCHAIN

Blockchain comprises of blocks that contain lists of transactions that have been generated in the network. Transaction can be any information about the exchange of any type of data. A block has a header and a body. Body of the block contains information regarding transactions. Header contains information that binds the block with other previous blocks, such as the hash of previous block and other information. Mainly in the block, it contains main data, a hash of past block, hash of current block, timestamp, and other data. Fig. 8.2 shows the structure of a block [25].

Main data: Contingent upon what administration is this blockchain applicate, for instance, exchange records, bank clearing records, contract records, or IoT data record [25].

Hash: At the point when an exchange was executed, it had been hash to code and after that communicate with each node. Since it could be contained, a considerable number of exchange records in each center point's block, the blockchain used Merkle tree function to create last hash esteem, which is moreover Merkle tree root. This last hash esteem will be recorded in the block header (hash of current block) by utilizing Merkle tree function, and data transmission and figuring resources can be drastically diminished [25] (Fig. 8.3).

Timestamp: Time of block created.

Other Information: Like the mark of the block, Nonce esteem or other data that client portrays.

8.1.5 ELEMENTS IN A BLOCKCHAIN

There are mainly four components in a complete blockchain infrastructure, and they are network of nodes, distributed database, shared ledger, and cryptography.

- *Network of nodes:* All nodes are connected together to form a network that validates and maintains all the transactions made in the network. Consensus protocol validates each transaction, and this eliminates the involvement of trusted third party. When the transactions are validated, they are recorded in the blockchain, and the process of recording into the blockchain is called as mining of the blocks. Block in blockchain can have zero or multiple number of transactions. Block generation in some blockchains are time based, that is, blocks are generated in particular interval of time, and all the transactions validated from the time previous block was generated to the current time are grouped together to form a block. If no transaction is being validated during this time, then the block will have no transactions in it.

8.1 INTRODUCTION

Hash of current block 1	Hash of previous block 2	Timestamp	Other information
Main data 1 ... Main data N			

Hash of current block 2	Hash of previous block 3	Timestamp	Other information
Main data 2 ... Main data N			

FIGURE 8.3

Structure of blockchain.

- *Distributed database:* Blockchain is a network and a database. It stores the data generated by the network. Unlike other central databases, blockchain resides in every node of the blockchain network. So, the copy of the blockchain is with every node. Each block will have list of transactions, timestamp, and information about the previous block. Some blockchains will have additional fields that help in reaching consensus. This linkage to the previous block makes blockchain immutable. If data in a block are tampered, then the hash of the block changes which in turn will make all other blocks after the tampered block invalid.
- *Shared ledger:* The ledger is shared among all nodes. If an attacker tries to tamper with the blockchain, then the attacker needs to make the change in every single node in the network.

This makes attacking difficult. To do so, the attacker needs the control at least 51% of the nodes in the network.
- *Cryptography:* Data in blockchain are cryptographically hashed. Hash function is a one-way function which means that hash can be generated from the plain text, but deriving the plain text from the hash is extremely difficult. Thus, tracking of information and unauthorized tampering of data cannot be done.

8.1.6 TYPES OF BLOCKCHAIN

Blockchain technologies can be generally partitioned into three sorts.

8.1.6.1 Public blockchain

A public blockchain network is totally open, and anybody can take part in the network by joining it. The network normally has a reward system to urge more members to join the network [26]. Everyone can check the transaction and confirm it and can likewise take part the way toward getting consensus. Bitcoin and Ethereum are both public blockchain. Fig. 8.4 demonstrates public blockchain.

Advantages of public blockchains are:

- *Trust:* The target of public blockchains since the beginning was to eliminate mediators of any structure, and all the more critically, it looks to evacuate the trusts put on them. Truth be told, members in the network shouldn't have to confide in one another for the network for

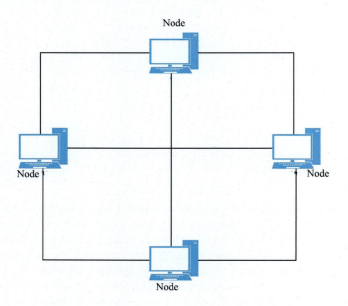

FIGURE 8.4

Public blockchain [2].

transactions to be handled and verified. Public blockchains are trustless, since everybody is boosted to make the best decision for the improvement in the network [27].
- *Transparent and open:* All information inside transactions are open for the public to check. The straightforwardness of a public blockchain is maybe a noteworthy element that pulls in a wide cluster of utilization cases, from casting a ballot to money-related transactions. Moreover, the legitimacy of transactions and the information can be confirmed by anyone in the network [27].
- *Secure:* The blockchain will be more secure with the increase in active participants and greater decentralization. With more number of nodes in the network, it will be a lot harder for any hacker to attack the environment. In a public blockchain network, anybody can become a full node or miner and add to the security of the framework. It is for all intents and purposes unimaginable for hackers to conspire and work together to oversee the agreement network [27].

8.1.6.2 Consortium blockchain

Consortium blockchains contrast to their public partner in that they are permissioned, and in this manner, anybody with a Web association could access a consortium blockchain. These kinds of blockchains could likewise be portrayed as being semi-decentralized. Command over a consortium blockchain isn't conceded to a solitary substance but instead a gathering of affirmed people. With a consortium blockchain, the agreement procedure is probably going to vary to that of a public blockchain. Rather than anybody having the option to share in the methodology, agreement members of a consortium blockchain are probably going to be a gathering of pre-endorsed nodes on the network. In this manner, consortium blockchains have the security includes that are inalienable in public blockchains, while likewise considering a more prominent level of authority over the network [28]. It implies the node that had specialist can be picked ahead of time, for the most part has organizations like business to business, and the data in a blockchain can be open or private and can be viewed as partly decentralized. Hyperledger, Corda [29], and R3CEV are some consortium blockchains. Fig. 8.5 demonstrates consortium blockchains.

8.1.6.3 Private blockchain

A private blockchain network requires a request and should be approved by either the network originator or the rules set up by the network originator. Organizations who set up a private blockchain will have a permissioned network. This makes confinements on who is permitted to partake in the network and some specific transactions. Members need to get an invitation or authorization to join. The entrance control system could fluctuate: Existing members could choose future participants; an administrative expert could issue licenses for interest; or a consortium could settle on the choices. When a node has joined the network, it will maintain the blockchain in a decentralized way. All nodes will take part in data transfer but only a few nodes will have special permission and have special access such as writing on the blockchain. Fig. 8.6 demonstrates private blockchain.

Advantages of private blockchain are:

- *Faster:* Private blockchains process higher transactions per second (TPS) when contrasted with public blockchains, since the presence of a limited number of approved members brings about essentially lesser occasions in securing an agreement for the network. This enables more transactions to be handled for each block; private blockchains can process thousands or even a huge number of TPS, contrasting with Bitcoin's seven TPS [27].

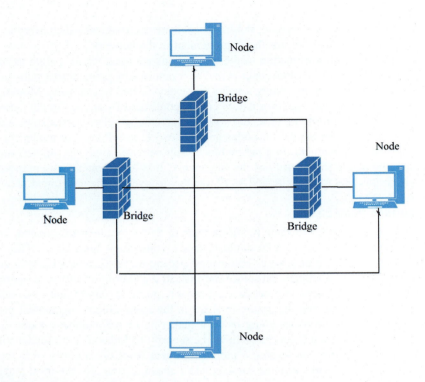

FIGURE 8.5

Consortium blockchain [2].

- *Scalable:* Since just a couple of nodes are approved and mindful to deal with the information, the network can support and process a lot of higher transactions. Not at all like a decentralized framework where accomplishing accord could require some investment, the basic leadership process in a private network is progressively incorporated and hence a lot more quicker. A similarity is that it requires some investment for 100 instructors to check your test paper (and concur that it is right/wrong) when contrasted with a solitary educator stamping it [27].

Disadvantages of private blockchains are as follows:

- Consensus algorithm is hard coded into the system, and validation of transaction uses trust-based model [30]. This limits the scalability of the network after deployment, and since the validation of transactions utilize trust among the nodes, this brings possibility of inside attacks within the network.
- Use of nonstandard programming languages for writing smart contracts makes it difficult to adopt. This creates confusion and errors among programmers. The open-source community working on Hyperledger is still small and inadequate documentation compared to other counterparts, and this makes deployment harder.
- Only deterministic transactions are supported by Hyperledger, and this makes it difficult in wide-spread adoption of the technology in IoT. IoT devices generate wide range of data set, and

8.1 INTRODUCTION

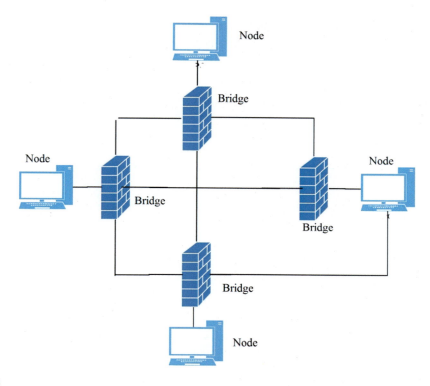

FIGURE 8.6

Private blockchain [2].

each device generate data different to other devices due to the various standards and protocols followed by manufacturers and programmers.
- The sequential execution of all transactions by all companions limit execution, and complex measures are expected to avoid denial-of-service assaults against the stage starting from untrusted contracts (Table 8.1).

Regardless of what kind of blockchain is, the two have an advantage. Sometimes we need public blockchain on the grounds of its accommodation; however, sometimes we possibly need private control like consortium blockchains or private blockchain, contingent upon what administration we offer or what spot we use it.

8.1.7 CONSENSUS

In blockchain, how to achieve consensus among the dishonest hubs is a change in the Byzantine Generals (BG) Problem, which was brought up in Lamport et al. [31]. In BG issue, a gathering of generals who order a bit of Byzantine armed force circle the city. A few generals like to assault, while different generals want to withdraw. Be that as it may, the assault would fall flat if just piece

196 CHAPTER 8 BLOCKCHAIN INTEGRATION WITH LOW-POWER IOT DEVICES

Table 8.1 Difference Between Public and Private Blockchains

Parameter	Public Blockchain	Private Blockchain
Read permission	All transactions are visible to public	Depends upon the blockchain architecture
Immutability	Tampering is difficult, as information is with all the nodes	Can be tampered
Efficiency	Latency is high	Low latency
Centralization	Decentralized	Partly or fully centralized depending upon the blockchain architecture
Consensus process	Everyone in the blockchain takes part in consensus process	Only permissioned nodes take part in consensus process

of the generals assault the city. Along these lines, they need to achieve a consent to assault or withdraw. Step-by-step instructions to achieve a consensus in conveyed condition is a test. It is likewise a test for blockchain, as the blockchain network is circulated. In blockchain, there is no focal hub that guarantees records on disseminated hubs are all the equivalent. A few protocols are expected to guarantee records in various hubs are steady.

8.1.7.1 Proof of work

Proof of work is a concord that has the principle objective of stopping cyberattacks, for example, an appropriated denial-of-service attack that has the motivation of debilitating the assets of a PC framework by sending different phony solicitations. The proof-of-work idea existed even before bitcoin; however, Satoshi Nakamoto connected this strategy to his or her digital currency changing the manner in which conventional transactions are set. The term "proof of work" was coined by Markus Jakobsson and Ari Juels in Jakobsson and Juels [32] in 1999 but idea of POW was first published by Cynthia Dwork and Moni Naor in Dwork and Naor [33] in 1993.

Bitcoin network uses POW. In bitcoin network, nodes compete with each other to validate transaction, or the miner can be chosen randomly. In POW, nodes solve a cryptographical puzzle and whoever solves it first gets the opportunity to add the block into blockchain. PoW is required for mining, which is a computational intensive task, usually a mathematical puzzle, in order to create a block of transactions [34]. In bitcoin network, every block will have a nounce. Nounce is a variable in the block that is used during mining process. All nodes compete against each other to find the hash of the block to be mined. The network sets a threshold value for the hash. Whichever node gets the least hash closer to the threshold value will get to add the block. To get the desired value of hash, nounce in the block is changed. There is a possibility of occurrence of side chains when two nodes mine the block at the same time. In such situation, the side chain with most number of blocks is chosen by the network. Mining using POW verifies the transactions, avoids double spending, and rewards miner with new coins by generating them during the process.

8.1.7.2 Proof of stake

While the general procedure continues same as of POW, the technique for achieving the true objective is altogether different, and consensus is achieved virtually in proof of stake (POS). In POW, miners unravel cryptographically hard riddles by utilizing their computational assets. In POS, rather than miners, there are validators. The validators deposit a portion of their crypto asset as a stake in the environment. Following that, the validators wager on the blocks that they feel will be added by the chain. At the point when the block gets included, the validators get a square reward in extent to their stake.

It is a vitality-sparing option in contrast to POW. Miners in POS need to demonstrate the responsibility for measure of cash. It is accepted that individuals with more monetary forms would be more averse to assault the network. The determination dependent on record parity is very unreasonable in light of the fact that the single most extravagant individual will undoubtedly be prevailing in the network. Therefore, numerous arrangements are proposed with the mix of the stake size to choose which one to manufacture the following block. Specifically, Blackcoin [35] utilizes randomization to foresee the following generator. It utilizes an equation that searches for the most reduced hash an incentive in mix with the size of the stake. Peercoin [36] favors coin age−based determination. In Peercoin, more seasoned and bigger arrangements of coins have a more noteworthy likelihood of mining the following block. Contrasting with POW, POS spares more vitality and is progressively powerful. Lamentably, as the mining cost is about zero, assaults may come as a result. Numerous blockchains receive POW toward the start and change to POS progressively. For example, ethereum is planning to move from Ethash (a sort of POW) [37] to Casper (a sort of POS) [38].

PoS is a safer alternative to PoW. It is energy efficient and reduces the chances of 51% attacks. In order to hack in the network, the attacker needs to possess 51% of the total coins in the network which is more expensive for the attacker.

8.1.7.3 Practical byzantine fault tolerance

Practical byzantine fault tolerance (PBFT) is a consensus algorithm used when there exists a partial trust between nodes. It is a replication algorithm to endure byzantine faults [39]. Byzantine fault tolerance (BFT) is the component of a network to achieve accord notwithstanding when a portion of the nodes in the system neglects to react or reacts with false data. The goal of a BFT component is to protect against the framework failures by utilizing aggregate decision-making, both right and flawed nodes, which intends to diminish to impact of the broken nodes.

Byzantine general's problem can be explained as follows. Envision that outside an adversary city, few divisions of the Byzantine armed force are staying outdoors, and every division is instructed by its very own general. Generals can speak with each other just by emissary. Subsequent to watching the foe, they should choose a typical game plan. In any case, a portion of the generals might be swindlers, attempting to keep the faithful generals from achieving an understanding. The generals must settle on when to assault the city; however, they need a solid dominant part of their military to assault simultaneously. The generals must have a calculation to ensure that every single steadfast general settles on a similar game plan, and few tricksters can't make the dependable generals receive an awful arrangement. The dedicated generals will all do what the calculation says they should, yet the tricksters may do anything they wish. The calculation must ensure a condition paying little heed to what the tricksters do. The dependable generals ought to achieve understanding yet ought to concur upon a sensible arrangement [31].

Hyperledger fabric uses the PBFT as its consensus algorithm, since PBFT could deal with up to one-third vindictive byzantine copies. Another block is resolved in a round. In each cycle, an essential would be chosen by certain guidelines. Furthermore, it is in charge of requesting the

198 CHAPTER 8 BLOCKCHAIN INTEGRATION WITH LOW-POWER IOT DEVICES

transaction. The entire procedure could be isolated into three stages: prearranged, arranged, and submit. In each stage, a hub would enter next stage on the off chance that it has gotten cast a ballot from more than two-third of all things considered. So PBFT necessitates that each hub is known to the network. Like PBFT, Stellar Consensus Protocol (SCP) [40] is likewise a Byzantine understanding convention. In PBFT, every hub needs to question different hubs, while SCP gives members the privilege to pick which set of different members to accept.

8.2 BLOCKCHAIN FOR INTERNET OF THINGS APPLICATIONS

Blockchain was initially utilized for financial transactions, where transactions are encoded and kept by all members (e.g., Bitcoins and different cryptographic forms of money). Blockchain can be connected to improve IoT security. As we would like to think, Blockchain is the missing bit of a riddle to unravel security and dependability defects in IoT. Since blockchain is decentralized, self-ruling, and trustless, it appropriate to be connected in a few distinct situations such as smart homes and smart transportation systems. For instance, the blockchain could keep an unchanging history of smart gadgets. In addition, it might empower a self-governing and eliminate central authority by the utilization of smart contracts. Moreover, blockchain can likewise make a protected route for smart gadgets trading messages with one another [41].

Hyperledger, IOTA, and Ethereum are three open-source implementations of distributed records. They all offer the major qualities of connecting among blocks, cryptographically secure for hashing of different transactions minimally in single blocks, hilter kilter cryptography, advanced marks, consensus, and smart contracts; the particular individual implementations change altogether. Moreover, the degree of relevance to IoT additionally contrasts among the three in how perspectives especially smart contracts are connected.

8.2.1 HYPERLEDGER

Hyperledger is a permissioned blockchain: It extraordinarily applies access control, chaincode-based smart contracts, variable consensus with a present implementation of viable byzantine fault tolerance (PBFT), and incorporates stays of trust to root testament experts as an improvement to the unbalanced cryptography and computerized mark highlights with SHA3 and ECDSA [42]. While most distributed records empower open enlistment, the permissioned parts of Hyperledger upgrade security by methods for avoiding Sybil assaults, an assault wherein consensus could possibly be undermined by a noxious substance making and selecting ill-conceived peers—or Sybil personalities—to antagonistically influence the network [43]. Besides, its implementation of smart contracts includes the chaincode implementation that can self-execute conditions, for example, resource or asset moves among peers in several milliseconds [44—46]. This inertness is low among relatively distributed record implementations. Hyperledger use of PBFT averts the probabilistic and computationally costly mining of hashes as in confirmation of work; be that as it may, it exchanges off prompt computational overhead with network usage. Numerous approving peers have communicated the transaction and merge upon a deterministic execution and, in this manner, a similar block [42]. To achieve this, the approving peers should likewise intercommunicate, causing more network

8.2 BLOCKCHAIN FOR INTERNET OF THINGS APPLICATIONS 199

overhead. In an IoT setting, the scale at which network use increments with the number of gadgets on a network must be examined and estimated further. Generally speaking, between applying chaincode smart contracts and a novel PBFT implementation that counterbalances computational overhead for expanded networking among peers, Hyperledger offers strong flexibility for IoT applications.

Hyperledger Fabric is the most widely used and first private blockchain. It is an open-source and modular system. Linux foundation hosts Hyperledger projects (www.hyperledger.org). It provides custom-use cases and modular consensus and is written in general-purpose languages. Hyperledger network does not rely on cryptocurrency to operate. Fabric is a permissioned model utilizing a versatile idea of enrollment, which might be coordinated with industry standard management. To help such adaptability, fabric presents a completely novel blockchain structure and patches up the way blockchains adapt to indeterminism, asset depletion, and execution assaults.

8.2.2 IOTA

IOTA is a one of a kind distributed record in that it doesn't use an express blockchain by any means; rather, it executes a coordinated acyclic chart of transactions — rather than blocks of numerous transactions that connects together, every transaction affirms and connects back to two different transactions. This is expected to be particularly lightweight, as consensus does not require a few peers intercommunicating or debilitating computational exertion approving increments to the coordinated acyclic diagram; rather, two transactions can be approved by single peers submitting a transaction themselves. This limits transaction time and overhead; nonetheless, this presents a couple of assault situations. With a few of Sybil personalities and enough processing force, an aggressor could possibly arrange concurrent, clashing transactions on the double [47]. The lightweight nature and one of a kind utilization of coordinated acyclic graphs make it among the most adaptable implementations of a distributed record; nonetheless, there are a few trade-offs. The lightweight and adaptable nature of IOTA counterbalance both computational and network overhead, making it particularly extensible toward a scope of IoT applications with certain trade-offs. A few components of heartiness, for example, an inherent smart contract system, are prohibited to keep IOTA as lightweight as could be expected under the circumstances. This restrains the full scope of distributed record abilities that could stretch out to IoT; moreover, because IOTA is among the latest of rising implementations, its inactivity and versatility still can't seem to be tried at scale. IOTA might be more up to date and not as hearty as other distributed records; be that as it may, its remarkable use of coordinated acyclic graphs and its lightweight methods for engendering transactions can possibly make it among the quickest IoT implementations.

8.2.3 ETHEREUM

Ethereum, albeit overwhelmingly proposed for cryptocurrency, is a versatile blockchain implementation with the implementation of smart contracts and a subsidiary of verification-of-work consensus known as Ethash. This likewise applies guided acyclic graphs to oversee probabilistic hash age to such an extent that it keeps potential maltreatment from particular hardware, and those other confirmation-of-work calculations are defenseless against Wood [37]. Notwithstanding actualizing smart contracts, Ethereum transactions can likewise store custom data. This expands the potential

200 CHAPTER 8 BLOCKCHAIN INTEGRATION WITH LOW-POWER IOT DEVICES

Table 8.2 Comparison of Hyperledger, IOTA, and Ethereum

Distributed Ledger Technology	Consensus Algorithm	Transaction Time	Smart Contracts	Other Characteristics
Ethereum	PoW	10−15 s	Yes	Less network usage and computationally heavy
Hyperledger	PBFT	0.05−100 ms	Yes	Network intensive but less computational cost
IOTA	N/A	120 s	No	Less network usage and computational cost

for auditability and unchanging nature of IoT data past cryptocurrency transactions. This permits strong extensibility for IoT applications with some presentation trade-offs. Due to Ethash being founded on evidence of work, Ethereum may require between 10 and 20 seconds to deliver a block. High recurrence and time-touchy IoT gadget tasks may not bolster such postponements [48]. While Ethash keeps maltreatment from potential specific hardware, it doesn't really improve fault tolerance. At scale, IoT gadgets would need to depend on trusted and computationally incredible peers to guarantee taking care of fault. Capacity additionally exhibits another issue; Ethereum requires all peers to store a blockchain that is likewise in the request of many gigabytes huge. The IoT gadgets will either need to intercommunicate with a proxy server that goes about as a friend in the Ethereum network or oblige huge capacity. Ethereum, in dynamic use as a cryptocurrency longer than most distributed record implementations, does as of now additionally have IoT prototypes. For instance, Ethereum is handling tokens and contracts for electronic lock sharing and supply chain affirmation prototypes [48]. To wind up down to earth, individual IoT gadget execution with Ethereum would be alluring in any case; moreover, smart contract security and development ought to be reassessed, as it has experienced huge changes since a robbery of assets because of a smart contract implementation imperfection [49]. By the by, Ethereum flaunts common sense, client acknowledgment, and the capacity to apply smart contracts—all of which stretch out toward strength in IoT potential (Table 8.2).

8.2.4 CHALLENGES IN INTEGRATING BLOCKCHAIN WITH INTERNET OF THINGS

The main challenge in integrating blockchain with IoT is the processing capabilities of end IoT devices. End IoT devices are low-power devices and cannot handle the processing power and memory capacity to handle blockchain. For comparison, current size of bitcoin blockchain is 120 GB, and a block in bitcoin network is around 1MB and contains roughly 1600 transactions [50]. The IoT devices have memory capacity of few megabytes maximum, and thus we can see that IoT devices cannot act as a full node or miner. To deal with the expanded traffic brought about by more clients and more transactions, more nodes are expected to process them along these lines, expanding the expense. Different systems are proposed to address the problem of adaptability such as Segwit, square size increment, Sharding, POS, off-chain state channels, and plasma. A ton of research should be done to propose a complete arrangement [41]. Given the fact that IoT ecosystems are very diverse and comprised of devices that have very different computing capabilities, and

8.3 PROPOSED ARCHITECTURE **201**

not all of them will be capable of running the same encryption algorithms at the desired speed, the time required to perform encryption algorithms for all the objects involved in blockchain-based IoT ecosystem are different.

8.2.5 RISKS IN INTERNET OF THINGS BLOCKCHAIN INTEGRATION

Vendor Risks: Practically speaking, most present organizations, looking to deploy blockchain-based applications, lack the required technical skills and expertise to design and deploy a blockchain-based system and implement smart contracts completely in-house, that is, without reaching out for vendors of blockchain applications. The value of these applications is only as strong as the credibility of the vendors providing them. Given the fact that the Blockchain-as-a-Service market is still a developing market, a business should meticulously select a vendor that can perfectly sculpture applications that appropriately address the risks that are associated with the blockchain.

Credential security: Although the blockchain is known for its high security levels, a blockchain-based framework is just as secure as the framework's passage. When considering a public blockchain-based framework, any individual approaches the private key of a given client, which empowers him or her to "sign" transactions on the public record, will successfully turn into that client in light of the fact that most present frameworks don't give multifaceted validation. Additionally, loss of a record's private keys can prompt total loss of assets, or information, constrained by this record; this risk ought to be completely evaluated.

Legal and compliance: It is another domain in all angles with no legal or compliance points of reference to pursue, which represents a significant problem for IoT makers and service suppliers. This test alone will drive away numerous organizations from utilizing blockchain innovation.

8.3 PROPOSED ARCHITECTURE

In this work, we propose Ethereum blockchain-based IoT architecture that will be used by low-power IoT devices to communicate with each other, validate transactions, and provide security. Due to the integration of smart contracts, we have chosen Ethereum. Another popular choice for the blockchain was Hyperledger and IOTA. But because of insufficient documentation and openness of the developers' community, we faced difficulty in using these blockchains. Ethereum has a well-defined documentation, and the open-source community is sufficiently large. Hence, most of the errors and bugs were able to rectify in time. Fig. 8.7 shows the basic architecture.

In this system, all the low-power devices communicate to the Ethereum blockchain, and all the nodes are connected to the same blockchain. Low-power IoT devices act as lite nodes and store only a portion of the blockchain. Local storages act as full nodes, where it stores all the data generated by the devices and stores transaction happening in the network separately. Full nodes do not mine the blocks but only store the chain. Miner nodes are other nodes in the network that validate the transactions and generate new blocks. In a real-life scenario, Internet service providers can act as miners and compete with each other to validate transcations. Thus, they receive incentives for validating transactions as ether coins (Fig. 8.8).

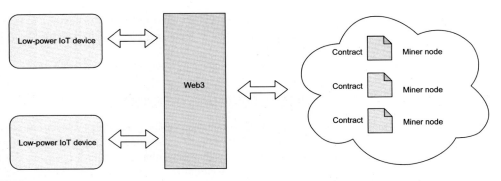

FIGURE 8.7

Proposed general block diagram.

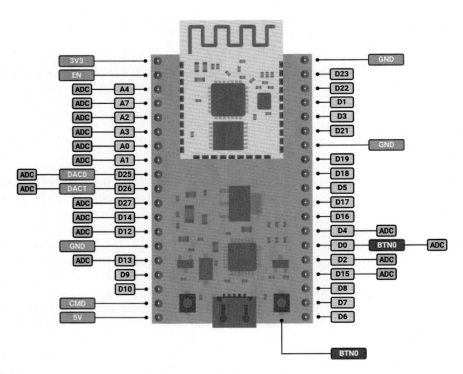

FIGURE 8.8

ESP32 DevKit [51].

8.3 PROPOSED ARCHITECTURE   203

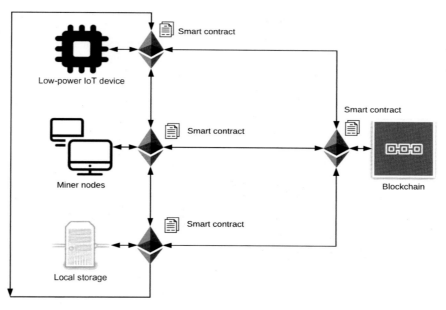

FIGURE 8.9

Ethereum-based architecture.

In our project, we used ESP32 as low-power IoT controller. ESP32 is a solitary 2.4 GHz Wi-Fi and Bluetooth combo chip planned with the TSMC ultra-low-control 40 nm technology. It is intended to accomplish the best power and RF execution, appearing flexible and with unwavering quality in a wide assortment of uses and power situations. ESP32 is designed to work with mobile applications and the IoT. It can be programmed using any of the programming languages such as C, embedded C, JavaScript, and Python. DHT11 temperature and the humidity sensor are interfaced with ESP32.

Temperature and humidity values are sent to the LAMP server in ubuntu machine. Both devices are connected to Ethereum blockchain using Ropsten private network and MetaMask. Ropsten is a test network for Ethereum applications. This works exactly the same as real Ethereum network but in a private network with fake eth cash. Ropsten network is used for developing and debugging DApps. MetaMask is a browser extension that goes about as a scaffold between Internet browsers, Ethereum, Dapps based on Ethereum, for example, MyEtherWallet. It empowers clients to execute Ethereum dApps in their Internet browser straightforwardly without running a full Ethereum hub. MetaMask enables clients to store, send, get, and encourage connections with the Ethereum organize (Fig. 8.9).

8.3.1 SMART CONTRACT

The main reason for the wide adoption of Ethereum blockchain is the inclusion of smart contracts. A smart contract, in basic terms, is a digitized type of a legitimate contract. It comprises of a lot of

204 CHAPTER 8 BLOCKCHAIN INTEGRATION WITH LOW-POWER IOT DEVICES

conventions, which the taking part elements ought to concur on and conditions causing the executions of those conventions. The public accessibility of the code on the BC makes a trust in taking the interest elements, and programmed execution eliminates the requirement for a TTP [41]. Smart contract on Ethereum allows developers to write program on blockchain that will do the computing for the network. Languages that can be used for smart contracts in Ethereum are Solidity, Serpent, and LLL, of which Solidity is widely used for writing smart contracts, and here in this project, we used solidity programming using Remix IDE and complier to compile and deploy code on the Ethereum test network. Smart contract provides autonomy and trust among the peers in the network.

In our project, we wrote two smart contracts, one for data logging and the other for authentication of valid accounts. Using these two, we can safe guard the data and if an attacker manipulates or tries to enter the network, the attacker is just ignored.

8.3.2 INSTALLATION OF LAMP SERVER

A LAMP Stack is a set of open-source software that can be utilized to make websites and web applications. LAMP server is a local server that can be deployed on the local machine, which can interact with the IoT nodes with/without active Internet. Since we need active Internet for Ethreum blockchain, we use it with Internet ON. This term is really an abbreviation which speaks to the Linux operating framework, with the Apache web server. The site data are put away in a MySQL database, and dynamic content is handled by PHP.

LINUX: It is an open-source operating system that is based on UNIX kernel. Debian, Ubuntu, Fedora, Mint, and RedHat are some of the Linux distributions widely used. Linux can be used for embedded system applications, and lightweight kernels and distributions are available for the same. In our project, we have used raspbian operating system, which is also a Linux distribution.

Apache HTTP Server: It is an open-source cross-platform web server software. It was released under the terms of Apache License 2.0. Apache is developed and maintained by an open community of developers under the auspices of the Apache Software Foundation.

MySQL: It is a free and open-source software under the terms of the GNU General Public License. It is a relational database and is written in C or C++.

PHP: Hypertext Preprocessor is a general-purpose programming language made developing web applications.

The Ubuntu machine used runs on 64-bit Ubuntu 18.04 Deskop with 4 GB RAM and 60-GB storage space. In a real-case scenario, most of the servers/cloud platforms run on ubuntu machines, and hence these steps can be used to deploy real cloud environment/server/virtual private server. Steps are as follows:

- Install an apache web server and update firewall
    sudo apt-get update sudo apt-get apache
- Allow all incoming requests from HTTP and HTTPS
    sudo ufw allow in "Apache Full"
- Install MySQL Server
    sudo apt install mysql-server sudo mysql secure installation

8.3 PROPOSED ARCHITECTURE 205

- Configure MySQL database creation by adding a password, creating MySQL user and updating privileges
- Install PHP

  sudo apt install php libapache2-mod-php php-mysql

8.3.3 DEPLOYING SMART CONTRACT ON NETWORK

For coding and debugging smart contract, we have used Remix IDE which is an online browser-based IDE with in-built editor and compiler. The programming language used is Solidity, and version of the compiler is v0.4.21. Account for smart contract and user has been created on Ropsten test net, and the accounts are imported on metamask. Both accounts are loaded with fake ether suing Ethereum Ropsten faucet so as to use it for sending transactions to each other accounts. In the decentralized network, to interact with the smart contract, we need to have access to one of the clients, and this can be done either by running a client by ourselves or by connecting to a remote node provided by services like Infura. Here, we use infura as a remote node that helps in connecting the node to Ethereum blockchain.

Steps involved in deploying a contract is as follows:

- Write the solidity code on Remix IDE.
- Create an account on the Ropsten network and load ether using Ethereum Ropsten faucet.
- Create a test node using infura.io. This would be the entry point to the Ethereum test network. Ethereum works on web3, and infura provides necessary API for it. Create a project in infura and copy the address of the project.
- Select web3 provider for deploying the contract and paste the address copied in the previous step.
- Click on run and a pop up from MetaMask extension will appear to create the transaction. Approve the transaction manually and thus the smart contract is deployed on the test network. Fig. 8.2 shows the contract creation snapshot of etherscan (Fig. 8.10).

8.3.4 INTERFACING ESP32 WITH ROPSTEN NETWORK

ESP32 software development is done using Zernyth Studio v2.1.1. Zerynth Studio provides an easy medium to interact with Ethereum blockchain, and software language used here is python. The following steps are done for interfacing ESP32 with Ethereum Ropsten network. Each transaction will cost 50 Wei. Wei is the smallest unit of ether. 1 ether is 1018 Wei.

- Write the code for fetching values from the DHT11 sensor and sending to a local server and initiating a transaction to the Ethereum network during each transaction.
- Import the smart contract to the project folder currently working on. The solidity code is converted into byte codes by the software.
- Specify the node address, contract address, network address, and other necessary details in config.py file. Other information includes the wi-fi credentials of the router through which the sensor interacts with the internet.

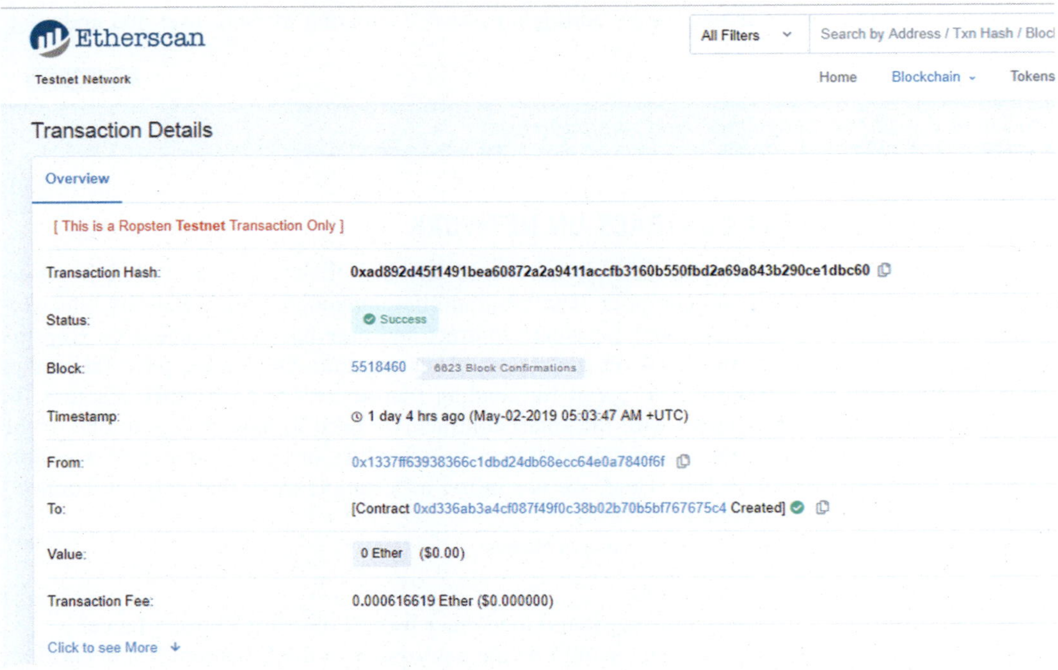

206 CHAPTER 8 BLOCKCHAIN INTEGRATION WITH LOW-POWER IOT DEVICES

FIGURE 8.10

Contract creation.

- Create a virtual machine for the ESP32 board connected. We use DOIT ESP32 Devkit v1 board.
- Connect the board with the computer and upload the program on to it.
- Once the code is successfully uploaded, reset the module and check MetaMask extension to see new transactions.

8.4 RESULTS AND DISCUSSION

Smart contract was deployed on the Ropsten network, and ESP32 was connected to the network. ESP32 generates, signs, and send the transaction inside the controller. Cryptographical algorithms take place inside the controller, and thus it removes the reliance of central gateways. The total size of the byte code was 72,705 bytes out of available 524,288 bytes. This is a very low memory requirement, and hence this method is ideal for low-power IoT applications. The code that was uploaded uploads the data to the local server and creates transaction each time. By validating this transaction, we can ensure that the data are unaltered.

This blockchain-based architecture is fully decentralized, and the transactions are automatically controlled by the smart contracts. Each IoT node will have a unique smart contract and will only

8.4 RESULTS AND DISCUSSION 207

communicate through the contract. Thus, we can check transactions periodically to know the nodes that accessed the data. Once the data are not stored in the cloud and is in local storage, the data are purely owned by the owner and not by third-party cloud providers. Even if we want to eliminate local server and integrate with other cloud platforms, this architecture provides a way for it. The cloud party will also be controlled by the contract. We can hardcode in the contract that how all can access the data. In this way, the data can be more secure. Since the contract is in every node of the network, even if the contract in the IoT device tampers, it will not affect the network. The IoT device will no longer be able to communicate with the network. Thus, network security is brought by the blockchain (Fig. 8.11).

The IoT has of late risen as a significant research theme. It gives the integration of various sensors and articles to communicate explicitly with one another without human interference. Besides, the prerequisites for the enormous scale sending of the IoT are increasing quickly with real security concerns. We displayed a far-reaching audit of the best-in-class IoT security dangers and vulnerabilities. We examined the IoT by presenting the scientific categorization of the present security dangers and vulnerabilities with regard to its communication. Also, we talked about the present cutting-edge IoT-enabling communication technologies. We talked about open-research issues and difficulties in IoT security (Fig. 8.12).

All things considered, different kinds of IoT applications may be absolute best with express characteristics among dispersed record executions. The IoT sensor and actuator data may not require incessant updates but instead require a gigantic level of trustworthiness affirmation by methods for consensus, and the capacity to incorporate custom data may warrant Ethereum. Time touchy sensor and actuator data including lively exchanges and not requiring additional convenience of smart contract authorization may be extensible toward IOTA and its coordinated

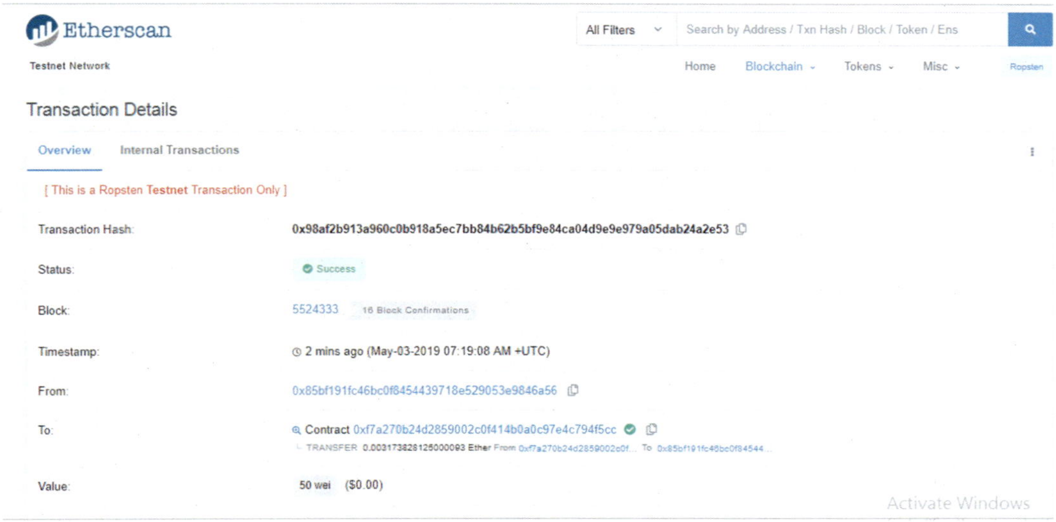

FIGURE 8.11

Transaction details.

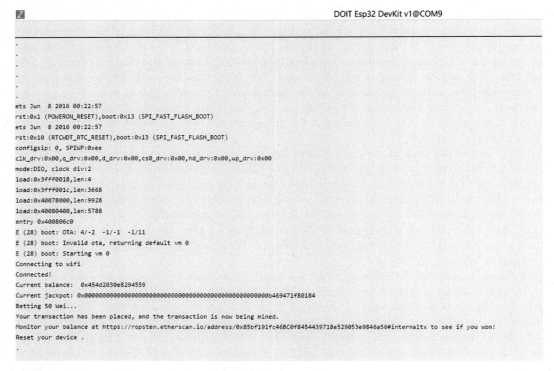

FIGURE 8.12

Snapshot of serial monitor of ESP32 while sending transaction.

noncyclic graph approach. For fault tolerance, quick exchanges, smart contract comparability, and high-data transfer capacity systems—Hyperledger—may be ideal. Overall, circulated records can ensure uprightness of IoT data in a powerful nice assortment of valuable applications; how definitely the introduction of new conveyed records and updates to existing executions may change the display and interoperability with IoT contraptions may be explained in future research.

A great deal of research should be done in explicit areas like smart energy and smart assembling. Being in the underlying stage, next to no exploration is finished with tending to the problem of versatility of the BC arrangement. Research is being completed with respect to agreement conventions to empower the adaptability of the combination. From the dissemination, we can see that a great deal of research has been done in the field of smart homes and smart urban communities. The following coherent advance lies in separating the most generally embraced stages in independent subsystems and after that building full-included stacks from institutionalized and pluggable parts. What's more, assurance against imaginative attacks like side-channel examination can be an intriguing endeavor. Another intriguing exploration heading can be the use of blockchain to take care of the problem of information trade and exchanging. With the pervasiveness of IoT gadgets and expanding creation of information, endeavors have begun to adapt the information bringing forth the Machine Economy. Blockchain can streamline the exchange forms, wiping out the requirement for a confided in mediator.

The architecture we proposed is suitable for low-power IoT devices and use less computational power in the end device side. But the network usage is intensive and is due to the resource-exhausting consensus mechanism employed by Ethereum. Transaction approval time is also not ideal for IoT applications. More research works are going on to reduce the transaction time in blockchain by implementing lightweight consensus. Hyperledger is a promising blockchain alternative for IoT, and the consensus used in the Hyperledger fabric is much faster than Ethereum. This along with other custom blockchain implementations marks a better future for IoT devices, making it more secure, reliable, and faster.

REFERENCES

[1] A. Reyna, et al., On blockchain and its integration with IoT. Challenges and opportunities, Future Gen. Comp. Syst. 88 (2018) 173—190.

[2] M.S. Ali, et al., Applications of Blockchains in the Internet of Things: A Comprehensive Survey. IEEE Communications Surveys & Tutorials, 21.2, 2018, 1676—1717.

[3] K. Wust, A. Gervais, Do You Need a Blockchain? 2018 Crypto Valley Conference on Blockchain Technology (CVCBT). IEEE, 2018, 45—54.

[4] F.A. Alaba, et al., Internet of Things security: A survey, J. Netw. Comput. Appl. 88 (2017) 10—28.

[5] J. Rivera, R. van der Meulen, Forecast Alert: Internet of Things—Endpoints and Associated Services, Worldwide, 2016. Gartner, 2016.

[6] C. Lu, Overview of Security and Privacy Issues in the Internet of Things, Washington University, 2014.

[7] H. Suo, et al., Security in the internet of things: a review, in: 2012 International Conference on Computer Science and Electronics Engineering, IEEE, 2012.

[8] K. Zhao, L. Ge, A survey on the internet of things security, in: 2013 Ninth International Conference on Computational Intelligence and Security, IEEE, 2013.

[9] P. Kasinathan, et al. An ids framework for internet of things empowered by 6lowpan, in: Proceedings of the 2013 ACM SIGSAC Conference on Computer & Communications Security, ACM, 2013.

[10] I.A.T. Hashem, et al., The role of big data in smart city, Int. J. Inf. Sci. 36 (5) (2016) 748—758.

[11] J. Han, M. Ha, D. Kim, Practical security analysis for the constrained node networks: Focusing on the dtls protocol, in: 2015 5th International Conference on the Internet of Things (IOT), IEEE, 2015.

[12] C. Simko, Man-in-the-Middle Attacks in the IoT. Globalsign, 2016.

[13] M. Vučinić, et al., OSCAR: Object security architecture for the Internet of Things, Ad Hoc Netw. 32 (2015) 3—16.

[14] P. Pongle, G. Chavan, A survey: attacks on RPL and 6LoWPAN in IoT, in: 2015 International Conference on Pervasive Computing (ICPC), IEEE, 2015.

[15] T. Kothmayr, et al., DTLS based security and two-way authentication for the Internet of Things, Ad Hoc Netw. 11 (8) (2013) 2710—2723.

[16] A. Botta, et al., Integration of cloud computing and internet of things: a survey, Future Gener. Com. Sy. 56 (2016) 684—700.

[17] L. Eschenauer, V.D. Gligor, A key-management scheme for distributed sensor networks, in: Proceedings of the 9th ACM Conference on Computer and Communications Security, ACM, 2002.

[18] S. Chakrabarty, D.W. Engels, S. Thathapudi, Black SDN for the Internet of Things, in: 2015 IEEE 12th International Conference on Mobile Ad Hoc and Sensor Systems, IEEE, 2015.

210 **CHAPTER 8** BLOCKCHAIN INTEGRATION WITH LOW-POWER IOT DEVICES

[19] Z. Zheng, et al., Blockchain Challenges and Opportunities: A Survey; Work Paper, Inderscience Publishers, Geneva, Switzerland, 2016.

[20] S. Nakamoto, Bitcoin: A Peer-to-Peer Electronic Cash System, Whitepaper (2008), 2009.

[21] Veena, Pureswaran, et al. "Empowering the edge-practical insights on a decentralized internet of things." Empowering the Edge-Practical Insights on a Decentralized Internet of Things. IBM Institute for Business Value 17 (2015).

[22] N. Szabo, Formalizing and securing relationships on public networks. First Monday 2(9) (1997).

[23] V. Buterin, Ethereum White Paper, GitHub repository, 2013, pp. 22−23.

[24] E. Androulaki, et al., Hyperledger fabric: a distributed operating system for permissioned blockchains, in: Proceedings of the Thirteenth EuroSys Conference, ACM, 2018.

[25] I.-C. Lin, T.-C. Liao, A survey of blockchain security issues and challenges, Int Journ. Net. Sec. 19 (5) (2017) 653−659.

[26] P. Jayachandran, The Difference Between Public and Private Blockchain, IBM Blockchain Blog, 2017, p. 31.

[27] Aziz, Public vs Private Blockchain Whats the Difference, Masterthecrypto Blog, 2018.

[28] B. Asolo, Consortium Blockchain Explained, Mycryptopedia Blog, 2018.

[29] R.G. Brown, et al., Corda: An Introduction. R3 CEV, 2016, vol. 1, p. 15.

[30] A. Singh, et al., BFT Protocols Under Fire, NSDI, 2008.

[31] L. Lamport, R. Shostak, M. Pease, The Byzantine generals problem, ACM Trans. Program. Lang. Syst. 4 (3) (1982) 382−401.

[32] M. Jakobsson, A. Juels, Proofs of work and bread pudding protocols, Secure Information Networks., Springer, 1999, pp. 258−272.

[33] C. Dwork, M. Naor, Pricing via processing or combatting junk mail, Annual International Cryptology Conference, Springer, 1992.

[34] A. Rosic, Proof of Work vs Proof of Stake: Basic Mining Guide. Blockgeeks Blog, 2017.

[35] P. Vasin, Blackcoin's Proof-of-Stake Protocol v2, 2014, p. 71. Available from: <https://blackcoin.co/blackcoin-pos-protocol-v2-whitepaper.pdf>.

[36] S. King, S. Nadal, Ppcoin: Peer-to-Peer Crypto-Currency With Proof-of-Stake, Self-Published Paper, August 2012, p. 19.

[37] G. Wood, Ethereum: A Secure Decentralised Generalised Transaction Ledger. Ethereum Project Yellow Paper, 2014, vol. 151, pp. 1−32.

[38] V. Zamfir, Introducing Casper "The Friendly Ghost," Ethereum Blog, 2015. Available from: <https://blog.ethereum.org/2015/08/01/introducing-casper-friendly-ghost>.

[39] M. Castro, B. Liskov, Practical Byzantine fault tolerance and proactive recovery, ACM Trans. Comp. Syst. 20 (4) (2002) 398−461.

[40] D. Mazieres, The stellar consensus protocol: a federated model for internet-level consensus, Stellar Develop. Found. (2015) 32.

[41] A. Panarello, et al., Blockchain and IoT integration: a systematic survey, Sensors 18 (8) (2018) 2575.

[42] C. Cachin, Architecture of the hyperledger blockchain fabric, in: Workshop on Distributed Cryptocurrencies and Consensus Ledgers, 2016.

[43] J.R. Douceur, The sybil attack, International Workshop on Peer-to-Peer Systems, Springer, 2002.

[44] M. Samaniego, R. Deters, Hosting virtual IoT resources on edge-hosts with blockchain, in: 2016 IEEE International Conference on Computer and Information Technology (CIT), IEEE, 2016.

[45] Z. Yan, P. Zhang, A.V. Vasilakos, A survey on trust management for Internet of Things, J. Netw. Comput. Appl. 42 (2014) 120−134.

[46] B. Smith, K. Christidis, IBM blockchain: an enterprise deployment of a distributed consensus-based transaction log, in: Proc. Fourth International IBM Cloud Academy Conference, 2016.

REFERENCES **211**

[47] S. Popov, The Tangle. cit. on, 2016, p. 131. http://www.iotatoken.com.

[48] K. Christidis, M. Devetsikiotis, Blockchains and smart contracts for the internet of things, Ieee Access 4 (2016) 2292−2303.

[49] C.K. Frantz, M. Nowostawski, From institutions to code: towards automated generation of smart contracts, in: 2016 IEEE 1st International Workshops on Foundations and Applications of Self* Systems (FAS* W), IEEE, 2016.

[50] M. Vukolić, The quest for scalable blockchain fabric: Proof-of-work vs. BFT replication, International Workshop on Open Problems in Network Security, Springer, 2015.

[51] E. Datasheet, Version 2.1.

CHAPTER 9

APPLICATIONS OF BLOCKCHAIN TECHNOLOGY

Chetna Laroiya, Deepika Saxena and C. Komalavalli
Jagan Institute of Management Studies, Rohini, New Delhi, India

9.1 INTRODUCTION TO BLOCKCHAIN TECHNOLOGY APPLICATIONS

Blockchain has received considerable hype, starting with "cryptomania" in the trading markets in the year 2017 to broad discussions about its potential impact across public and private sectors and in society. The "Blockchain" is one among the most exalted technology today which has a pervading impact on all industries, specifically in Banking and Financial Services.

Blockchain combines the principles of cryptography, peer-to-peer networking, and game theory. Blockchain evolved as the formal name for tracking the database underlying the cryptocurrency, that is, bitcoin, but now it is referred as distributed ledger with software algorithms to record transactions as a chain of blocks with trustworthiness and anonymity. Blockchain also uses the concept of smart contracts where business rules are implied by an agreement that are embedded in the blockchain and executed with the transaction.

Blockchain uses digital signature to guarantee the provenance of the transaction. The key advantage of blockchains resilient architecture which protects the distributed ledger. Other benefits of blockchain include a reduction in time, complexity, and cost. Blockchain has the promising feature to make financial and trading processes efficient, improve regulatory control. Decentralized consensus mechanism makes transactions immutable and updatable only through consensus among peers over the network. This design protect displace traditional third-party functions in a transaction. Blockchain distributed ledger will offer consensus and immutability about the transfer of assets within a business networks.

The Banking and Financial Services Industries are gravely looking at this technology. The Central Banks in India have formed committees to evaluate the extent of adoption of the blockchain technology (BCT), to address some of the problems that the industry is trying to overcome over many years. Although institutions appreciate its potential, they are still working to figure out whether BCT provides a cost cutting or represents a margin-eroding threat which could lay them out of business.

Any public blockchain is a decentralized system open to everyone and where the distributed ledger is updated by n number of anonymous users, whereas a private blockchain is used within a bank or an institute premices, where the organization controls the entire blockchain system. Hybrid

---

Handbook of Research on Blockchain Technology. DOI: https://doi.org/10.1016/B978-0-12-819816-2.00009-5
© 2020 Elsevier Inc. All rights reserved.

214 CHAPTER 9 APPLICATIONS OF BLOCKCHAIN TECHNOLOGY

Blockchain is an amalgamation of both public as well as private implementations, which is open to a labeled group of trusted and verified users that have rights to update and maintain the network collectively.

The technology is getting commercialized, and several industry groups are coming out with the use cases portraying that the technology could be proven suitable for across different industry vertical.

Numerous Blockchain applications and platforms are widely known, starting with Bitcoin, followed by Ethereum, which act as a platform for building decentralized applications using smart contracts and inspired a whole new concept of "token economy." Emerging applications in voting, digital identity, banking, and health sector illustrate how blockchain can potentially be used to address global business challenges. There is now also emerging new line of thought about block chain's potential to global efforts of advance environmental sustainability.

9.1.1 RELEVANCE OF BLOCKCHAIN TECHNOLOGY APPLICATIONS

As per May 2019, there are 44% organizations globally who have implemented blockchain. Technology provides trusted online transaction to the industry. People and industries are trying to use blockchain because of the following benefits it offers.

- *Transparency*

    Transparencies offered in the network make it most agreeable technology. The distributed ledger is shared among all the participants over the network. All can keep an eye on every ongoing transaction leaving behind any chance of discrepancy.
- *Security*

    In today's time, hackers are using all their tactics to hack the device. Blockchain promises its users to offer vigorous security. Each block creates a hash based on the data in the last block thus making it fully interconnected with each other. It is very pricey to hack the blockchain network in terms of time. Also, it is worthless to hack a Blockchain since the technology is developed such that even if one block is altered, all information of the block gets spoiled.
- *Inexpensive*

    Blockchain rules out the brick and mort model office to facilitate traditional financial transaction which is expensive, even it is not required to pay huge commission to avail financial services.

Let us assume that you are a part of the music industry where a singer is paid at last after intermediaries. One of the most intelligent part of the BCT is the smart contract which is highly munificent. Nowadays, the giant music companies have started using these smart contracts in their newly developed songs. This technology helps the artists and musicians in selling their songs directly to the consumers digging out the requirement of intermediaries wholly. The artists are now able to get dues easily and smoothly as royalties and agreements are executed automatically. The success of the giant music companies such as Apple, Sony, etc., will depend completely on how they adopt this particular technology in future.

- *Secure platform*

  Blockchain provides a new digital platform with ensured security to all the intellectual property. The digital certification and certification of ownership is some of the reason to trust this technology. Several new start-up music companies have blockchain applications which allow artists to register their work get their payment first. These blockchain applications are also allowing artists to upload their work with digital signature on their work. This way artist get proper value for their work, and their work is not lost in the crowd.

- *Creating a better contribution economy*

  The BCT provides a better sharing platform, It provides suppliers and buyer a trusted network to trade. Without intermediaries or third-party involvement, traders can bit lower price and can even earn more profit. Thus, the technology creates a trusted and transparent marketplace and also better economy.

- *Prevents payment scams*

  BCT helps to prevent online payment scams. Both buyers as well as sellers can use smart contracts to purchase and sell products. Basic reason of why Blockchain provides powerful security is because if a coin is spent, it cannot be used for next payment. This is the way it stops corruption. Since everything including traders and payment amount is accounted and kept under the track, it is really difficult for discrepancies or corruption. Another reason of popularity of the technology is that if a transaction occurs between two parties, it has digital signature from both parties which prevents any fraud.

- *Transactions in minutes*

  One can send or receive money or financial documents in minutes, thus it saves time. Traditional payment system involving third party directs every document to the clearing house for approval which causes down time in the transaction.

9.2 BACKGROUND OF BLOCKCHAIN

9.2.1 WHAT IS BITCOIN?

The most admired representation of BCT in the world is the cryptocurrency bitcoin. In modern monetary systems, a currency is not denoted by a physical commodity but is popularly known as fiat money. Fiat money preserves its value and is approved by a government or an institution to approve it as a legitimate resource of currency.

Bitcoin does not necessitate any validation from any bank or government. Both the parties involved in bitcoin have public key. Separate private key is with both the parties for digital signature as a proof of ownership. Sender party issues the bitcoin of desired amount. After the transaction is complete, digital signature and the time stamp of the transaction are broadcasted for validation to all the parties over the network. Once the nodes over the network confirm the validity of the transaction, a block having the information is added to the chain of blocks. After this point, transaction is irreversible and is open for anyone over the network to view and verify that it did take place (Fig. 9.1).

216 CHAPTER 9 APPLICATIONS OF BLOCKCHAIN TECHNOLOGY

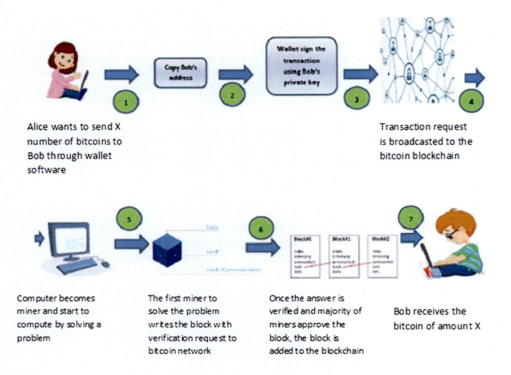

FIGURE 9.1

Bitcoin transaction.

9.2.2 EMERGENCE OF BITCOIN

Bitcoin concept came into existence with the Bitcoin white paper authored by a fictitious name Satoshi Nakamoto, whose identity is still a mystery. This volatile cryptocurrency has fought through years of controversies and success. With the release of the white paper, cryptocurrency mission got a dimension which was left an indelible mark across many industries.

Bitcoin.org domain was recorded on August 18, 2008 by an anonymous entity. This came to knowledge of people with the publishing of the Bitcoin white paper "Bitcoin — A peer-to-peer electronic cash system," on October 31, 2008. On January 3, 2009, Nakamoto successfully created the Genesis Block of the Bitcoin blockchain. The first-ever known bitcoin transaction of 10 BTC took place on January 12, 2009 between Nakamoto and late Hal Finney. On October 5, 2009, first-ever bitcoin exchange rate was set against dollar. That time, $1 equaled 2300.03 BTC.

The very first transaction of bitcoins for the purchase of physical goods was on May 22, 2010. Two bitcoin Pizza were bought for 10,000 BTC. On June 12, 2011, bitcoin hiked to value $10.25. BTC price continued to rise into 2013, and on April 9, the cryptocurrency reached to $200. By November 2013, the bitcoin reached to $1000.

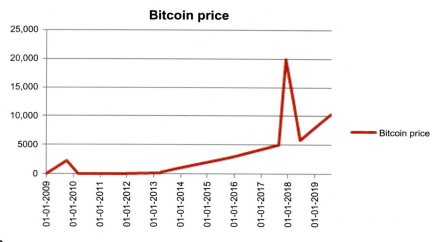

FIGURE 9.2

Bitcoin price.

On December 11, 2014, a great news that Microsoft began accepting bitcoin payments, made a big milestone on the path for bitcoin approval by companies globally. On October 31, 2015, cryptocurrency appeared on the front page of Economist. On June 11, price crossed the $3000 mark and ignited the debate on issues affecting bitcoin.

On August 1, 2017, concept of bitcoin cash to existance, still bitcoin surpassed the $5000 mark on September 2, 2017. The cryptocurrency surpassed $10,000 in value on 29 November, and then reached the $11,000 mark a few hours later. Bitcoin finally reached to historic hike of $20,000 mark by November 2017. Cryptocurrency dropped as low as $10,000. In March 2018, Twitter followed Facebook and banned cryptocurrency advertising. Also Google announced to stop cryptocurrency. The value of bitcoin was lowest at $5868 on June 24, 2018 (Fig. 9.2).

9.2.3 WORKING OF BITCOIN

Bitcoin client or wallet application is a program which allow user to join bitcoin network. User can participate in bitcoin network in different capacities. A transaction is the process to control and proof the creation and movement of bitcoin.

Step 1: User start a transaction using the wallet application to spend or transfer a certain token from them to the trader or receiver.

Step 2: This transaction is broadcasted through wallet application to all the miners over the blockchain network. Transaction will wait till it is chosen up by a miner. Till the time it is not taken up, it will remain in a "puddle of unconfirmed transactions."

Step 3: Miners or nodes on the blockchain network select transaction from the pool. Every miner creates his own block of transactions. Multiple miners simultaneously might select the same transaction to verify for block creation. Before the miner can add the transaction to their

218 CHAPTER 9 APPLICATIONS OF BLOCKCHAIN TECHNOLOGY

block, every miner has to check the validly of the transaction to be executed according to the blockchain history. If sender's wallet has enough balance funds to complete transaction according to the active blockchain history, the transaction is marked as valid thus can be added by the miner to the block. Miners will assign priority to transactions based on transaction fee set (higher the fee of transaction greater is the priority).

> For example, Alice has 10 bitcoins and she wants to give 5 bitcoins to Bob. Five bitcoins for Bob and another 4.90 bitcoins for Alice where the remaining 0.10 bitcoin will be claimed by the miner as transaction fee.

Step 4: After block is added by the miners, next this block of transactions is appended to the chain of blocks for which the block is to be first signed. This signature can be created by the miner after solving a very complex mathematical problem. This problem is unique to each block of transactions. This process of signing the block is termed as mining. To solve these problems, lots of electricity is required because lots of computational power is used.
Step 5: (Proof of Work - consensus algorithm): The complex mathematical problem, which every miner is solving before it could add one block to the blockchain, is to calculate a hash value for the data in its block. This hash value has to start with a definite number of consecutive zeros.

9.2.4 RISK IN BITCOIN

Now a days, bitcoin is the most successful cryptocurrency, but with the introduction of any new dimension, there are bound some obstacles too. There are many severe risks when it is to invest in bitcoin.

1. *The volatile and fluctuating market*
   The price of bitcoin is changing dramatically. To be noted that on November 6, 2018, each bitcoin was worth $6461.01. The bitcoin market ripples back and forth and is one of the unpredictable markets. To avoid a huge loss, small investment for long term is recommended.
2. *Cybertheft*
   Cryptocurrency or bitcoin is technology-based, many reports reveal that there had been losses during mining or exchange. Hackers have an eye on exchanges and even the wallet application is protected.
3. *Fraud*
   Apart from hacking, there is a possibility of fraud in the bitcoin trading. Some of the bitcoin exchanges can be fake. The lack of security causes risk for big investors. Although systems exist to deal with online fraud, security is a big issue.
4. *No regulation*
   The bitcoin market is running without any regulations. The government has not formed a clear stance on bitcoin or cryptocurrency. Trading in cryptocurrency is not taxed, which can make it attractive investment option. State of the bitcoin market in near future is unpredictable.

9.2 BACKGROUND OF BLOCKCHAIN   **219**

5. *Technology reliance*

Bitcoin are digitally mined, exchanged, and generated via smart wallet. It has no physical collateral to back it up. Unlike bitcoin, with the purchase of gold, real estate, or mutual funds, we own something as a proof that can be exchanged.

6. *Limited use*

Bitcoin definitely a footstep toward a new monetary exchange, but only a few companies accept it.

7. *Financial loss*

Bitcoin investment creates bubble economy. Whenever the bubble bursts, bitcoins are useless. There might be many people holding cryptocurrency and willing to sell. There is no return on investment resulting into a painful financial loss.

8. *Currency or investment opportunity*

Buyers buy bitcoins the way investment is done in stocks. Although bitcoin potentially payoff, this investment is to be done with caution. Small investments at small steps will give good results.

9.2.5 LEGAL ISSUES IN BITCOIN

The legal status of cryptocurrency bitcoin varies a lot from state to state. It is still ambiguous and changing from time to time. Majority of countries have announced bitcoin illegal and do not use it. It is money (or commodity) that is also debatable and varies with different regulatory implications.

Each country has formed its own rule to deal with local and foreign currencies. Foreign currencies are controlled by the enactment of special legislation. Citizens can make transactions in any foreign currency subject to rules and regulations. The rules are related to other legislations, that is, investment caps and corporate law. Bitcoins are not issued or regulated by any bank. They are produced through mining, a computer-generated process. Although cryptocurrency is unrelated to government, it is a one-to-one payment system, as it does not have a physical form. Till date, there are no standardized international laws to regulate bitcoin.

Canada

Canada has a bitcoin-friendly status, but makes sure that cryptocurrency is never to be used for money laundering. Bitcoin is considered as a commodity by the Canada Revenue Agency (CRA). This means that income generated by bitcoin transaction is considered as business income.

The European Union

European Union (EU) has not made any decision on legality of cryptocurrency, acceptance, or regulation. Without any central guidance, countries in EU have developed their individual bitcoin regulations. In Finland, bitcoin is exempted from value-added tax (VAT) by the Central Board of Taxes (CBT). Bitcoin is treated not as currency but as commodity in Finland.

The Federal Public Service Finance of Belgium has also assigned a VAT exempt status to bitcoin.

220 CHAPTER 9 APPLICATIONS OF BLOCKCHAIN TECHNOLOGY

United Kingdom
The Financial Conduct Authority (FCA) in the United Kingdom needs to develop the regulatory background to support the digital currency.
Australia
Australia treats bitcoin as a currency and allows parties to trade and buy it.

9.3 APPLICATIONS OF BLOCKCHAIN TECHNOLOGY

9.3.1 APPLICATIONS OF BLOCKCHAIN TECHNOLOGY IN FINANCIAL SERVICES

BCT has been applied to banking and financial services in various ways and getting numerous benefits. Smart contract service helps in conducting financial transactions without an intermediary. It has the potential to manage securities, deeds, settlements, and claims in an automated manner.

Few applications and use cases in financial transactions are as follows:

1. *Cross border transactions*

This is one of the most popular application of BCT in the world. Movement of funds have always observed to be slow and expensive using the traditional modes of transaction in the centralized system. With the use of decentralized ledger system, verification and processing of cross border transactions are a matter of seconds across different time zones.

2. *Smart bonds*

Smart contract technology enables investors to hold smart bonds. Smart bonds are the bond contracts which are automated and uses BCT for its registration services. These also allow instant settlement of transactions.

3. *Point of Sales systems*

Point of Sales system using BCT allows the merchants and users to accept cryptocurrency as payment. This helps in removal of expensive merchant services and costly card transaction fee. Also, it can assist the cash management and its analysis.

4. *Lending and borrowing*

BCT helps banking and financial institutions to lend and borrow money in a better way using distributed ledger system. Interbank borrowing and lending can also be made possible, speedy, transparent, smooth, reliable, verifiable, and secure.

5. *Securities trading*

BCT helps in the reduction of cost of securities trading in stock exchanges and provides an altogether new and innovative way to exchange assets in digital platform without any intermediary. Smart contracts are used to generate buying and selling of security for the investor without the involvement of any third party. The present trading mechanism needs documentation, duplicate copies, databases, their reconciliation, etc. Blockchain helps to remove all such issues present in the trading of securities.

9.3 APPLICATIONS OF BLOCKCHAIN TECHNOLOGY 221

6. *Clearing and settlements*

The present clearing and settlement system is a typical three-day cycle which can be transformed using the BCT. The present clearing and settlement cycle is extremely complicated and needs to have a match of balances, their reconciliation and resolution in domestic and international trading system. The distributed ledger technology (DLT) can make this process smooth, automated, efficient, and frictionless.

7. *Bookkeeping and auditing*

Using BCT, auditors will be of great benefit, as it will help them to verify the relevant and material information and data behind the financial statements and will help to save cost and time. The DLT can make it possible to prove the integrity of electronic data and files. One of the most innovative mechanisms which can bring the auditing process into real time is the hash string concept. The hash string of file represents its digital fingerprint along with time stamp of writing it. In order to check the integrity of the records, auditor can create fingerprint and compare the same with the one stored on the blockchain. Matching fingerprint will prove that data and records are authentic and have not been modified. These audits can also be conducted real time, not having to wait for weeks and months to complete.

8. *Hedge funds*

A group of investors maximize their return and alleviate the risk involved in investment in the stock market with the help of a fund manager. Now-a-days, hedge funds are traded in cryptocurrencies, some investors investing centrally and some in a decentralized hedge funds. The decentralized hedge funds allow investors invest without the need of a fund manager/ intermediary or a single controlling entity. This kind of mechanism provide investors an open platform to attract investors and thereby minimizing the investment risk.

9. *Credit score reports*

Credit score report is an essential requirement, now-a-days whenever credit is granted to any individual/entrepreneur/organization etc. BCT provides a platform which saves and stores the data and information related to credit score reports in an immutable ledger and at the same time, protects the personal details of the user. This information cannot be bought/sold/leaked/ hacked due to immutable nature of the distributed ledger system. Using BCT platform, small businesses or new loan applicants find it easy to get credits and their approval in secure and transparent manner.

Last year, the Barclays Bank put themselves ahead of others by implementing the security and transparency features of BCT in their processes. This bank announced the blockchain-based credit transactions between Ornua and Seychelles Trading Company. It has included the first-trade documentation which was encrypted and managed on a blockchain network.

The use of a decentralized ledger technology for sending the documents have saved time and money in transaction processing else the process would have taken 10 days for Barclays Bank if done through traditional channels. Accenture has predicted that the international financial sector may save up to $10 billion by implementing BCT for storage and processing of clearing and settlement system.

222 **CHAPTER 9** APPLICATIONS OF BLOCKCHAIN TECHNOLOGY

9.3.1.1 Use cases of applications of blockchain technology in banking

Trade Finance, Cross Border Payments, FX trading, capital market operations, consortium accounts, etc., are the areas in which BCT can be applied in banking sector (IDBRT, 2017).

- *Trade finance*

  In trade finance, companies are implementing BCT for replacing paper-based letters of Credit into distributed ledgers. This enables all the parties involved, that is, exporters, importers, and banks in the transaction to share information in their network. The trade deal can be executed automatically without the intervention of third party. It reduces the time from days to hours.

- *Cross border payments*

  International payments or cross border payments processing involves a series of steps, and in this course of action, it becomes very difficult to escape from the cyberattacks. Ripple technology under blockchain is a distributed ledger which banks have been using for not only making international payments easier and faster but also safe and secure.

- *FX trading*

  In the existing banking system, various records of currency trade are required to be created for sellers, buyers, clearers, brokers, and various third parties, and continuous reconciliation is required across multiple systems. Using BCT, multiple trade records can be removed for all these participants and can present a shared view of trade which frees up back and middle-level resources leading to continuous reconciliation across multiple systems. Due to the complexity involved in the existing system, FX market participants have to incur various expenses such as license fee, ticketing fee, staff costs, IT overheads, etc. BCT helps in providing immediate efficiency benefits and cost-reduction benefits by integrating seamlessly with all trading sources and venues.

- *Capital market operations*

  In Capital market trading, different parties like exchanges, central counter parties, Central Securities Depositories (CSDs), brokers, custodians, and investment managers are involved, and they have to maintain their ledgers based on the exchanged messages between them. For completing a transaction, the ledgers must be up-to-date and need intermediate beneficiaries for cash management. It might lead to delay for the final settlement and involves additional costs. BCT is playing a vital role in each and every stage of the trade such as pretrade and posttrade. BCT system facilitates for Know Your Customer (KYC) check and avoids multiple numbers of same checks again and again. It provides transparency and verification of holdings and reduced credit exposure. In the trade stage, it ensures the real-time transaction in more transparent and secure way and provides the automatic reporting. In posttrade, BCT removes the concept of central clearing needed for real-time cash transactions. Various other benefits are provided using BCT in the areas of Custody and Securities Servicing, Pre-Initial Public Offer (IPO) shares allotment, Loan Syndication, Bond Trading, Supply Chain Financing, etc.

- *Monitoring of consortium accounts*

  The major concern of banks today is the prevention of diversion of funds. The borrower moves funds from one bank to the other, and the end usage of funds is not known to the banks. Due to the nonexistence of the central entity, it has not been operationally feasible to securely and reliably trail the movement of money between various accounts maintained across various banks and financial institutions by the borrower; thus it has become one of the challenging area for banks. An integrated approach among banks and financial institutions is required which enable them to monitor money movement and detect anomalies in the process. BCT can help resolve the problem by assisting the banks and financial institutions to have visibility on the money movement and tracking the end use of borrowed funds thus, BCT will help strengthening the monitoring mechanism.

- Know Your Customer

  Regulatory compliance committee have been enforcing antimoney laundering and KYC for every bank. KYC process takes so much of time for collecting the data and uploading the data individually in the system. This might lead to the false entry and duplication of the data. BCT stores this data in a central repository and generates a reference number which is shared among all banks and financial institutions in near real time. Banks can access the same data for due diligence related to any customer's request for any other service in the same bank or with other banks. This helps in removing the efforts of collecting and checking KYC information again and again. Since the data are stored in encrypted form, security is maintained.

9.3.2 BLOCKCHAIN APPLICATIONS IN INSURANCE SECTOR

The BCT is bringing transformation in the insurance sector by bringing optimization in the business processes and sharing the information with better efficiency, security, and transparency. It is bringing the policy shift into the insurance system from manual to automated using the smart contracts on the peer-to-peer networks and thereby eliminating the traditional processing system. There are numerous benefits which insurance companies and individuals seeking insurance can take using BCT. With the help of decentralization, insurance sector will get streamlined in underwriting, payments, claims, and reinsurance processes. This technology will provide higher security, as the data is not stored at any centralized place and do not have any single entity control and thereby provides higher level of protection and cost saving.

Blockchain technology provides benefits to various insurance verticals:

1. *Health insurance*

   This technology not only improves health insurance but also can transform the services of healthcare providers. Health insurance is in direct connection with the medical institutions and patients using advanced data analytics. All such processes and operations can be done through DLT in an efficient, secure, immutable, and transparent manner.

2. *Auto insurance*

   Auto insurance industry can be benefitted in terms of reducing the level of paperwork, making underwriting easier, storage of data related to previous repairs, and damages to a vehicle in immutable and distributed manner, resulting in getting affordable quotes and faster resolution of accident claims.

3. *Life insurance*

   A lot of improvement is required in the existing life insurance system in terms of removal of paperwork, efficiency, and transparency in death claims and funds transfer to beneficiaries.

The DLT can be used to connect various cords in life insurance, namely, insurance companies, insuree, funeral homes, beneficiaries, government, etc. With the help of smart contract technology, insurance companies can automate the whole processing and can operate smoothly with better efficiency with reduced time and money on operations.

4. *Travel insurance*

Travel insurance could also be a vertical which use BCT to protect the traveler in case of flight delay without the need to make repeated calls to airline's company. Operational efficiency in this vertical of insurance would increase international coverage as traveling to different countries is not a big deal for individuals across the globe.

9.3.2.1 Use cases for blockchain technology applications in insurance

1. *Claims settlements*

Under insurance sector settlement of claims is the biggest challenge. These claims settlement become simpler with the help of custom smart contract code which takes various parameters of insurance policy and processes the operation automatically through trustless identity verification mode. In distributed ledger system, smart contract processes the funds for claims settlement, and controlling is not done exclusively by either policyholder or insurance company. Funds can be directed to the genuine party automatically after the verification using the digital contract on BCT. Smart contracts can settle the insurance claims in a faster and speedy manner without the requirement of any paper documents, photocopies, and complicated web portals.

2. *Reinsurance*

Reinsurance refers to the situation when several insurance companies purchase insurance policies for the purpose of offsetting possible losses which arise due to an incident or disaster. Blockchain can be very fruitful for reinsurance purpose also, as it helps in automating all calculations, balances, and reconciliation. This technology can track the funds available for settlement of claims and help the insurance companies in assessing financial risks and improve upon the reinsurance strategy in totality and simultaneously benefitting in terms of minimizing cost and time.

3. *Customization*

BCT helps in attracting the customers with lesser cost and further customized and easy-to-use interfaces. It is true that personalization or customization of insurance policy is difficult, and getting it at a reasonable rate is even more challenging. With improved transparency under the public distributed ledger, customers can easily upload and share their data and information even with more security.

4. *Real-time claims settlement and automation of payments*

Personalized and customized payment plans and insurance policies can function effortlessly for both the entities such as insurance company and insurance policy holder using an event-triggered smart contract technology. The real-time data from various systems work together in order to process the insurance claims automatically and make payments to claimants or collect the insurance premium payment from the policy holder. This brings an improved customer experience and at the same time prevents the insurance companies from losses and improves cost savings.

5. *Underwriting*

The process of underwriting involves technical skill in an individual in order to calculate the coverage amount on the policy for policy holder and the annual premium chargeable from him or her. It is presently very time-consuming process and need high level of data analysis. With the help of blockchain, storage of data and analysis can be done automatically using its data storage management and tools for analysis. It can help the underwriters to reduce the risk liability and automate insurance policy price determination process, which can result in cost-efficient model of insurance and better experience for the customer. Transparency in the underwriting process may lead in the direction of building trust between customers and insurance companies.

> Accenture - With the goals to boost the productivity in the insurance sector, a blockchain solution is built for the insurance clients by Accenture. The initiative was taken to support key insurance industry processes on to blockchain system.

9.3.3 BLOCKCHAIN TECHNOLOGY AND FINANCIAL TECHNOLOGY

With the digital innovation in financial sector, BCT examples have already been implemented in international payments which are benefitting banks in terms of reducing costs and also to shorten processing times. Also originally, cryptocurrency was invented to side-step principal controlling mechanisms worldwide. Moving to blockchain for financial solutions will open up possibility for innovative legalized landscapes and mechanisms and even more customer-centric and user-friendly trade and industry models.

> One more domain where blockchain can make positive impact is in the Middle East. According to the World Bank Global Findex Database, only 14% of the surveyed citizens in the Middle-East are reported to own a financial account. Using BCT combined with the mobile banking system and fintech, Middle Eastern digital innovators will have a much higher chances of penetration of the consumer base as compared with their traditional predecessors.

9.3.4 APPLICATION OF BLOCKCHAIN IN HEALTHCARE

Distributed ledger technology which stores the data in an immutable manner and update the information in real time have been reshaping the healthcare sector in totality. The traditional models in this landscape are proved to be highly inefficient in terms of delivering quality healthcare which is affordable in nature by the individuals. BCT-based healthcare applications are ready to be used and transform the healthcare institutions across the world. As the BCT works for the improvement in terms of transparency and efficiency, various parties are associated with the healthcare system, and patients get benefitted. The regulatory mechanism in the healthcare system and auditing has become now easier to manage. Improvement can be seen in the business operations using the innovative technology. The traditional or existing healthcare system is slow and expensive and also involves various intermediaries into the system; all such issues get resolved with the help of this innovative technology.

226 CHAPTER 9 APPLICATIONS OF BLOCKCHAIN TECHNOLOGY

> Healthcare industry is most excited for switching to BCT. Deloitte is disclosing its investment in millions in this domain. Transition to a blockchain supported healthcare system would cut costs and improve privacy and interoperability of health reports. The blockchain healthcare real examples and cases use smart contracts which they could use in order to process the surgery receipts easily and efficiently, and movement of hospital bills among the hospital, patient, and the insurance provider. A patient can interact with a blockchain-based healthcare system in order to view all his claims, medical history, and overdue payments easily and in a better manner. With the help of smart contracts, the individuals can also use BCT platform to plan and schedule appointments and engagements with their staff, which can be started as soon as the registration amount is paid, and doctor confirms availability.

Many countries in the world have centralized healthcare system, national healthcare system, and government administered programs. The main issues and concerns with the centralized healthcare system are: information sprawl to multiple parties, insecurity of data and information as the centralized servers is prone to hacking and data theft, inefficiency in operations and processes, very high cost of administration, extremely expensive and overpriced tests and medications, slow processing, duplicate treatments, poor patient outcomes, dissatisfaction of the patients, opaque operations and pricing, involvement of unrequired intermediary, etc.

The benefits of introducing decentralization in the healthcare are numerous. Deploying the BCT can drastically improve supply chain in the healthcare sector in number of ways. Healthcare system is extremely data heavy, and in case of critical information gets lost or altered may impact the life of the patient.

Various ways in which blockchain technology can transform the healthcare sector are:

- *Interoperability*

    Various healthcare systems can work together and in cohesion across the organizational boundaries in order to provide more advanced and effective healthcare services. The DLT is an ideal platform for this sector, as it can help resolve the issues of patients' data and information sprawl.
- *Better data storage and analytics*

    BCT helps in creating "patient-first" atmosphere in which the individuals can manage and store their data and information on a "permissioned" category of blockchain. Identity details can also be removed from the actual healthcare data and information to develop statistics and at the same time provides real-time, alter-proof data analytics for better healthcare services.
- *Immutability*

    Blockchain provides immutability feature in which data and information become unalterable. When every patient has an immutable document, the data are alter proof, accuracy is definite. When data and information are more accurate, data analysis for the medical practitioners and researchers become easier and almost error proof. Immutability also aids in securing documentation and stop falsified behavior and malpractices.
- *Stricter security*

    DLT is secure by nature itself. Data and information storage in decentralized manner, and the payment platform used are inevitably more secure. BCT data and blocks are difficult to

9.3 APPLICATIONS OF BLOCKCHAIN TECHNOLOGY   227

hack as it is based on cryptography and based on asymmetrical private key systems in order to safeguard transactions and data. The encrypted signatures on the block of data and information are difficult to manipulate. Also, when data is decentralized, not stored at single place, but it creates its multiple copies and shares that on peer-to-peer network which is intrinsically more difficult to hack.

- *Reduced costs*

    BCT has been working toward streamlining the expensive multistep intermediary prone processes. Eliminating the intermediaries and unwanted processes and steps which leads to delay in receiving timely and real care for the patients helps to reduce the cost.

- *Faster services*

    The DLT helps the healthcare industry to reduce delays in the processes for both service providers and patients. When the data and information are available and shared on a distributed ledger, the process between the doctors, specialists, health insurance providers becomes shortened, faster healthcare services can be provided to the patients.

- *Transparency*

    The BCT will provide transparency in the healthcare system when all the medical practitioners have access to accurate and unaltered data and information. Various areas where Blockchain applications could provide transformation and improve upon the overall healthcare delivery system across the globe are health insurance, pharmaceuticals, medical research and development, private healthcare providers, national healthcare systems, nursing homes, dentistry, healthcare administration, etc.

> Government and hospitals want to deliver complete care which is reasonable to manage and control and easy for the general public to access and approach. Now with the help of BCT, various tools are available which make dream a reality. Small start-ups and larger companies have been trying and figuring out ways to cut down the overhead cost, deliver improved care, restructuring insurance coverage processes, and thereby improving the overall quality of life and lengthen life expectancy for the larger population.

9.3.4.1 Use cases for blockchain applications in healthcare

- *Electronic medical records*

    Electronic Medical Records using BCT is the first and foremost use case in healthcare. This is because a single, longitudinal, and alter-proof record of patients can be made possible with the help of DLT. This will maintain all data related to vaccines, results of lab tests, treatments availed, and prescription history on ledger which is stored on a decentralized peer-to-peer network.

- *Tokenized healthcare*

    With the help of tokens, community members are motivated to improve the public health outcomes which not only help in creating and building a better society in terms of health but also become an income brook for the participants and develop the economy. The user can share, absorb, and earn with the help of their personal medical records and data. Prevention and treatment can be monetized and incentivized for the patients by tokenization.

228 CHAPTER 9 APPLICATIONS OF BLOCKCHAIN TECHNOLOGY

- *Medicine prescription compliance*

  Millions and millions are spent every year on medical prescription due to inappropriate patient compliance. Medication is costly for individual patients. Incentive can be provided for any improvement in medication through aplication program interfaces (APIs) which will gamify the medical prescription-taking process. Information which can stored on the blockchain will be accessible to both doctors and patients.
- *Automated health insurance (claims adjudication)*

  Using smart contract on the blockchain, verification of claims can be done over peer-to-peer network, and claim processing can be executed automatically after claim verification. This does not involve a biased third-party authority. Claims fraud might be prohibited through the trustworthy blockchain environment, and this can speed-up the claims process.
- *Personalized care*

  Personalized treatment can be there when easy-to-share data are available over time in distributed ledger. Medical practitioner can import medical history of family members to make clear understanding of the medical condition of an individual.
- *Medical product supply chain tracking*

  Pharmaceutical companies that provide medical care equipment or medicines to patients have to browse through a complex supply chain. When medication is to be recalled, it becomes difficult to trace the medicine back to their supplier. When blockchain applications are used to record the transaction related to the exchange of medicines, medical equipment, and services. A volatile audit trail is maintained. The transaction history can be used to decrease counterfeiting of medicines with harmful side effects.
- *Telehealth provider credentialing*

  Blockchain can provide telehealth which can become a suitable remote-care option. Blockchain can automatically send updates to the patient. This help to reduce telehealth. Whenever a new medical practitioner is available to accept patients, notifications could be sent to patients.
- *Patient consent management*

  Patient consent papers verification can be done on blockchain, rather than relying on office person for authenticate. Before going to the doctor, patients can record his/her symptoms and consent to treatment through the blockchain portal. These time stamped documents can settle malpractice by doctors.
- *Blockchain payment platforms*

  Health insurance can be implemented on the blockchain using smart contract, which can automatically issue funds held in a smart contract as soon as a condition is met (a patient gets discharged) occurs. Users can lock their funds in a smart contract for medical emergency for automatic payments.

Medical chain — It is the first healthcare services provider which has used BCT to aid the storage and employment of automated health records for delivering an innovative and altogether different telemedicine experience. They are the actual practicing doctors and specialists in UK healthcare structure and want to alter the system from inside.

> *MedRec* — To provide secure access to patients' records to the medical practitioners, MedRec uses BCT. This technology helps them to save money, time, and efforts and avoids duplication in procedures and processes in various facilities and services. Patients can also be grant access to their medical records and also to anonymous people to be used for further research.
> *Nano Vision* — Beholding to fling of medical field innovation and transformation with traditional data silos and irreconcilable records systems, Nano Vision combines the power of BCT with artificial intelligence (AI) in order to collect molecular-level data on Nano Tokens. AI then scrutinizes the data in order to identify various trends and patterns and thus analyze the networks and connections that may result into medical revolutions and innovations.
> *Gem* — With the objective of providing patients control and access of their medical information and genomic data with the help of BCT, the company Gem has partnered with "Centers for Disease Control and Prevention" in order to do experimentation of deploying BCT in order to monitor infectious and communicable diseases.
> *Simply vital health* — This particular platform uses the BCT and authorizes the providers and patients to have control and access of their medical and healthcare records and can also share their healthcare data.

9.3.5 APPLICATION OF BLOCKCHAIN IN VOTING

Three quarter of countries in world are democratic and dependent on voter consensus to elect officials. Unfortunately, the voting systems at present are inefficient and manipulation prone. BCT can improve the system to identify the rationality of the individual citizens. The decentralized ledger to store voting data through BCT correspondingly means that the result is not managed by a centralized authority thus eliminating the menace of voting result manipulations.

Another factor of supremacy which may be transformed using BCT is notary services. These administrative time stamps actually validate an action that happens in a person's life including birth and death details, documentation for new identity, receiving educational certificate, or transfer of ownership titles. As of now, many of these practices are done on secluded databases or through brick-and-mortar offices, which are generally prone to errors. Due to the encryption of the data and information stored in a blockchain, all these recorded data will be stored safely and will be only observable to the owner or the permitted parties.

> Technocratic states like Dubai are trying to transfer their entire governmental infrastructure on a blockchain. Also they are preparing to use smart contracts to reduce heavy documentation for movement of goods within the state.

230 CHAPTER 9 APPLICATIONS OF BLOCKCHAIN TECHNOLOGY

Projects promoting voting using blockchain systems:

- *BitCongress* is developed using the Ethereum platform. Its idea is that every voter has access to one "votecoin." This enhances him to cast vote only once.
- *Remotengrity* provides every vote with a cryptographic code to verify the authenticity of the vote.
- *AgoraVoting* uses the bitcoin network for blockchain-based voting.

Blockchain-based voting system has increased reliability and the convenience to offers to the voters.

9.3.6 APPLICATION OF BLOCKCHAIN IN REAL ESTATE

Blockchain could help in creating a new business models by connecting buyers and sellers. This technology will lower the barrier of real estate investment. New model of property ownership and rental contracts will come up with this shift in the real estate business.

Blockchain real estate platform will cut out inspection costs, registration, and loan fees and also property taxes, through smart contract. Blockchain could be used to change rental property payments system. Using blockchain all stakeholders including owners, tenants can interact together in secure way.

United States, Vermont, and Arizona have already recognized smart contract as format point of reference for any real estate transaction.

BitProperty

Using BCT and the smart contract concept, BitProperty application wants to develop a decentralized society where anyone, anywhere in the world (except the US and Japan due to some regulatory concerns) is allowed to invest in the real estate.

Deedcoin

Deedcoin is reported to run on less commission than the traditional real estate commission, and will hopefully be the innovative mode for home buyers and sellers to join with real estate agents.

Ubiquity

This Software-as-a-Service (Saas) blockchain implementation that offers a simpler interface to securely record property data and information so as to confirm a clean record of ownership.

Real estate is totally paper-based industry and thus with the help of BCT, the real estate sector will be significantly benefitted in terms of operational efficiency, data storage, and record

9.3 APPLICATIONS OF BLOCKCHAIN TECHNOLOGY 231

maintenance. The real estate industry can perform its important operations such as payment, title/ownership transfer, escrow, etc. with the help of blockchain and DLT to create extraordinary efficiency and cost savings. The most beneficial area which gets benefitted in real estate is the reduction of fraud, faster transactions, and enhanced privacy. BCT play an important role in title management of the property and provides better ownership record tracking.

The DLT has exceptional potential to decrease friction in paper-based transactions and replacing the same with automation. The transactions in the real estate generally involve intermediary/third-party which makes these expensive and time consuming. From rental property to larger commercial dealings, smart contract technology helps in making the real estate transactions smoother.

9.3.6.1 Problems in existing real estate sector

The real estate sector industry faces a lot of problems across the globe. Some of the major problems which affect this sector are: *frauds* at various levels from lower to higher. Security of data and information is the major problem seen in the centralized electronic funds transferring system, leading to stealing of data, information and funds. Inaccurate market data are another issue faced by this industry. Various popular platforms such as Zillows have failed to provide the real-time data and information to information seekers (investors and tenants). There is a lack of reliable sources of data providers in real estate sector, therefore creating a gap of authentic information for professionals and buyers to rely on. Total dependency on pen and paper format creates complexities within the system and becomes time intensive process.

Benefits of decentralization in real estate sector

- The trustless environment created by DLT with full automation and security using smart contracts is the major benefit gained by real estate sector.
- The chances of getting the data and information hacked on the blockchain platform is almost nil, therefore, decentralization provides higher level of data security.
- Immutability is one the important factor for the success of the DLT. Therefore, data stored in it are more reliable compared with the data stored on centralized server/database.
- Real-time availability of data is another benefit which real estate industry has not taken so far using traditional modes, it is possible with the help of BCT. Decision making by professionals and buyers can be now done in a better manner.
- Removal of intermediaries from the process can bring drastic change in the operations of real estate industry. It will bring transparency and reliability with speedy transactions, leads to lesser chances of errors and lesser involvement of cost.

9.3.6.2 Use cases for blockchain applications in real estate

- *Managing ownership titles*
  BCT helps to enhance traceability using the feature of transparency and immutability in the data and information. These features and benefits makes the investigations easier for the professionals and players in the real estate. Ownership risks get reduced with the help of DLT.
- *Tokenization of real estate platforms*
  Tokenization of real estate assets refers to buying the property through investment in digital currency. Tokens can increase its value in a larger real estate ecosystem through its use. These tokens can be exchanged or liquidated for various other cryptocurrencies; also buyers of

232 CHAPTER 9 APPLICATIONS OF BLOCKCHAIN TECHNOLOGY

property can receive in crypto as an investment. This benefit of tokenization in real estate can be consolidated in all transactions and processes and results into elimination of closing costs and brokers' charges. The transaction can be completed within a day as compared with several weeks in traditional modes.

- *Real Estate Investment Trusts*

    Real Estate Investment Trusts (REIT) can be benefitted from end-to-end with the help of decentralization. Smart contracts can help implement an event on the basis of the predetermined conditions. REIT can be crowd funded with the help of IPOs. Investors of the property can gain funds in timely fashion and do not have to wait for REITs to execute process on paper contracts.

- *Smart contract escrow*

    The most viable use for blockchain applications is the smart contract escrow. The smart contracts create a secure and safe repository for the funds which can released to verified parties in the system at the triggering of event information. The tenants have multisignature transactions with the help of "public private key" cryptography. The security deposit can also be deposited in escrow for the whole duration of lease and only be returned at the end of tenure of lease after the validation of private key.

- *Blockchain notarization*

    The notary is required for all kind of paperwork in the real estate sector. Both the parties in the transaction, that is, buyer and seller sign transaction agreement which can be recorded on smart contract. This agreement after all formalities receives a designated hash (code). Notarization can be done with the smart contract address and signing of the final documents. With the help of the private key, notary can mark the deal which is executed on blockchain. In future, with the acceptance of authorities, the notary can be removed from the entire process.

- *Tracing the property history*

    There has always been dearth of transparency in the property history. With the help of BCT, tracking of history of property is easier and in a transparent manner. This will help the buyers to invest in property with full confidence and trust.

9.3.7 APPLICATION OF BLOCKCHAIN TECHNOLOGY IN SUPPLY CHAIN

Global commerce is a reaction to the changing needs rather than an organized expansion. With manufacturing process taking place across the globe and the need of transparency between suppliers, supply chains are must. Blockchain is perfectly suitable supply chain management, with real-time tracking of goods and especially appealing to companies having multiple supply chains.

With the help of BCT, all inefficient and incompetent supply chain will be eliminated. Businesses are getting transformed with the help of blockchain-based supply chain solutions which offer end-to-end decentralized processes through DLT and digital public ledger.

Supply chain cannot be discussed without logistics industry, freight, trucking, shipping, and all other modes of transportation which we use to transport goods. There is a strong need to streamline and make its system transparent and that can only be done with DLT.

9.3 APPLICATIONS OF BLOCKCHAIN TECHNOLOGY   **233**

Walmart in association with IBM has started working on the Hyper-ledger Fabric block-chain, in order to track the food staples from the suppliers to retail stores shelf. Giant tech companies such as IBM have already realized the potential for blockchain supply chain management and have web 3.0 solutions in development or pilot program stages.

Blockverify

To enforce transparency in supply chains, Blockverify focuses on the solutions which use BCT to verify counterfeited products, stolen stock and merchandise, and deceitful transactions.

OriginTrail

Already is in use by food industry which is a platform which allows consumers discern where their product came from and how it was produced.

Problems in the traditional supply chain management

The existing supply chain management system is outdated and unable to match the pace of changes happening across the globe. Problem of *transparency* related to supply of goods from one place to other, their genuineness as "real" and "certified." Identifying the true value of transaction, expensive and inefficient systems are the major problems which have been face by the industry in supply chain management. The speed of existing supply chain is extremely slow, risk of counterfeiting and fraud always exists, lack of trust, unreliability, and insecurity in data are major problems which have been observed.

Benefits of decentralization of supply chain management

There are numerous benefits of decentralization of supply chain management. Major benefits are that it will bring traceability and transparency into the system. Real-time tracking of data is possible which helps to locate the items and their conditions, resulting in reduction of human error. There would be a change in the speed of transactions and efficiency level will also enhanced. Since the BCT is trustless chain therefore provides more security and eliminates the chances of fraud and errors. Other benefits of decentralized supply chain are improved inventory management, lower courier costs, less paperwork, faster issue identification, happier customers, and more time to innovate better products. Proper implementation of the distributed ledger could also prove to be valuable for pharmaceutical giants, which by law have to maintain the chain of custody over every tablet.

9.3.7.1 Use cases of blockchain application in supply chain management

- *Provenance tracking*

    For big multinational companies, it is difficult to keep track of all transactions and records. This also creates questions of company's reputation and reliability. Blockchain-based supply chain solutions provide answers to such issues and make provenance tracking possible with easy access to product information using embedded sensors and RFID tags.

234 **CHAPTER 9** APPLICATIONS OF BLOCKCHAIN TECHNOLOGY

- *Inventory management*

  Many companies have now started using BCT for inventory management. This also helps them to maintain real-time records in a distributed and transparent manner, and also provides better data analytics.

- *Identity verification*

  BCT also helps in identity verification, and it is a very popular use case for this technology. With the help of universal blockchain identity solution, the company do not require an intermediary to handgrip the international business relations and arrive at contracts and agreements. It helps not only in the verification of transactions, goods, and services but also the person involved in such transactions. The immutable nature of distributed ledger can help to validate the identity involved in the supply chain transactions.

- *Shipping logistics*

  In order to operate globally, the shipments are required to move on time. Blockchain carries the real-time tracking of all the participants in the supply chain in order to view the accuracy, alter-proof information and data. DLT helps making the best platform for the management of each and every aspect of shipping logistics.

- *Payments efficiency*

  BCT uses smart contract technology in order to connect the logistics related to delivery and its payments into digital contracts and bring efficiency in the payment mechanism in the supply chain management.

- *Food supply chain*

  Food-borne diseases can be easier to prevent and cure using the BCT which acts as tracing platform. Identifying the source of food and their further distribution can easily be done with DLT. It can be possible to find the good distributed to communities are contaminated or not. Tracing the food source can prove to be lifesaving for individuals.

- *Automotive supply chain*

  Supply chain based on BCT can prove to be a boon for the automotive industry. The whole vehicle history can be stored on distributed ledger with immutable feature which allows the buying of used vehicles trustless and reserve the resale price of the currently used new vehicle. Information on the public ledger will help the future buyers know the exact value of the vehicles and also help the owners to receive the correct value for their vehicles. This technology will also help in eliminating the counterfeiting in the automotive industry.

OriginTrail

Already is in use by food industry which is a platform which allows consumers discern where their product came from and how it was produced.

De Beers

De Beers Mines Company is planning to use a blockchain ledger technology to track diamonds from the mine to the purchase. This kind of transparency will definitely help the industry and the one who wishes to confirm diamonds.

9.3 APPLICATIONS OF BLOCKCHAIN TECHNOLOGY   **235**

AidCoin

As per the research, 43% of people do not trust charities. AidCoin is to increase that trust using distributed ledgers, smart contracts, and cryptocurrencies will make the nonprofit sector transparent.

Guts

It is a blockchain implementation solution to eliminate ticket fraud and secondary ticket market by making ticket allocation transparent.

Warranteer

It is a blockchain application which is used by the customers to access the information and feedback about the products and the services. Customers can complain in case of product malfunction or delay in issue.

9.3.8 APPLICATION OF BLOCKCHAIN TECHNOLOGY IN MUSIC INDUSTRY

BCT has transformed the music industry and empowered the musicians. It helps in streamlining the ownership rights of the music and helps in providing fair payment to the musicians for their work in a transparent manner.

9.3.8.1 Problems in the existing music industry

The major issues with the existing system in the music industry is the lack of transparency, ownership clarity, distribution of royalty, and the struggle to monetize digital music files. The data with its reliability and accuracy are extremely important to ensure that music creators and owners get the right and fair payment for their work. Complexities in the royalty distribution and collection process, copyright issues, etc. creates problems in the music industry.

BCT using the smart contracts will facilitate the creation of inclusive, reliable and accurate database of music files, music rights, and total transparency in the system with real-time distribution of royalties to cowriters, technology partners, producers, publishers, etc.

Integration of web 3.0 solutions will open up the potential to develop a totally fresh decentralized system which can support scenarios in which fans could pay for the amount of song which they have to listen or make payments in real time on a micro scale. Most significantly, BCT's smart contracts ensure that each time a payment is generated for a given work, the money is automatically get split as per the preset terms. Each party's account would be reflected instantly with the additional revenue and that too without the requirement of a third-party fund distributor.

9.3.8.2 Use cases for blockchain applications in music industry

- *Revenue sharing*

  BCT provides the means for artists to share revenue using smart contract. The smart contract will facilitate the parameters of the contract for song/album and release funds accordingly and shares revenue between artists, managers, etc. in a transparent manner and in real time.

236 CHAPTER 9 APPLICATIONS OF BLOCKCHAIN TECHNOLOGY

- *Tokenized fandom*
    Blockchain can bring transformation in the relationship between the fans and the artists. Fans will be able to participate in the tokenized ecosystem in order to buy music directly and can also participate in various polls, contests and interact with the artists in a meaningful way.
- *Media ecosystems*
    Decentralization will immensely benefit the music industry as there will not be any central repository where the data and content will be stored. It will help provide a new ecosystem to the music industry eliminating the intermediaries and third parties in the process, facilitating direct distribution which empowers the musicians in multiple ways. It will provide decentralized music streaming platforms in the industry.
- *Blockchain-based digital rights management*
    Copyright issues have always been one of the biggest hurdle in growth of companies, music industry, and the artists. BCT provides means to validate and authenticate copyright with a better level of transparency and provides rights management in digital and transparent manner.

9.3.9 APPLICATION OF BLOCKCHAIN TECHNOLOGY IN IDENTITY MANAGEMENT

BCT also provides transformation in the way identity management is done across the globe. It helps tracking and managing the digital identities in a secure and efficient manner which results in reduction of data leakage and fraud. Identity verification and authentication are required in every industry be it healthcare, insurance, banking, national security, online retailing, citizenship documentation, entry to a bar, or anywhere else.

9.3.9.1 Problems in the existing identity management system

Due to the lack of common platform of identity management, individuals every time need to verify and authenticate their identities at all places wherever they go or whatever facility/service they avail. Advancement of technology has brought biometric identification which is password-based and stored in an unsecured system which are highly prone to data theft and hacking. The centralized systems are identity centric and people get one social security number containing all details. Misuse of this information may result in disastrous acts such as frauds in banking, purchasing, emailing, fake identity creation, terrorism, etc. Moreover, companies also sell the personal information for commercial purposes and generate revenues.

The decentralized mechanism of identity management helps resolves all such issues by providing altogether new model of identity management using BCT infrastructure. Cryptography is used to separate data from the identity of individuals for better security. With the help of separate data management companies can possess and obtain data which is of their use and preserving the individuals' privacy at the same time. This will create a win—win situation for government, individuals', and companies.

9.3.9.2 Use cases for blockchain application in identity management

- *Data collection and its analysis*
    Accuracy of data is of utmost importance for any country, state, or company. The real-time data storage using blockchain can be analyzed and improved in numerous industry practices. This will also bring faith in terms of security of data. Analysis and analytics can be applied in a

better manner for the research and development, decision-making, and further betterment of the society.

- *E-Residency*

  The identity management done using BCT can help individuals to vote, file their income tax returns, perform various other processes in a more efficient and secure manner. It can also help virtual residency authentication due to Government verified ID's maintained using DLT. All the government transactions can be moved to blockchain using e-residency identity management in order to streamline all interactions among individuals and government.

- *Immigration and identity*

  Digital identity card can be linked to the details on the blockchain, and this will act as a temporary ID card when entering in new country. This will help immigration services to get smoother. Travelers can also link their debit/credit cards and can monitor their account activity, at the same time access can only be given to account holders to prevent theft and fraud.

- *Self-sovereign identities*

  The individuals can have a better life with the secure, transparent, reliable, and accurate identity management system provided by BCT. This will eliminate the involvement of third parties for digital/manual identity management, eliminate identity sprawl, and identity theft.

Around one-sixth population in the world does not have the documented evidence of their existence. Blockchain can help in establishing the immutable identity record for this one-sixth population and giving them access to various other services such as education, banking, mobile communication, etc. Biometric data, when used in combination with the BCT, helps to create immutable identity records on the distributed ledger. BCT enabled immutable identity can be lifesaving in refugee camps and allow more families to unify amid displacement.

9.4 BLOCKCHAIN CHALLENGES AND CONCERN

Major challenges in blockchain implementation are because of the consensus protocol which demands very high computation support. More the number of nodes in blockchain means better blockchain network, but computational power increases in direct proportion to the network size. Many questions are to be still answered, that is, When businesses should implement private blockchain? When public blockchain make sense for an organization? How to provide integration among them? There is a direct relation between the blockchain size and security. Different application domains have different concerns while blockchain is implemented.

- *Millitary application*

  With enough number of nodes in a blockchain system, the information transfer between the sensors at the battle field and the command authorities, the system ensures to become almost unhackable. Only challenge with increasing number of nodes would be the processing time taken to validate an information transfer might exceed the duration within which a response or decision is needed to a legitimate threat reported by sensors.

238 CHAPTER 9 APPLICATIONS OF BLOCKCHAIN TECHNOLOGY

- *Enterprise resource planning (ERP)*
  When blockchain is combined with ERP, a significant amount of infrastructure support will be needed for cooling and maintaining the systems or nodes. A solution could be Proof-of-stake consensus algorithm instead of Proof-of-work.
- *Shipping industry*
  TradeLens is a implementations case in supply chain where shipping organization can simply access data related to goods. Furthermore, this solution runs of IBM's Hyperledger fabric which supports peer-to-peer interaction. Moreover, all the nodes can access any data through the ledger. In short, Company A can view data relating to Company B.
- *Banking*
  People use VISA for their cross-border transactions. Banks use SWIFT for cash movement in bulk. Such structures have so much entrenched in the current system that it will definitely take a long time to introduce new mechanisms. This is what BCT is doing, changing the nature of operations.
- *Healthcare*
  Using smart contract concept, patient data will be encrypted, and permission is granted to access this information. Data are digitally signed, and medical officials would have to access to this data if patient cannot provide it, for example, in case of unconscious patients. This means that the patient has to entrust his or her private key with someone else, thus creating a good security gap. There also exist interoperability challenges among hospital systems, which results into lack of coordinated data, leaning fragmented health records.

9.5 FUTURE CASES

9.5.1 BLOCKCHAIN TECHNOLOGY IN MILITARY

Introduction of digital technologies have changed the warfare. Nowadays, war fighters make use of connected devices for air strikes, drones in the battle place too are controlled from very far places. In earlier years, hackers used to take control of the operator's terminal and could be seen, in real time, whatever the operators used to see on their screens. Hackers could now compromise the system and can send a pop-up on user's screen. Later, a London-based, non-governmental organization warned in a report in January 2018 that nuclear weapons systems are becoming increasingly vulnerable to cyberattacks. The Nuclear Threat Initiative (NTI), US nonprofit organization published a report on cyber threat to nuclear weapons. They concluded that there is a high possibility that US nuclear weapons systems will be compromised. Considering these cyberattacks, a new paradigm is needed to address the vulnerabilities of defense systems. It is believed that blockchain be a key role player in rectifying these weaknesses.

The potential benefits of blockchain to protect defence system against cyberattack can be presented as distinct use cases.

- Defending critical weapons systems
- Managing automated, swarm systems
- *Defending critical weapons systems*

The operator in system receives data from a lot many sensors. This data notify command authorities about the incoming threat. Command authorities then direct the weapon to respond to threats. In the centralized system, there is one points of vulnerability and that can be breached by external bad actors. So, command authorities of the weapon system may receive deceptive information. This might lead to either illegal use of weapons or even failure in response to a legitimate threat. Alternatively using blockchains, data transmissions from sensors to the operators are validated using a consensus system. As transaction is approved by at least most of the nodes within the blockchain network, any hacker would have to hack all nodes in the chain simultaneously. The computing power needed to hack such a system is magnificent.

- *Managing automated, swarm systems*

  A swarm robotics is a way to coordinate many robots as a system. It is to implement a desired combined behavior from the connection between many robots and also interaction of robots with surroundings. The dependence of robots for communication and interaction open a loophole for hackers. Blockchain proposes a mechanism to protect intraswarm coordination. In this system, each robot of the swarm will act as a node in their blockchain. In such implementation, the swarm can exchange information and protect itself from cyberattack.

9.5.2 BLOCKCHAIN TECHNOLOGY IN INTERNET OF THINGS

Internet of Things (IOT) is defined as a network of connected objects through internet for collection and exchange of data. In IOT, billions of connected devices are raising the concern of security, storage, cost and cloud attacks, and privacy. Storing large volume of data raises the concern of storage and security in cloud platform.

With ongoing innovations in BCT shall bring the new transformations in the IOT industry. Blockchain helps in tracking of billions of connected devices, processing of transactions and coordination among devices in IOT. The decentralized approach of blockchain eliminates the single point failure, thus ensures the more resilient system in IOT industry. IOT industry manufactures can be benefitted with substantial cost savings because of integration of BCT and IOT. Blockchain empowers IOT devices with the enhanced security and transparency in IOT systems. BCT driven with cryptographic algorithms become enable for security challenge of IOT industry.

Cryptographic signatures enhances the security feature of IOT. Immutable and time stamped transaction assures the tamper proof data. Storing IOT data in a distributed fashion save the cost of IOT by preventing monopoly of service providers and damage cost of hackers. Distributed ledger provides the trust between parties and devices of IOT and automated services by smart contract. Blockchain enables the autonomy of smart devices and removes intermediate parties completely.

- *Smart home:* Home appliances are smarter nowadays, and they are able to connect with the internet, other appliances, and mobile phones. All appliances are chipped with the sensor, and sensed data are stored in the central server or cloud storage nowadays. This raises the concern of security and privacy. These data could be stored in blockchain, thus ensures security and privacy.

240 CHAPTER 9 APPLICATIONS OF BLOCKCHAIN TECHNOLOGY

- *Automobile industry:* Automobile industries are more benefitted by automating the vehicles. Vehicles integrated with the sensors enable the communication between vehicles and exchange information among users. Automobile combined with BCT enable the easy way of fuel payment, autonomous cars, automatic traffic control, and smart parking. Since the vehicles are IOT enabled, it could easily update the blockchain-based ledges for the movement of vehicle record. These records are immutable and transparent. It enhances the transparency in automobile industry and keeps track of parts movement also. Blockchain systems would help in transforming the vehicle manufacturing, distribution, selling also.

9.6 CATEGORIES OF BLOCKCHAIN DEVELOPMENT

9.6.1 APPLICATION OF BLOCKCHAIN TECHNOLOGY AS A DEVELOPMENT PLATFORM

BCT is used to create distributed applications which have potential much more than just allowing a digital currency. Such type of applications is called as Crypto 2.0, Blockchain 2.0. Ethereum was introduced in July 2015. It is most well-established, decentralized software platform which supports smart contracts and distributed applications to be run without any flaw and control from a third party. Ethereum is not only a platform. It is a programming language too helping developers to publish distributed applications.

9.6.2 APPLICATION OF BLOCKCHAIN TECHNOLOGY AS A SMART CONTRACT

Smart contracts are contract which execute on their own when the terms of the agreement set between buyer and seller are satisfied. A smart contract could act as notary, as it can be implemented in real estate for registry transfer. Also a user can create his/her will on this using blockchain smart contract feature, and the contract will execute on its own once he/she is death, without any intervention by third party (notary or judge) to validate it. Smart contract can be helpful for betting, that is, users put their money on distributed digital ledger account. The virtual contract defines the conditions of winning and losing. Once the result is out, whosoever wins or loses, the contract gets executed the terms, and money is transferred to the winner's account.

9.6.3 APPLICATION OF BLOCKCHAIN TECHNOLOGY AS A MARKETPLACE

Decentralized marketplaces are where there is no middle man between buyer and seller. The conditions of trade are transparent and immutable. No tampering of financial data which was a part of the transactions is possible. Some more key features are transaction does not require any third-party payment system, BCT creates the global marketplace, the infrastructure is not centralized, so cannot be hacked, anyone's personal data are with no centralized owner, so there is no threat that personal sensitive data are at stake and can be sold.

9.6.4 APPLICATION OF BLOCKCHAIN TECHNOLOGY AS TRUSTED SERVICE APPLICATION

BCT could be put to use in much of the government or business scenarios where trust service is the requirement in the business applications. Permissioned blockchains will support the implementation of a trustworthy ecosystem in which required services can be developed in the long run. Emphasis is majorly on trust and confidentiality of transactions.

9.7 SUMMARY

In fact, interconnected blockchain application based on the concepts of scalability, smart contract, trustworthiness, privacy, and interoperability can serve many benefit to stakeholders. A shared ledger, among automotive manufacturers, automotive dealers, insurance providers, vehicle leasing organization, buyers, and sellers will provide a higher degree of transaction transparency and trust in almost every vehicular transactions, preventing disputes, and lowering down the overall cost of maintenance. Also, it could significantly streamline processes, especially those that rely on regulatory and compliance approvals from third party.

This chapter consists of various sections and subsections. The first section introduced applications of blockchain and relevance of BCT. The second section discussed the background of the BCT applications, emergence of bitcoin, and working of bitcoin. This section also discussed various legal issues involved in bitcoin. The third section gave elaborated discussion on various applications of BCT. This section has also covered real-time use cases, companies developing these real-time applications, their benefits to the organizations and society at large. The fourth section mentioned various concerns and challenges in the implementation of the BCT in various domains. Fifth section covered various future use cases which are under implementation across the globe. Sixth section of the chapter focused on various categories under which the application of BCT can be done for future development.

REFERENCE

[1] IDBRT, 2017. Applications of Blockchain Technology to Banking and Financial Sector in India Available from: http//www.idrbt.ac.in/assets/publications/Best%20Practices/BCT.pdf.

FURTHER READING

R. Aggarwal, Blockchain and financial inclusion. Digital currency group, 2017. Available from: <http://finpolicy.georgetown.edu/sites/finpolicy.georgetown.edu/files/Blockchain%20and%20Financial%20Inclusion%20120417.pdf>.
Available from: <https://www.coindesk.com/information/applications-use-cases-blockchains>.
Available from: <https://www.investinblockchain.com/blockchain-transform-industries/>.
Blockchain Applications. Available from: <https://blockgeeks.com/guides/blockchain-applications-real-world/>.

242 CHAPTER 9 APPLICATIONS OF BLOCKCHAIN TECHNOLOGY

Blockchain Technology. Available from: <https://www.blockchaintechnologies.com/applications/>.

M. Gebert, Application of blockchain technology in crowdfunding. New Eur, 2017. Available from: <https://www.researchgate.net/publication/318307115_application_of_blockchain_technology_in_crowdfunding>.

Y. Guo, C. Liang, Blockchain Application and Outlook in the Banking Industry, (n.d.).

M. Gupta, IBM Blockchain for Dummies, second ed., A Wiley Brand. M. Young, The Technical Writer's Handbook. University Science, Mill Valley, CA, 1989.

HyperledgerSawtooth. Available from: <https://www.hyperledger.org/projects/sawtooth>.

Introduction to Blockchain. Available from: <https://www.javatpoint.com/blockchain-introduction>.

S.K. Johansen, Working paper on a comprehensive literature review on the blockchain technology as a technological enabler for innovation, 2018. Available from: <https://www.researchgate.net/publication/312592741>.

A. Kharpal, Blockchain Revolution: Everything You Need to Know About the Blockchain, CNBC, 2018.

D.E. Krause, V. Velamuri, T. Burghardt, D. Nack, M. Schmidt, T.M. Treder, Blockchain technology and the financial services market — state of the art analysis. A joint report by Infosys Consulting and HHL Leipzig Graduate School of Management, 2016. Available from: <https://www.infosysconsultinginsights.com/insights/blockchain-technology-and-the-financial-services-market/>.

A. Krishnan, 24 industries that blockchain will radically transform, 2018. Available from: <https://www.investinblockchain.com/blockchain-transform-industries/>.

M.H. Miraz, M. Ali, Applications of blockchain technology beyond cryptocurrency, Ann. Emerg. Technol. Comput. 2 (1) (2018).

S. Nakamoto, Bitcoin: a peer-to-peer electronic cash system, 2008. Available from: <https://bitcoin.org/bitcoin.pdf>.

A. Narayanan, J. Bonneau, E. Felten, A. Miller, S. Goldfeder, Blockchain technology in India: opportunities and challenges. 3Princeton University report on "Bitcoin and Cryptocurrency Technologies" by Arvind Narayanan, Joseph Bonneau, Edward Felten, Andrew Miller, Steven Goldfeder, Deloitte, 2017. Available from: <https://www2.deloitte.com/content/dam/Deloitte/in/Documents/strategy/in-strategy-innovation-blockchain-technology-india-opportunities-challenges-noexp.pdf>.

J. Naughton, Is blockchain the most important IT invention of our age? 2016. Available from: <http://www.theguardian.com/commentisfree/2016/jan/24/blockchain-bitcoin-technology-most-important-tech-invention-of-our-age-sir-mark-walport?CMP = share_btn_fb>.

M. Peck, Freelance technology writer, a white paper on reinforcing the links of Blockchain, in: IEEE Spectrum Magazine Special Edition "Blockchain World", 2017.

Proof of Authority. Available from: <https://github.com/paritytech/parity/wiki/Proof-of-Authority-Chains>.

S. Seebacher, R. Schüritz, Blockchain technology as an enabler of service systems: a structured literature review, in: International Conference on Exploring Services Science, Italy, 2017.

T. Shah, S. Jani, Applications of blockchain technology in banking and finance, 2018. Available from: <https://doi.org/10.13140/rg.2.2.35237.96489>.

M. Swan, Blockchain: Blueprint for a New Economy, Vol. 3, no.3, O'Reilly Media, 2015, pp. 38−72.

L.J. Trautman, Is disruptive blockchain technology the future of financial services? in: The Consumer Finance Law Quarterly Report 232, 2016. Available from: <https://papers.ssrn.com/sol3/papers.cfm?abstract_id = 2786186>.

H.T. Vo, A. Kundu, M. Mohania, Research directions in blockchain data management and analytics, in: 21st International Conference on Extending Database Technology (EDBT). OpenProceedings.org, 2018, pp. 445−448.

G.R. White, K. Brown, Future applications of blockchain: toward a value-based society, in: Conference: INCITE, At Amity University, India, 2016. Available from: <https://www.researchgate.net/publication/308916112_Future_Applications_of_Blockchain_toward_a_value-based_society>.

FURTHER READING

J. Yli-Huumo, D. Ko, S. Choi, S. Park, K. Smolander, Where is current research on blockchain technology? A systematic review, PLoS One 11 (10) (2016) e0163477. Available from: https://doi.org/10.1371/journal.pone.0163477.

Z. Zheng, S. Xie, H. Dai, X. Chen, H. Wang, An overview of blockchain technology: architecture, consensus, and future trends, in: Proceedings of the 2017 IEEE BigData Congress, Honolulu, HI, 2017, pp. 557−564.

CHAPTER

BLOCKCHAIN-POWERED SMART HEALTHCARE SYSTEM

10

Rashmi G. Shukla[1], Anuja Agarwal[1] and Shekhar Shukla[2]

[1]Technology Management Area, MPSTME, NMIMS, Mumbai, India [2]Information Management Area, S.P. Jain Institute of Management and Research, Mumbai, India

10.1 INTRODUCTION

The advent of blockchain derives its roots from a white paper introducing decentralized peer-to-peer electronic cash system called bitcoin by Satoshi Nakamoto owing to the financial slowdown in the year 2008. Blockchain, which is developed from bitcoin, has revolutionized every industry in the past decade and has become one of the path-breaking technology in today's world. This technological offering is based on trust and transparency being offered in a decentralized environment [1]. Since then, companies and researchers have been trying to implement the technology in many industries like marketing, finance, human resources, operations and supply chain, and so on [2]. The market of blockchain is expected to rise from 2.2 billion USD in 2019 to 23.3 billion USD in 2023 [3]. Healthcare which faces many problems with respect to privacy and security has great deal of scope for implementing blockchain technology [4,5].

Initially, blockchain was developed to be an alternative to currency. A secure and decentralized currency that can be used as a medium of exchange across the globe. A peer-to-peer connected network would facilitate the online payments to be sent directly from one party to another without any intervention of a third-party vendor [6]. The timestamps by network would add another layer of security by avoiding any data infringement in the ongoing transaction. This property makes blockchain far ahead of other technologies and could be used to perform any exchange with an added advantage of trust, transparency, and traceability across different industries.

After few years of release of bitcoin, the code was released as open source enabling researchers to utilize the same which resulted in creation of varied blockchain-based applications and prototypes. Researchers began to realize the potential of blockchain and started exploring the technology in other industries outside the realm of nonfinancial industries. Blockchain in itself is a decentralized, peer-to-peer distributed ledger which is capable of recording all the transactions that are happening over the network. This property makes blockchain useful for any kind of exchange like data, currency, information, and so on. This realization stimulated a huge investment and research in blockchain attempting to revolutionize the technology for use in applications like healthcare, insurance, operations and supply chain, and many more.

The recent developments in blockchain technology are aimed at nonfinancial applications of blockchain. Hence, the research is going on to extend this technology to other industries like human

---

Handbook of Research on Blockchain Technology. DOI: https://doi.org/10.1016/B978-0-12-819816-2.00010-1
© 2020 Elsevier Inc. All rights reserved.

245

resources, identity management, supply chain management, and so on [7,8]. With the growing passion for blockchain across different industries, healthcare is one of the most used arena where a number of applications and prototypes are developed in recent times. Healthcare practitioners suffer with fragmented and delayed data, extended and late communications, and maintaining patient records. Blockchain-based framework is also presented to gather insightful information through predictive techniques embedded in the framework [9]. With the implementation of blockchain technology in healthcare, all the stakeholders of healthcare can be brought under a single window and exchange information in a much better way.

The most obvious advantage of blockchain is that it eliminates the need of a third party by enabling two parties to transact with each other in a distributed environment. The transaction speed is increased due to elimination of third party, and the transaction cost also is mitigated. Basically, blockchain uses cryptography to derive most of its properties. Each participant on the node is categorized as a node and has a pair of public and private keys. The public key acts as a public address of the participant, and private key is used for authentication. When a transaction is created, it will have the public key of the sender, public key of the receiver, and the transaction message. Then the transaction is cryptographically signed using the private key and transmitted over the blockchain network. This completes one transaction. A block is a collection of many such valid transactions which are sent over the network within a specified time limit. Validation of transaction ensures that the transaction is legitimate and generated from a valid and connected node. The special nodes on the network that are responsible for checking the validity of the transaction are called miners.

With time, there have been variations in the development of blockchain. Hence, many types of blockchain have been developed. In some implementations, any node is free to join the network and can become a miner. Such blockchain implementation is called public blockchain. Permissioned blockchain is yet another implementation in which each node wanting to join the blockchain network needs to be authorized and have permission to join the network. However, only some nodes can become miners in a permissioned blockchain. Hence, the permissioned blockchains are small and are comparatively faster and secure. If only one node is permitted to be a miner, it is a private blockchain. But the private blockchain loses the decentralization property to a great extent as the control is given only to a single node. If more than one node acts as a miner, it is called consortium blockchain.

One of the most important characteristic of blockchain that is advantageous to healthcare is decentralization making it possible to implement distributed healthcare network not dependent on a single authority. Thus blockchain protects the potential loss of data or data corruption in the healthcare network. It is very important to maintain the integrity and validity of patients' records to ensure their wellness. The immutability property of blockchain which makes it impossible to change any record helps in maintaining the integrity of data in healthcare.

The fundamental property of blockchain lies in its Information Technology architecture and the chain of data entries that allow secure and open transactions which is unbreakable. Blockchain can also increase interoperability among different systems which is important for improvement in the current healthcare system. Blockchain has the ability to replace the current obsolete Information Technology systems existing in healthcare with a single interoperability system. Apart from offering interoperability, blockchain also gives the advantage of being cryptographically secure and irrevocable transactions thus ensuring privacy between parties involved in transactions.

Healthcare system opens up a plethora of opportunities to blockchain for implementation. There are many stakeholders involved like patients, hospitals, doctors, clinical researchers, policy insurers, and drug suppliers to name a few. A lot of data exchange takes place across different stakeholders in the industry. Hence, it becomes important that the data are not being misused by any of the entity by sharing the records illegally, tampering of the records, or any other means. The data tampering between exchanges of any of stakeholders may affect the treatment of the patients and thus pose severe danger to their life. One of the most important applications of blockchain is monitoring the drug supply chain [10].

Healthcare industry faces major problem with respect to storage and data exchange risking the data privacy and security [11]. The researchers are developed a blockchain system called BlocHIE that will store and exchange the information between different stakeholders involved in the process ensuring that the data are not tampered, and patient data are kept private [12]. Besides storage and information exchange, daily data collection is also an important factor in the healthcare industry. Blockchain can thus handle the entire data system in healthcare where all the stakeholders can access the data from a single point safely and securely.

Blockchain can have a greater impact on the clinical research field of healthcare since blockchain allows for storing, sharing, and tracking of data [13]. This blockchain system can increase the credibility of clinical research which has been repeatedly maligned due to various scandals in recent years (another citation). Blockchain technology can act as a positive catalyst for the improved clinical research methodology and will be a step closer to better transparency and trust in the network. It can also enhance the safe communication between research and patient communities and improve trust within the research communities.

A major problem with respect to healthcare systems today is absence of secure links which connects all the independent healthcare systems together to create an end-to-end system ensuring data protection and privacy [14]. Eventhough there are data standards maintained across systems for data exchange, it comes with a maintenance cost. Thus a more transparent discipline which is secure and private would be more apt for the system. A more sufficing solution to these problems comes with the implementation of blockchain, enabling transactions via decentralization. Complex system of healthcare industry poses additional challenges to implementing blockchain technology.

There are many blockchain implementations available for various functional aspects in healthcare. The major benefit of blockchain is to enable efficient data sharing among the healthcare stakeholders keeping privacy and security intact [15]. However, the discrete blockchain applications for different functionalities of healthcare system would add additional complexity to the existing system. This chapter proposes a holistic model of healthcare system in order to develop a smart healthcare system which is particular to the context of smart cities. Other technologies like Internet of Things (IoT), Analytics, and artificial intelligence (AI) will mitigate any challenges that are faced during blockchain implementation.

10.2 HEALTHCARE SYSTEM

Healthcare industry comprises several sectors which are dedicated for providing health services and products. United Nations International Standard Industry Classification categorizes it as an industry consisting of hospital, medical, and dental activities under the supervision of nurses, physiotherapists, doctors, pathologists, and other allied health professions. Healthcare mainly concentrates on maintenance and improvement of health by prevention, diagnosis, treating diseases, illness, injuries, and other physical and mental ailments. Healthcare availability to public may vary across the countries, communities and is largely driven by social and economic environment and also on the government policies. However, healthcare organizations are established to meet the health needs of general public.

The healthcare delivery mainly depends on trained professionals coming together as interdependent teams. It includes teams from medicine, physiotherapy, psychology, dentistry along with public health practitioners, community workers who work together to deliver preventive and rehabilitation health services. Although the basic goal of every healthcare practitioner is the same, they are categorized in three different types.

Primary healthcare: It consists of those professionals who act as first point of contact and consultation for patients in the system. This category provides the widest scope for people with all age groups, all socioeconomic backgrounds, and all types of chronic diseases. Hence, the primary healthcare practitioner is expected to have broad knowledge in many areas. Primary healthcare often plays a role in local community.

Secondary healthcare: This segment includes intense care for the illness which is serious but has a brief time period. It can be considered as the synonym for the hospital emergency department. Based on the healthcare policies and organization rules, patients sometimes may be required to visit a primary practitioner for referral before consulting secondary care.

Tertiary healthcare: This sector is a specialized unit consulted mainly on the basis of referral from primary or secondary care. This sector requires more care and time to heal the illness and is majorly referred only when the patient is expected to have chronic diseases.

Healthcare industry is functional due to involvement and contribution of different set of professionals working for it. These professionals are called as stakeholders. In other words, any person who is directly or indirectly affected by the operation or who contribute to the functioning of the industry are called as stakeholders. They are the entities who are involved completely or partially in the industry and are largely affected by the functioning of the system. Major stakeholders of the healthcare industry are patients, physicians, pharmaceutical companies, insurance companies, and government. However, the interrelationship among the stakeholders in the industry is quite complex. Some of them are publicly owned, and some are individual workers. Hence, the regulations and policies vary for each relation among the stakeholders.

Following schematic diagram in Fig. 10.1 shows the relationship between different stakeholders in the industry.

10.2.1 INSURANCE INDUSTRY

Health insurance is the coverage plan that provides whole or part of the expenses that are incurred for medical purpose by a person or group of persons. According to Health Insurance Association of America, health insurance is an entity that covers payments of beneficiaries as a result of sickness or injury. The insurance company develops a routine finance structure based on the number of people covered for the plan. Such benefit is administered by a central body-like government, private organization or nongovernmental organizations. In recent times, it has become important to have an

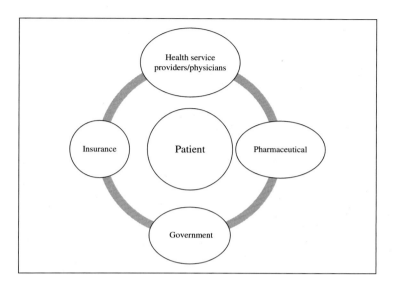

FIGURE 10.1

Schema of patient-centric healthcare system.

insurance for oneself and family since the medical care has become expensive and the chronic diseases have become more prevalent in the world.

Insurance companies deal with lot of paper work back and forth and are prone to human errors. There are also chances of scams and bankruptcies due to mismanaged data and information tampering. Blockchain technology can overcome these documentation shortcomings and privatize the data concerned with each individual. The blockchain technology for insurance sector will speed up the business process in the industry client registration to policy issuance, reduce the operational costs, make client data more confidential and accessible only to authorized public [16].

10.2.2 PHARMACEUTICAL INDUSTRY

The pharma companies play a vital role in the healthcare industry, since the patients are dependent on the products manufactured by them. The industry discovers, develops, produces, and sells the medical drugs which are used as medications to patients. The companies are bound to various laws and regulations governed by private bodies or government. The drug expenses are rising with time and are becoming unreachable to the common people. The increasing price is due to research investment of the companies in the drug discovery, the marketing and supply chain expenses, and many more.

The pharma industry, in recent times, has become victim of counterfeiting at global level [17]. The consequences of this problem pose serious after math problems to patients involved. This phenomenon is very common in the areas where surveillance and regulation need an improvement [18]. Drugs move across the supply chain and involve many vendors in between before finally reaching the end user. A regulated body tests a batch of the drugs in order to know the quality of

250 CHAPTER 10 BLOCKCHAIN-POWERED SMART HEALTHCARE SYSTEM

the drug, but this system fails as it is not possible to track down the origin or authenticity of every drug involved. The blockchain framework developed can actually trace and track the drug products. It also enables traceability to the source of the drug [17].

10.2.3 PHYSICIANS

Physicians play a vital role in ensuring that the patients receive proper care. They have to play a balance between the patient and other stakeholders of the healthcare industry. Physicians are alone capable of taking any decisions in regard to the patient for any of his clinical complexity and uncertainty, thus understanding the patient's case with his knowledge and experienced clinical judgement. The doctors should maintain and promote population health based on clinical study and experience. The role of doctor is changing in healthcare and with the advancement in medical science, his experience-based errors and mere guessing of the diseases based on the symptoms that has reduced.

With the blockchain technology implementation in healthcare industry, the job of the physicians has become comparatively easier. Due to exploding population, the physicians would not have enough time to look at individual patient's charts and give prescription accordingly. With the help of blockchain, where all patients' records are made available on a single platform and summarized reports help them in better interpretation in lesser time.

10.2.4 PATIENTS

Patient is one of the most important stakeholder and is the end user in the healthcare industry. The healthcare programs and policies should be made keeping patients in mind. Sadly, the concentration of policy makers and healthcare organizations have shifted from patient-driven to profit-driven. In today's world, the patient possess a greater knowledge than before but still are given very less or no opportunity for contributing to policy making and decision-making system. The interests of organizations, medical professionals, and other healthcare providers are represented through various government bodies or unions. However, no such organization or body exist to represent the patient community or to regulate the principles from their point of view.

The blockchain technology is going to benefit the patients in a larger perspective. The medical records will be made available on a single platform which helps the patients to access the records from any location [12]. With the help of such framework, it is not only going to help the patients with access but also protects the privacy and security of the data. The patients will be able to control who has access to their data and can also protect their records from tampering.

10.2.5 GOVERNMENT

The role of government in healthcare has expanded over a period of time and has a great influence on the power and political discourse in the healthcare industry [19]. The government has played a vital role in promoting the use of preventive measures for chronic diseases and also contributes to healthcare efficiency and cost savings. The government is also responsible for making budget and other planning activities related to expenditure in healthcare industry. Still, the government has a large area for market influence and policy reforming in the healthcare sector for achieving better

10.3 ISSUES AND CHALLENGES FOR A HEALTHCARE SYSTEM **251**

quality and value. It is government's responsibility to protect and advance the society interests by delivering high-quality healthcare and also serve the interests of patients by supplementing the gaps where there are gaps and inefficiencies [20].

Healthcare has broadened its wings ever since its existence and with increasing time, it is becoming more complex [21]. Aging phenomena and number of prevalent chronic diseases have become driving forces for creating increasing demand for the healthcare industry. There will still be an increasing demand for the healthcare industry in the times to come. Due to complex communication between different stakeholders in the industry, the system is more prone to errors and scams costing life of patients involved. Industry should not only take care of the patient but also consider on protecting the privacy and security of the data [11]. It is driven by record maintenance, compliance, and regulative principles [22]. Considering the importance of healthcare industry to people around the world and for national economies, it becomes increasingly important to manage the industry in an efficient manner.

10.3 ISSUES AND CHALLENGES FOR A HEALTHCARE SYSTEM

Healthcare today finds a constant challenge in order to find ways to improve the quality of care, reduce costs, and also increase revenue [23]. Although there are many technologies that have come up in order to cope up with the problems of the industry, it still faces lot of challenges with respect to privacy and security. Despite several improvements, inefficiency still exists in the industry, sometimes causing serious threats to life of the patients involved [24]. Although the advent of wireless sensor networks in healthcare industry is gaining popularity [25], its blockchain applications in healthcare industry will create revolution with respect to the way the industry works.

Since healthcare organizations gather highly sensitive patient data, the industry is a major target to the cybercrime [26]. This problem will continue to persist since the healthcare organizations are very slow in responding to such issues. The existence of centralized systems to maintain data makes it more vulnerable to cyberattacks. When a data breach happens, it not only leaks the important information of the patients but also violates the rules and regulations of the organizations thus causing serious threats to organization and the patients involved. Apart from preventing the cybercrime, healthcare organizations should adapt a technology that is more robust and safe in nature as a full-time solution.

Interoperability is the ability of heterogenous information systems and software technologies, like Electronic Health Records (EHRs), to be able to communicate and exchange the data [27]. Permitting the information to work together within and across the organizational limits is important for effective care and safety to individuals and communities [28]. For example, the interoperability allows healthcare providers to share the patient data securely with one another irrespective of locations and trust relationship between two parties involved in exchange [29].

Secure data sharing is vital to provide effective collaborative treatment and care to the patients. Data sharing assists in improving the diagnostic accuracy [30] by collecting confirmations and opinions from different experts. It will also prevent inefficiencies and errors in treatment schedule and medication [31,32]. Despite the importance of data sharing in the healthcare systems, today's healthcare systems require patients to collect and share their medical records with physicians by hard copies or electronic copies. This process is inefficient since it is slow, insecure, may be incomplete, and lacks context. The ineffective data sharing mechanism leads to lack of trust among the providers and lacks interoperability between the health systems and applications today.

The healthcare industry is taking a paradigm shift from volume-based to value-based. Earlier, the providers were offered incentives in order to provide more treatments because the payment was related to quantity of care. The value-based care facilitates patient-centered care with utmost quality. This system involves patients in the decision making and are kept well informed. It also allows the data to be collected on the digital platform and gives them easy access thus reducing information fragmentation and inaccuracy due to communication [33].

In addition, it becomes necessary that the patients can control when and to whom their medical data are being shared and should also be able to choose to what extent are they ready to share their information. However, healthcare systems today do not provision such flexibility. Once a provider gets his hands on any particular patient, it will be with him permanently. The current system does not allow the patient to revert access given to a particular provider. Hence, if a patient visits several providers in his lifetime, his sensitive medical data are available permanently at several sites. This increases risk of data theft because it only causes a single provider who is not up-to-date with the latest security practices.

Medical data possess huge amount of data in different formats like images, scan reports that may be shared across different stakeholders. These large set of data is difficult to share on an electronic platform due to restrictions on the firewall settings or bandwidth. Also, there is no single platform or infrastructure to retrieve, store, and share the data from various sources. Loads of data are being generated on a regular basis, it is scattered across different sources and parties like payers, patients, and physicians. Hence, there is no single point of access for providers in order to optimize the patient experience.

Because of safety regulations, the providers should be able to trust each other even before any kind of communication takes place with respect to patient data [33]. A mutual trust relationship must exist between two parties who are ready for exchange. It can be in-network providers, health organizations, patients, and so on. However, it is difficult to establish such relationship when the receiver of data does not comply with the security system of the sender. In such cases, the security standards are compromized.

The rising cost of the healthcare is a major challenge posing to the industry. This phenomenon is observed across different categories and services of the healthcare industry [34], be it pharmacy, the physician consultation, clinical tests, and so on. The reason for such an inflation is the absence of a proper tracking infrastructure for the drugs and other materials. Hence, the providers are at the benefit of selling the goods at a higher price. When a patient can track the source and manufacturing of the drugs, he will be able to know the pricing value of each commodity. The healthcare industry should implement strategies to address the rising value of costs in the industry.

The current patient billing management systems are complicated and are sensitive to manipulation by the service providers. In recent times, 50% of the healthcare billings are fraudulent thus leading to excessive billing or billing the patients for the services that are not performed [35]. For instance, America has caused a loss of $30 million due to Medicare fraud in 2016 [36]. It is expected that the blockchain-powered healthcare systems will provide realistic solutions to such frauds and minimize medical billing related frauds. It is also assumed that blockchain will eliminate any middle man involved in the payment process and reduce the administrative costs as well [35].

Clinical trials form an important part of healthcare industry in today's world. The intensity of dependency on clinical trials has increased over a period of time due to chronic diseases coming up in recent times. It starts with initial human testing in order to know the intensity and status of the disease and establishes safety and efficacy. However, it is estimated that more than 50% of clinical trials go unreported. Hence, it created crucial safety concerns for the patients involved and may cause serious danger to their life. But with the advent of blockchain in the clinical research area may address these issues since blockchain facilitates time stamping and immutability. It may also help in active participation from the participants of the healthcare industry leaving no communication gap due to existence of a single platform.

10.4 THE BLOCKCHAIN TECHNOLOGY: CONCEPT AND APPLICATION AREAS

Blockchain, which gained popularity as distributed ledger technology through bitcoin in the year 2008 [6], is a public ledger which is capable of storing the immutable record performed as transactions on a peer-to-peer network [37]. Ever since the advent of blockchain technology in white paper in 2008, its scope has been broadened across different industries. Researchers and experts are exploring the new avenues for the implementation of this technology in different domains. This section briefly discusses about the technology, its journey from beginning of cryptocurrencies and beyond, its architecture and how it is face of change in the healthcare industry.

The ideation of blockchain technology was first demonstrated in the research work by Haber and Stornetta in their pioneer work [38]. They presented a cryptographically secured chain of blocks with data storage capabilities where tampering was not possible with the combination of timestamps. The idea was upgraded using the concepts of Merkel Tree for improved efficacy and storage capability of the blockchain system [39]. However, it was until 2008, when the concept of bitcoin blockchain came up, and then the technology really took up and moved fast forward. Satoshi Nakamoto [6], whose identity is not yet clear, gave the research about bitcoin and the underlying blockchain technology. He explained how blockchain technology enhances digital trust given the decentralization aspect that meant nobody would ever be in control of anything.

Blockchain is a distributed ledger that was developed by Satoshi Nakamoto in 2008 as basic concept for cryptocurrency bitcoins [6]. Using blockchain and cryptocurrency, the transactions between two people can be directly performed without the involvement of an intermediary party. Blockchain is facilitated with cryptographic properties to ensure immutability, sequence, and integrity of the involved transactions [40]. The purpose of this technology was to eliminate the intermediary parties involved ensuring the interests of the transacting parties is intact while costs involved in intermediary parties are mitigated.

The most obvious benefit of blockchain lies in the fact that it removes the need for any centralized authority in distributed applications. Blockchain mitigates the problem of single point of failure which would exist in case of centralized authority when two or more parties are involved in transactions. Thus it improves the transaction by bypassing the centralized third party and reduces the transaction cost. Consensus mechanism is used to avoid any discrepancies. The following shows the difference between centralized and decentralized systems. In centralized system, though there

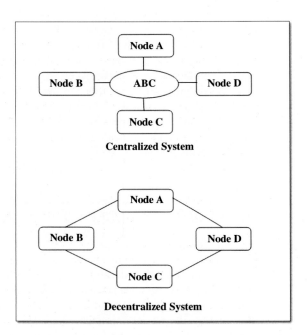

FIGURE 10.2

Schema of centralized and decentralized systems.

are multiple ledgers, all records are held in one single place. System ABC as denoted in Fig. 10.2 maintains state of the ledger. In case of any discrepancies on the state of the ledger, the central authority ABC is consulted for final arbitration. On the other hand, in decentralized system, there is a single ledger, but all the ledgers hold the copy of records and same access to its contents. When the nodes agree on a particular state of the ledger, it is called consensus. There are different ways to achieve the consensus.

Although blockchain is dominating in cryptocurrencies, its transparent ability to maintain transaction history has widened the scope of blockchain application in other domains as well [41]. The ability of blockchain to incorporate smart contracts has made the technology disruptive. Smart contracts are nothing but a logical code executed in blockchain and imitate the regular contracts. A smart contract regulates the conditions that are to be met and also administers the behavior if these conditions are not met. Hence, they can be used to automate payments, documents, and so on. The ability of smart contracts to eliminate intermediaries and establish trust between transacting parties has created disruption of technology in many domains [41].

Before tracing back to the evolution of blockchain technology, it is important to know the functionality of the technology. Bitcoin is the classic example of first ever public blockchain [42]. Blockchain is a peer-to-peer connected network between computational resources which are called as nodes. Each of these nodes carry out the function of processing transaction blocks. Since these nodes are rewarded to compete with each other to mine new blocks, there exists a mechanism to

10.4 THE BLOCKCHAIN TECHNOLOGY 255

agree upon which block must be added to the blockchain. This process is called consensus protocol. With respect to bitcoin, it is called Proof-of-Work (PoW) [6]. As the blockchain grows bigger, the computing difficulty to mine new blocks becomes difficult. PoW implies that miners must come to a hash code difficult to compute but easy to verify [43]. The integrity of the blockchain is compromised only when an attacker tries to modify a block and then he has to spend resources on modifying all the blocks that follow up to current block. PoW along with rewards makes sure that its impractical to hack a public blockchain for an attacker [6,42]. However, the computational intensity of blockchain consumes lot of time and is a major drawback. Thus a public blockchain has poor performance and high latency [42,44].

Later, a need for identity-based blockchain was conceived to overcome the drawbacks of public blockchain. Permissioned blockchain filled this gap. Permissioned blockchain detaches itself from classes PoW and forms a network with only those identified participants and have permission to be part of the blockchain network. Hence, all the participants can know each other and establishes accountability in the network [45]. Digital certificates of each participant are verified in order establish identity of the participants. Hence, the permissioned blockchain is more efficient as compared with public blockchain on all the parameters mentioned in the previous paragraph [46]. This category of blockchain is more useful for enterprises, consortia, and governments. The practical Byzantine Fault Tolerance algorithm (PFT) is used to come to consensus in permissioned blockchain.

A permissioned blockchain is further classified into private and consortium blockchain. The special nodes in the blockchain that run the consensus algorithm are called miners. The difference between private and consortium blockchain lies in the number of miners in both the blockchains. If there is a single miner in the blockchain, it is referred as private blockchain. However, if only one node acts as a miner, the decentralization property of the blockchain is lost. Consortium blockchain allows one or more nodes to be miners. Hence, the consortium blockchain carries decentralization property along with privacy and security.

The chaining of blocks in blockchain is achieved by cryptographic primitive called hash functions. A hash function takes a message of particular length and changes it into an output of fixed length called as message digest or digital fingerprint. The hash function being collision resistant where no two different messages produce same output adds to integrity of messages. This forms the building block for chaining the blocks. The hash of previous block is included in the new block header in order to chain the blocks. Thus the last block of blockchain contains hash of transactions in the previous block and also carries hash of transactions of the next block and so on. The following Fig. 10.3 depicts the chaining of blocks.

Blockchain has evolved beyond the application in cryptocurrency. As per the report by Price Waterhouse Cooper (PwC), the technology has brought the maximum revolution in financial industry. The adoption of technology by different industries shows that, financial services lead all the domain with a percentage of 46%. Other industries have adopted the technology with industrial products and manufacturing (12%), energy and utilities (12%), healthcare (11%) and government (8%). Finally, retail and consumer goods (4%) and media (1%) complete the survey [47]. Some notable work with the help of blockchain include creating digital identities for refugees from Syria to ensure a nutrition supply to them, a project that is being supported by UN's World Food Program [48]. The most notable blockchain application has been in supply chain management apart from cryptocurrencies. Particularly, the agricultural supply chains, from farm to fork have been

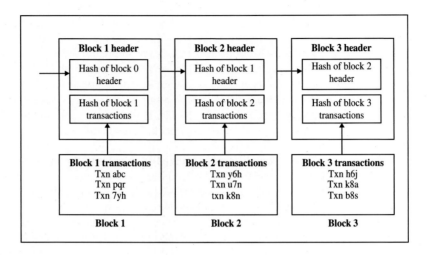

FIGURE 10.3

Act of digital chains among the blocks by using concept of hashing and cryptography in the blockchain.

implemented using blockchain. Walmart and IBM have collaborated in order to increase the food traceability [49].

Another industry in which blockchain has created revolution is healthcare industry. Medical records of patients are highly confidential and sensitive [50]. The continuity and integrity of these historical medical records is important to properly diagnose and treat the patients. Blockchain's property of immutability, facilitating the easy data sharing securely, and data integrity help in maintaining the integrity of the medical records. An end-to-end blockchain solution has been for this purpose [51]. The ability of smart contracts is of importance in the financial industry. They help in speedy settlements, provide a trail for all financial transactions which can be audited in future and are immune to data tampering [40]. Blockchain ensures that medical data are protected and also maintain access control history of data so that auditing can be done in future.

10.5 BLOCKCHAIN APPLICATIONS IN HEALTHCARE

It is important and preliminary to understand the relevant areas of applications where blockchain suitably finds its use. This premise can be well understood by analyzing the research papers or practice-based work which is conducted in the same realm of research. However, the analysis of these research papers indicates that not all the conceptual ideation of blockchain technology in healthcare has been transformed into working prototype. It is therefore both relevant and critical to understand the real-time prototype of blockchain working among the existing use cases of blockchain technology in healthcare [52]. This helps and becomes a guiding principle to understand what the potential research gaps are and where the research of blockchain technology applications

10.5 BLOCKCHAIN APPLICATIONS IN HEALTHCARE 257

in healthcare is headed towards. The analysis of existing prototype also directs us to the understanding of potent challenges and limitations of blockchain technology in healthcare.

Since the ideation and implementation of bitcoin blockchain, there are several tweaks to the original model of technology in order to get better equipped and efficient format of blockchain technology. It is also an important aspect to understand by the analysis of different application of blockchain technology in healthcare, to assimilate the present trend and format of blockchain technology adoption. Now, let us systematically visualize the application of blockchain technology in different dimensions of applications in healthcare through the table presented (Table 10.1; Fig. 10.4).

One of the most popular areas of healthcare applications with blockchain technology is found in the arena of EHRs. Patient data are often very sensitive in terms of both maintaining the sanctity of records for future reference and also sharing across stakeholder with due access permissions. Blockchain technology has inherent feature of transparency, trust, and traceability as mentioned in the previous sections, which is realized through the synthesis of cryptography and hashing in a timestamped setting transactions [6,11,50,54,55]. Blockchain technology applications for healthcare data exchange provide a great implementation insight to regulatory frameworks like General Data Protection Regulation (GDPR) [60] for safely exchanging healthcare data across stakeholders and maintaining the data sanctity. The blockchain technology is considered as a robust mechanism to ensure secure data exchanges between stakeholders appropriately [11]. Liu [78] presented a framework based on Big Data and blockchain and presented different challenges and possibilities in healthcare. MIT Digital Media Lab presented [51] a unique application of blockchain technology called Medrec for handling sensitive health data appropriately. Smith and Dhillon [4] presented a Value-Focused Thinking-based approach to analyze the features of blockchain that assist in maintaining the heath data sanctity.

Drug supply chain is another important area where applications of blockchain are sought as an important innovation. To avoid fake drugs reaching the end customer, a traceability mechanism powered through blockchain can be a very good premise to ensure genuine drug items [7,10,64,65]. There are advanced prorotypes developed on the same theme based on IBM Hyperledger [7] and ways to maintain quality control and sanctity of drug supply chain.

Another application of blockchain technology in healthcare is around the area of biomedical research especially from the point of view of clinical trials and maintaining the data sanctity and anonymity of individuals in this process [13,66—70]. Another aligned application is also in the area of healthcare training and inferencing through the data and insights generated from the blockchain platforms. An Etherium-based blockchain [67] platform was developed as a prototype to demonstrate the process of maintaining data sanctity in clinical trials.

Blockchain technology also facilitates the Remote Patient Monitoring by using the synergies of IoT and blockchain technology. These IoT devices can gather patient and biomedical data which can be stored, analyzed, and maintained on the Blockchain Network. Different approaches to maintaining biomedical data through remote sensing by IoT devices deploy platforms like Hyperledger, Etherium, and so on, to develop patient-centric, advanced intelligence applications of blockchain in healthcare [4,71—73]. These applications act as rapid, reliable, and robust source of deploying Remote Patient Monitoring and also smoothening the process of health data record gathering, analyzing, and maintaining.

258 CHAPTER 10 BLOCKCHAIN-POWERED SMART HEALTHCARE SYSTEM

Table 10.1 A Detailed Review of Blockchain Applications in Healthcare.

References	Use Case Specification in the Realm of Healthcare
[53]	Electronic health record
[11]	Electronic health record
[50]	Electronic health record
[54]	Electronic health record
[55]	Electronic health record
[56]	Electronic health record
[57]	Electronic health record
[14]	Electronic health record
[58]	Electronic health record
[59]	Electronic health record
[60]	Electronic health record
[61]	Electronic health record
[62]	Electronic health record
[51]	Electronic health record
[63]	Electronic health record
[7]	Drug supply chain
[64]	Drug supply chain
[65]	Drug supply chain
[11]	Drug supply chain
[10]	Drug supply chain
[13]	Biomedical research
[66]	Biomedical research
[67]	Biomedical research
[68]	Biomedical research
[69]	Biomedical research
[70]	Biomedical research
[71]	Remote patient monitoring
[72]	Remote patient monitoring
[73]	Remote patient monitoring
[4]	Remote patient monitoring
[74]	Health insurance
[53]	Health insurance
[11]	Health insurance
[55]	Health insurance
[75]	Health data analytics
[76]	Health data analytics
[77]	Health data analytics

10.6 CHALLENGES TO BLOCKCHAIN APPLICATIONS IN HEALTHCARE

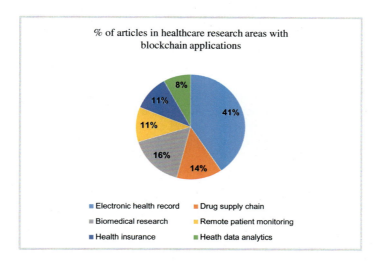

FIGURE 10.4

Healthcare research areas with blockchain applications.

The features of blockchain technology like auditability, traceability, and transparency also provide a very great application in the area of health insurance. Processing policy claims, ensuring right accountability and maintaining the data sanctity provides both the insurers and insurance companies a background of open and trusted environments [11,53,55,74]. Various applications based on the platform of Etherium has been developed as a prototype. An illustration of deployment of blockchain in health segment is MIStore which is a biomedical insurance system. Poditok is another organization that has partnered with Intel to generate more advanced implementation of blockchain technology.

An advanced thought of blockchain technology is in the area of health data analytics where the blockchain technology is interfaced with business intelligence, cognitive capabilities and analytical power in order to better process the data. A very interesting work has been demonstrated in terms of health records and precision medicine by Moamosihna [75]. Saxena et al. [76,77] designed an intelligent blockchain framework that helps stakeholders to harness the power of AI, Machine Learning (ML), and Natural Language Processing (NLP) due to the cognitive intelligence inherent in the technology.

10.6 CHALLENGES TO BLOCKCHAIN APPLICATIONS IN HEALTHCARE

On the basis of the analysis of research articles and practice-based developments in the realm of blockchain applications in healthcare, we identified some important key challenges that are present in the sphere of applications in this area. These challenges are either at theoretical or conceptual level or application level. Few prominent challenges that we identified are:

- There is a clear bias in many applications in terms of the stakeholder of healthcare that they are trying to serve [14,60,79].
- Blockchain platform has to be made more robust in terms of working at the interface of other technological advancements like IoT, analytics, and so on, in order to fully harness the utility of the platform [75−77].
- The involvement of stakeholders and their interactions should be more and more simpler and exhaustive in terms of comprising their behavioral data as close to reality as possible [73,75].
- Most of the applications are still at conceptual level and robust testing and usage has to be done in real-time environment. This will help us to gauge the actual practical usage of blockchain technology [51,76,78].
- Most important and critical factor is working upon the interoperability and integration of different blockchain applications that have developed in the realm of healthcare in terms of usage and having an end-to-end system design on blockchain [61,66,80].
- Other important challenge is exploring the scalability aspect of blockchain technology applications in the realm of healthcare [66].
- One of the most prominent challenge is ensuring the engagement of stakeholder on a blockchain platform of healthcare system. Say engagement of elderly and children in the blockchain application [70].

Some illustrations on the similar challenges are deployment of Etherium blockchain [37] for high volume biomedical data where transactions needs validations from all peer nodes which might not be necessary as well as induce latency in the system. Another application developed on Etherium Platform by Griggs [79] is clearly not operable with the application developed by Liang on the Hyperledger Platform [81]. This is the main issue with most of the applications that are developed. Although they might sound working well in the silos but hardly have they complied with each other in terms of integrated functioning.

Let us visualize some of the applications developed on Etherium and Hyperledger platform as shown in Table 10.2.

However, irrespective of all these developments across varied platforms of blockchain, there are hardly the ones who claim that they are interoperable blockchain with integration support. Smart contracts that can facilitate this type of exchange among different blockchain platforms will play a critical role in addressing the issue of interoperable blockchain [83]. PwC estimates that most of the blockchain systems that exist today might be replaced by 2022 and 2023 to mitigate the challenge of interoperable blockchain and get end-to-end system integrated blockchain.

Some suggestions on the aspect of blockchain applications scalability in the realm of healthcare is thought to be addressed through selective data maintenance of blockchain platform. Say, only the encrypted health data be maintained on blockchain and whenever required can be computational

Table 10.2 Development of Healthcare Applications Across Platforms of Blockchain.

Blockchain Platform	Applications
Etherium	Medrec [51], Ancile [61], MIStore [79], GHN [50], Medium.io [65]
Hyperledger	Healthchain [82], Medical Chain [14]

decrypted to increase the efficacy and scalability of blockchain in terms of data storage [14,57,84]. Furthermore, access control of blockchain platform can be controlled through the infrastructure of permissioned blockchain and suitable reversal mechanism needs to be established in order to reverse the fraudulent transactions [58,82].

The benefits that blockchain technology renders have brought a lot of attention of the research and practice-based communities to improvize on the fallouts of the technology in order to facilitate a better and robust version of blockchain technology. The rapid pace at which this is being conducted shows positive signs that soon these drawbacks of the technology would be mitigated and healthcare segment would be no exception in that case.

10.7 AN INTEGRATED BLOCKCHAIN-POWERED SMART HEALTHCARE FRAMEWORK: CONCEPTUAL MODEL AND ANALYSIS

The analysis of various research articles and practitioner documents give a clear understanding to us about the usage of blockchain technology in the realm of healthcare. However, we also understand that the technology is still in its nascent stage for large scale adoptions. It becomes really critical and important to analyze the issues with present dimension of blockchain technology applications in healthcare and imply them and formulate a plan of action to mitigate these challenges. Few prominent challenges that we identified during the analysis of relevant research material as presented before are:

- *Feasibility issue*: There are many applications that deploy blockchain technology as a solution to problem space in healthcare. However, it becomes critical to look them from the lens of feasibility and analyze the suitability of technology in those areas. For instance, cognitive intelligence in blockchain is still a far dream as claimed by many conceptual models.
- *Scalability analysis*: Another critical issue is to scope the deployment of blockchain technology in terms of scalability. Many applications see that although blockchain technology is a feasible and appropriate technology but it not scalable in terms of adoption. An illustration is the area of healthcare data analytics.
- *Convergence of technologies*: Blockchain can become magnanimously useful if it interfaces well with upcoming technologies like AI, ML, IoT, Cloud, and so on. However, there is an immense focus of research yet needed to make this aspect mature. An illustration is business intelligence and analytical practices in healthcare.
- *Interoperability dimension*: This forms the most challenging part as most applications of blockchain in healthcare systems are developed on different platforms and work in silos. A mechanism is needed to make these blockchain interoperable. An example is MedRec on Etherium and Drug Chain on Hyperledger.
- *Stakeholder neutral*: The applications are mostly cantered around benefit of one stakeholder; however, they should have smart contracts to facilitate unbiased behavior to all stakeholders.
- *Technology adoption*: A strong push is needed for technology adoption in terms of enhancing usage as well as knowledge. The utility of blockchain should reach to healthcare stakeholders.
- *Incorporation of cognitive intelligence*: The applications should be propagated and developed in a manner that maturity of blockchain applications in healthcare start driving behavioral

262 **CHAPTER 10** BLOCKCHAIN-POWERED SMART HEALTHCARE SYSTEM

intelligence to incorporate perspectives of stakeholders and understanding healthcare norms possibly by effective usage of Deep Learning Techniques.

- *Real-time applications*: There is a dire need of moving quickly and robustly from the phase of Ideation to Implementation. Conceptual models should be tested on prototypes and prototypes after suitable testing should transform to real-time applications.

After understanding some prominent issues in the adaption of blockchain technology in the realm of healthcare, it also becomes very important to think on lines of a possible solution. The zone of possible solution lies around developing and implementing an integrated framework that can help different stakeholders in the healthcare domain to integrate blockchain applications that are developed in silos in a holistic and integrated solution. The main premises that ought to be considered while developing this framework is the interoperability of distinct blockchain applications in terms of their smart contracts and mechanisms that facilitates the exchange of information, regulations, and interactions in a specific manner. We present an integrated blockchain framework for the healthcare domain considering the challenges identified earlier in order to pave the way for future applicability.

The proposed framework mainly consists of the following prototypes of the blockchain applications:

- Electronic Health Record
- Remote patient monitoring
- Healthcare analytics
- Drug supply chain
- Biomedical research
- Healthcare insurance

The aspects around which this framework is intended to integrate the applications are:

- Real-time functionality
- Interoperability
- Cognitive intelligence

These aspects are presented as layers of different blockchain applications and facilitate an integrated and synergetic ecosystem of healthcare domain with varied blockchain applications. It becomes important to strategize the functioning of these layers in terms of different smart contracts facilitating the execution of these layers. Since any inherent logic in blockchain setting can be incorporated using the mechanism of different layers. Thus the structuring of these layers in terms of setting a special set of smart contracts for prioritizing which layer to be invoked when is also an important ingredient. Thus from a superlative view the type of smart contracts needed for execution of this framework can be organized into following categories:

- Layer Functionality Smart Contracts (LFSC)
- Layer Coordination/Cognitive Intelligence Smart Contracts (LCSC)
- Layer Interoperability Smart Contracts (LISC)

Any other important aspect of smart contracts that can be thought as an integral component of integration process can be incorporated as per the requirements. Fig. 10.5 given below presents an

10.7 AN INTEGRATED BLOCKCHAIN-POWERED SMART HEALTHCARE

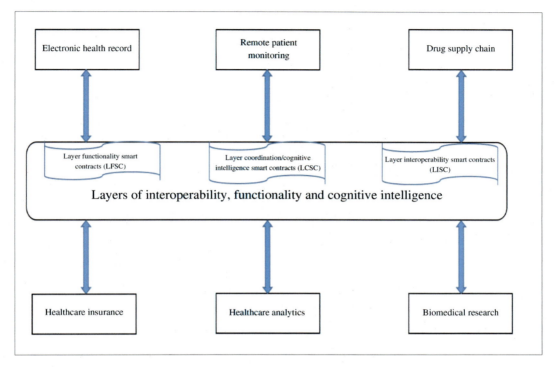

FIGURE 10.5

An integrated blockchain-powered smart healthcare framework.

integrated framework for integrating blockchain applications in healthcare domain for a synergetic and optimal usage of blockchain technology:

The integrated framework presented above addressed the concerns of real-time functionality, interoperability, and involvement of cognitive intelligence in the realm of healthcare applications for blockchain. Let us briefly understand the role of each type of smart contract that facilitates the functionality of this integrated framework across applications of blockchain in healthcare domain:

- *LFSC*

    LFSCs are essentially meant to facilitate the functioning of each of the layer. An illustration to understand this is that for the layer of say interoperability or cognitive intelligence, it performs a check on appropriate regulations on the formation of the layer in terms of facilitating each layer to function well in its defined domain.
- *LCSC*

    LCSC takes care of how to access information and perform cognitive intelligence techniques like AI, ML, and NLP and so on, in an appropriate fashion. This makes the blockchain applications more robust not only in terms of the sanctity of data but also in terms of sanctity of analytical results that form action parameters for the healthcare stakeholders.

264 CHAPTER 10 BLOCKCHAIN-POWERED SMART HEALTHCARE SYSTEM

- *LISC*

    LISC takes care of one of the most prominent and challenging situation of present paradigm of blockchain applications. It is very important to formulate common consensus protocols and regulations in order to facilitate the mechanism of handling interoperability of blockchain applications. Once blockchain applications are well integrated with another, this forms a basis of synergetic and integrated smart healthcare system with an end-to-end compliance and utility on blockchain applications

    Some other important concepts around blockchain applications like scalability, prototype development, and creating technology awareness and utility require both components of research in development with respect to the technology and other dimensions as well. However, the promising nature of blockchain technology to facilitate trust and transparency in the present systems actually makes the research endeavors in the direction of blockchain technology improvement a worthy effort. With improved attention toward blockchain technology improvements mainly in terms of integrating applications from interoperability point of view will impact the adoption and usage of this technology to the fullest. Applications developed on varied platforms like Hyperledger, Etherium, Coda, and so on, are pity close to the point of getting a universal acceptance and utility when integrated. The proposed framework can act as a driving principle in this realm by utilizing the parameters of smart contracts that facilitate most of the logic in blockchain technology.

10.8 CONCLUSION AND IMPLICATIONS

Blockchain technology has evolved over the timeframe since its inception. From being an underlying technology for the revolutionary cryptocurrency bitcoin, blockchain has now taken shape as an enabler of mechanism like trust, transparency, traceability, authentication, auditability, and so on, for an applications in varied domains of business and society. In this chapter, we presented a broad overview of this technology in the realm of healthcare. We started the discussion with understanding the present status of healthcare systems, the issues and challenges they face. We presented the concept of blockchain technology and then highlighted the prime applications of blockchain technology in healthcare. We also pointed at the challenges faced by these blockchain technology applications in the realm of healthcare and how to mitigate these challenges. We also presented an ideation of interoperable and integrated blockchain system that can be functional across end-to-end system. On the basis of our analysis of more than 75 research or practitioner articles through this chapter, we have following important conclusive insights for the healthcare systems in terms of blockchain technology applications (Table 10.3).

10.9 FUTURE RESEARCH DIRECTIONS

Blockchain is an evolving technology and its applications in the realm of healthcare are still at innovation stage. As we have already discussed, in order to make blockchain applications success in healthcare systems, there is a dire need that the researchers now move very hastily from the

10.9 FUTURE RESEARCH DIRECTIONS 265

Table 10.3 Implications From Research and Analysis.

Agenda of Discussion	Implications From Research and Analysis
Feasibility analysis	• Movement from ideation to prototype development • Blockchain is not a solution for all problems • Analyze from a feasibility point of view if there is actually a need of blockchain technology deployment.
Scalability analysis	• Peep into the future and see if the present dimension of technology suits the use case, especially in case of healthcare systems where there is ample and varied data • Ideate and experiment the models for scaling up the blockchain technology
Convergence of technologies	• Blockchain can find its best use when the data reside on blockchain and can be procured, stored, analyzed, and maintained with the interface of other technologies • Healthcare systems which are very sensitive in terms of their data can utilize these features more aggressively • The interface of IoT, AI, ML (machine learning), and Cloud is the next big thing for blockchain technology to aspire for
Interoperability dimension	• There are ample applications in healthcare systems developed across different platforms • However, there is a serious need to think about integration of these silo blockchain and make them interoperable • This will facilitate an end-to-end integrated Blockchain Network for Healthcare System.
Stakeholder neutral	• Most applications are patient centric • However, because data as an asset is important to every stakeholder of the healthcare system, there should be mechanism to smoothly access the data in permissioned and permission less settings.
Technology adoption	• The technology adoption of blockchain is yet at a very slow rate when compared with technology evolution • Stakeholders should be educated and encouraged for the usage of technology in terms of the benefits it provides • Effective usage and knowledge are the most assured way of technology adoption and evolution.
Incorporation of cognitive intelligence	• Blockchain should be empowered with power of AI/behavioral sciences in order to facilitate the intelligence of understanding system regulation and interpreting stakeholder transactions better.
Real-time applications	• There is a strong need for movement from ideation to implementation • There are yet very less number of applications that are developed as prototype, as most of the applications are still at the phase of highlighting conceptual models.

phase of ideation to the phase of execution or implementation. Without having actual prototypes and visualizing end-to-end functioning of the blockchain applications, it is really difficult to gauge their actual usage. Another important and critical issue is catering interoperability issue in the blockchain. Research should bring out the prospects of standard protocols in terms of smart contracts and consensus mechanisms that can facilitate the easy exchange of data among the blockchain built on different platforms. Other aspect of scalability should be taken care by developing

266 **CHAPTER 10** BLOCKCHAIN-POWERED SMART HEALTHCARE SYSTEM

suitable computational protocols of data storage and processing so that blockchain models become scalable in nature. Once these challenges are suitably mitigated, the chances of blockchain fostering healthcare system will increase substantially. With the suitable adoption of blockchain technology, many challenges to the present dimension of healthcare systems would be mitigated and thus lead to more stabilized and powered healthcare system.

REFERENCES

[1] S. Olnes, J. Ubacht, M. Janssen, Blockchain in Government: Benefits and Implications of Distributed Ledger Technology for Information Sharing, Elsevier, 2017.

[2] R. Beck, J. Stenum Czepluch, N. Lollike, S. Malone, Blockchain—The Gateway to Trust-Free Cryptographic Transactions. ECIS 2016 Proc., 2016, pp. 1—15.

[3] Statistica, Global Market for Blockchain Technology 2018—2023 | Statistic [WWW Document], Available from: <https://www.statista.com/statistics/647231/worldwide-blockchain-technology-market-size/>, 2019.

[4] K. Smith, G. Dhillon, Blockchain for digital crime prevention: the case of health informatics, Inf. Syst. Secur. Priv. 1 (2017) 1—10.

[5] H.D. Zubaydi, Y.-W. Chong, K. Ko, S.M. Hanshi, S. Karuppayah, A review on the role of blockchain technology in the healthcare domain, Electronics 8 (2019) 679.

[6] S. Nakamoto, Bitcoin: A Peer-to-Peer Electronic Cash System. Available from: <http://bitcoin.org/bitcoin.pdf>, 2008.

[7] E. Androulaki, A. Barger, V. Bortnikov, C. Cachin, K. Christidis, A. De Caro, et al., Hyperledger Fabric: A Distributed Operating System for Permissioned Blockchains, in: Proceedings of the Thirteenth EuroSys Conference. ACM, 2018, p. 30.

[8] C. Wood, B. Winton, K. Carter, S. Benkert, L. Dodd, J. Bradley, How blockchain technology can enhance EHR operability, ARK. Invest. Res. 1 (2016) 1—13.

[9] T.-T. Kuo, L. Ohno-Machado, Modelchain: decentralized privacy-preserving healthcare predictive modeling framework on private blockchain networks, ArXiv Prepr 1 (2018) 1—13.

[10] J.-H. Tseng, Y.-C. Liao, B. Chong, S. Liao, Governance on the drug supply chain via gcoin blockchain, Int. J. Environ. Res. Public. Health 15 (2018) 1055.

[11] M.A. Engelhardt, Hitching healthcare to the chain: an introduction to blockchain technology in the healthcare sector, Technol. Innov. Manag. Rev. 7 (2017).

[12] S. Jiang, J. Cao, H. Wu, Y. Yang, M. Ma, J. He, Blochie: a blockchain-based platform for healthcare information exchange, in: 2018 IEEE International Conference on Smart Computing (SMARTCOMP), IEEE, 2018, pp. 49—56.

[13] M. Benchoufi, P. Ravaud, Blockchain technology for improving clinical research quality, Trials 18 (2017) 335. Available from: https://doi.org/10.1186/s13063-017-2035-z.

[14] W.J. Gordon, C. Catalini, Blockchain technology for healthcare: facilitating the transition to patient-driven interoperability, Comput. Struct. Biotechnol. J. 16 (2018) 224—230. Available from: https://doi.org/10.1016/j.csbj.2018.06.003.

[15] A. Dubovitskaya, Z. Xu, S. Ryu, M. Schumacher, F. Wang, Secure and trustable electronic medical records sharing using blockchain, AMIA. Annu. Symp. Proc. 2017 (2018) 650—659.

[16] M. Raikwar, S. Mazumdar, S. Ruj, S.S. Gupta, A. Chattopadhyay, K. Lam, A blockchain framework for insurance processes, in: 2018 9th IFIP International Conference on New Technologies, Mobility and Security (NTMS). Presented at the 2018 9th IFIP International Conference on New Technologies,

Mobility and Security (NTMS), 2018, pp. 1–4. Available from: <https://doi.org/10.1109/NTMS.2018.8328731>.

[17] P. Sylim, F. Liu, A. Marcelo, P. Fontelo, Blockchain technology for detecting falsified and substandard drugs in distribution: pharmaceutical supply chain intervention, JMIR Res. Protoc. 7 (2018) e10163. Available from: https://doi.org/10.2196/10163.

[18] F.C. Coelho, Optimizing disease surveillance with blockchain, bioRxiv 1 (2018) 1–8. Available from: https://doi.org/10.1101/278473.

[19] B.M. Straube, A role for government: an observation on federal healthcare efforts in prevention, Am. J. Prev. Med. 44 (2013) S39–S42.

[20] N. Tang, J.M. Eisenberg, G.S. Meyer, The roles of government in improving health care quality and safety, Jt. Comm. J. Qual. Saf. 30 (2004) 47–55.

[21] M.T. Taner, B. Sezen, J. Antony, An overview of six sigma applications in healthcare industry, Int. J. Health Care Qual. Assur. (2007). Available from: https://doi.org/10.1108/09526860710754398.

[22] W. Raghupathi, V. Raghupathi, Big data analytics in healthcare: promise and potential, Health Inf. Sci. Syst. 2 (2014) 3. Available from: https://doi.org/10.1186/2047-2501-2-3.

[23] J. Emanuele, L. Koetter, Workflow opportunities and challenges in healthcare. BPM Work. Handb. vol.1, 2007, p. 157.

[24] R. Thakur, S.H.Y. Hsu, G. Fontenot, Innovation in healthcare: issues and future trends. J. Bus. Res., 65, 562–569, 2012. Available from: <https://doi.org/10.1016/j.jbusres.2011.02.022>.

[25] H.S. Ng, M.L. Sim, C.M. Tan, Security issues of wireless sensor networks in healthcare applications, BT Technol. J. 24 (2006) 138–144. Available from: https://doi.org/10.1007/s10550-006-0051-8.

[26] MailMyStatements, 5 Major Challenges Facing the Healthcare Industry in 2019 [WWW Document]. Medium. 2018. Available from: <https://medium.com/@MailMyStatement/5-major-challenges-facing-the-healthcare-industry-in-2019-60218336385f>, (accessed 19.07.19).

[27] G. Fanjiang, J.H. Grossman, W.D. Compton, P.P. Reid, Building a Better Delivery System: A New Engineering/Health Care Partnership, National Academies Press, Washington, DC, 2005.

[28] B. Middleton, M. Bloomrosen, M.A. Dente, B. Hashmat, R. Koppel, J.M. Overhage, et al., Enhancing patient safety and quality of care by improving the usability of electronic health record systems: recommendations from AMIA, J. Am. Med. Inform. Assoc. 20 (2013) e2–e8.

[29] P. Zhang, D.C. Schmidt, J. White, G. Lenz, Blockchain technology use cases in healthcare, Advances in Computers, Elsevier, 2018, pp. 1–41.

[30] C. Castaneda, K. Nalley, C. Mannion, P. Bhattacharyya, P. Blake, A. Pecora, et al., Clinical decision support systems for improving diagnostic accuracy and achieving precision medicine, J. Clin. Bioinforma. 5 (2015) 4.

[31] R. Kaushal, K.G. Shojania, D.W. Bates, Effects of computerized physician order entry and clinical decision support systems on medication safety: a systematic review, Arch. Intern. Med. 163 (2003) 1409–1416.

[32] G.D. Schiff, O. Hasan, S. Kim, R. Abrams, K. Cosby, B.L. Lambert, et al., Diagnostic error in medicine: analysis of 583 physician-reported errors, Arch. Intern. Med. 169 (2009) 1881–1887.

[33] G. Hripcsak, M. Bloomrosen, P. FlatelyBrennan, C.G. Chute, J. Cimino, D.E. Detmer, et al., Health data use, stewardship, and governance: ongoing gaps and challenges: a report from AMIA's 2012 Health Policy Meeting, J. Am. Med. Inform. Assoc. 21 (2014) 204–211.

[34] J. Pennic, Top 10 Challenges, Issues and Opportunities for Healthcare Executives in 2019. Available from: <https://hitconsultant.net/2018/09/28/challenges-issues-opportunities-healthcare-executives/>, 2018.

[35] K. Rabah, Challenges & opportunities for blockchain powered healthcare systems: a review, Mara Res. J. Med. Health Sci. 1 (2017) 45–52.

[36] R. Das, Does Blockchain Have A Place In Healthcare? [WWW Document]. 2017. Available from: <https://www.forbes.com/sites/reenitadas/2017/05/08/does-blockchain-have-a-place-in-healthcare/#44f0fdb71c31>.

[37] J. Yli-Huumo, D. Ko, S. Choi, S. Park, K. Smolander, Where is current research on blockchain technology?—a systematic review, PLoS One 11 (2016) e0163477.

[38] S.A. Haber, Jr, W.S.S., Method for Secure Time-Stamping of Digital Documents. US5136647A, 1992.

[39] D. Bayer, S. Haber, W.S. Stornetta, Improving the efficiency and reliability of digital time-stamping, in: R. Capocelli, A. De Santis, U. Vaccaro (Eds.), Sequences II, Springer, New York, 1993, pp. 329—334.

[40] M. Peterson, Blockchain and the future of financial services, J. Wealth Manag. 21 (2018) 124—131. Available from: https://doi.org/10.3905/jwm.2018.21.1.124.

[41] M. Crosby, P. Pattanayak, S. Verma, V. Kalyanaraman, Blockchain technology: beyond bitcoin, Appl. Innov. 2 (2016) 71.

[42] I.-C. Lin, T.-C. Liao, A survey of blockchain security issues and challenges, J. Netw. Secur. 19 (2017) 653—659. Available from: https://doi.org/10.6633/IJNS.201709.19(5).01.

[43] D. Mingxiao, M. Xiaofeng, Z. Zhe, W. Xiangwei, C. Qijun, A review on consensus algorithm of blockchain, in: 2017 IEEE International Conference on Systems, Man, and Cybernetics (SMC). Presented at the 2017 IEEE International Conference on Systems, Man, and Cybernetics (SMC), 2017, pp. 2567—2572. Available from: <https://doi.org/10.1109/SMC.2017.8123011>.

[44] J. Park, J. Park, Blockchain security in cloud computing: use cases, challenges, and solutions, Symmetry 9 (2017) 164.

[45] M. Vukolić, Rethinking permissioned blockchains, in: Proceedings of the ACM Workshop on Blockchain, Cryptocurrencies and Contracts, ACM, 2017, pp. 3—7.

[46] Z. Zheng, S. Xie, H. Dai, X. Chen, H. Wang, An overview of blockchain technology: architecture, consensus, and future trends, in: 2017 IEEE International Congress on Big Data (BigData Congress). Presented at the 2017 IEEE International Congress on Big Data (BigData Congress), 2017, pp. 557—564. Available from: <https://doi.org/10.1109/BigDataCongress.2017.85>.

[47] Price Water House Coopers, Blockchain is here. What's your next move? [WWW Document]. PwC. Available from: <https://www.pwc.com/gx/en/issues/blockchain/blockchain-in-business.html>, 2019.

[48] R. Zambrano, A. Young, S. Verhulst, Connecting Refugees to Aid Through Blockchain-Enabled ID Management: World Food Programme's Building Blocks, GOVLAB, 2018.

[49] D. Galvin, IBM and Walmart: blockchain for food safety, PowerPoint Present (2017).

[50] M. Mettler, Blockchain technology in healthcare: the revolution starts here, in: 2016 IEEE 18th International Conference on E-Health Networking, Applications and Services (Healthcom), IEEE, 2016, pp. 1—3.

[51] A. Azaria, A. Ekblaw, T. Vieira, A. Lippman, Medrec: using blockchain for medical data access and permission management, in: 2016 2nd International Conference on Open and Big Data (OBD), IEEE, 2016, pp. 25—30.

[52] C.C. Agbo, Q.H. Mahmoud, J.M. Eklund, Blockchain technology in healthcare: a systematic review, Healthcare, Multidisciplinary Digital Publishing Institute, 2019, p. 56.

[53] S. Angraal, H.M. Krumholz, W.L. Schulz, Blockchain technology: applications in health care, Circ. Cardiovasc. Qual. Outcomes 10 (2017) e003800.

[54] T.-T. Kuo, H.-E. Kim, L. Ohno-Machado, Blockchain distributed ledger technologies for biomedical and health care applications, J. Am. Med. Inform. Assoc. 24 (2017) 1211—1220.

[55] J.M. Roman-Belmonte, H. De la Corte-Rodriguez, E.C. Rodriguez-Merchan, How blockchain technology can change medicine, Postgrad. Med. 130 (2018) 420—427.

REFERENCES 269

[56] H. Kaur, M.A. Alam, R. Jameel, A.K. Mourya, V. Chang, A proposed solution and future direction for blockchain-based heterogeneous medicare data in cloud environment, J. Med. Syst. 42 (2018) 156.

[57] Q. Xia, E. Sifah, A. Smahi, S. Amofa, X. Zhang, BBDS: blockchain-based data sharing for electronic medical records in cloud environments, Information 8 (2017) 44.

[58] Z. Alhadhrami, S. Alghfeli, M. Alghfeli, J.A. Abedlla, K. Shuaib, Introducing blockchains for healthcare, in: 2017 International Conference on Electrical and Computing Technologies and Applications (ICECTA), IEEE, 2017, pp. 1−4.

[59] V. Patel, A framework for secure and decentralized sharing of medical imaging data via blockchain consensus, Health Inform. J. (2018). 1460458218769699.

[60] Patients and Privacy: GDPR Compliance for Healthcare Organizations—Security News—Trend Micro DK [WWW Document], 2018. Available from: <https://www.trendmicro.com/vinfo/dk/security/news/online-privacy/patients-and-privacy-gdpr-compliance-for-healthcare-organizations>.

[61] G.G. Dagher, J. Mohler, M. Milojkovic, P.B. Marella, Ancile: privacy-preserving framework for access control and interoperability of electronic health records using blockchain technology, Sustain. Cities Soc. 39 (2018) 283−297.

[62] A.F. Hussein, N. ArunKumar, G. Ramirez-Gonzalez, E. Abdulhay, J.M.R. Tavares, V.H.C. de Albuquerque, A medical records managing and securing blockchain based system supported by a genetic algorithm and discrete wavelet transform, Cogn. Syst. Res. 52 (2018) 1−11.

[63] K. Peterson, R. Deeduvanu, P. Kanjamala, K. Boles, A blockchain-based approach to health information exchange networks, in: Proc. NIST Workshop Blockchain Healthcare, 2016, pp. 1−10.

[64] T.K. Mackey, G. Nayyar, A review of existing and emerging digital technologies to combat the global trade in fake medicines, Expert. Opin. Drug. Saf. 16 (2017) 587−602.

[65] T. Bocek, B.B. Rodrigues, T. Strasser, B. Stiller, Blockchains everywhere-a use-case of blockchains in the pharma supply-chain, in: 2017 IFIP/IEEE Symposium on Integrated Network and Service Management (IM), IEEE, 2017, pp. 772−777.

[66] M.N.K. Boulos, J.T. Wilson, K.A. Clauson, Geospatial blockchain: promises, challenges, and scenarios in health and healthcare, BioMed. Cent. (2018).

[67] T. Nugent, D. Upton, M. Cimpoesu, Improving data transparency in clinical trials using blockchain smart contracts, F1000Research 5 (2016).

[68] P. Mytis-Gkometh, G. Drosatos, P.S. Efraimidis, E. Kaldoudi, Notarization of knowledge retrieval from biomedical repositories using blockchain technology, Precision Medicine Powered by PHealth and Connected Health, Springer, 2018, pp. 69−73.

[69] Z. Shae, J.J. Tsai, On the design of a blockchain platform for clinical trial and precision medicine, in: 2017 IEEE 37th International Conference on Distributed Computing Systems (ICDCS), IEEE, 2017, pp. 1972−1980.

[70] I. Radanović, R. Likić, Opportunities for use of blockchain technology in medicine, Appl. Health Econ. Health Policy 16 (2018) 583−590.

[71] D. Ichikawa, M. Kashiyama, T. Ueno, Tamper-resistant mobile health using blockchain technology, JMIR MHealth UHealth 5 (2017) e111.

[72] J. Zhang, N. Xue, X. Huang, A secure system for pervasive social network-based healthcare, Ieee Access 4 (2016) 9239−9250.

[73] M. Weiss, A. Botha, M. Herselman, G. Loots, Blockchain as an enabler for public mHealth solutions in South Africa, in: 2017 IST-Africa Week Conference (IST-Africa), IEEE, 2017, pp. 1−8.

[74] V. Gatteschi, F. Lamberti, C. Demartini, C. Pranteda, V. Santamaría, Blockchain and smart contracts for insurance: is the technology mature enough? Future Internet 10 (2018) 20.

[75] P. Mamoshina, L. Ojomoko, Y. Yanovich, A. Ostrovski, A. Botezatu, P. Prikhodko, et al., Converging blockchain and next-generation artificial intelligence technologies to decentralize and accelerate biomedical research and healthcare, Oncotarget 9 (2018) 5665.

[76] M. Saxena, M. Sanchez, R. Knuszka, Method for Providing Healthcare-Related, Blockchain-Associated Cognitive Insights Using Blockchains, 2018a.

[77] M. Saxena, M. Sanchez, R. Knuszka, Providing Healthcare-Related, Blockchain-Associated Cognitive Insights Using Blockchains, 2018b.

[78] P.T.S. Liu, Medical record system using blockchain, big data and tokenization, in: International Conference on Information and Communications Security, Springer, 2016, pp. 254−261.

[79] K.N. Griggs, O. Ossipova, C.P. Kohlios, A.N. Baccarini, E.A. Howson, Hayajneh, Healthcare blockchain system using smart contracts for secure automated remote patient monitoring, J. Med. Syst. 42 (2018) 130.

[80] G. Carter, H. Shahriar, S. Sneha, Blockchain-based interoperable electronic health record sharing framework, in: 2019 IEEE 43rd Annual Computer Software and Applications Conference (COMPSAC), 2019, pp. 452−457.

[81] R. Liang, A linguistic intuitionistic cloud decision support... Google Scholar [WWW Document]. 2019. Available from: <https://scholar.google.co.in/scholar?hl = en&as_sdt = 0%2C5&q = A + Linguistic + Intuitionistic + Cloud + Decision + Support + Model + with + Sentiment + Analysis + for + Product + Selection + in + E-commerce&btnG = >.

[82] T. Ahram, A. Sargolzaei, S. Sargolzaei, J. Daniels, B. Amaba, Blockchain technology innovations, in: 2017 IEEE Technology & Engineering Management Conference (TEMSCON), IEEE, 2017, pp. 137−141.

[83] M. Swan, Blockchain Economics: "Ripple for ERP" [WWW Document]. Eur. Financ. Rev., 2018. Available from: <https://www.europeanfinancialreview.com/blockchain-economics-ripple-for-erp-integrated-supply-chain-ledgers-to-free-3-9-trillion-in-capital/>.

[84] C. Esposito, A. De Santis, G. Tortora, H. Chang, K.-K.R. Choo, Blockchain: a panacea for healthcare cloud-based data security and privacy? IEEE Cloud Comput. 5 (2018) 31−37.

CHAPTER

11

INTERNET OF THINGS AND BLOCKCHAIN: INTEGRATION, NEED, CHALLENGES, APPLICATIONS, AND FUTURE SCOPE

Deepak Kumar Sharma[1], Ajay Kumar Kaushik[1], Aarti Goel[1] and Saakshi Bhargava[2]

[1]*CAITFS, Department of Information Technology, Netaji Subhas University of Technology (Formerly known as Netaji Subhas Institute of Technology), New Delhi, India* [2]*Department of Electronics and Communication, Banasthali Vidyapith, Vanasthali, India*

11.1 INTRODUCTION: BLOCKCHAIN

There is an exponential increase in the Internet of Things (IoT) [1] devices in the last 20 years, and it has been observed that the data with the devices have also increased, and it is expected that the total data which will be produced by IoT devices will touch more than 700 Zettabytes per year by 2022 [2]. It has been foreseen [3] that within the next few years, the Machine-to-Machine (M2M) connections will increase to greater extent that will deal with wide-ranging spectrum of applications in various domains, for example, transportation, automation [4], defense, agriculture, augmented reality, wearable, etc.

To achieve the proper linkage between the two technologies (IoT and blockchain) and also to uphold data as well as focusing on its security is the major concern. Although solutions are working properly, whenever there is an increase in the number of devices, a new technology has to be proposed for it. While working on the idea of proposing a better technology, the concept of decentralized architectures [5] were suggested for the creation of peer to peer (P2P) wireless sensor networks, but there were certain blocks of puzzle that were still missing and were the main factors for better working of devices called privacy and security. This problem was dealt very easily with the arrival of blockchain technology [6].

Before the commencement of blockchain, there was no central authority ruling the whole scenario of data entry and retrieval. Then distributed databases made by blockchain had a profoundly diverse digital mainstay that proves to be an important feature of blockchain technology. With the centralized server, a user with consents allied with its account is capable of changing the records, for example, Wikipedia. Every time a user accesses the Wikipedia page, they get a restructured version of the master copy of the Wikipedia record. The control of the data and page will be in the hands of Wikipedia administrators, where the allowance and accessibility are maintained solely by

Handbook of Research on Blockchain Technology. DOI: https://doi.org/10.1016/B978-0-12-819816-2.00011-3
© 2020 Elsevier Inc. All rights reserved.

271

272 **CHAPTER 11** INTERNET OF THINGS AND BLOCKCHAIN

the central authority running it. But, in blockchain, every single node in the network is approaching to the conclusions of similar type, each one is updating the record independently. Transactions that are made are broadcasted, and every single node is generating their own updated type; this difference makes blockchain technology very useful.

11.1.1 BASICS OF BLOCKCHAIN

Blockchain is simply a correlation of three technologies namely [6]:

1. Internet
2. Private key cryptography
3. Protocol governing incentivization

Blockchain is a decentralized, distributed, shared, and immutable (data cannot be erased or altered) database ledger where block data contain a list of all transactions and a hash to the previous block [7]. Moreover, blockchain has a full history of all transactions and provides global trust [7].

In current banking system, there are various issues such as double spending, centralized power, prone to hacks, and private ledgers. This arises a demand for safe P2P network transaction [7]. Nowadays, blockchain is being used highly for transaction management and replacing the current transaction management system.

While dealing with IoT devices, the main concern is to define the digital trust between the different parties operating in different network areas. Determining trust often requires two major things, authentication and authorization, so that nobody disrupt or compromise or hacks the data of the devices as intruding into someone's device may break the digital trust over the technology. Using blockchain technology, it is not possible for a person or a group of persons to directly modify a transaction part of the blockchain. This is one of the key reasons why blockchain is highly being used in today's payment platform.

To provide security, elliptic curve cryptography (ECC) and SHA 256 hashing algorithms are used for data authentication and integrity [7].

11.1.2 WHY DO WE NEED BLOCKCHAIN?

Ever wondered whether there is an easier way to complete transactions without having to deal with [8]:

- online wallets,
- banks, and
- third party applications.

So, the answer is yes, it is definitely possible, thanks to blockchain.

Imagine four friends Jack, Ted, Sam, Phil and they meet up for dinner after they are done, Jack pays the bill and all of them decide to split the expense among each other. Phil transfers the money to jack online money transfer and transaction goes through without a hitch but when Ted and Sam send their respective share to Jack, their transaction do not go through. The failed transactions are the ones which have some issues from the bank, that is, when Jack comes to know about the many ways of bank transactions that could fail, it can be due to technical issues, accounts hacked, daily transfer limits exceeded, and sometimes transfer charges.

To solve these problems, the concept of crypto currency comes into the picture. "Crypto currencies are a part of digital or virtual currency that uses a technology called blockchain".

11.1.3 CRYPTO CURRENCY

Crypto currencies are immune to counterfeiting, do not require a central right, and are secured by robust and compound encryption systems. There are around 1000 crypto currencies in the market right now such as Etherium, Zeecash [9], and so on, one reins the realm, that is, Bitcoin [10].

Returning to the previous example, now let us say they sent 2 Bitcoins each to Jack. First Phil sends 2 Bitcoin to Jack; the record is created in the database. The transaction details between them are permanently inscribed in the block. This record also holds the number of Bitcoins in each block. So, after phil's transaction, Jack has 7 Bitcoins and phil has 1. Following this Sam and Ted send 2 Bitcoins to Jack, the record is then created for each of these, and this account holds transaction details as well as how many Bitcoins Jack holds, Sam and Ted hold. These blocks are linked to each other as each of them takes the reference from previous one for the record of Bitcoin. Each friend owns this chain of the records or blocks, this is called a Ledger and this Ledger is shared among all the friends. Jack's ledger is a distributed ledger; this forms the basis of the blockchain.

If Phil has one Bitcoin and he tries to send, then the transaction would not take place as all his friends have this ledger. All these friends have this link. A hacker will not be able to alter the data in blockchain because each user has a copy of the ledger. The data within the blocks are encrypted by complex algorithms. All of this is made possible by blockchain technology.

11.1.4 TYPES OF BLOCKCHAIN

Fig. 11.1 depicts the types of blockchain. It can be permissioned or permissionless depending on the transmission process it will undergo, if every node is able to create a new block in the network, then it becomes permissionless. In other words, it does not require any permission from the user to make the changes to the database created. On the other hand, if some specified nodes are allowed to make the changes in the network by creating new blocks, then it is called as permissioned blockchain. With transmission process, it is meant that while transferring any sort of data through blockchain, blocks get created. Once they get into the system, final confirmation of the transmission has to be sent to the system which is mainly done through the consensus algorithm [11].

The field of blockchain is not only restricted to these types but is also further classified as private and public based on how it relates to the access of blockchain data. Public blockchain is permissionless and lacks in privacy, whereas private blockchain is permissioned and has better security features.

Blockchain can be described as follows [12]:

- Assembly of accounts
- Related with each other
- Sturdily resilient to amendments
- Secured by Cryptography

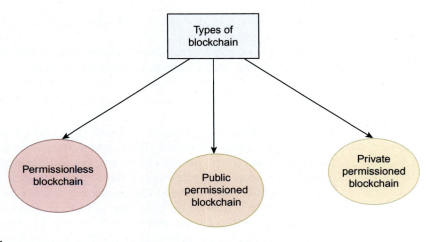

FIGURE 11.1

Types of blockchain.

11.1.4.1 Mechanism of Bitcoin transaction

In this section, the mechanism which is used for Bitcoin transaction is discussed. Every user in the Bitcoin network has two keys:

1. Public key: for example, e-mail address of user.
2. Private key: for example, password of an e-mail account

First, phil chooses the number of Bitcoins he wants to give to Jack with their wallet address with hashing encryption algorithm; all this is a part of transaction details, these details are encrypted using encryption algorithms and using Phil's private key. This is done to digitally sign the transaction, ensuring that these are obtained from phil, these transactions are then transmitted across the world using Jack's public key, this message or transaction can only be decrypted using Jack's private key. Different crypto currencies use different hashing algorithms: Bitcoin uses SHA256 [13] algorithm and ETHERIUM uses ETHASH.

These types of transactions are taking place all around the world; they are first validated and added block by block; the people who validate it are called miners. For the blocks to be validated and integrated into the blockchain, miners need to solve complex mathematical problems. The miner who solves it first adds the block to the blockchain and is rewarded with 12.5 Bitcoins. The process of solving complex mathematical problems is called validation of work and the process of integrating a block into a blockchain is called mining. With this, Phil's and Jack's wallets and every person in the network who made the transaction are updated. Even the world famous U.S Company "WALMART" uses blockchain for their customers to improve the services they provide.

Initially, blockchain was not used in the WALMART until when they faced problems in delivering quality products to its customers; they were facing high return rate and large amounts of refunds due to their products bad quality. They were unable to determine the point of failure in the supply chain which started from farm to storage to transportation to processing all the way to the customers

11.1 INTRODUCTION: BLOCKCHAIN

to the distribution center. Then WALMART adopted blockchain technology; with blockchain, the quality of the block with each step were inscribed. If a customer reports the product was damaged, it can be identified at which point was it damaged going through the supply chain, helping WALMART to identify the problem areas and fixing them.

This is one of the several ways blockchain is used in different real-life problems. Blockchain has also impacted industries by adding a layer of security and transparency into their processes.

11.1.5 APPLICATIONS OF BLOCKCHAIN

This section explains applications of blockchain in different sectors such as banking, health care, real estate, supply chain, government, cyber security, social media, and artificial intelligence (AI) as shown in Fig. 11.2.

1. *Banking*: Success of Bitcoin has led us to expect blockchain to replace many core functions of banking system. Due to its decentralized architecture, it is expected from blockchain technology to provide core banking facilities like payment with minimal or no fees with fast and secure transfer among the clients. However, for blockchain to be adopted by financial institutions, blockchain developers need to win the trust of customers and financial institutions. Major challenges involved are to provide security with respect to financial sector requirements. Time is not far when blockchain technology will be adopted in financial institutions to take advantage of smart contracts and unified currency.

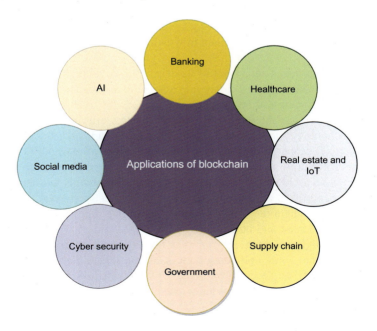

FIGURE 11.2

Applications of blockchain.

276 CHAPTER 11 INTERNET OF THINGS AND BLOCKCHAIN

2. *Health care*: Let us consider the life of a patient and his/her health conditions as different states of a network then anyone can map the health and condition of a patient at different instances of time perfectly with blockchain architecture. By its definition, blockchain keeps record of different transactions over the network with zero alteration policy by anyone in transaction data once it is committed. Efforts need to be put in to make the health care work better. This problem is more prevalent in developing countries or poor countries where hardly any attempt is made by health care sector to record the patient lifetime data in an organized manner. Onus to keep a record of patient data is largely the responsibility of patient in such countries. Patients too, most of the time due to unawareness, do not keep the records of their medical condition over a period of time. It is here where blockchain can strengthen the patient while providing them the facility to control their own medical records. It will also help doctors and other medical staff to retrieve the previous medical record of a patient. Such an easy and integrated access of patient data across the different stakeholders of health care system like doctors, pharmacist, and other medical students and researchers will lead to more effective health care system.

3. *Real estate:* With the increase in IT applications and introduction of regulators, such as real estate regulatory authority (RERA) in India, real estate market has shown some commitment to increase the trust in selling, buying of lands. According to Paul Clitheroe, "Before you start trying to work out which direction the property market is headed you should be aware that there are markets within markets" [14]. Above statement clarifies that things are not that simple and transparent in real estate sectors. Not so long ago, it was a common practice adopted by greedy sellers to sell a piece of land umpteen numbers of times. To do so, most of the time, they were able to alter the chain of property records. Blockchain provides a clarification to this problematic situation by giving a mechanism in which it is difficult to transfer the existing property records.

   Blockchain technology takes advantage of its decentralized architecture and increase the resistance to fraud. It is also supposed to provide a streamlined process that not only reduces the time involved in verification of records but also reduces the manpower involved in doing so. The main aim is to remove the intermediaries by storing, verifying band record transfer using blockchain technology. It can also support a user in searching for the desired property. Most of the commercial property Web sites in India like Magic bricks and 99 acres do not guarantee the property data to be accurate as anyone can list their property on the above platforms. All the above problems can be fixed when property records will be listed on a single blockchain-based decentralized database. It can provide prospective buyers a reliable data that are authenticated by blockchain technology. It can also help in digitally signed smart contracts rent agreement between tenant and landlord.

4. *IoT*: For IoT, a private lightweight architecture can be developed on technology that abolishes the overhead of blockchain while upholding both of its confidentiality and secrecy benefits. IoT can be benefited a lot from the decentralized nature of blockchain to make data storage and device connectivity trustworthy through nodes, without the involvement of any centralized authority.

5. *Supply chain management (SCM):* SCM involves the shipping and logistics industry; transportation means to transfer goods from one place to another. So good managerial skills are needed for SCM to be effective and to keeps the chain functioning. Terms such as decentralized

process, distributed, and public ledger are the attributes of blockchain technology that has the power to modify the technique related to: how business is done across SCM? In SCM, blockchain has inbuilt chain of commands as records of a chain which practically cannot be erased completely and one can verify the chain of records, thanks to the distributed nature of blockchain ledger

6. *Government*: Government in its several welfare schemes get benefited by adopting blockchain technology. It can improve its delivery services, eliminate the middleman, and prevent tax fraud cases. Most of the government functions and blockchain technology are essentially the same. Government launches welfare schemes that can be considered as a network involving several stakeholders'. Similarly blockchain maintains a distributed ledger.

7. *Cyber security*: Government also focuses on cyber security across several frontiers of digital world. Lots of efforts have already been put to strengthen cyber security. Beauty of blockchain is that data are very difficult to alter as every alteration in data can be tracked easily. One can easily verify its data by looking at previous versions of a block. It means that to tamper with network, one needs to destroy the entire network. So with the increased size of blockchain-based network, chances of penetrating a network decrease. Keys of verification by distribution of data to every other node can be removed. Any change in any data node gets reflected across all the nodes.

8. *Social media*: There was a time not so long ago when social media was not considered as a serious place. Its primary functions were entertainment and social connectivity. But with the passage of time, both the popularity and size of social media have increased considerably. Nowadays, it is sharing an equal amount of space with primary media like news and print. Issues in social media are spreading of fake news and provoking information leading to communal riots. Blockchain-distributed ledger system can prevent the fake information and news. Blockchain technology cannot easily authenticate the content of data on social network, but can easily identify and verify the source of the data. Need is to develop some new business models based on blockchain for social media.

9. *Artificial intelligence*: Although blockchain technology is a decade old, little attempt has been made to create real world applications that can utilize the concepts of both blockchain and AI. But scenario is changing very fast; nowadays, there are considerable and significant attempts to combine blockchain with AI. Traditionally, Bitcoin working is based on hashing algorithm which is based on brute force approach. It examines every candidate for the solution. AI algorithm can tackle the situation in a more intelligent and efficient way. A large blockchain network can pass on large amount of data for AI algorithm to work on. Integration of machine learning and AI will enhance blockchain underlying architecture.

11.1.6 POTENTIAL AREAS OF RESEARCH IN BLOCKCHAIN

1. Due to complex and connected ledger calculations, blockchain calculations can be a time consuming as compared to traditional payment system. Effort needs to be put towards reducing the processing time for blockchain calculations.
2. Regulations need to be more and standardized.
3. Some models need to be made to ensure the transparency and flexibility to view the information stored on ledgers by stakeholders of an enterprise.

278 CHAPTER 11 INTERNET OF THINGS AND BLOCKCHAIN

4. Models of blockchain can improve the existing supply chain management of an enterprise and can make it more transparent.
5. Nowadays most organizations use data and are data driven. Models can be developed to deal with data theft in an organization.
6. Like cloud services, blockchain can be used by organizations whenever and wherever required as per their needs. Such organizations instead of developing their own blockchain network can opt for some contract with a vendor to avail blockchain as a service.
7. Existing blockchain models are generally useful in making smart contracts where an electrical code ensures the smart management of contracts.
8. Existing models can be improved to ensure that online advertisements are viewed by human only.

11.1.7 CHALLENGES WITH BLOCKCHAIN

1. *Storage capability and scalability*: Information and communication can be secured if they are stored as transactions of blockchain [15,16]. But blockchain is not designed to store large amounts of data. Unfortunately, as the size of the blockchain increases with the number of transactions, performance also gets affected, and synchronization time increases [15].
2. *Security*: Although blockchain technology has several advantages for financial sector. It has the potential to streamline the digital payments. But there are many crypto currencies (> 2000) available in the market. People are losing their hard-earned money in trading with most crypto currencies. Government can regulate, but too much regulation by the government or any centralized agency will nullify the advantages given by decentralized blockchain-based architecture. Currently, one needs to educate the masses about crypto currencies. Blockchain in isolation is very much secure but new security concerns arise with the merger of blockchain with IoT and machine learning algorithms. Data generated by malicious IoT devices are a challenge.
3. *Data privacy*: In blockchain every single transaction can be patterned, inspected, and copied from the system's very first transaction [15]. Dash [17] was the first crypto currency that focused on privacy [17]. Mix coin [18], coin shuffle [19], coin swap [20], and blind coin [21] are some proposed techniques to increase privacy [15]. Therefore, measures should be taken for data privacy while using blockchains.
4. *Smart contracts:* Smart contracts are conceptually similar to triggers in database management system. Triggers are executed automatically in the happening of an event. Similarly smart contact is the code which is executed automatically when the conditions of a predefined agreement are made. Benefits of such an arrangement are that all the parties involved in pre-defined agreement cannot deny if the conditions in the contract happened. All the stakeholders in agreement are well aware about the procedure and the consequences if the conditions in smart contracts are met. For example, a car buyer who wants to get their car finance can get rid of many intermediaries. According to the smart contract, blockchain will streamline the process by providing the verification details of the buyer to the lenders. This will automatically speed up the time to pass/reject a loan. Accordingly, a smart contract will be prepared between buyer, dealer, and banks. So from an implementation point of view, blockchain-based smart contracts

are similar to decision-making statements like switch case statement. Some of the advantages of smart contracts are:

1 reduced processing time as lot of intermediaries is removed.

2 every stakeholder is bounded by the pre-defined rules, so it increases the trust among them.

3 due to the interlinked nature of blockchain, transaction security is increased.

5. *Legal issues:* With the emergence of upcoming technologies like blockchain and IoT, law and order of a country needs to be revised, controlled, and updated. New laws and regulations can ease centralization [15].

6. *No specified software*: There is no standard software till now on which we can connect the two technologies, that is, Blockchain and IoT. As a result of which each company has to build its own software and the entire set of use cases and most importantly they will have to build the entire infrastructure from scratch. This procedure of proper establishment will take years.

7. *Rapid change in technology*: Existing blockchains are immature, and they get outdated very quickly. No organization wants to commit to the technology that gets outdated in 6 months or does not work properly for the use cases due to the hardware constraints or the slow validation times with IoT devices. Till now, they are not able to find which Distributed Ledger Technology can be used for better results.

11.2 INTERNET OF THINGS

When various types of smart devices gets associated to each other and starts communicating via Internet, then IoT comes into the picture. The only factor that holds the roots of IoT is interconnectivity of the devices. Initially, devices collect information and then exchange it through embedded software, cameras, sensors, and so on. These sensors can sense different forms of energy from the environment and then act accordingly for the desired output. IoT is also known as a system of physical entities or publics called to be "things" which when entrenched with software, sensors, electronics, and network will allow these things to receive as well as collect data and then convert it. The aim of IoT is to widen to Internet connectivity from typical devices such as mobile, computer, tablet, Smart TV to relatively dumb devices such as fan, microwave, air conditioner, etc. IoT has a capability of making everything "smart" virtually, by improving facets of human life with the supremacy of data collection, AI algorithm, and networks [22]. The thing in IoT can be anything may it be a person suffering from diabetes and to keep a record to monitor implant or it can be an animal with a tracking device on it for keeping track of its location, etc.

By the year 2020, there may be almost 20 billion IoT smart devices and worldwide expenditure on IoT technologies is predicted to touch $1.3 trillion approximately, while the global market for blockchain development is predicted to be worth $3 trillion by 2024. Companies such as Google, Amazon, Cisco Systems, Intel, general electric, IBM, Dell, Microsoft, HP, Bosch, and many more are adopting IoT technology. General electric (GE) established GE digital in 2015 with the aim of incorporating IoT to predict conservation needs. They produced predix, software that receives generated data from machines and associates it with outdated data. GE also has IoT for hospitality production, the airline diligence, and utility enterprises. Roll Royce is another fortune company which practices smart sensors in combat jets, civil aerospace jet engines, and other vehicles of flight.

Cisco smart grid technologies are used for efficacy services and energy regulator (oil and gas). Cisco also analyzes and collects data related to energy convention and circulation, application supervision, wireless network, and cyber security

11.2.1 ARCHITECTURE OF INTERNET OF THINGS

There are four fundamentals as shown in Fig. 11.3 that governs the IoT system:

1. *Sensors/devices*: The key components ruling the IoT system that will help us to manifest live data from environment are sensors or devices. The data that have been collected from surrounding may have number of complexities in it, depending on the type of input the sensor will collect to the type of output anyone desire. For example, it could be a simple temperature monitoring sensor or a form of the video feed. It might be possible for a device to have a variety of sensors in it, which can perform numerous tasks together while sensing. Example, a smartphone is a device which has multiple sensors like GPS, camera, and fingerprint sensor so that a customer can efficiently interact with its environment.
2. *Connectivity*: The data that are collected from different devices are sent to cloud infrastructure, for doing so the sensors should be linked to the cloud with the help of different mediums of communications. These communication mediums include mobile phones or satellite networks, bluetooth, zigbee, Wi-Fi, WAN, etc.
3. *Data processing*: once the collected data gets linked to the cloud, a software is needed that will help in processing of the gathered data; this can be as simple as checking the temperature, reading devices like AC, ovens, heaters, etc, to a complex task of identifying objects and using computer vision on video.
4. *User interface*: It is a way to provide the information to the end user. Some way either by triggering alarms or by sending a text message or e-mail to the user on their phones or computers. The user sometimes might need an interface for keeping a check on their IoT systems. For example, camera installed in home can allow the user to access video recording and also allows him to check all the feeds with the help of a web server. However, it might not be a one-way communication. Depending on the IoT application and complexity of the system, the user may also perform actions that may lead to cascading effects. For example, suppose a user detects changes in the temperature of the refrigerator and wants to adjust the temperature of the refrigerator sitting far away from it, this can be easily done using the mobile phone.

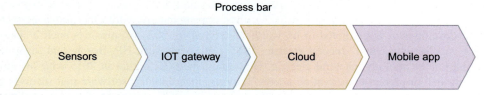

FIGURE 11.3

Architecture of Internet of Things.

11.2.2 HOW A DEVICE BECOMES SMART DEVICE?

A device can be transformed into a smart device by following two steps. First, the capability to operate and interact with physical world by processing collected data using open wireless communication technologies such as bluetooth, RFID, and Wi-Fi. Second, is the capability to collect data through sensors or actuators. For example, smart watch; an ordinary watch connected via bluetooth is capable of recording calories burnt, heart beat rate, steps walked, messages, weather forecast, and sleep monitoring along with date, day, and time.

Smart devices are capable of generating big data that can be stored (using cloud computing), processed, and analyzed (using genetic algorithm, neural network, and AI) to make decisions. To ensure that the data are not accessed by unauthorized access, transparent and reliable devices are needed [23].

11.2.3 APPLICATIONS OF INTERNET OF THINGS

Fig. 11.4 shows applications of IoT in different sectors:

FIGURE 11.4

Applications of Internet of Things.

282 CHAPTER 11 INTERNET OF THINGS AND BLOCKCHAIN

11.2.3.1 Smart home

Use of Internet and Wi-Fi in home computerization has grown over the years with multiple electronic devices like TVs, mobile phones, tablets forming part of a personal network with increasing use of these devices and their connectivity to the Internet. In this context, IoT has allied individuals and devices in the unchanged home, in the similar building, and even in the local area network. On a larger scale, this takes the form of smart cities and smart society. In homes, IoT can manage energy depletion, provide collaboration among home applications, spot predicaments, ensure security, etc. On the larger scale of smart cities, it can be used for witnessing and regulating decent air quality, recognizing emergency paths, etc. Various components are smart sports, cycling, logistics, tourism, environment, road condition monitoring, traffic management, accidental measures, etc. [24–27].

Google home voice controller enables user to control the lights, manage timer and alarm, TV, and speakers by their voice. In addition to this, Amazon echo plus voice controller is capable of running songs, making calls, and messaging, asking questions, providing information, checking weather, managing to do, and shopping list. On the other hand, Amazon dash button allows users to order products quickly by reducing the search time frame for searching required product. But this product is fully functional for Amazon Prime members only.

Other IoT invention is August doorbell cam which is an effective IoT product that constantly checks door and capture motion changes. This innovation allows user to answer the doorbell from anywhere, while August smart lock allows user to know about each and every person coming and going into the house. Additionally, there is no fear of stolen keys as it provides unlimited digital keys. It also has an auto unlock system which gets open as the user arrives near the door.

Kuri mobile robot is kind of home robot and most popular product of IoT, which interacts with user and captures moments all around the house daily. It has sensors, HD camera, microphone, speaker, charging pad, and processor.

11.2.3.2 Wearables

In today's world, wearable's are not only easy to use but are portable at the same time, and it is one of the hottest trends in IoT, some of the examples of wearable that have come across are Apple, Jawbone, Samsung, and many more, all of them are surviving in a highly competitive market and meeting customer requirements.

Even many start-ups have also been initiated in the market for wearable's and have extended this domain creating a niche for Wearable's IoT tech in the future. The large domains have an array of devices included in it like fit bits, hologram watches, fitness belt, etc. This application of IoT has basically covered the fields like fitness, health, entertainment, and lifestyle.

The requirement of IoT in this application is to have a high energy efficiency and it should be of small size that is easier to carry. Some of the examples of wearable's in the market are:

1. Motorola Moto 360 Sport
2. Fitbit ChargeHR
3. Jawbone UP2, etc.

11.2.3.3 Smart city

Footbot air quality monitor helps in measuring indoor pollution, humidity, and temperature. This leads to an increase in the lifespan of user, improved air quality in the house, workplace, and indoor activity space. Other very useful innovation of IoT is nest smoke alarm that can think, speak, and alert mobile phone about any unwanted emergency at home. Security of the house is the most important aspect; nowadays, people tend to focus on this part more than anything else and when a home is under smart surveillance, it not only becomes safer but user can have a control of the processes taking place. IoT can benefit a city in other ways like smart transportation, smarter dynamism management systems, and environmental intensive care.

Due to population outburst, people living in cities have to face a lot of troubles day in and day out, such as increasing pollution, shortages of energy supply, and poor infrastructure. Our aim is to make better use of the public resources, increasing the quality of the services offered to the citizens, while reducing the operational costs of the public administrations [28]. Hence, smart cities can be the real considerable solutions for the problems people usually face.

Some of the examples of IoT devices at work are:

1. Libelium-Metiora Smart Parking Sigfox Kit [29]
2. CitySense-Smart Street Lighting [30]
3. Bigbelly Smart Waste and Recycling System [31]

11.2.3.4 Smart grid

Although the smart grids possess the capability of having an optimistic influence on efficiency and the sustainability of the market, it also has to face some dares related to the secrecy of the customer. One of the use circumstances of subtle data is for the forecast prediction [32,33].

Smart Thermostats: Supports you to protect resources on heating bills by determining your usage arrangements. The sources being used for the generation of electricity today are quickly depleting. Growing public responsiveness in this regard may result in shifting toward renewable infrastructure and resources. In such a scenario, the need of a smart and reliable electric grid responsive to power distinctions through monitoring resources linked to electrical energy foundations, such as loading and sinks, such as energy load and storage, with proper reconfigurations, is undeniable. The backbone of such applications will be grounded on IoT ideas [26].

11.2.3.5 Industrial Internet

Industrial Internet is leading in an innovative way of industrialized mutiny from end-to-end networking, industrial equipment, calculating, and users. This is creating different ways to link various sensors and actuators that are inserted in the machines with the help of Internet. The gigantic quantities of data are produced via sensors and are supplementary analyzed to comprehend the functioning and conditioning of the apparatuses and industrial systems, and even authorize independent operations and controlling the complete systems. It is going to influence a wide industrialized spectrum that includes trade, health care, conveyance, power, oil and gas, operation, smart buildings, and cities. The topics that are generally included in this are: industrial cyber-physical systems, industrial communication and schmoozing technologies, industrial cloud and edge computing, industrial robotics and autonomous systems, data management and analytics, security,

284 CHAPTER 11 INTERNET OF THINGS AND BLOCKCHAIN

trust, real-time, privacy, reliability, and safety, industrial big data and AI and applications, test beds, and case studies.

11.2.3.6 Connected car

Automobile companies use IoT to handgrip billing procedures, insurance, parking, and other associated stuff inevitably. Parking sensors identify real-time accessibility of parking places on customer's handset which can be easily done by IoT technology. Most popular applications of IoT in the field of transport include ceaseless highway toll, law execution, checking rule breaking by vehicles and may soon include declining environmental pollution, assuaging traffic, etc. [27] Google's self-driving cars are well acknowledged by everybody now. IoT is made things like "connected cars" a prospect but slowly and steadily it has become an entrenched fact that any upcoming technology always takes no less than a few years to cultivate in conventional automotive manufacturing.

Companies and start-ups in the market are announcing ingenious technologies and have anticipated the concept of connected car platforms for the future. Electric vehicle with sensor fusion is the new technology taking birth in different parts of the world which will not only make the cars smart but will help the owner to communicate with the environment inside and outside of the car in a well versed way.

Here are some examples of the work that is going on in this field:

1. Caterpillar's newest equipment
2. Latest locomotive from GE

11.2.3.7 Connected health

Health care can be met using RFID sensors and barcode. These measuring devices can be used by hospitals to pathway cloud computing system that can store all medical records on demand. More significantly, health data are extremely personal for the patients and needs protection from both the unauthorized entities and cloud servers that can excerpt surreptitious information of personal health information (PHI) from patient and health service provider interactions [34]. Lack of appropriate security and privacy may lead to leakage of PHI which may be exposed to various kinds of threats: eavesdropping, modifying, and injecting messages that may result in wrong medical diagnosis or treatment, spoofing patient's identity or location to mislead the medical rescue. Therefore, it is crucial to guarantee both the patients personal health data security and location privacy in cloud. For data concealment and privacy, key management strategies have delivered a convincing result to protect wireless communication [34]. The concept of associated health upkeep scheme simplifies real-time health checking and patient care. It benefits in making better curative decisions depending on patient's health data. IoT finds application in health care for monitoring metrics accompanying with health, handling medicines management in inventory, etc. Biopatch, a biomedical sensor device takes decision of calling inaccessible physician, hospital, emergency center, test clinic, and supply chain medicine retailers [24−27].

IoT examples in this domain are many, some of them are:

1. Philips' medication dispensing service [35]
2. Future Path Medical's UroSense [36]

11.2.3.8 Smart retail and supply chain

SCM may be defined as the flow of belongings and services while counting the movement and storage of raw materials, work-in-process inventory, and finished goods from point of source to point of consumption. IoT can be used to deliver eminence-based perishable stuff in a shrewder way. Future Stores by Metro Group have implemented commodity-centered retail funding to the customers by incorporating RFID on top of State-of-the-Art enabled SCM system. Supplier, constructor, and trader information-focused servers interconnect with loading and inventory workforces through pre-installed database system. This annexation of sensor-based system into perishable goods can lower down carbon footprint in retail.

The current equipment in manufacturing sector either has no competency to observe, record, or collect information about its use and potential faults or has some capability for this purpose but data capacity is limited and the records are isolated and not readily available when required. With the use of Industrial IoT (IIoT), machine logs are gathered centrally as well as analyzed and reported consistently. As an instance, sensors may be used to observe humidity circumstances during vehicle painting and permit real-time monitoring by regulating ventilation systems [24,25,37,38].

11.2.3.9 Smart farming

IoT is instrumental in recording and controlling various environmental variables like, NH3, PM2.5 concentration, soil situations, including soil moistness and temperature, habitat environment containing temperature, humidity, illumination, CO_2 and tilt, and finally growth circumstances such as tree diameter and sap flow in different types of environments like henhouse system, agriculture environment, wild vegetation environment, domestic waste treatment and disposal management, etc. Various researches have been conducted and many models are proposed, each relating to different aspects of different environments [24,25].

Various sensors can be used to find out different land requirements and appropriate action be taken as per need. Some useful cases are smart packaging of seeds and fertilizers according to specific environmental situations. Smart farming will support in countering climate instability by adopting better farming practices. This has the potential to drastically enhance farming productivity [27]. Agriculture segment needs very accustomed as well as extremely compatible technology as its clarifies IoT applications that can be delivered to the farmers. There are many new systems like "Smart irrigation system"; "Smart farming systems" [39] are developed which can help farmers and lead a good production of the crops and then on the collection of data by the systems it is sent to the cloud for keeping track of every parameter which is involved in farming [40]. CleanGrow's Carbon Nanotube Probe [41] and The Open IoT Phenonet Project [42] are the two projects which were undergone in the field of Smart agriculture and on the collection of data by the systems

11.2.4 ADVANTAGES OF USING INTERNET OF THINGS TECHNOLOGY

- With IoT technology, a lot of technologies have been improved and made better basically leading them to technical optimization. Example, a manufacturer can collect the data from various sensors in a car and then can analyze them for the improvement of its design for an efficient and effective model.

286 CHAPTER 11 INTERNET OF THINGS AND BLOCKCHAIN

- The way the data has been collected traditionally had some limitations, but with IoT, data can be collected efficiently which can facilitate immediate actions on it.
- With lots of data in this environment, one can also have to face a problem of waste data, but with IoT, real-time information can lead us to effective decisions and proper management of the resources; similarly, one can easily reduce waste from the systems.

11.2.5 CHALLENGES WITH INTERNET OF THINGS SYSTEMS

- *Security*: IoT is an ecosystem of devices connected to the network. Nevertheless, the system may present little authentication control despite providing adequate amount of security to the systems.
- *Privacy*: The usage of IoT can sometimes render a significant sum of peculiar data, without the user's lively involvement. This results in loads of confidentiality-related issues.
- *Flexibility*: At hand is a massive apprehension concerning the flexibility of system. It is largely about assimilating with another diverse system which can be convoluted in a process.
- *Complexity*: The designing of the IoT system is relatively complex. In addition to this, its disposition and upkeep is also a tough task.
- *Compliance*: IoT systems have their individual set of policies. Nevertheless, as it is complex, its task of compliance is also challenging.

11.3 INTEGRATION OF INTERNET OF THINGS AND BLOCKCHAIN

The focus of this section will be on the collaboration between IoT and blockchain referred to as blockchain-based IoT (BIoT). Integration of these two above-mentioned technologies will be a decentralized approach that will solve many issues. By adoption of a consistent P2P communiqué, the user can process hundreds and billions of dealings between the devices that will apparently lessen the cost. Blockchain allied with IoT networks will allocate computational and storing needs across numerous devices. This will hence prevent entire network from failure.

During establishment of P2P communication, number of challenges such as security, scalability, privacy, etc, comes into the picture. When both the technologies combine together, a solution which will uphold privacy as well as safekeeping in gigantic networks can be proposed. This combination will not compromise any form of validations and agreement for the transactions so that the activities like spoofing and theft can easily be avoided.

To accomplish the utilities related to customary IoT solutions without decentralized tactic, it will be difficult to manage transactions. So, it must keep three foundational functions:

1. Peer-to-peer messaging;
2. Disseminated file sharing;
3. Autonomous device harmonization.

Safekeeping needs to be constructed as a basis of IoT structures, with severe legitimacy authorizations, substantiation, and data authentication. After that, data need to be encoded in a better way for enhancing the security system [43]. At the solicitation level, software development

11.3 INTEGRATION OF INTERNET OF THINGS AND BLOCKCHAIN 287

administrations need better coding skills that are not only stable but also robust and dependable, with enhanced code improvement standards, menace study, and analysis. When systems interrelate with every single device, it is really crucial to have an approved interoperability standard, which is innocuous and legal. Without having a concrete bottom−top structure, it may craft more number of intimidations with all devices that are connected to the IoT network. Our aim is to secure and safeguard IoT devices. This can be done with the help of blockchain as it supports cryptographic keys and decentralized network.

IoT devices are generating lots of data. Need of the hour is to provide data security and reliability for the content generated by millions of heterogeneous IoT devices. Blockchain technology can help in increasing the data security. Decentralized nature of blockchain technology eliminates the drawbacks associated with a centralized architecture-based system. For example, mobile phones and smartphones; users want device size to get reduced without any compensation with device security or computational feature. Similarly in case of IoT, people prefer small size devices which can track their health or financial data without compromising with any kind of loss due to high vulnerability to security threats.

11.3.1 NEED FOR INTEGRATION OF INTERNET OF THINGS AND BLOCKCHAIN

IoT generates gigantic large network using machine-to-machine interactions. Digital sensors, camera, or smartphones are all machines. Our challenges are to increase the trust in such an environment where each machine has access to data and they will verify whether there is any tampering by any device or node. BIoT integration can increase the data sharing and the stakeholders of a supply chain management. Smart cities and smart vehicles need to share the data which must be reliable. Reliable data bring more nodes in the system and will ultimately increase the performance. Blockchain technology brings autonomy by creating an environment for BIoT where devices or machines are able to interact. BIoT will enable IoT for digital payments with its traceability features. While low computational speed of blockchain nodes is a serious issue and it also reduces IoT performance. However, BIoT will also provide fast calculations. This increase in computing power is in fact the result of BIoT.

Such an integration of blockchain technology and IoT in fact allows users to enjoy the advantages of both the technologies in a more efficient and powerful manner. However, there are many issues and challenges in implementing integration of blockchain and IoT which will be discussed in the next sub-section. Some of the advantages of BIoT are as follows.

1. Increased computational speed
2. Increased trust
3. Data security
4. Increased data generating capability of IoT nodes
5. Increased power of ML algorithms.
6. To tackle scalability issues of blockchain nodes in a seamless manner.
7. To increase the traceability of any transactions
8. To address the addressing issues
9. To make machines smart
10. To collect data from source and processing

288 CHAPTER 11 INTERNET OF THINGS AND BLOCKCHAIN

11.3.2 CHALLENGES IN INTERNET OF THINGS AND BLOCKCHAIN INTEGRATION

BIoT brings the best out of blockchain technology and IoT. But there are some challenges which needs to be addressed:

1. Currently consensus-based protocols pick the miner with some degree of randomness. But in the absence of network consensus, they may become victim of blockchain forks. Therefore, there is a need to develop IoT focused consensus protocols for BIoT system. Such protocols can handle the ever increasing communication complexity.
2. Scalability of blockchain is also one of the important issues.
3. Security of IoT devices and protocols against network threats such as malware also need to be addressed. Blockchain in isolation is able to counter such attacks very effectively but it becomes an issue with IoT devices.
4. Privacy of data in blockchain nodes is a very critical issue which needs to be dealt in a very effective manner. For example medical records of a patient or smart quotation for smart contracts are some use cases of BIoT which demands privacy.
5. Data generated from large IoT network in real time brings a serious limitation for BIoT.

11.3.3 BENIFITS OF INTEGRATION

By integrating blockchain technology with IoT, we expect to have the following benefits:

1. Decentralized architecture of blockchain increases the security of highly vulnerable IoT devices. Hacker can modify or alter only local blockchain as it is very difficult to make changes in every block of blockchain.
2. Enhanced security of blockchain using encryption and cryptographic algorithms of the blockchain.
3. Smart contracts are stored in blockchain network that not only reduces the processing time of a transaction but helps in several kinds of fraud detection.
4. BIoT combination has the potential to eliminate the need of intermediaries.
5. BIoT combination leads to the framework where machine learning algorithms can be applied to the real-time data generated from blockchain-based IoT devices and can enable the strengthening of supply chain management.

11.3.4 INDUSTRIAL INTERNET OF THINGS

IoT network consists of many devices like sensors, camera, laptop, RFIDs, machines, etc. Such devices are interconnected in one way or another. Due to this interconnection, such devices are also vulnerable to several attacks. Several models are proposed for security of IoT devices. However, when the term IoT is said most of the time, it is meant for non commercial usage. For industrial usage, it can be referred to as IIoT.

IIoT demands more stringent storage and computational requirements in comparison with IoT. This stringent demand arises from the fact that IIoT devices are more likely to deal with real-time big data. Such kind of infrastructure leads to the challenge of transmission of this big data among networks in real-time. It also demands more sophisticated computing with the need to improve the

11.3 INTEGRATION OF INTERNET OF THINGS AND BLOCKCHAIN 289

time complexity of existing algorithms. It also demands the need to strengthen the existing security and privacy protocols for the same. IIoT rides upon the merger of embedded intelligence and network connectivity of IoT devices. For an industry, sometimes a little change in the existing infrastructure may be enough to avail the IoT benefits. However, for big industries, a lot needs to be done to avail the maximum benefits from the intersection of people, data, and intelligent devices. Keen interest is shown by certain industry segments as they have responded by doing large investment in IoT technology. They are collecting analytical data from devices like gas turbines and locomotives. Situation can improve further when the principles of cognitive computing are applied. Such merger of IoT with cognitive capabilities will help in cost reduction and can improve the existing business models.

11.3.5 INTERNET OF THINGS ON CLOUD PLATFORM

Most of the IoT devices are storage constrained. IoT devices need significant storage for data by computation [44]. Cloud computing can provide storage as a service feature for several of IoT devices connected. Such arrangement in fact not only solves the storage issues of IoT devices but also provides a strong foundation for IIoT network by providing the storage needed for real-time computation of big data for the smart devices involved in IIoT. IoT devices generates a lot of data may be in the range of terabytes and pet bytes. Need is to capture and process such data. Cloud computing provides an environment to do real time analysis of this data. It is useful in making applications like an intelligent refrigerator that automatically orders an item whenever it is less in quantity.

Clouds can store data for providing smart phone services and can handle multi-structured data. Cloud-based services will enable IoT to be deployed as a platform as a service. It is needed to link IoT applications, services, and data and to work on blockchain challenges related to IoT such as smart contracts, scalability, licensing, privacy, and security issues. IoT is just a point in the long list of potential use cases of blockchain.

11.3.6 CREATING AN ENVIRONMENT

It has been observed that blockchain, IoT, and cognitive computing are working fine in isolation. But the need of the hour is the integration of blockchain, IoT, and cognitive computing. Convergence of these technologies can be achieved by creating an environment. The constituent parts of an environment are as follows:

1. Stand alone isolated customer who expects the real-time digital services from the available devices like digital camera, sensors, smartphones, etc, with minimum delay.
2. IT developers think about a lightweight architecture and protocols that can deal efficiently with the information coming from many digital devices.
3. There are cloud service providers willing to offer storage as a service for many storage constrained IoT devices.
4. There are industries willing to excel upon the information collected not only from the devices like smartphones, cameras, and sensors but also from machines and locomotives.
5. There are algorithms based on AI, machine learning, data sciences, and cognitive computing

290 CHAPTER 11 INTERNET OF THINGS AND BLOCKCHAIN

The main aim is to take care of the requirements of customers, developers, and industry. It is needed to create an environment where these constituents can be integrated in a hassle free manner. It is expected to build an environment where linear and permanent-indexed records of blockchain is used by IoT to increase the trust in the IoT network which otherwise is connected to vulnerable public networks. Blockchain is supposed to provide a solution to the typical client/server-based paradigm of most IoT networks. As mentioned in above sections that many IoT devices are storage constrained, the cloud computing storage facilities can enable IoT devices to fulfill their storage needs. Once the IoT devices are enabled to store big and real data from many IoT devices (blockchain decentralized architecture based), it provides an excellent platform for the buildup of IIoT. IIoT as explained in the above section thrives on the fully connected devices, machines sending data at a rapid rate which eventually will be utilized by machine learning algorithms. Ml- and AI-based algorithms can replace the existing time-consuming blockchain mining algorithms. Such integration will lead toward the development of new business models. Currently, such an environment is helping in fraud identification. Machine learning algorithms are able to evaluate the facts from sensors to assess whether the defect claimed in a vehicle is genuine or not.

11.4 DISCUSSION

Blockchain technology was initially supposed to be the architecture behind the crypto currency but now it is being used in many diverse fields. Efforts need to learn from the experience of its usage in Crypto currency and move toward its applications in diverse fields like health care, economics, and government welfare schemes [45]. An awareness to increase knowledge about blockchain Technology in organizations is needed. Areas need to be identified, where blockchain technology can solve the problem of intermediation. IoT although gaining popularity suffers from security and privacy vulnerabilities [46]. Traditional ways to provide security and privacy are inappropriate for IoT due to decentralized topology. Problem becomes more severe as in IoT many devices have resource constraints. Millions of devices which form the backbone of any IoT network are also likely to be attacked by hackers leading toward data theft. Effort is needed to look for a seamless integration for establishing the relationship among the objects and their environments.

IoT further can be applied in several industries leading to the term IIoT. Combination of blockchain and IoT need to be used so as to overcome the limitations of blockchain. Such combination is able to reduce the computationally expensive mining algorithms of blockchain technology. Thus, with this combination, one can opt for lightweight IoT architecture. To maximize the benefits of blockchain and IoT combination, some new models should be created to support different distributed IoT applications. For successful combination of blockchain, IoT requires to solve the issues like scalability and reducing the undesirable overhead traffic. Database designers must work upon to improve transaction throughput of blockchain databases. With the rapid increase in the number of transactions, the convolution of mining algorithm increases, which makes it more difficult for power constrained IoT devices to do substantial computing?

Blockchain is analogous to the Internet; the Internet that is known today is nothing like it was back then, the same is for blockchain. Blockchain when merges with IoT can add a whole fresh aspect to the Internet, this will not only introduce new standards of data transparency but also will result in better P2P communication. With the rapid development in this technology, you will find that this too is becoming very slow. Devices now cannot only transfer data but can also monitor values, can open up brand new markets which have the worth of trillions in the future.

11.5 **FUTURE WORK**

Combining the concepts like IoT and blockchain allows users and companies to directly amass the billion times more information that can be generated by approximately 30 sensors in the car, the sensors incorporated in engines of planes that measure more than 5000 elements every second and other sensors that one use in every small part of our daily life which measures different parameters that affect human being such as weather conditions, pollutants, fuel, humidity, temperature, sound, moisture, vibration, resistance, electricity, weight, and many more factors that affect our lives.

Renowned companies like GE and Cisco have started working in this field, and it has been estimated that by 2020, 1 trillion sensors data will be present in the world.

Sensors make system smart and help in the proper interaction with the surrounding and the data that they produce has to be incorporated somewhere or another. With this scenario, one can lead to the way of endless possibilities; this can even lead to opening up of more than gazillions of data points. They are to be measured each day by the number of sensors and the track of data has to be kept which can be easily done via blockchain technology that was specially designed to sell and buy the data that comes through different private and public marketplaces where the sensors are placed.

On realizing the need of data incorporation on its increase with time, very limited amount of organizations in market have started testing the basic use cases for the technology which will slowly deploy with the rapid increase in data, but till now, no company is using BIoT solutions today, and many new blockchain projects that include "The supply chain sector" will be only used for blockchain part of BIoT to track the goods of the company.

Blockchain developers can stabilize and ensure guarantee of verification then certainly users will be using more of blockchain-based products. IoT can be clubbed with cloud computing platform to counter the limited storage and computational capabilities of IoT devices. BIoT will eventually collaborate or they will continue to simply co-exist, with the subsequent increase in data and the need of security.

The solution to the problems related to BIoT technology can be an embedded wallet for the IoT, and it is basically constructing a "Ledger Nano" for the machines. This will allow all the IoT devices to get connected with the distributed ledger and as it is not being the IoT gateway, it will retain all the security benefits of the blockchain. Also it is not the full node; hence, it will reduce the load on an IoT device, thereby alleviating the efficacy of the entire system. This can be one of the ways to move toward the desired system which has scalable infrastructure for the autonomous machine.

REFERENCES

[1] K. Karimi, G. Atkinson, What the Internet of Things (IoT) Needs to Become a Reality. White Paper, FreeScale and ARM, 2013.

[2] Gartner, Report: Forecast: The Internet of Things. Worldwide, 2013.

[3] Cisco Systems, White Paper: Cisco Visual Networking Index: Global Mobile Data Traffic Forecast Update, 2016−2021, 2017.

[4] M. Suárez-Albela, P. Fraga-Lamas, T.M. Fernández-Caramés, A. Dapena, M. González-López, Home automation system based on intelligent transducer enablers, Sensors 16 (10) (2016) 1−26. no. 1595.

[5] Block chain technology report to the US Federal Advisory Committee on Insurance. Available from: <https://www.treasury.gov/initiatives/fio/Documents/McKinsey_FACI_Blockchain_in_Insurance.pdf> (accessed 10.04.18).

[6] M. Swan, Block Chain: Blueprint for a New Economy, first ed., O'Reilly Media, 2015.

[7] M.A. Khan, K. Salah, IoT security: review, block chain solutions, and open challenges, Future Gen. Comput. Syst. 82 (2018) 395−411.

[8] K. Christidis, M. Devetsikiotis, Block chains and smart contracts for the Internet of Things, IEEE Access 4 (2016) 2292−2303.

[9] E. Karafiloski, A. Mishev, Block chain solutions for big data challenges: a literature review, in: Proceedings of the IEEE International Conference on Smart Technologies, Ohrid, Macedonia, 6−8 July 2017.

[10] A. Dorri, M. Steger, S.S. Kanhere, R. Jurdak, Block chain: a distributed solution to automotive security and privacy, IEEE Commun. Mag. 55 (12) (2017) 119−125.

[11] Bitfury Group; J. Garzik, Public Versus Private Block Chains Part 1: Permissioned Block Chains, 2015, 1−23. Available from: <https://bitfury.com/content/downloads/public-vs-private-pt1-1.pdf>.

[12] S. Raval, Decentralized Applications: Harnessing Bit Coin's Block Chain Technology, first ed., O'reilly Media, 2016.

[13] I. Makhdoom, M. Abolhasan, H. Abbas, W. Ni, Block chain's adoption in IoT: the challenges, and a way forward, J. Netw. Comput. Appl. 125 (2019) 251−279.

[14] https://blog.hubspot.com/sales/real-estate-quotes.

[15] A. Reyna, C. Martín, J. Chen, E. Soler, M. Díaz, On block chain and its integration with IoT. Challenges and opportunities, Fut. Gener. Comp. Syst. 88 (2018) 173−190.

[16] G. Prisco, Slock. It to introduce smart locks linked to smart ethereum contracts, decentralize the sharing economy, 2016. Available from: <https://bitcoinmagazine.com/articles/slock-it-to-introduce-smart-locks-linked-to-smart-ethereum-contracts-decentralize-the-sharingeconomy-1446746719/> (accessed 01.02.18.).

[17] Dash, 2017. Available from: <https://www.dash.org/es/> (accessed 20.10.17).

[18] J. Bonneau, A. Narayanan, A. Miller, J. Clark, J.A. Kroll, E.W. Felten, Mixcoin: anonymity for bit coin with accountable mixes, International Conference on Financial Cryptography and Data Security, Springer, San Juan, Puerto Rico, 2014, pp. 486−504.

[19] T. Ruffing, P. Moreno-Sanchez, A. Kate, Coinshuffle: practical decentralized coin mixing for bit coin, European Symposium on Research in Computer Security, Springer, Heraklion, Crete, Greece, 2014, pp. 345−364.

[20] G. Maxwell, CoinSwap: Transaction Graph Disjoint Trustless Trading, 2013.

[21] L. Valenta, B. Rowan, Blindcoin: blinded, accountable mixes for bit coin, International Conference on Financial Cryptography and Data Security, Springer, San Juan, Puerto Rico, 2015, pp. 112−126.

[22] John Stankovic, Research directions for the Internet of Things, IEEE Inter. Things J 1.1 (2014) 3−9.

[23] A. Ravishankar Rao, D. Clarke, Perspectives on emerging directions in using IoT devices in block chain applications, *Internet of Things*, (2019).

REFERENCES 293

[24] P.P. Ray, A survey on Internet of Things architecture, J. King Saud Univ.—Comp. Inform. Sci. 30 (3) (2018) 291−319.

[25] P. Asghari, A.M. Rahmani, H.H.S. Javadi, Internet of Things applications: a systematic review, Comp. Netw. 148 (2019) 241−261.

[26] S. Albishi, B. Soh, A. Ullah, F. Algarni, Challenges and solutions for applications and technologies in the Internet of Things, Procedia Comp. Sci. 124 (2017) 608−614.

[27] R. Mehta, J. Sahni, K. Khanna, Internet of Things: vision, applications and challenges, Procedia Comput. Sci. 132 (2018) 1263−1269.

[28] H. Schaffers, N. Komninos, M. Pallot, B. Trousse, M. Nilsson, A. Oliveira, Smart cities and the future Internet: towards cooperation frameworks for open innovation, Fut. Inter. Lect. Notes Comput. Sci. 6656 (2011) 431−446.

[29] http://www.libelium.com/smart-parking-surface-sensor-lorawan-sigfox-lora-868-900-915-mhz/.

[30] https://economictimes.indiatimes.com/citysense-indian-origin-designer-chintan-shahs-smart-street-lamps/articleshow/25636569.cms.

[31] bigbelly.com.

[32] A. Ahmad, M. Hassan, M. Abdullah, H. Rahman, F. Hussin, H. Abdullah, et al., A review on applications of ANN and SVM for building electrical energy consumption forecasting, Renew. Sustain. Energy Rev. 33 (2014) 102−109.

[33] Commission for Energy Regulation, Electricity smart metering customer behaviour trials (CBT) findings report. Technical Report CER11080a, 2011. Available from: <http://www.cer.ie/docs/000340/cer11080 (a)(i).pdf>.

[34] J. Zhou, Z. Cao, X. Dong, N. Xiong, A.V. Vasilakos, 4S: a secure and privacy-preserving key management scheme for cloud-assisted wireless body area network in m-healthcare social networks, Inform. Sci. 314 (2015) 255−276.

[35] https://medium.com/@medipense/2017-the-year-of-the-iot-automated-pill-dispenser-ca1d41f0592b.

[36] https://www.bioenterprise.com/companies/future-path-medical/.

[37] K. Alexopoulos, S. Koukas, N. Boli, D. Mourtzis, Architecture and development of an Industrial Internet of Things framework for realizing services in Industrial Product Service Systems, Procedia CIRP 72 (2018) 880−885.

[38] A.R. Sfar, E. Natalizio, Y. Challal, Z. Chtourou, A roadmap for security challenges in the Internet of Things, Dig. Commun. Netw. 4 (2) (2018) 118−137.

[39] N. Gondchawar, R.S. Kawitkar, IoT based smart agriculture, Int. J. Adv. Res. Comp. Commun. Eng. 5 (6) (2016). 2278−1021.

[40] T. Baranwal, Nitika, P.K. Pateriya, Development of IoT based smart security and monitoring devices for agriculture, in: 6th International Conference—Cloud System and Big Data Engineering, IEEE, Noida, India, January 14−15, 2016, 597−602.

[41] https://internetofthingswiki.com/title/cleangrows-carbon-nanotube-probe-compressor/.

[42] http://www.openiot.eu/.

[43] S.N. Swamy, D. Jadhav, N. Kulkarni, Security threats in the application layer in IoT applications, in: 2017 International Conference on I-SMAC (IoT in Social, Mobile, Analytics and Cloud) (I-SMAC), IEEE. Palladam, India, October 5, 2017, pp. 477−480.

[44] P. Corcoran, The Internet of Things: why now, and what's next? IEEE Consum. Electron. Mag 5 (1) (2016) 63−68.

[45] K. Rose, S. Eldridge, L. Chapin, The Internet of Things: An Overview, Understanding the Issues and Challenges of a More Connected World, Whitepaper released by Internet Society, 15 October 2015.

[46] A. Dorri, S.S. Kanhere, R. Jurdak, Block Chain in Internet of Things: Challenges and Solutions, 2016, arXiv preprint arXiv:1608.05187.

294 CHAPTER 11 INTERNET OF THINGS AND BLOCKCHAIN

FURTHER READING

Al-Fuqaha, M. Guizani, M. Mohammadi, M. Aledhari, M. Ayyash, Internet of Things: a survey on enabling technologies, protocols, and applications, IEEE Comm. Surveys & Tutorials 17 (4) (2015) 2347−2376.

M. Banerjee, J. Lee, K.-K.R. Choo, A block chain future to Internet of Things security: a position paper, Digital Commun. Netw (2017). <https://doi.org/10.1016/j.dcan.2017.10.006>. <http://www.sciencedirect.com/science/article/pii/S2352864817302900>.

Bit coin is a fraud that will blow up, says JP Morgan boss, 2017. Available from: <https://www.theguardian.com/technology/2017/sep/13/bitcoin-fraudjp-morgan-cryptocurrency-drug-dealers> (accessed 01.02.18).

Bit coin could be here for 100 years but it's more likely to 'totally collapse', Nobel laureate says, 2018. Available from: <https://www.cnbc.com/2018/01/19/bitcoin-likely-to-totally-collapse-nobel-laureate-robert-shiller-says.html> (accessed 01.02.18).

Bit coin could hit $100,000 in 10 years, says the analyst who correctly called its $2,000 price, 2017. Available from: <https://www.cnbc.com/2017/05/31/bitcoin-price-forecast-hit-100000-in-10-years.html> (accessed 01.02.18).

E.B. Centralny, Virtual currency schemes—a further analysis, Luty, 2015. Available from: <https://www.ecb.europa.eu/pub/pdf/other/virtualcurrencyschemesen.pdf> (accessed 01.02.18).

C.K. Elwell, M.M. Murphy, M.V. Seitzinger, Bit coin: questions, answers, and analysis of legal issues, Congressional Research Service, 2013. Available from: <https://fas.org/sgp/crs/misc/R43339.pdf> (accessed 01.02.18).

https://www.bgr.in/news/digital-india-narendra-modi-flags-cyber-security-concerns-says-india-can-play-big-role/.

B. Hammi, R. Khatoun, S. Zeadally, A. Fayad, L. Khoukhi, IoT technologies for smart cities, IET Netw. 7 (1) (2018) 1−13.

J. Lopez, R. Rios, F. Bao, G. Wang, Evolving privacy: from sensors to the Internet of Things, Future Gener. Comput. Syst. 75 (2017) 46−57.

J.S. Kumar, M.A. Zaveri, M. Choksi, Task based resource scheduling in IoT environment for disaster management, 7th International Conference on Advances in Computing & Communications, ICACC-2017.

R. Roman, J. Lopez, M. Mambo, Mobile edge computing, fog et al. a survey and analysis of security threats and challenges, Future Gener. Comput. Syst. 78 (2018) 680−698.

R. Roman, J. Zhou, J. Lopez, On the features and challenges of security and privacy in distributed Internet of Things, Comput. Netw. 57 (10) (2013) 2266−2279.

J. Zhou, Z. Cao, X. Dong, A.V. Vasilakos, Security and Privacy for Cloud-Based IoT: Challenges, IEEE Commun. Mag. 55 (1) (2017) 26−33.

CHAPTER

BLOCKCHAIN AND INTERNET OF THINGS: AN OVERVIEW

12

A. Sherly Alphonse[1] and M.S. Starvin[2]

[1]DMI Engineering College, Aralvaimozhi, India [2]University College of Engineering, Nagercoil, India

12.1 INTRODUCTION

The growth of Internet of Things (IoT) technology has a significant role in developing a sophisticated world that has a machine to machine communication. The machines communicate with the other machines to save cost and energy enabling full automation and monitoring of devices. This leads to transparent and quality communications. IoT connects all the devices to the internet like mobiles, air conditioners, washing machines, coffee makers, health monitoring devices, and so on. The privacy is lost as all the information is communicated through the internet. As such, the data should be protected through an encryption process. In health-monitoring applications, the protection of the patient's information is very much needed to ensure that the medical information is not leaked to the hackers. As all the industrial machines, home appliances, water, and power supply devices are connected through the internet, it should be ensured that the confidential information is not leaked to hackers. The conventional approaches used for maintaining security cannot be applied in an IoT-based application because of its huge resource constraints.

Blockchain and Bitcoin are used to provide security to the architectures that are similar to the ones used for IoT. For applications having innumerable transactions blockchains can be used. However, blockchains have huge computational complexity and suffer from huge overhead and delays. There are various lightweight architectures in literature that reduce the overhead without compromising its advantages. These architectures can be used in various smart devices that employ IoT. There are different types of blockchains for different network hierarchies. Qualitative estimation is used for evaluating the architecture to ensure its effectiveness. Blockchain helps to improve productivity without sacrificing security and scalability. It functions as a distributed ledger having secured information communication via encryption technologies. Each block in a blockchain is a collection of data. A blockchain has a chain of such blocks. The first block of the chain is named as the Genesis block. The amount of bitcoin is kept track of as the blockchain has a ledger, that is a digital file that helps to track all the transactions. All the nodes in the network may receive the message and copy the needed transaction to the ledger, manipulating the account balances. The technology permits the digital information for distribution and not for copying. The data are constantly submitted to the database, and is stored in various locations and manipulated instantly.

Handbook of Research on Blockchain Technology. DOI: https://doi.org/10.1016/B978-0-12-819816-2.00012-5
© 2020 Elsevier Inc. All rights reserved.

295

Bitcoin-like cryptocurrencies create more digital innovations. Applications of the blockchain in the IoT are very vast. This chapter provides a detailed overview of how the blockchain affects the IoT. The Internet of Everything (IOE) has benefits like fast-tracking. The emerging technologies have not only changed the way the internet behave but has also changed the way the devices and sensors connect to the internet. The devices, gateways, and sensors connect to the network without any human intervention. In a world with smart homes and smart factories, the devices communicate with each other without any human intervention. There will be good monitoring of devices and fewer security breaches. The systems are more transparent. IoT connects all the systems that we use in day-to-day life to the internet. To monitor the health and safety of humans, wearable devices connected to the internet can be used in our daily life. The data communicated are encrypted through some cryptographic techniques. The main problem is that the conventional security approaches used for other applications cannot be applied for an IoT-based application.

The artificial intelligence (AI) techniques have helped to gather only useful information. The application of blockchain technology has helped the IoT to be used in different organizations like the government, private companies, and industries. The blockchains have made the IoT-based systems not only affordable but also more secure. The contributions of this chapter are given below:

- The chapter gives a detailed overview of IoT.
- The applications of IoT in different sectors are briefly described.
- The chapter describes the challenges faced while implementing IoT in day-to-day life.
- Blockchain technology is explained in detail.
- The recent researches that help in overcoming the drawbacks are also discussed.

There are a series of blocks in a blockchain. Blockchain uses cryptocurrencies like Bitcoin, Litecoin, and Ethereum, and so on. The data depend upon the type of blockchain used for the particular application. Each block has the data about the sender, receiver, and details about the bitcoins. The genesis block is the first block of a blockchain. The blockchain functions like a ledger that keeps track of all the transactions. The systems connected as a network process the transaction, and store the data as a block in the blockchain. Although the data are stored at different locations, it is manipulated to maintain consistency. There is distribution of information and no copying. Even without knowing the identity, money can be transferred. Blockchain and Bitcoin provide security to the architectures in literature that are similar to the ones used for IoT. In the case of applications involving numerous transactions, blockchain techniques can be used. But these techniques suffer from computational complexities. There are different types of blockchain architectures for different networks. There are also lightweight architectures that help the blockchain to be applied in different applications [1,2].

Section 12.2 gives an overview of the IoT technology. It also gives information about the different applications of IoT. It explains the architecture in detail. It explores the different challenges faced in IoT-based applications and gives an overview of the existing security approaches. Section 12.3 presents an analysis of blockchain technology. This section presents some of the properties of blockchain. It also explains some of the researches going on in the field of blockchain technology. The section also explains certain challenges faced. Section 12.4 explains the advantages of using Blockchain technology with IoT. Section 12.5 explains the convergence of IoT with blockchain technology and the uses of AI in decision making. Section 12.6 concludes the chapter.

12.2 OVERVIEW OF INTERNET OF THINGS

The IoT converts the objects existing in the real world into the intelligent objects of the virtual world. IoT gathers all the objects into one so that there is better control of the objects and updates about the current state of the objects. It helps in better accumulation of information. IoT can be used in many applications. This allows tracking the objects through codes so that the industries may become efficient, flexible, and rapid functioning ones. 'The Internet of things' phrase has both internet and things. The internet uses transmission control protocol (TCP)/internet protocol (IP) protocol to connect the devices. The internet consists of lots of private networks, public networks, networks from governmental organizations that are connected by a wide array of wireless, electronic, and optical network techniques. More than 100 nations are interconnected today. The internet can also be used in space through Cisco's Internet Routing in Space (IRIS) program in the coming four years (http://www.cisco.com/web/strategy/government/space-routing.html). A thing is distinguished as an object or a person in the real world. These objects are not only the electronic gadgets, but also the food, furniture, clothing materials, merchandise, parts of equipment, landmarks, specialized items, monuments, and works of art, items used in commerce, culture, and sophistication. Also, living things like a person, calf, dog, pigeon, animal, cow, rabbit, and so on, plants like mango tree, jasmine, banyan and nonliving things like a chair, fridge, tube light, curtain, plate, and so on can also be considered as real-world objects.

12.2.1 APPLICATION OF INTERNET OF THINGS

IoT technology has been used in day-to-day applications to military applications as Internet of medical things, Internet of vehicles, and Internet of military things. This section discusses some of the applications of IoT as follows [3].

12.2.1.1 Smart home

The IoT Analytics company database for smart home comprises of 256 companies and some startups are Nest, AlertMe, Philips, Haier, and Belkin.

12.2.1.2 Wearable devices

Wearables are used by consumers like Apple's new smartwatch. There are lots of other innovations like the Sony Smart B Trainer, the Myo gesture control, and the LookSee bracelet.

12.2.1.3 Smart city

Smart city has traffic management, water distribution, waste management, urban security, and environmental monitoring. Smart City solutions will alleviate the problems of people. It will reduce noise traffic, congestion problems, and pollution to make cities safer.

12.2.1.4 Smart grids

Smart grid uses details about the electricity suppliers and the consumers in an independent fashion that improves reliability, efficiency, and economics.

298 **CHAPTER 12** BLOCKCHAIN AND INTERNET OF THINGS: AN OVERVIEW

12.2.1.5 Industrial Internet of Things
The industrial internet is one of the vital applications of IoT. Many types of research from Gartner or Cisco show that the potential hasn't reached smart homes yet.

12.2.1.6 Automated car
The connected car in the automotive industry should be developed in the coming two to four years. Large automakers and some brave startups have started their research on the working of connected car solutions.

12.2.1.7 Telemedicine
This is the connected system for healthcare and medical devices that has significant benefits.

12.2.1.8 Smart retail
Immediacy-based advertising is a division of smart retail.

12.2.1.9 Smart supply chain
Supply chains when smarter offer solutions for monitoring goods and inventory information.

12.2.1.10 Smart farming
Smart farming is a great application of IoT because it has health, mobility and industrial benefits. The isolation of farming techniques and the huge number of live-stock can be monitored. IoT can transform the way farmers normally work.

12.2.2 ARCHITECTURE, CHALLENGES, AND EXISTING SECURITY TECHNOLOGIES USED FOR INTERNET OF THINGS
There is a lack of publicly available data sets that can be used in research. There should be certain standards for sharing the IoT data sets among the research and development communities.

12.2.2.1 Architecture
The IoT architecture [4] is given in Fig. 12.1. It has five layers as follows:

1. *The perception layer*
   In the architecture used for IoT, the perception layer collects the data from the sensors.
2. *The network layer*
   This layer transfers the data to the information processing systems.
3. *The middleware layer*
   This layer does the management of services. This layer uses protocols. It has various communication protocols such as Radio Frequency Identification (RFID), 3G/4G, WiFi, Bluetooth Low Energy (BLE), infrared, ZigBee, and so on.
4. *The application layer*
   This layer manages the applications based on the information obtained from middleware.

12.2 OVERVIEW OF INTERNET OF THINGS

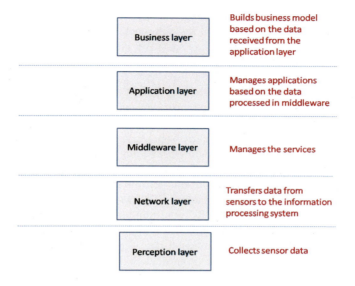

FIGURE 12.1

Architecture of Internet of Things.

5. *The business layer*

    This layer manages all the activities. It forms the business model, graphs, and models future activities.

The perception layer collects the sensor data. The network layer is responsible for transferring the data from the sensors to the information processing layer. The middleware layer is responsible for managing the services. The application layer manages the applications based on the data obtained from the middleware. Based upon the data obtained from the application layer the business model is built.

Apart from the benefits of IoT, there are several challenges that affect the implementation of IoT in various areas. The conventional security techniques applied in other areas are not applicable to the systems using IoT. The subsequent section discusses some of the security threats and the techniques that can be used to prevent security breaches.

12.2.2.2 Challenges of Internet of Things and the security technologies used in literature

All the existing Intrusion Detection System (IDS) and Intrusion Prevention System (IPS) systems are capable to identify the attempts to unauthorized access and Distributed Denial of Service (DDoS) problems. The author, Alsunbul et al. [5] proposed a network defense system for the detection and prevention of unauthorized accesses and attempts. They have dynamically created a new protocol to substitute the standard protocol used. The number of scanning attempts is being confused. There are many cyber-attacks like Mirai, Ransomware, Shamoon-2, and DuQu-2 on industrial systems and IoT which shows that the existing technologies are more vulnerable to attacks.

To avoid unauthorized access the network path is changed occasionally to avoid traffic scanning. But there are excessive numbers of packets in this approach. Also, among the documents that are being uploaded, one-fifth of the documents has sensitive information. Most of the cloud service providers assure the security of data while transmission.

In the method proposed by Zitta et al. [6], Raspberry Pi 3 was used to make safe the Ultra High-Frequency (UHF) RFID readers that run the Low-Level Reader Protocol (LLRP). Also, Fail2ban and Suricata are the solutions for their working and scalability. Fail2ban can be used in complex architecture, and it can be used for deployment in a cloud environment along with multiple sensors and servers. The IoT devices require some cryptographic security measures with the proper key management system and there should be revoking and updating of the keys. In the case of data security, there should be proper enrollment, authentication, ID management, and authorization. Suricata has good performance than Snort and permits multithread processing needed for the multicore CPU of Raspberry Pi 3. The performance of Snort and Suricata when dealing with DoS attacks [7] concluded that Snort has lower CPU consumption. However, there was a better single and multicore detection performance in multithreaded Suricata.

A network-based IDS is better than a host-based IDS. Saracino et al. [8] proposed a multilevel behavior-based anomaly detector for use in Android devices. It was used for analyzing and correlating the features at different Android levels like kernel, user application, and package. This detector finds and blocks the threats by detecting the behavior of certain security threats. The security risks are checked by requesting permission and reputing metadata during the installation of metadata.

Chen et al. [9] presented an ASIC design for a VoIP IPS that has a hierarchical architecture for statistical and stateful technique-based protocol anomaly detection. But the performance of statistical technique is poor. Although it differentiates normal and abnormal traffic, the throughput is poor. When statistical technique is used with stateful protocol-based technique they complement each other and the processing performance increases significantly. The profile analysis module reduces the false-positive rate by updating the threshold.

Freudiger et al. [10] have proposed protocols for privacy-preserving and measuring data quality matrices using the homomorphic encryption technique of completeness, uniqueness, consistency, validity, and timeliness. The client discovers the quality metric in the case of a semi-honest party. The assessment of data quality assures that poor quality data are rejected and this reduces the overhead of cleaning the data.

Sharma et al. [11] presented a hybrid encryption technique using RSA to attain high throughput and security and reduced overheads in MANETs. The technique has Secure Ad hoc On-Demand Distance Vector (SAODV) routing which is experimented using the NS-2 network simulator.

The Stochastic Activity Networks (SAN) [12] was used to model the communication between employees and the organization's security policy. The approach possesses a number of problems. Designing a representation from the attacker and administrator view is very difficult. Human behavior follows descriptive theory that is difficult to confine with mathematical models. The validation results are difficult to achieve. They proposed an agent-based model that is used as a substitute approach, and the system is considered as a group of independent agents capable of calculating the present situation and creating the decisions.

Shi et al. [13] have presented a hierarchically formed cloudlet mesh and a remote cloud platform. The objective was to perform intrusion detection among multiple Wi-Fi-enabled cloudlets by using cloud services through Wi-Fi networks. There is filtering of malicious attacks through trusted

12.2 OVERVIEW OF INTERNET OF THINGS 301

clouds. Security analytics were used for scanning malware signatures and spam removal. The clouds have the capability of using security service to all end users. This approach was used on EC2 cloud along with MapReduce and was experimented on 1 TB of data from Twitter. A hybrid IDS uses a cloudlet mesh for detecting the malware and data-coloring techniques [14].

The main difference between conventional networks and IoT is the resource levels available. IoT uses RFID and sensors. IoT devices require low computing power and low battery life. The conventional networks include supercomputers, servers, and mobiles that have sufficient resources. These networks have highly complex security protocols. But those security protocols cannot be applied in the IoT systems. The IoT networks require lightweight algorithms and there should be a balance between resource consumption and security. IoT devices uses communication media like 802.15.4, LoRa, ZigBee, 802.11a/b/g/n/p, SigFox, and NB-IoT.

The end media uses fiber digital subscriber line (DSL)/asymmetric digital subscriber line (ADSL), Wi-Fi, optics, 4G, and long term evolution (LTE). Also, a traditional network device uses the same operating systems and data but IoT uses different formats and operating systems. Hence, conventional networks use firewalls and IDS/IPS. The security approach uses for devices in a conventional network cannot be applied to the resource constraint IoT devices There is a lack of software updates, lack of access control measures, security patches, cross-device dependencies, and lack of IoT-focused attack signatures.

The conventional methods cannot prevent the attacks from both insiders and authorized persons in case of an IoT application as they are web based. There is a lack of consistency and standardization of IoT protocols.

There are different security requirements as follows:

1. Distributed computing
2. Decentralization
3. Data security
4. Device security
5. Data authentication
6. Data privacy
7. Key management
8. User security
9. Integrity
10. Access control
11. Controlled access to user data
12. Restricted network
13. Trustless environment

Blockchain is like a shared record. The records are controlled by a central authority and the digital files are owned by a database administrator. The records are governed by the central authority. The replicated form of blockchain is present in various locations. Thus it is a distributed record. It is not owned by a single participant. Blockchains are immutable as the history of the entire blockchain is stored in the first entry known as the genesis block. Each entry in a blockchain is responsible for creating the identity of the next entry. Each block is linked to all the previous blocks in a blockchain. Therefore it is impossible to change the contents of a blockchain and it is immutable. Thus the records are tamper-proof and protected against attack by hackers.

When blockchains are used for an enterprise it acts as a data storage. The traditional databases can also be used for data storage and retrieval. They can also be searched for data by using various querying techniques. The blockchains do not support faster queries. Also, once a block is added the details can't be updated. The operations performed by a traditional database cannot be performed by a blockchain. But, the main advantage of using a blockchain is that it avoids the usage of third parties like a banking service. The two end parties involved in the transaction are having a secured transaction as the data added to the blockchain can't be changed. The contract among large corporations can be done using a blockchain and without the involvement of a third party as the details can't be changed.

There are two types of blockchain, public and permissioned. The public blockchain is opened to all and the persons using the blockchain are not known to each other. Anyone can download the blockchains and read the transactions. But, these types of blockchains are not relevant.

The other type of blockchains are permissioned ones that are open to only limited participants. But, the identities of all the participants are all known. Private blockchains and semiprivate blockchains are the two types of blockchains. Private blockchains are used within an organization and semiprivate blockchains are used between two organizations.

The semi-private blockchains are used by most companies as every company has to interact with several other companies. The consortium like R3 and Hyperledger are using permissioned blockchains. Financial service is one such sector that has several applications for the blockchain. The data flow in IoT application is depicted in Fig. 12.2. There are various devices interacting in an IoT application. These devices interact with each other through the internet. If the devices need to communicate, they will do that through a central server. Therefore all the data flow will happen through the central server.

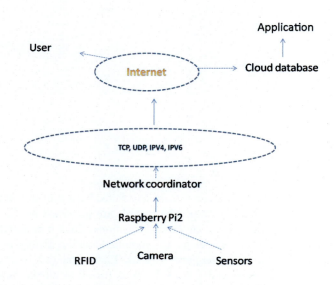

FIGURE 12.2

Data flow.

The data collected through the input devices like RFID tags, cameras, and sensors are taken by the Raspberry Pi to the network coordinator. The network coordinator communicates to the internet through the TCP/IP protocols. The cloud storage is used for storing the data. The internet communicates with the user through some communication device. The entire data flow is given in Fig. 12.2. The IoT application should be capable of operating in a trustless environment. Security against unauthorized sharing and usage of data is needed. The devices must be secured from physical and cyber-attacks. All the devices must be authenticated. There should be integrity among data that is shared and the code used. There should be protection against unauthorized access to data. In the case of the cryptographic security techniques used, the keys should be revoked and updated whenever needed [2]. There are various advantages of using blockchain technology in an IoT application.

12.3 OVERVIEW OF BLOCKCHAIN TECHNOLOGY

Blockchain has a list of records called blocks. All the blocks are linked using cryptographic techniques. The cryptographic hash of each block is computed and placed in its next block along with a timestamp, and Merkle tree which is a transaction data (Fig. 12.3).

The modification of data is restricted in a blockchain. Blockchain is an open and distributed ledger that records transactions among two parties very efficiently and in a demonstrable and everlasting method. The blockchain was invented by a set of people who were called by the fictitious name, Satoshi Nakamoto. This technology is in a recent trend that has created a significant development of IoT applications.

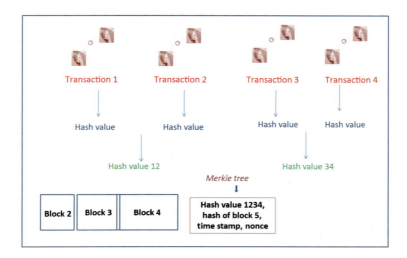

FIGURE 12.3

Blockchain technology.

304 CHAPTER 12 BLOCKCHAIN AND INTERNET OF THINGS: AN OVERVIEW

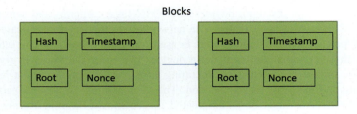

FIGURE 12.4

Data in a bitcoin.

Each block has the hash of the previous block. Also, the root hash value for all the transactions is stored. The timestamp at the time of the creation of block is also stored. The Nonce value ensures that the current hash value is below the target as in Fig. 12.4.

Blockchain technology takes the internet to the next stage by allowing distribution of the content without copying. It was initially devised for the usage of the bit currency bitcoin. But it has tremendously helped in improving the security of a lot of applications. In this section, the short introduction to the blockchain and its significant properties has been explained. Blockchain is like a ledger for all transactions. It cannot only be used for financial applications but also can be applied in different fields.

A blockchain is a series of immutable records of data that is time-stamped and managed by a group of computers that do not belong to a single entity. All these blocks of data are very secured and connected to each other with cryptographic techniques that form a chain. The network of blockchain has no transaction cost and central power. But it is shared among people so that each and everyone can see the information. So, any application built using blockchain works in a transparent manner. Blockchain technology can be used for the secure sharing of data sets and the usage of secured IoT systems.

The advantages of blockchain technology are as follows:

1. Data are tampering proof
2. Communication to a trustless person
3. Security from failure
4. Reliable
5. Data privacy
6. Recording of activities
7. Data recording
8. Proper direction of tasks.
9. Distributed access
10. Distributed storage
11. Economic benefits
12. Trust
13. Faster communications

The transaction or the distribution of information among two parties start by passing a block between X and Y. Either X or Y can create the block. This particular block is checked by several

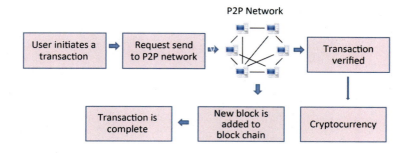

FIGURE 12.5

Blockchain and transactions.

other computers on the internet. This block is also added to a chain and distributed on the internet. Therefore corrupting one block is equivalent to corrupting the whole chain on the internet which is impossible. Bitcoin uses this technology for monitoring the transactions. While booking air tickets through the credit card, the company charges for processing the transaction. But this money can be saved when the entire transaction is being done using blockchain technology. The ticket becomes a block that has been added to the blockchain as in Fig. 12.5.

Blockchain transactions are free. There is no need to subscribe to an annual magazine, but a person can pay the fee for only the article read using bitcoin technology. The transactional information can be encoded using bitcoin ensuring its safety. E-books can also use blockchain code. All the money can be transferred to the author itself. This can be done on a book review website. The market places like Flipkart are unnecessary.

In transactions involving money, the involvement of blockchain is tremendous. Blockchains modify the transactions involved in different applications. This would almost remove the tasks done by banks. The financial offices that work mainly based on commissions would no longer exist. Data on a blockchain are like a shared and continually submissive data set. This uses the network with significant benefits. The data set is not a particular location, and the records are public and easily provable. No centralized storage of this information will be there for a hacker to easily corrupt.

12.4 PROPERTIES OF BLOCKCHAIN

Some properties of blockchain make it superior over other technologies. They are as follows:

- Decentralized
- Transparent
- Immutable

Previously the centralized networks were commonly used in which the information was stored. The usage of bitcoins and blockchains has removed the centralized versions as in Fig. 12.6. For example, in case of money transactions, all the transactions are centralized around a banking

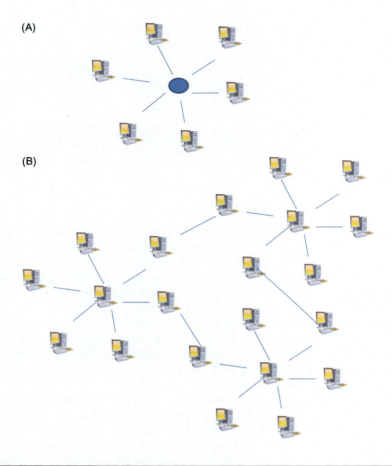

FIGURE 12.6

Types of network. (A) Centralized network. (B) Decentralized network.

database. The client–server network communication is mainly used in a centralized network as in Fig. 12.7.

- In the case of centralized communication, as all the information is stored in a central hub, it is probably attacked by attackers more frequently.
- If the software within the central system is in problem, then the entire network would cease.
- Then no one can use the information stored.
- If data get corrupted the whole system would stop.
- This paves the way for a decentralized system

Here the data are not only stored in the central system, but it is also spread throughout the network. Also, one communicates with the other systems in the network without interacting with a third party. This is the main benefit of bitcoins. There is no need to communicate the bank for the transaction of money to other persons.

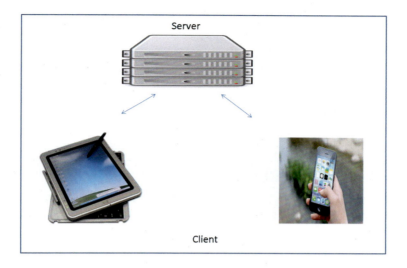

FIGURE 12.7

Client-server technology.

12.4.1 TRANSPARENT

Blockchain offers both transparency and privacy. The identity of a person involved in the transaction will be hidden using cryptography so that the name is not revealed while sending money to another person. So, as the public addresses are used for the transactions, the level of transparency is high. With cryptocurrency, by using the public address of big companies, a browser would help in getting all the transactions of them.

12.4.1.1 Immutable

Information once entered into blockchain cannot be tampered. The bitcoin technology uses SHA-256 that produces a fixed-length output for each given input. So, the output is always of 256-bit length.

A peer to peer network is used so that the communication is faster when compared to client-server technology. There will be no problem in the transaction and no errors when blockchain technology is used. Blockchain is the future for IoT.

12.4.1.2 Blockchains in data set sharing

There are some conditions to be maintained while sharing a data set [2].

- For maintaining integrity, a Reference Integrity Metric (RIM) for the data set is used in blockchain.
- Before using the data set, integrity can be checked using the RIM.
- In this approach, there is a central hub that maintains the references of member repositories where the data sets are actually stored and distributed.
- The membership information is shared among all the members.

12.5 CHALLENGES

- Limitation of storage
- Shortage of skilled persons
- Legal problems
- Scalability
- Processing time

12.6 RESEARCH ON BLOCKCHAIN

The idea of blockchain was initiated in 2008. It was devised as a decentralized method used initially for cryptocurrency. The technology has its own advantages like integrity and security along with some challenges and issues too. The interest in blockchain technology has been increasing since the idea was coined in 2008. This technology can also be used in smart contract systems. There are many challenges of blockchain technology and several solutions are also existing as a result of research. But their effectiveness still needs to be evaluated. Some challenges like latency, scalability, and throughput have still to be researched. [15]

Blockchain has a distributed database that has an enormously growing list of records. As this is a decentralized approach, the information on all transactions is send to all the nodes. The security techniques used in blockchain should be very strong and the data integrity should be well maintained so that the blockchain technology will be more robust. The computational power should also be high when using this technology. As the number of people using this technology is exponentially increasing the conventional currency usage like KRW, EUR, and USD, should be altered in several applications. Bitcoin makes use of the public key technology for ensuring security [16]. Bitcoin has a public and a private key. The address part has the public key. The authentication and security of the user using the technology are well maintained using the private key. A bitcoin will have the public keys of both the sender and receiver and also the value transferred. Within a few minutes, a block having the transaction information will be created. Then it will be linked to the previously created block in a blockchain.

In the nodes, some disk storages will be there in a decentralized network. The nodes make use of the disk storage for storing the information. The integrity of the data within a block is ensured by checking the previous blocks. The correctness of information present in the blocks are periodically ensured by a technique called mining. When the correctness of a block is ensured, all the blocks are connected together as a blockchain. This blockchain acts as a public ledger for the bitcoin.

There are some limitations of the blockchain technology. The number of transactions per second is very low in case of bitcoin technology. Other current transaction processing networks like Twitter has a very high throughput. So, the number of transactions per second or the throughput should be increased for the effective use of blockchain technology. The bitcoin network takes 10 minutes for a transaction. As the blocks are checked for their integrity and security before adding to the blockchain, the time taken is more.

The size of the blockchain created is very high resulting in high throughput. This is a big challenge, limiting the usage of bitcoin technology in various applications as in Fig. 12.8. Blockchain is

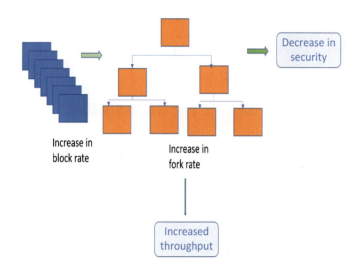

FIGURE 12.8
A general bitcoin system.

very much prone to attacks. This issue also needs to be solved through security keys so that it can be used in more critical applications. Also, bitcoin uses a huge amount of resources. The Application Programming Interface (API) is not very developer-friendly. This issue also needs to be solved in the case of blockchain technology. The chains are also needed to be split for administrative purposes. Although there are many advantages of blockchain technology, some challenges and limitations are also there that interrupts the vast usage of this technology. All these issues create more scope for research in the area of blockchain technology.

The transaction confirmation in a bitcoin -base system happens under 60 minutes. The average duration for a bitcoin transaction confirmation is 20 minutes. If a transaction is not confirmed it disappears from the blockchain. A bitcoin transaction cannot be undone as the information in a blockchain is immutable. But the transaction is highly secure as the information regarding the transaction cannot be changed. Through some ways, a bitcoin can also be converted to money. There are some exchanges for cryptocurrency like Coinbase and Kraken. Through these exchanges, the bitcoin can be converted to cash. Also, a bitcoin ATM can also be used. The Costa Rican Central Bank made an announcement against cryptocurrencies. But they are not considered illegal. Some merchants too accept bitcoins.

12.7 SCOPE FOR RESEARCH

There is scope for more research in the following areas:

- More challenges are needed to be identified and more solutions are needed to be implemented.
- The scalability issues still need to be concentrated on in research as most of the present research focus on the privacy and integrity of data.

310 CHAPTER 12 BLOCKCHAIN AND INTERNET OF THINGS: AN OVERVIEW

- Current blockchain technology is based on the bitcoin. More research should be done focusing on other cryptocurrencies too.
- The performance of the algorithms should be analyzed using some metrics.

12.8 SOME RECENT ADVANCES IN THE TECHNOLOGY

Some recent advances can also be seen in the literature as stated below.

12.8.1 PAYMENT NETWORKS WITH BITCOIN

Bitcoin does not scale because the blockchain technology limits the maximum size of the blockchain. There are off-blockchains that enable the usage of a large number of transactions between two users. These channels create a payment network. This creates the transaction to be done between two users with high security and less delay in service [17,18]. Bitcoin protocols, time locks, and shared accounts are a part of these off-blockchains. Once the payment network is established there is the usage of micropayments. This micropayment is incremented as the transaction proceeds between the sender and the receiver. As the limit for the micropayment is crossed the entire transaction will be closed between the sender and the receiver.

A shared account is set between the sender and the receiver for enabling a micropayment network. Hashed Timelock Contracts are used in a payment network. If a person needs to get a payment from another person, he needs to reveal a secret to receive the payment. There are duplex micropayment channels that offer faster service by establishing micropayment channels between the sender and the receiver than sending bitcoins between the sender and the receiver.

12.8.1.1 Bitcoin-NG—a scalable bitcoin network

This is a bitcoin network that is scalable as it has a new protocol. This is a fault-tolerant protocol that has the same benefits as that of a bitcoin. The experiments [18] confirm that the scalability of the bitcoin is optimal. The security and efficiency are evaluated by using different metrics. Bitcoin-NG serializes the transactions just like the bitcoin but has good bandwidth and a reasonable delay.

The ledger entries have key blocks and micro blocks. The header part refers to its predecessors, a hash value created by using cryptographic techniques. The key blocks refer to its leader and they act a leader. For a key block to be valid the cryptographic hash should be smaller than its limit just like a conventional bitcoin. The key block has a public key that will be used in other micro blocks. The key block is generated by iterating through the nonce values.

The leader is responsible for serializing the transactions until the next leader is chosen. The key blocks are used for leader election. They arrive at an interval of 10 minutes. The transactions are initiated by the key blocks. The micro blocks have the transactions. They are always signed by the leader's private keys.

12.8.1.2 GHOST—a secure technology

Bitcoin is gaining momentum as a new cryptocurrency. When there is a high volume of data, there occurs the problem of scalability. When there is high throughput the weak attackers will reverse

12.9 BLOCKCHAIN AND INTERNET OF THINGS   311

throughput. This security problem is reported by GHOST which is a change to the way the nodes may be built and the blockchains are constructed with a distributed data structure. GHOST is developed as a part of the Ethereum project which is a platform of second generation.

12.9 **BLOCKCHAIN AND INTERNET OF THINGS**

The (private) blockchain can be extended into the cognitive IoT using IBM blockchain. It is a grouping of IoT, AI, and blockchain that results in interesting IoT applications. Many blockchain platforms concentrate on IoT in the industries. IOTA is one of the blockchain platforms with IoT. There is a layer for data transfer. Several blockchain platforms focusing on IoT are emerging as the industry gets bigger. One of the first blockchain IoT platforms is IOTA.

It was created for the IoT and resulted in a settled transaction and data transfer layer for devices that are connected. When IoT is used to interconnect the devices the number of hackers affecting the security will also increase. So, a security measure is very essential. In blockchain, the blocks are secured using cryptographic techniques to form a blockchain. When blockchain technology is used with IoT then a better secured IoT application is developed which is less prone to hackers. The decentralized nature of hackers helps in its better combination with the IoT devices. Because in the case of critical applications of IoT like health monitoring and safety monitoring, hackers can induce serious threats by hacking the transactions and communications. Blockchain technology helps in removing these threats and also reduces the cost of communication.

Blockchain ensures security without the intervention of authority. Some of the other limitations of IoT has been overcome by Big data and Cloud computing. A Japanese startup Nayuta has released its lightning implementation "thunder bird" concentrating on IoT. The implementation has the code for the interconnected devices that are connected using networks. The Nayuta interconnects the bitcoin network, allowing real bitcoin transactions. Till now, the Nayuta has released its fourth implementation as Acinq, Blockstream, and Lightning Labs. The name "thunder bird" is given relating to the most promiscuous results.

The micropayments are an important component of devices using bitcoin technology and IoT technology. Many industries are involved in the process of integrating IoT and blockchain techniques for interconnecting the devices. The small payments are thus enabled in IoT devices. The small payments are done using bitcoin. These networks have the following properties:

1. a transaction using small amounts
2. border-less transactions
3. real-time communication
4. larger transaction per second

Because of the combination of bitcoin and IoT, there is a new view of marketing. Nayuta had a partnership with a Japanese electric company for making the payments for recharging electric cars. This is actually a vision of companies like Qualcomm Ventures. But these technologies have not been yet developed on a large scale. Nayuta is testing the idea. They have a device called a shield. Arduino are smaller computers used to create robots. These shields also have motors to enable the movement of robots. Using these motors, the robots can spin and move around. Because of these facilitie, the number of people who are developing IoT applications with Lighting networks is increasing exponentially nowadays. Research and development are going on in developing a small lightning network Ptarmigan that uses small hardware. This experimentally uses SPV wallet mode that works on an independent node in Raspberry Pi Zero [19].

The secured online transactions help in making a better living. These cryptocurrency systems power the IoT. Rather than using a washing machine connected to the internet and using a credit card, a bitcoin can be used to buy washing liquid for washing. Theoretically, a cryptocurrency system could be used to power the IoT. For instance, instead of hooking your internet-connected washing machine to a credit card, it could use bitcoin to buy new detergent when needed. Also, our fridge can buy milk using a bitcoin system if IoT technology is linked with blockchain technology. But all these techniques will affect the survival of a bank if the devices buy the things they use by themselves.

Maintaining security in IoT is a very difficult task due to the widely distributed nature of the networks of IoT. When blockchains are used, they provide security and privacy in IoT applications due to its decentralized nature. But there is a considerable expenditure of energy and significant delay and a high number of computations. This affects the IoT devices which have resource constraints. The lightweight version of a BC is also applied in IoT by eliminating the technique of coins and the proof of work. This paper [19,20] has the concept of a smart home with cloud storage, smart home, and overlay. This paper explains the various components involved in a smart home. The miner in a smart home is responsible for handling all the communications happening in a smart home. The communications are secured with the help of blockchain. The privacy, integrity, and availability are very much maintained in the smart home because of the blockchain concept. The overall traffic and energy consumed are also analyzed just like security and privacy concerns. These devices include lots of safety-critical information. So, there will be lots of cyber-attacks too.

As the networking devices have to spend more energy these devices find it difficult to provide better security. Also, security methods are very expensive. Most of the security methods are more centralized making it difficult to be implemented in a networking device. To provide privacy and security, most of the IoT applications provide incomplete information. Blockchain provides a lightweight and secure system for distributed services. The Public key system is used by the users for security and a new block is created.

As the block is full, it is attached to the end of a Blockchain. If the miners want to mine the block, they need to solve a puzzle which is called the proof of work and it is a high resource-consuming process. This paper [20] eliminated the resource-consuming proof of work and used a hierarchical structure for the distributed nature of IoT. The smart homes have cloud services for storage and also use mobile phone services to communicate with the user.

The overlay network brings the distributed network into the architecture. The nodes in the overhead are grouped into clusters and each cluster has a cluster head. Blockchain has two key lists. The list of the overlay users is called as the requestor key list. The list of keys allowed to be accessed is called the requestee key lists. The cloud storage is used to store data.

A private blockchain is used for providing secured services as in Fig. 12.9. There is also a set of time-ordered transactions between the consumer and the service provider. The security is affected by diverse features like devices that can't be directly accessed and complex transactions. Symmetric encryption reduces the complexity of the process. The security approaches in the smart

12.9 BLOCKCHAIN AND INTERNET OF THINGS 313

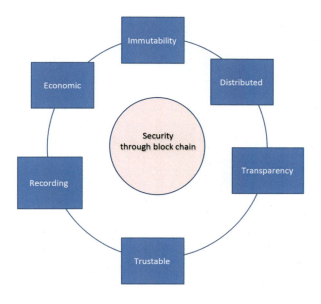

FIGURE 12.9

Security through blockchain.

home should achieve integrity, confidentiality, and availability. There are security attacks like linking attacks and DDOS. This would reduce the overhead of the entire process [21].

12.9.1 COMPONENTS

The communication between the devices is called transactions. The store transactions help the devices to store data. Access transaction helps to access the data from cloud storage. A monitor transaction helps to monitor the service. Adding new devices is known as a genesis transaction. A key is shared in the communication as in Fig. 12.10. Lightweight hashing detects the changes in the transaction.

All of the aforementioned transactions use a shared key to secure communication. Lightweight hashing is employed to detect any change in transaction content during transmission. All the transactions are stored in a private blockchain through which the outgoing and incoming transactions are monitored. The transaction of the devices is held together as an immutable ledger. Each block header has the hash of the previous block. Fig. 12.10 shows the functions of an electronic cash system where each block has a sign created using the private key of the previous transaction.

The IoT has exponential growth prospects along with drawbacks in security. Because of the decentralized topology, conventional security techniques are applicable. Fig. 12.11 indicates the different mechanisms in a blockchain-based transaction.

314 CHAPTER 12 BLOCKCHAIN AND INTERNET OF THINGS: AN OVERVIEW

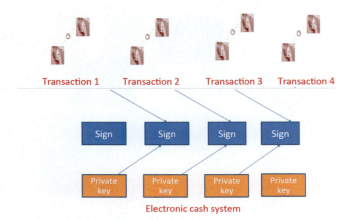

FIGURE 12.10

Electronic cash system.

FIGURE 12.11

The mechanisms of a blockchain-based transaction.

There are different components for a transaction.
- The requester public key is the requester parameter.
- The second component is the policy header that stores data locally on cloud storage and gets access to the stored data.
- The third component indicates the ID of a device present inside the home.
- The last component is the action that must be done for the transaction.

12.9.2 USE CASES

12.9.2.1 Smart home

The smart home architecture has the following functions as in Fig. 12.12.

- A person must be able to access the temperature of his or her home.
- The application should be able to store data on storage devices.
- Local blockchain is created.

The different appliances used in a home like the fridge, air conditioner, and washing machine, and so on are monitored as in Fig. 12.12. Blockchain can ensure secure communication between the home appliances [22–25]. The records of the appliances can be entered by the house owner in a distributed ledger created by a blockchain. Each device has a unique identifier to identify it. The house owner can have a security key to maintain the security of the home which they can access using a mobile phone.

Through this security key, the transactions are managed and protected. There are different access levels for each device. For example, the kiosk can be used at a home to accept the delivery of an order. The kiosk can accept or decline an order depending upon various parameters.

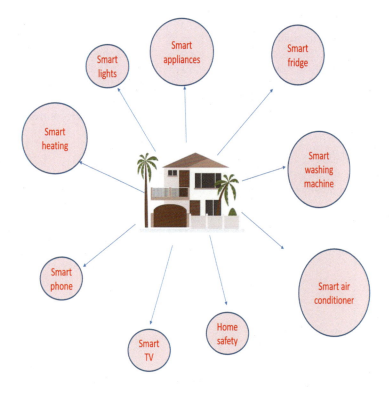

FIGURE 12.12

Smart home.

316 CHAPTER 12 BLOCKCHAIN AND INTERNET OF THINGS: AN OVERVIEW

By connecting their smart appliances to a portal, they can be easily managed. Through the usage of IoT and blockchain techniques, this can be achieved.

Walmart has filed several patents for certain applications like:

- Blockchain enabled smart package system.
- A market place for reseller.
- An electricity grid solution.

These applications will be under usage within a few years. The smart devices can communicate with each other and also with a central authority managing the home. Each device may communicate with another device for some service. For example, the light bulb communicates with the motion sensors present within the home. This enables us to turn on the lights automatically when there is some motion inside the home. A shared key is used to enable communication between devices. The miner asks the owner or checks the policy header to allocate the shared key. As long as the key is valid there will be communication among devices. If the miner marks the key as invalid there will not be any communications based on the key. Thus the miner has the list of the devices used in the home. It behaves as the central authority for managing the keys.

The transactions occur between devices [26]. For a particular function to be done in a smart home there are several transactions equipped with blockchain. Devices use transactions to store data. To access the cloud service another transaction is needed. To monitor a device another transaction is needed. A new device can be added or an old device can be removed using the transactions. All the transactions will involve keys. Lightweight hashing is used to detect any change in transactions. All the transaction information is stored in a blockchain. The smart miner manages the transactions and integrates the device with the internet.

For each smart home, there is a private blockchain. The transactions are stored as a distributed immutable ledger. Starting from the genesis transaction, each block has a block header and a policy header. The hash of the previous block is stored in the block header. User controls are managed by the policy header.

12.9.2.2 Smart security

These devices lock the cyber-physical devices using deadbolts which are electronically measured by mobile devices. There are two categories of attacks. There are several defense systems that mitigate these attacks.

12.9.2.3 Background of smart lock systems

There are five locks namely:

1. August
2. Danalock
3. Kevo
4. Okidokeys
5. Lockitron

There are three components:

1. An electronically amplified deadbolt in a door.
2. A mobile device to control the lock.
3. A remote web server.

12.9 BLOCKCHAIN AND INTERNET OF THINGS

FIGURE 12.13

Smart lock.

The lock using a local wireless channel, such as BLE. Smart locks do not have a straight connection to the internet. They make use of mobile phones as an internet "gateway" that transmits information to and from the servers. This architecture is presented in the Device-Gateway-Cloud (DGC) model as in Fig. 12.13. The second architecture is Lockitron that has an internet connection between the smart lock and the server. The lock has a Wi-fi modem which gets connected by itself to a Wi-fi network. An "owner" key can lock and unlock the smart lock any time. Atonomi has a new security protocol and a new infrastructure that enables lots of IoT devices to have trusted service with interoperability [23].

It's based on a new ledger, the Tangle, which overwhelms the inefficiencies of blockchain designs and paves a new way for a decentralized system. A decentralized network in IoT is powered using a privacy-centric blockchain. There are also some emerging technologies as follows.

- The Watson IoT Platform has a built-in capability that lets you add selected IoT data to a private blockchain.
- Ambrosus is another IoT network-enabled with blockchain for pharmaceutical enterprises with secure dialogue among sensors, with distributed ledgers and thereby databases that optimize the supply chain discernibility and quality assurance.
- Waltonchain is a secure business system with data sharing and absolute transparency. It is formed as a group of RFID and blockchain.
- OriginTrailis Protocol is based on blockchain for unified supply chains.
- Streamr is creating tradeable data streams.

Fig. 12.14 indicates the identification of security breaches during communication in blockchain-based IoT. In different areas like cyberspace, geo-space, aqua-space, the ecosystem benefits from fast-tracking and there is a convergence of blockchain and AI. The existing systems alter things linked to the internet. The devices and gateways in a distributed network request an action without any disturbances. AI obtains data from a decentralized network and gets meaningful information from it. An automated and man-to-machine (M2M) networks shape the systems. So, IoTs define and develop a connected and secure system.

The change from centralized to distributed system and automated system to the autonomous system causes a different era of technology. It creates challenges for securing the ecosystem and for doing everyday tasks like sensing, storing, communicating, and taking timely actions.

318 CHAPTER 12 BLOCKCHAIN AND INTERNET OF THINGS: AN OVERVIEW

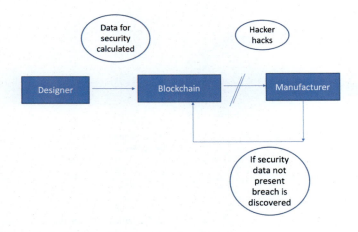

FIGURE 12.14

Security in blockchain-based Internet of Things.

The Internet of Nano Things (IoNT) and IoE create billions of connections. when objects in space become surrounded by sensors and communicate across space. The concept of IoE is the machines with unique identifiers and the ability to communicate data over a centralized and decentralized network. This communication would be done without any HCI (Human−Computer Interaction) requiring new techniques along with the merging of blockchain and AI/ML. The IoE ecosystem can be helpful in evolving technology across nations.

12.9.2.4 Smart contracts

The smart contracts are stored as scripts in a blockchain [27,28]. For the deployment of IoT and blockchains in the sector of smart contracts there are several issues that need to be addressed. While using blockchain there will be high throughputs and greater latencies. Each participant participates using a key. All the transactions are done in open. If a private blockchain is used, the same blockchain cannot be used for all the transactions.

The transactions included in a smart contract is always deterministic. The same input produces the same input. The code of a smart contract resides in the blockchain so that all the participants can verify it. All the transactions are done using signed messages so that the security is enforced in transactions. These transactions using bitcoin helps in transferring of assets among parties that do not trust each other. The usage of digital signatures is ensured throughout the transactions. The transactions are done even between two parties that do not trust each other because they do not deny the information as all the data stored in a blockchain are immutable.

It is preferable that all the participants belong to a private network. Private networks are good for establishing contracts between the stakeholders. The cryptocurrency is transferred between the parties. Using cryptocurrency each device has an account in the bank and by permitting its usage to other users it can compensate its resource usages over the network. This can also be done using micropayments. Also, the owner of a house can permit their house for rent by giving access to the

12.10 THE CONVERGENCE OF BLOCKCHAIN, IOT, AND AI

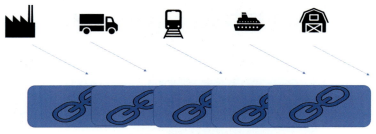

When the product reaches each destination a signed smart contract is sent via blockchain

FIGURE 12.15

Smart contract.

smart locks only for a permitted period. A party interested in the rented house can contact the owner, sign the contract via digital signature, and can use the house for the permitted period.

Also, when a product is manufactured in a factory it is transported via railways and ships. Then it is taken to the distributor and to the retailer. On the way, it gets verified by different stakeholders. The stakeholders have to maintain a database of their own to store the asset information. Using blockchains, they can keep track of the product and update the information via smart contracts as in Fig. 12.15.

Özyilmaz and Yurdakul [28] proposed that the exponential growth of IoT applications needs a secure and scalable network technology for data storage and processing of the transactions. Blockchain helps in stabling a distributed and safe storage of data. But the end devices using the IoT applications cannot handle the heavy computations occurring due to blockchain. The authors proposed some resource-constrained IoT devices suitable for blockchain architecture. Based on this objective, an IoT gateway was shaped as a node of blockchain. It used an event-based mechanism for passing the messages between the low-power end devices.

12.9.3 STATIC SENSORS AND AUTONOMOUS MACHINES

The connection of sensors and machines using AI forms the IoT. The sensors measure the real value and communicate the information with the devices, thereby extracting the information using AI. The devices also respond based upon the information gathered using AI. There is data collection, extraction of useful information from the gathered data using AI, and machine learning. The decisions are obtained and based upon that the devices respond accordingly. Thus these static sensors and automated devices perform without any human intervention.

12.10 THE CONVERGENCE OF BLOCKCHAIN, INTERNET OF THINGS, AND ARTIFICIAL INTELLIGENCE

The IoT system is based upon a centralized server/client model in which all the devices communicate and get authenticated through the cloud servers that support the processing making the system

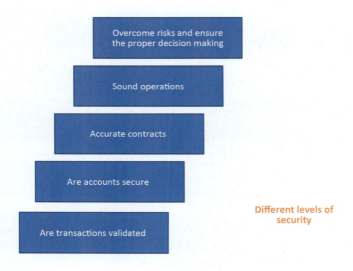

FIGURE 12.16

Different levels of security.

more vulnerable to the attacks. The applications have tasks that are more sensitive by nature and it impacts the IoT ecosystems.

Blockchain technology helps to create a decentralized environment in IoT systems [29]. So, a distributed system like blockchain can create a distributed ledger using blocks. All the IoT devices are tagged, and thereby, monitoring how the devices communicate. Thus the blockchain technology creates secured networks interconnected as a mesh.

Fig. 12.16 indicates how different levels of security are achieved in a blockchain-based IoT system.

12.11 CONCLUSION

Day-to-day money transactions are more centralized and this paves the way for hackers to attack the information communicated. Blockchain technology is a decentralized technology that uses bitcoin and removes the hurdle of using a centralized bank for transactions. The usage of the proof of work which solves a cryptographic puzzle helps to maintain security, privacy, integrity, and authenticity. Blockchain has a distributed ledger that has blocks added to it. It avoids the usage of the third party in any transaction or communication. This saves a lot of money spent as commission to third parties. The IoT system has facilitated our human life in many ways. The usage of sensors to monitor the real data have automated the system of monitoring our health, home, and several other day-to-day life activities. But, in the case of critical applications, the IoT system is affected by hackers. This can be avoided by the usage of a security technique. Blockchain technology is very helpful in maintaining the security of an IoT system.

There are some backlogs like high computational overhead and processing power. But several types of research are going on in this area that can make a lightweight blockchain. The proof of work phenomenon in blockchain increases the computational overhead. By changing this phenomenon, the computational overhead can be reduced. The public key cryptography technique is used to provide additional security along with Diffie Hellman Key exchange process. The IoT ecosystem and the IoNT are seriously evolving and changing the world. The advantages of blockchain are its decentralized and tamper-proof properties. The problem of scalability is addressed by a new technology called GHOST.

The application of this blockchain technology with IoT has its usage in several areas like smart homes, smart locks, automated cars, and smart retail management, and so on. The machines with unique identifiers paved the way for IOE. There are no human—computer interventions in this system and with the help of the IoE ecosystem, the nations can be connected.

REFERENCES

[1] S. Madakam, R. Ramaswamy, S. Tripathi, Internet of Things (IoT): a literature review, J. Comp. Commun. 3 (05) (2015) 164.

[2] M. Banerjee, J. Lee, K.K. Choo, A blockchain future for internet of things security: a position paper, Dig. Commun. Netw. 4 (3) (2018) 149—160.

[3] https://iot-analytics.com/10-internet-of-things-applications/.

[4] R. Khan, S.U. Khan, R. Zaheer, S. Khan, Future internet: the internet of things architecture, possible applications and key challenges, in: Proceedings of the IEEE10th International Conference on Frontiers of Information Technology (FIT), 2012, pp. 257—260.

[5] S. Alsunbul, P. Le, J. Tan, B. Srinivasan, A network defense system for detecting and preventing potential hacking attempts, in: 2016 International Conference on Information Networking (ICOIN), Kota Kinabalu, 2016, pp. 449—454.

[6] T. Zitta, M. Neruda, L. Vojtech, The security of RFID readers with IDS/IPS solution using Raspberry Pi, in: 2017 18th International Carpathian Control Conference(ICCC), Sinaia, 2017, pp. 316—320.

[7] W. Park, A. Seongjin, Performance comparison and detection analysis in Snort and Suricata environment, Wirel. Personal. Commun. 94 (2) (2017) 241—252.

[8] A. Saracino, D. Sgandurra, G. Dini, F. Martinelli, MADAM: effective and efficient behavior-based android malware detection and prevention, in: IEEE Transactions on Dependable and Secure Computing, 15 (1) 83—97.

[9] M.J. Chen, C.C. Wen, H.C. Lin, Y.S. Chu, ASIC design and implementation for VoIP intrusion prevention system, in: 2016 International Conference on Applied System Innovation (ICASI), Okinawa, 2016, pp. 1—4.

[10] J. Freudiger, S. Rane, A.E. Brito, E. Uzun, Privacy preserving data quality assessment for high-fidelity data sharing, Proceedings of the 2014 ACM Workshop on Information Sharing &Collaborative Security (WISCS '14), ACM, New York, 2014, pp. 21—29.

[11] A. Sharma, D. Bhuriya, U. Singh, Secure data transmission on MANET by hybrid cryptography technique, in: 2015 International Conference on Computer, Communication and Control (IC4), Indore, 2015, pp. 1—6.

[12] W.H. Sanders, J.F. Meyer, Stochastic activity networks: formal definitions and concepts, in: E. Brinksma, H. Hermanns, J.-P. Katoen (Eds.), Lectures on Formal Methods and Performance Analysis, Ser. Lecture Notes in Computer Science, vol. 2090, Springer Berlin Heidelberg, 2001, pp. 315—343.

[13] Y. Shi, S. Abhilash, K. Hwang, Cloudlet mesh for securing mobile clouds from intrusions and network attacks, in: 2015 3rd IEEE International Conference on Mobile Cloud Computing, Services, and Engineering, SanFrancisco, CA, 2015, pp. 109–118.

[14] K. Hwang, D. Li, Trusted cloud computing resources and data coloring, IEEE Internet Co. 14 (2010).

[15] J. Yli-Huumo, D. Ko, S. Choi, S. Park, K. Smolander, Where is current research on blockchain technology? A systematic review, PLoS One 11 (10) (2016) e0163477. Available from: https://doi.org/10.1371/journal.pone.0163477.

[16] R. Housley, Public Key Infrastructure (PKI), John Wiley & Sons, Inc, 2004. Available from: https://doi.org/10.1002/047148296X.tie149.

[17] C. Decker, R. Wattenhofer, A fast and scalable payment network with bitcoin duplex micropayment channels, Symposium on Self-Stabilizing Systems, Springer, Cham, 2015, pp. 3–18.

[18] I. Eyal, A.E. Gencer, E.G. Sirer, R. Van Renesse, Bitcoin-NG: a scalable blockchain protocol, in: 13th {USENIX} Symposium on Networked Systems Design and Implementation ({NSDI} 16), 2016, pp. 45–59.

[19] Blockchain for IoT Security and Privacy: The Case Study of a Smart Home Conference Paper, March 2017.

[20] A. Dorri, S.S. Kanhere, R. Jurdak, Blockchain in Internet of Things: Challenges and Solutions, 2016. arXiv preprint arXiv:1608.05187.

[21] G. Ho, D. Leung, P. Mishra, A. Hosseini, D. Song, D. Wagner, Smart locks: lessons for securing commodity internet of things devices, in: Proceedings of the 11th ACM on Asia Conference on Computer and Communications Security.

[22] H. Gross, M. Holbl, D. Slamanig, R. Spreitzer, Privacy-aware authentication in the Internet of Things, Cryptology and Network Security, Springer International Publishing, 2015, pp. 32–39.

[23] https://www.postscapes.com/blockchains-and-the-internet-of-things/.

[24] https://www.forbes.com/sites/cognitiveworld/2019/07/05/a-changing-internet-the-convergence-of-blockchain-internet-of-things-and-artificial-intelligence/.

[25] https://www.forbes.com/sites/geraldfenech/2019/01/21/the-combination-of-blockchain-and-smarthomes/#7b447a0aa2c4.

[26] A. Dorri, S.S. Kanhere, R. Jurdak, P. Gauravaram, Blockchain for IoT security and privacy: the case study of a smart home, 2017 IEEE International Conference on Pervasive Computing and Communications Workshops (PerCom Workshops), IEEE, 2017, pp. 618–623.

[27] K. Christidis, M. Devetsikiotis, Blockchains and smart contracts for the internet of things, IEEE Access 4 (2016) 2292–2303.

[28] K.R. Özyılmaz, A. Yurdakul, Work-in-progress: integrating low-power IoT devices to a blockchain-based infrastructure, 2017 International Conference on Embedded Software (EMSOFT), IEEE, 2017, pp. 1–2.

[29] https://www.toptal.com/insights/innovation/blockchain-applications-create-enterprise-solutions.

CHAPTER

CRYPTOCURRENCY MECHANISMS FOR BLOCKCHAINS: MODELS, CHARACTERISTICS, CHALLENGES, AND APPLICATIONS

13

Deepak Kumar Sharma, Shrid Pant, Mehul Sharma and Shikha Brahmachari

CAITFS, Department of Information Technology, Netaji Subhas University of Technology (Formerly known as Netaji Subhas Institute of Technology), New Delhi, India

13.1 INTRODUCTION

This section introduces various concepts and mechanisms in cryptocurrency and blockchain technology.

13.1.1 INTRODUCTION TO CRYPTOCURRENCY

The cryptocurrency was developed based on the collaborative open-source principles to help with strong and secure financial transactions. With the rise of bitcoin, this has been associated with speculators as well. Cryptocurrency is similar to cash in various ways—the value is determined and can fluctuate, both can directly be used for transactions without a need for a middleman.

Cryptocurrencies are a medium of exchange, just like any other currency, but these are only digitally available. Cryptocurrencies usually use a decentralized control, through technology like blockchain, and strong cryptography to enable secure transactions. There are several cryptocurrencies available along with bitcoin, like Ethereum, Dash, Monero, etc., which are maintained through distributed consensus and a system is used to keep an outline of the virtual currency units and their possession. The cryptocurrency network, like any normal banking system, performs transactions by changing the entries in a database. The intermediaries that change these entries in the public database are a decentralized network rather than a private network of people working on private databases as in a bank. The public database (blockchain) is a transparent method of keeping a score of the cryptocurrency tokens. It also ensures the anonymity of the user. The use and trade of cryptocurrency vary from country to country and the concerning laws keep changing. Some countries, like the United States, treat tokens as a property which can be taxed.

13.1.2 CONTEMPORANEOUS ADVANCEMENTS IN BLOCKCHAINS

The introduction of the bitcoin cryptocurrency led to the first universal use of the blockchain technology. To this day, bitcoin is arguably the most popular application of this blockchain technology.

Handbook of Research on Blockchain Technology. DOI: https://doi.org/10.1016/B978-0-12-819816-2.00013-7
© 2020 Elsevier Inc. All rights reserved.

323

324 **CHAPTER 13** CRYPTOCURRENCY MECHANISMS FOR BLOCKCHAINS

Other blockchain-based cryptocurrencies like Ethereum, NEO, Ripple, and Steller soon followed after the initial success of bitcoins, leading to a transformation in the economic sector. While the applications of blockchain technologies range from e-healthcare and smart contracts to the Internet of Things (IoTs), the focus will remain on the field of cryptocurrencies in this chapter [1].

The majority of cryptocurrencies have, in one way or another, been the clones of some existing virtual currencies and have plainly featured different parameter values such as issuance schemes, currency supply, or different block time. These cryptocurrencies have shown extremely little to no innovation in technology or methods and are generally termed as "altcoins," for example, Dogecoin and Ethereum Classic. In contrast, a few virtual currencies have been introduced which, while borrowing a few concepts from other cryptocurrencies, provide innovative features that display considerable differences. These include the development of novel consensus models (e.g., proof-of-capacity) and/or decentralized platforms with "smart contract" capabilities providing unprecedented functionality. These blockchain and cryptocurrency innovations can be categorized as (1) new blockchain systems which present their own blockchain technology (e.g., Zcash and Peercoin) and (2) others which are built on existing blockchain systems. (e.g., Augur and Counterparty) [2,3].

13.1.3 **MOTIVATIONS FOR THE USE OF BLOCKCHAINS IN CRYPTOCURRENCIES**

Blockchain is known to be the underlying database of various existing cryptocurrencies. It is a peer-to-peer (P2P) and decentralized transaction system without third party involvement along with an efficient data management solution. A cryptocurrency ecosystem is based on the blockchain architecture, which contains a totally ordered back-linked list of blocks. These blocks are developed using proofing methods like Proof-of-Stake (PoS), Proof-of-Elapsed-Time (PoET) along with other existing proofs. Each block created or added to the chain contains the output transaction and cryptographic hash, previous block's hash is also stored guaranteeing integrity and determinism. Mining is defined as the activity for appending transactions to the existing chain. The objective of the distributed network is to achieve consensus on the next block/transaction to be appended to the blockchain by means of a Consensus Protocol. The validation of the transactions and the verification of ownership is ensured by public-key cryptography and algorithmic rules. Blockchain also plays an important role in reducing the problem of double-spending, which is one of the primary concerns associated with cryptocurrencies in which the transaction of the same cryptocurrency is carried out twice.

Cryptocurrencies are emerged to be the first generation blockchain-based applications, having a finite ability to bear programmable transactions. The second-generation (Ethereum) aims to deliver a non-specialized programmable infrastructure, whose programs are known as smart contracts [4].

13.1.4 **INTRODUCTION TO CRYPTOCURRENCY MECHANISMS FOR BLOCKCHAINS**

The blockchain is the core mechanism for most cryptocurrencies. A cryptocurrency uses a distributed ledger in a P2P system accounting for transactions in chains of blocks to overcome the limitations of conventional financial objects. Central features like decentralization, persistence, anonymity, and verifiability in blockchains enabled by numerous principal technologies like distributed consensus mechanism, cryptographic hash, and asymmetric cryptography-based digital signature ensure affordable and efficient transactions in cryptocurrencies. The consensus in a distributed system faces numerous challenges including message delays, corrupted messages, failures of

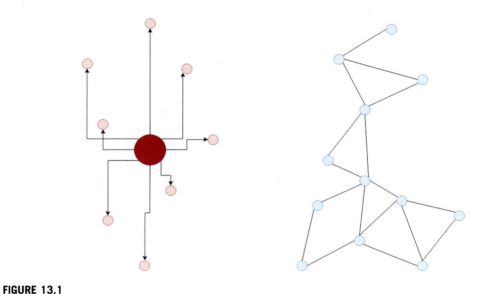

FIGURE 13.1

Graphical representation of centralized (left) and decentralized (right) networks.

participants and division of the whole network. Therefore, consensus models work to update the distributed ledger in a secure fashion. Many consensus mechanisms like PoS, Proof-of-Work (PoW), proof of capacity (PoC), and others are evolving to overcome their inherent issues. In a blockchain network, achieving consensus is to bring about an agreement on a constant global state of the public ledger. Furthermore, the cryptographic model used by virtual currencies cater high data integrity, confidentiality, and non-repudiation services. Hence, cryptocurrencies use public-key cryptography to verify transactions among the nodes and also use digital signatures to achieve the aforementioned services.

In Fig. 13.1, the centralized network (left) consists of a central node, which is connected to its edge nodes. In centralized distribution (left), particular nodes have complete control over the functioning of the network. However, in decentralized networks (right), there are no central administrations, and all the participants have equal rights.

While cryptocurrencies have their limitations in both the technological and economic contexts, the rapid research and development of various mechanisms have shown immense possibilities in mass use.

13.2 PRINCIPAL DISCIPLINES OF CRYPTOCURRENCY MECHANISMS FOR BLOCKCHAINS

13.2.1 BLOCKCHAIN ARCHITECTURE

Blockchain is decentralized distributed network consisting of a series of blocks, each series representing committed transactions. Every block has a parent block associated with it and the header of the block contains a hash value of previous block. In the blockchain, the initial block is termed as genesis block. As new transactions are generated and approved by all the members of the public ledger, the block is appended to the chain. The blockchain system has a very comprehensive and efficient architecture. Below is the architecture of a blockchain in detail.

13.2.1.1 Block
A block is the basic unit of blockchain. It can also be termed as a committed transaction and has many functionalities. Block body and block header are the two major components of the block. Block header comprises of the below components:

1. *Parent block hash:* This signifies the previous block's hash value, which is 256-bit.
2. *Version of block:* Signifies the block verification guidelines to be followed.
3. *Merkle tree root hash:* All agreements in the block are associated with a hash value described by merkle tree root hash.
4. *Timestamp:* Refers to the present timestamp in terms of seconds since 1970-01-01T00:00 UTC.
5. *Nonce:* For every hash calculation, nonce is a 4-byte field, starting with a zero and raises with every computation of hash value.
6. *nBits:* A compact format of the current hash target.

Another major component of a block consisting of transactions and its counter is the block body. The magnitude and the block size of every transaction are responsible for the large amount of transactions the blocks contain.

In Fig. 13.2, each of the blocks from $block_0$, $block_1$, ..., $block_n$ consists of their body and head, which further contain timestamps, nonce, and other relevant information. This clearly describes the various components of the blocks in the system.

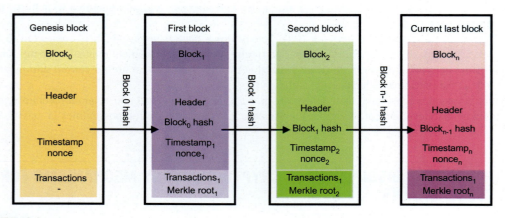

FIGURE 13.2

Components of blocks in a blockchain network.

13.2.1.2 Digital signature

In a deceitful environment, an asymmetric cryptographic mechanism is used to verify the credibility of the transaction. Digital signature works on the principle of asymmetric cryptography. Each transaction member has a private and public key. A private key is stored confidentially because it is used for signing negotiations. The transactions that are signed are transmitted across the distributed network and the public keys are used for their accessibility. There are two levels involved with digital signature: verification phase and signing. During the phase of signing, the encryption of the data is carried out using the private key by the sender. Encrypted result and native data are delivered, which are sent to the receiver of the transaction. In the verification stage, for the validation of the received value by the receiver, the public key is used and the data are checked if it has been meddled. Elliptic Curve Digital Signature Algorithm (ECDSA) is used for blockchains in the digital signature mechanism.

In Fig. 13.3, the process of assignment of the digital signature is described. The private key is used to sign and record various messages, while the digital signature is validated using the public key.

13.2.1.3 Blockchain features

- *No central authority:* A central trusted agency is required for any transaction to be processed in a conventional centralized system, thereby causing issues in performance and the cost of congestion at the middle servers. Blockchain is a P2P system with no influence from any third party. In distributed network, data consistency is maintained by consensus algorithms in blockchain.
- *Persistency*: Throughout the network transactions are distributed; thus, validated by all the nodes and invalid transactions are discovered immediately and not appended to the chain of blocks. Once a transaction has been introduced to the blockchain after consensus from all the members in the ledger, it is not possible to delete or rollback transactions.
- *Anonymity*: With a generated address, the interaction with blockchain is possible by all the users. The users real identity is not revealed by the address generated, thereby eliminating the

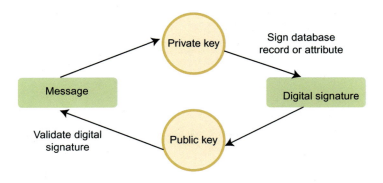

FIGURE 13.3

Digital signature assignment process.

necessity of a central agency supervising users' private information. Transaction privacy in the blockchain is preserved by this mechanism.
- *Auditability*: Previous records can be easily tracked and verified, as each transaction is stored and validated with a timestamp. This improves the transparency and traceability of the transactions.

13.2.1.4 Taxonomy of blockchain system
Blockchain is classified into three major categories (Fig. 13.4):

I. *Public blockchain*: All records in this blockchain system are broadcasted and all the members and everyone is involved in the process of confirming and validating the transactions.
II. *Private blockchain*: This is also referred to as a centralized network, as the complete control is under one organization and only the members of the organization are involved in the process of concluding to consensus
III. *Consortium blockchain*: This blockchain system is monitored and regulated by several bodies and is moderately decentralized as few pre-selected group of nodes would be selected to contribute to the decision-making process.

The comparison among different blockchain system is listed below:

- *Consensus determination*: Each node in public blockchain participates in the agreement process. Whereas, for validation of the block in consortium blockchain, only a few nodes are involved. Private blockchain system is under control by a single organization which also responsible for the validation.
- *Visibility*: In a public blockchain, transactions are transparent and available to the nodes, whereas the control is under one organization in the case of a private blockchain or a consortium blockchain.
- *Flexibility*: In public blockchains, the transactions are validated and checked by all the nodes in the public and it is not possible to tamper transactions. On the contrary, agreements in private or consortium blockchain can be meddled easily as it involves only a set of people.

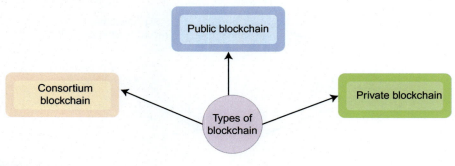

FIGURE 13.4

Types of blockchain.

- *Efficiency*: With limited number of nodes in consortium and private blockchain, the transactions are efficient as it is not propagated throughout the network. In public chain network, a large amount of time is required to deliver the transactions and blocks among the nodes. This results in limited transaction throughput and high latency.
- *Centralized*: Consortium blockchain involves partial centralization, public blockchain is decentralized, and private blockchain is completely centralized as the control is under a single unit.
- *Consensus Process*: Anyone can actively participate in the consensus process associated with public blockchain. Private and consortium blockchain nodes both being decentralized require permission from the regulatory organization.

13.2.2 BLOCKCHAIN CONSENSUS MODEL

The blockchain consensus between untrustworthy nodes is an extension of the Byzantine Generals Problem. A set of protocols need to govern the consistency of the distributed nodes in the public ledger. A blockchain-based structure can only be as reliable and functioning as its consensus algorithm. Several approaches are presented in the following subsections for consensus mechanisms in blockchains.

13.2.2.1 Proof-of-work model

PoW is a consensus strategy applied to Ethereum (called Homestead, in its current version) and bitcoin. PoW has its own variation for different cryptocurrencies—EthHash, Hashcash PoW, Merkle Tree-based, and others. In PoW, by requiring a set of work from the service requester, denial of service (DoS) and other abuses/attacks are deterred. In blockchains, this model is used to verify transactions and, thus, generate new blocks for the chain [5,6].

Fig. 13.5 represents the chain selection in blockchains. A6, A7, …, A12 represent nodes of one chain, while W8 and W9 represent nodes of another chain. The process of chain selection is as described below—on the basis of the longest chain rule.

In bitcoins, hashcash is used as a part of the mining algorithm. To add nodes in bitcoins, each participant must obtain a hash value lower than a particular numeral by solving computational puzzles set by the blockchain network. The dynamically tuned difficulty level presently ensures that each block is added every 10 minutes. The first node to obtain the winning hash takes the mining prize and is added to the proposed blockchain. When more than multiple nodes find the winning

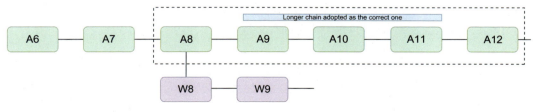

FIGURE 13.5

Selection of chain.

hash simultaneously, each winning node attaches the proposed block to the network and transmits this over the blockchain generating a temporary fork. However the mechanism that establishes the longest branch (greatest PoW) gets included and the others get discarded, as more blocks are attached to these forks. Fig. 13.5 represents this process. Therefore the state of the blockchain will be consistent. Even though bitcoin-PoW wastes large sums of power during the mining process in the computation of hashes, it operates in a completely decentralized fashion and has high scalability in terms of nodes participation. Thus, PoW in bitcoins proves to be efficient in multiple ways despite some inherent flaws.

Fig. 13.6 represents blocks *XYZ1*, *XYZ2*, and *XYZ3* as a part of a longer blockchain. Each block holds its structures in the form of header and body. The transaction counter consists of the list of transactions *Tx1*, *Tx2*, ..., *TxN*.

Ethereum uses its own PoW consensus model called EthHash, which allows quick verification time and builds Application-Specific Integrated Circuit (ASIC) reference to combat abuses/attacks that the BTC PoW is vulnerable to. EthHash was, in part, designed to combat the monopoly of the computing power in the blockchain by powerful entities, which might create mining pools. EthHash uses techniques called Memory Hardness and Greedy Heaviest Observed Subtree (GHOST) to combat this mining centralization. Memory Hardness is the ability of a machine to move data around in memory, as opposed to rapidly performing calculations. As this cannot be achieved efficiently on ASICs, but is well-suited for general purpose computer hardware, it prevents undue influence over the network. GHOST, on the other hand, is a substitute for the greatest chain rule for formalizing consensus among the PoW-based networks, and intends to remove the effects of stale blocks, that is, blocks that were transmitted to the blockchain and accepted by a few nodes to be legitimate, but eventually were discarded as a longer chain achieved dominance (forking). Ethhash, like bitcoins, makes use of the idea of obtaining an input that can create a hash value under a particular complexity. The Ethereum Network regulates this complexity level to process a block every 15 seconds. In the future, Ethereum is expected to advance into a PoS model (Serenity) [7–9].

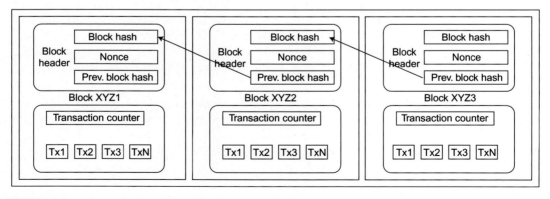

FIGURE 13.6

The structure of the PoW model.

13.2 PRINCIPAL DISCIPLINES OF CRYPTOCURRENCY MECHANISMS   331

13.2.2.2 Proof-of-stake model

PoS algorithms, primarily, were created to resolve the limitations of the PoW algorithms, with respect to excessive energy utilization during mining operations. An alternative approach involving users' stakes of the virtual currency is used to replace the traditional mining operation of PoW algorithms. Therefore, simply put, instead of investing a certain amount in, say, mining equipment, the user could use that amount to invest in the cryptocurrency share to buy equivalent block-creation opportunity by advancing as a validator. The pseudo-random selection of validators for block-generation ensures that no validator can determine its chance beforehand. However, inefficient implementation of PoS could lead to the nothing-at-stake problem, which do not incentivize nodes to vote on the correct blocks. Multiple versions of PoS have been used by PeerCoin, BitShares, Tendermint, and NXT.

Casper, ethereum's PoS algorithm, uses the concepts of "stakes" to achieve consensus, and is arguably the most advanced PoS algorithm. Nodes, which are bonded with the Ethereum system on staking their security deposits become bonded validators that are initially tracked by the Casper contract. Thereafter, the-formed initial bonded validator list might grow on the basis of nodes binding to and retiring from the system. The probability of selection in the "pseudo-random selection" of active validators is linearly weighted by the validators' deposits. Block rewards correspondingly equivalent to the ether in the active validator set are provided when validators produce blocks that get included in the chain. However, if the produced block is not included in the chain, the validator forfeits the security deposit equivalent to the block winnings. This protocol attempts to resolve the issues pertaining to nothing-at-stake [10,11].

13.2.2.3 Proof-of-elapsed time model

Developed and thereafter open-sourced by Intel, IntelLedger/Intel SawtoothLake uses a consensus model called PoET, to be executed in a Trusted Execution Environment (TEE), like Intel's Software Guard Extensions (SGXs).

The PoET model uses both permissionless and permissioned blockchains in a probabilistic transaction finality. It has no cost of participation and is highly scalable. PoET essentially follows the following strategy: each node in the blockchain waits for an arbitrary amount of time, and the first node to conclude becomes the new leader. This random leader election model, based on SGXs, is used by PoET to randomly select the next leader to finalize the blocks. Trusted Execution Environment (TEE) is used to assure the randomness and safety of selecting the leader. This helps to deal with the open-ended participation of nodes and untrusted nodes in the consensus algorithm. A validating node has to prove having the least wait time and must wait for some specific time (set by the protocol) before mining another block. All validating nodes must execute the TEE using Intel SGX. The details of working are provided as follows.

Random waiting times

As mentioned earlier, in PoET model, each node is expected to wait for a specific time before generating another block. This waiting time follows a probability distribution F, which is marked by a two-tier approach. A formula is used by each node to produce an output as its interim time, which could be used to develop other blocks until it is revised. Particularly, when nodes have produced their blocks with the interim time, the generation of the next block using this waiting time is also

332 CHAPTER 13 CRYPTOCURRENCY MECHANISMS FOR BLOCKCHAINS

decided randomly, that is, using a particular probability p, they recreate new waiting times, otherwise they continue to use the previous waiting time. The details are as follows.

Registration. Nodes must register the following to the system: their public/private key pair and interim waiting times. While the private/public key pair remains unaltered, the waiting times are subject to upgrade.

Calculation of the waiting time

The equation specified below is used by nodes to calculate the waiting time, *wait_time*:

$$wait\_time = minimum\_wait - local\_average\_time.log(r) \tag{13.1}$$

In the aforementioned equation, $r \ \varepsilon \ [0, \ 1]$ is obtained with the hash value of the participant's earlier certificate. *minimum_wait* is a set variable. For *local_average_wait*, nodes check the updated *sample_length* blocks to approximate the active nodes in the system by checking the interim waiting time data on these blocks, and multiplies this by a constant value to get *local_average_wait*.

Block verification

Any newly generated block is verified by noes in the blockchain network before it can be accepted into the system. Many attacks, ranging from novice to sophisticated ones, are detected by using statistical tests. Z-test is used to check whether nodes are generating blocks too quickly. The z-score may be computed as follows:

$$z = \frac{win\_num - mp}{\sqrt{mp(1 - p)}} \tag{13.2}$$

where *win_num* is the number of blocks that have been successfully created by the node [12].

This approach has two major advantages: (1) *Efficiency*. PoET does not require expensive computation workload by participating nodes to create new blocks; (2) *Fairness*. PoET fulfils the objective of "one CPU one vote". While the randomness ensures that leaders are randomly distributed among the mining nodes, the algorithm's reliance on specialized hardware is a drawback. However, the generation of random wait times by the use of a trusted execution environment, instead of computational effort to solve cryptographic puzzles (like BTC), is a more energy-efficient approach.

13.2.2.4 Proof of capacity model

PoC is another consensus model similar to PoW. An ecofriendly protocol, it was developed as a substitute to make a resource scared to prevent its abuse. PoC, differs from PoW as it reckons on the hard disk capacity of the miners (nodes present in the distributed network responsible for calculating hashes are referred to as miners) instead of relying on the computing power of the miners. It consumes disk space as a resource. It is significantly more energy-efficient than ASICs mining used in PoW.

Miners in PoW, keep altering the block header and hash for observing and calculating the solution whereas PoC, utilizing Shabal algorithm, generates all the random solutions, referred to as plots, and stores it on the hard drive. This level is called plotting, and depending on the storage capacity of the drive, it might take days or weeks for its completion. Next level involves miners matching their solutions to the most recent puzzle and the fastest solution producing node gets to

13.2 PRINCIPAL DISCIPLINES OF CRYPTOCURRENCY MECHANISMS   333

mine the block. It takes 4 minutes to create a block in PoC. Mining is relatively cheap as it does not depend on any specialized hardware, rather can be performed on any hard drive. Upgradation of the equipment is not a necessity, data can be stored on older disk. This model supports and provide a more decentralization by allowing users to use the free disk on phones to participate in the network. It takes 4 minutes to create a block in PoC.

Burstcoin, PermaCoin, and SpaceMint are cryptocurrencies using PoC. Burstcoin, a decentralized cryptocurrency and payment system is a major implementation of PoC. It primarily depends on disk space as its mining resource. The mining can be even performed on mobile devices making it inexpensive and energy efficient. This is supposed to increase the scalability of blockchain while decreasing the transaction confirmation time. Spacemint is another conceptual cryptocurrency based on PoC model. Miners rather than computation commit to disk space and are fairly rewarded according to their contribution to the network, thus promoting more decentralization. Spacemint alleviates the problem of large energy consumption in most cryptocurrencies and hard drive consumes very less power. Spacemint has several advantages: (1) *Ecological*: Mining cost is marginal and cheap. (2) *Economical*: Readily available free disk space, and the cost would be small for spacemint. (3) *Egalitarian*: Bitcoin mining is mostly dependent on large mining farms and ASICs. SpaceMint are not influenced by specialized hardware. PoC requires more P2P interactivity thereby leading to network congestion.

13.2.2.5 Byzantine fault tolerance and variants-hyperledger fabric

Developed by Linux foundation, the most approved permissioned blockchain platform designed for consortiums is the hyperledger fabric. Byzantine fault tolerance (BFT) finds its application in resolving the problem of reaching consensus when nodes could generate arbitrary data. Fabric provides a flexible architecture supporting smart contracts on blockchains referred to as chaincode.

A category of BFT, Castro, and Liskov recommended Practical BFT (PBFT), a replication algorithm. PBFT is used as the consensus algorithm by hyperledger fabric to tolerate byzantine failures.

For state changes, the concept of replicated state machine and voting by replica are used. Encryption and signing of messages interchanged between clients and replicas; reduction in magnitude and quantity of messages interchanged are some of the important elaborations provided by PBFT. For tolerating "f" failing nodes, "3f + 1" replicas are required by this algorithm. It has the potential to manage up to 1/3 hostile byzantine replicas and is also accountable for the ordering of the transactions.

PBFT comprises of two kinds of nodes, a leader node and some validating peers. In every round, a new block is determined, this also results in the selection of the primary. The PBFT is executed in three phases, known as commit, pre-prepared, and prepared. Two of the three votes from all the nodes is the minimum requirement to enter another phase. Broadcasting of the recommended block by the leader to the peers is carried out in the pre-prepare phase. Peers receive the block and store them locally. Prepare and commit phase is to ensure the similarity of the block received from the leader. The consensus achieved is faster and more economical. PBFT also finds its implementation in IOT networks, which require high throughput, low latency, and low computational overhead. It is limited only to small IOT networks as its high network makes it unscalable for large networks.

Sieve is a variation in PBFT and specializes in handling non-deterministic chaincode execution. Different output can be produced when different replicas are involved in the execution in a distributed network when the chain codes are non-deterministic in nature. Chaincode is treated like a bloack box. All the operations are initially executed separately and then the output is compared across the replicas; this leads to the sieving out of diverging values.

Cross-fault tolerance (XFT) is a protocol that makes attack model more comprehensive and ensure the feasibility of BFT feasible, making it efficient for practical situations. An adversary is assumed to be powerful when it has the potential of controlling the compromised nodes along with message delivery of entire network. XFT reduces these powerful adversaries and simplifies the state machine replication problem. If the major replicas are correct and can interact with each other concurrently, XFT provides correct services [13–15].

13.2.2.6 Federated byzantine agreement

Due to the decentralized nature, one of the major challenges is the trust that can be put in every participant of the transaction. The federated byzantine agreement consists of two protocols—Ripple and Stellar which aim toward providing payment protocols to ensure better transaction experiences. The two protocols are blockchain-based platforms that work on node participation and making it as open-ended as possible. Before describing both the protocols in detail, it is necessary to understand how the transaction happens. The participants of any transaction are the market makers who help generate work for a gateway that will eventually make the payment on their behalf. Banks are one of the most prominent gateways present at the moment. These complete the transaction by referring to the account balances of the user and the process is completed in a few days. A similar course is taken through the transaction with cryptocurrency but it is completed within a matter of seconds as opposed to days. While using cryptocurrency, the end user must be able to trust the gateway to handle the transaction. Transaction also occurs in a similar way by making changes in the users' balances in the blockchain. But as these gateways might not be as trustworthy as a bank, certain mechanisms are used to ensure the safety of the information and money.

Ripple and Stellar protocols, although derived from BFT, have various features that help them stand apart. Ripple consensus protocol algorithm works toward turning the data collected by the node into a valid block. The process begins with a bunch of transactions that need to be finished. Each node collects data based on the level of trust it has on other nodes in the ripple network. This data by each node is known as the Unique Node List and is collected in the candidate set. The list must have a 40% convergence with the other nodes in the network. Each node votes on the transaction which are broadcasted, and based on this, the ledger is made.

In Fig. 13.7, the different circles represent nodes and the gray outline around certain nodes depicts a subset of a quorum, which can persuade other nodes for consensus. The quorum slice works on the mechanism as described in the texts below.

The Stellar system is similar but the network is divided into quorum slices where each node votes on a selection of statements and can only accept a different statement if the slice it is a part

FIGURE 13.7

A quorum slice.

13.2 PRINCIPAL DISCIPLINES OF CRYPTOCURRENCY MECHANISMS

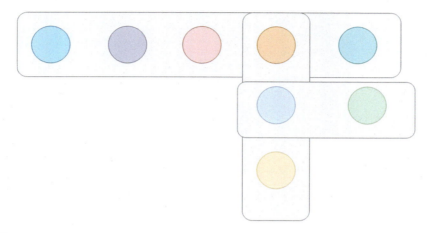

FIGURE 13.8

Quorum intersections.

Table 13.1 Comparison of Some Blockchain Consensus Models					
	PoW	PoET	PoS	Federated BFT	BFT and Variants
Trust model	Untrusty	Untrusty	Untrusty	Semi-trusted	Semi-trusted
Transaction finality	Probabilistic	Probabilistic	Probabilistic	Instantaneous	Instantaneous
Transaction rate	Slow	Medium	Rapid	Rapid	Rapid
Cost of participation	Present	Absent	Present	Absent	Absent
Scalability	Large	Large	Large	Large	Low
Token requirement	Yes	No	Yes	No	No
Type of blockchain	Permissionless	Both	Both	Permissionless	Permissioned

BFT, *Byzantine fault tolerance;* PoET, *Proof-of-elapsed-time;* PoS, *Proof-of-stake;* PoW, *Proof-of-work.*

of, has accepted that one. As one node can be part of various slices, the ratification happens where the complete quorum agrees on certain statements. These two protocols ensure that there is open-ended participation from the nodes [16,17].

In Fig. 13.8, the different circles represent nodes, and the gray outline on them represents quorums. The working of the mechanism of quorum intersection is described in the texts above.

In Table 13.1, the rows represent certain characteristics, while the columns represent the blockchain consensus models. It shows a comparison of the popular consensus algorithms [18].

13.2.3 BLOCKCHAIN CHALLENGES AND OPPORTUNITIES

This subsection provides insight on the challenges pertaining to the technicalities, implementation, and acceptability, and the opportunities this miraculous technology provides in reference to cryptocurrencies.

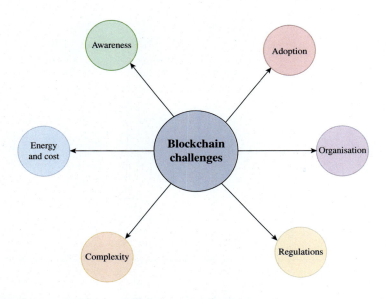

FIGURE 13.9

Challenges in blockchain.

In Fig. 13.9, the outer circles in the concentric plane represent different challenges pertaining to technicality, implementation, and acceptability. It represents the various challenges faced while using the blockchain technology.

13.2.3.1 Technical challenges

As blockchain is an upcoming technology, it faces several challenges.

1. *Space*: One of the major challenges faced by blockchain is the lack of space. The chain keeps growing and thus requires more and more resources and affects performance negatively. Due to millions of transactions requiring attention, the blockchain becomes heavy. It is necessary to store all previous transactions to validate the new ones which increases the need for a bigger capacity. This creates another problem—miners prefer transactions with a higher cost, which results in a delay in the small transactions. A few of the proposed solutions are as follows.
    a. Storage optimization can be done by deleting older data which frees the nodes from holding all previous transactions.
    b. The blockchain could be redesigned where the block is divided into two sections—one to hold the transactions and the other for the miners to compete to become the leader who generates the microblocks.
2. *Security*: There are various vulnerabilities discovered in the protocol.
    a. *Miner selfishness*: The majority (51%) of the miners could affect the blcokchain and even change the transactions that have occurred. This poses a serious threat to the security of the users and the blockchain. Recent researchers have found that even without the majority, the miners could seriously affect the blockchain.

13.2 PRINCIPAL DISCIPLINES OF CRYPTOCURRENCY MECHANISMS **337**

    b. *Double spend attack*: Since it requires a transaction of a certain depth before it can be confirmed, which can take 20—40 minutes on average, it is possible for malicious users to spend the same coin again. This could also be done with the help of a miner. Thus, it poses a serious risk on the transactions.

3. *Coin loss*: As the cryptocurrency is only present in online wallets, it is possible that a user forgets the password of their account over time which leads to the loss of those coins as there is no method to operate with such coins.

4. *Privacy:* A major feature of all cryptocurrencies is that all transactions are transparent. As the information is available to the public, it has been found that it could be used to reach the users involved in the transaction. Each user can be identified through the nodes it connects to and this information could be used to find the beginning of a transaction. This violates the privacy of the user which was ensured by through anonymity of the user. There is a method available to link the user's pseudonym with their IP address even if they use firewalls. This is a breach of the privacy of the user. A few proposed methods to tackle this are as follows.

    a. *Mixing:* This refers to performing transactions through multiple input and output addresses. This would make it difficult to find a relationship between the two participants. Intermediaries could be involved to ensure even more privacy. But if the intermediary node is selfish, it could reveal the participants' information or even keep the money to itself. An easy solution would be to encrypt the data so that the theft could be identified.

    b. *Anonymous:* This refers to the idea of completely anonymous transactions where the miners do not have any information about the transaction and the user information is encrypted.

    c. *Off chain*: Sensitive data are not stored on the blockchain and can only be accessed only by authorized personnel. This also solves the problem of space as some of the information is stored in a different location.

In Fig. 13.10, the outer circles in a concentric plane show the different challenges discussed in the previous paragraphs. It focuses on the technical challenges faced by blockchains.

Therefore, due to an increase in privacy and anonymity, cryptocurrencies could be used in illicit activities which is another threat to the technology [19—23].

13.2.3.2 Implementation and acceptability

In the recent years, cryptocurrencies are increasingly becoming accepted by payment processors (Square and Paypal), traditional vendors (Expedia, Dell, Microsoft, and Whole Foods), and companies like Tesla, Wikipedia, Reddit, and WordPress. Modern payment systems based on cryptography, that allow the ability to transfer and store funds in a secure and trustless manner, have created a shift away from the traditional, government-based currencies, and financial systems.

However, questionable network scalability, heightened transaction times, and large fees have raised serious questions about the practicality of many virtual currencies for everyday use. In fact, as more businesses start to accept virtual currencies, there will be immense scope for offenders. Thus, it is imperative to apprehend the use of prevalent cryptocurrencies in illegal activities like Ponzi schemes, money laundering, hacking, and drug trafficking. Silk Road, AlphaBay and Hansa were large Darkweb markets that used cryptocurrencies like XMR, BTC, and ETH for illegal transactions. Their transactions involved a variety of products and services such as firearms, stolen identification cards, human trafficking, sex trafficking, drugs, and other illegal services. Given the

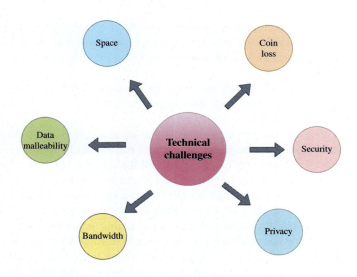

FIGURE 13.10

Technical challenges in blockchains.

anonymity, cross-border accessibility, and heightened security of cryptocurrencies, they provide opportunities numerous of these criminal activities.

Furthermore, the development of state-backed virtual cryptocurrencies by nations such as Russia, Iran, Venezuela, and North Korea have led to more concerns. Many experts believe that the state-backed virtual currencies lead to further security challenges because nations might use cryptocurrencies to bypass sanctions or use them for other fraudulent activities. Sweden, Marshall Islands, Turkey, Thailand, Tunisia, and many countries in the Eastern Carribean are investigating the feasibility of developing state-backed virtual currencies. The state-backed cryptocurrencies, for many of these countries, are a method to gain freedom from the reliance on U.S. fiat currency and create financial security. Nations that are assessing blockchain systems to develop their own virtual currencies are using various variations in centralization and control, from purely government-based virtual currencies to central bank–issued virtual currencies in coalition with private organizations.

Therefore strong and consistent regulations need to stay in place, the absence of which could obstruct the legitimate utilization of virtual currencies and support fraudulent services [24–29].

13.2.3.3 Opportunities in cryptocurrencies

The development and evolution of cryptocurrencies has brought about numerous opportunities for financial institutions, businesses, and customers alike. Cryptocurrencies are one of the most important subsets of digital currencies. Unlike other digital currencies that are tied to fiat currencies or organizations, centrally distributed, or limited within a geographic location, cryptocurrencies have different characteristics. The blockchain technology used by these cryptocurrencies uses an open distributed ledger for recording transactions. Decentralization allows faster settlement, better security, and increased capacity. Most of these features resolve the limitations of traditional financial systems.

13.2 PRINCIPAL DISCIPLINES OF CRYPTOCURRENCY MECHANISMS 339

The market of cryptocurrency has brought about unprecedented opportunities in business, banking, finance, and numerous sectors. Their inherit value and nature make them tradable commodities, and are, in fact, actively traded in the worldwide market. Hundreds of cryptocurrencies resulting in billions of dollars in transactions available for investors to invest in. The wide assortment of features provided by these virtual currencies encourage investment, which in-turn boosts innovation. Furthermore, these virtual currencies have driven a diverse group of start-ups and attracted a huge flows of venture capital to invest them for developing and advancing value within and between this technology. To co-create and seize values through organized business models, cryptocurrency companies seek to modularize the components as they are rooted within a deeply connected but decentralized innovation environment. Therefore, for companies, to help achieve their business goals, cryptocurrencies can be developed as mediating technological artifacts. Additionally, some countries have begun to develop their own versions of cryptocurrencies to facilitate trustworthy and stable transactions. Many proposals regarding taxation and regulation of these virtual currencies aim to provide a safer and more efficient experience for communities.

Information technology has, indeed, vastly changed the way institutions compete and interact. This technology-driven transformation of the traditional markets has led to unprecedented innovation and opportunities [30–33].

13.2.4 CASE STUDY: BITCOIN

Bitcoin is believed to be developed by Satoshi Nakamoto and open-sourced in 2009. It is a P2P decentralized virtual currency for financial transactions. It uses PoW consensus model and digital signature to validate and verify transaction history.

The bitcoin network functions to ensure decentralization and efficient verifiability. All recent transactions are broadcast to all participants in the system. These nodes gather the transactions in the form of blocks. The nodes work to discover a difficult PoW for their blocks, and broadcast the block on finding the PoW. The nodes in the network demonstrate their approval of the block by using the hash of the approved block to create new blocks. The participating nodes in the system select the longest chain, and work on extending it. In the event of a fork, the tie between multiple chains is broken by PoW, and one branch lengthens. The participants of the other branch will then turn to the longer one. This model of introducing new bitcoins into the network is termed as bitcoin mining. The primary incentive for bitcoin mining remains the rewards and transaction fees associated with it.

Fig. 13.11, graphically visualize bitcoin's distributed ledger system showing nodes $N1$, $N2$, ..., $N7$ inter-connected, each one having a copy of the public ledger with them.

The practicality and implementation of bitcoins has been a major topic of discussion, involving opportunities and limitations pertaining to bitcoins. Bitcoin is one of the most successful and popular cryptocurrencies. Many problems of accessibility, inflation, fraud prevention, and others can be resolved with the adoption of bitcoins. While lack of regulations and technological infrastructure, along with inherent faulty nature of the bitcoin infrastructure might hinder its implementation in many countries, it still remains a tour-de-force in the modern digital currency world. Bitcoin has the ability to transform the current payment methods, and immense incentive for individuals by becoming more private and less influenced by regulatory supervision [34,35].

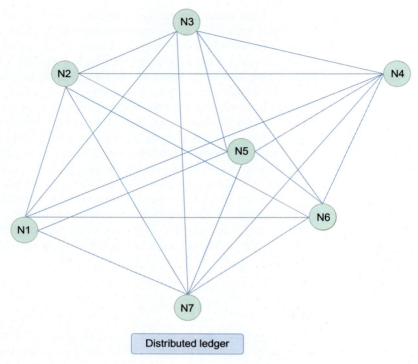

FIGURE 13.11

Bitcoin's distributed ledger system.

Fig. 13.12 depicts a proposed block going through the process of combination and hashing to produce a hash humber. This hash number is compared with the target value, which determines whether PoW has been achieved or not. The nonce is increased and the process is retried if the result is negative. Else, the block is approved and a reward is obtained.

13.3 COMPARATIVE ANALYSIS WITH OTHER MECHANISMS

This subsection provides information regarding various new technologies developed which are different from blockchain but aim toward the same goal.

13.3.1 THE IOTA TANGLE

IoTs is an extension of the internet to control everyday things. IOTA was designed specifically to control transactions using IOT. The IOTA tangle was an improved ledger, where there was no distinction between the users and miners. Bitcoin first used blockchain but created a distributed ledger. The tangle works toward removing this distribution to ensure that there is no conflict between the

13.3 COMPARATIVE ANALYSIS WITH OTHER MECHANISMS 341

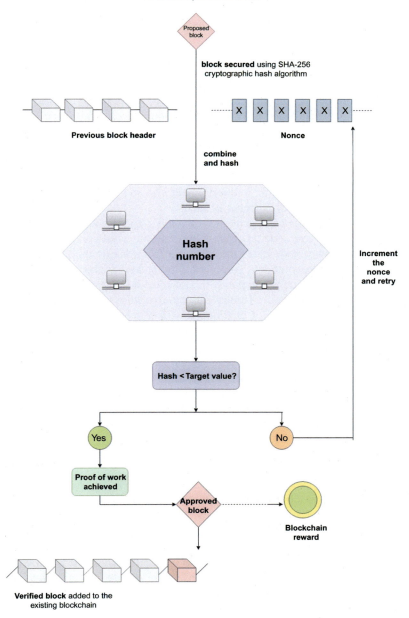

FIGURE 13.12

Flowchart for bitcoin's PoW model.

FIGURE 13.13

Blockchain architecture.

two groups. The users are the participants of a transaction who benefit through a faster transaction and less fees. On the other hand, miners who work toward providing the cryptocurrency and completing the transactions between users, rely on the transaction fees. Thus, there is a possibility of conflict which could then affect the network.

The following diagrams for Blockchain Architecture and Tangle show the inherent differences they have.

Fig. 13.13 represents blockchain architecture for comparison with IOTA Tangle. The rectangles represent participating nodes that form a linear chain. As the rectangles represent anonymous nodes, they have not been individually labeled as such.

To resolve this, the tangle uses an acyclic and directed graph for its distributed ledger technology. In this, the nodes need not provide any transaction fees but must contribute toward other transactions. This ensures that every node helps the other and can thus be used for IoTs. The tangle is made from the transaction between nodes. To transact, the node must approve of two other transactions which must not be conflicting. Along with this, it must be able to solve an encrypted puzzle similar to ones in the bitcoin blockchain. It must run the algorithm to choose coinciding transaction multiple times to check the percentage of confidence with which it was approved. As the tangle is acyclic, it is possible that there are conflicting transactions in the tangle. One of these could be indirectly disapproved by incoming transactions. If a node does not participate in other transactions, it could be considered too lazy and then dropped by the network which ensures that all nodes participate even if they do not have any transaction of their own to perform.

In Fig. 13.14, the diagram represents tangle (directed acyclic graph). The different rectangles represent participating nodes connected in a nonlinear chain. As the rectangles represent anonymous nodes, they have not been individually labeled as such [36–43].

13.3.2 HEDERA HASHGRAPH

Hedra Hashgraph is another public distributed ledger developed by Leemon Baird, it goes beyond blockchain in creating fast and secure applications. It provides with all the advantages of blockchain along with a better consensus mechanism without its limitations. Hedra hashgraph is based on the graph like structure where all the nodes communicate their information to each other, and their communication is reported by building a graph of connections. All the information is stored in hashes, describing certain events. The consensus mechanism adopted involve Gossip about Gossip and Virtual Voting. In Gossip about Gossip protocol all the information by each node on the network pertaining to which node, when and with whom it communicated is shared. Each device shares its own data on the hashgraph. All the transactions ever performed on the hedera hashgraph network are available in exact order. Each node is aware of the decision of the other node and can predict in advance without the need of active, effective decision. This prediction leads to virtual

13.3 COMPARATIVE ANALYSIS WITH OTHER MECHANISMS 343

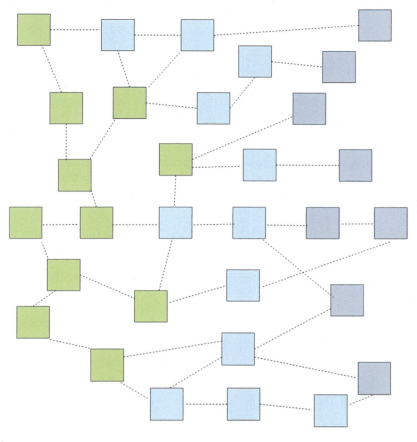

FIGURE 13.14

IOTA tangle.

voting, where the prediction of all the votes of the nodes is predicted in advance. These consensus algorithm use bandwidth efficiently and have the potential of processing thousands of transactions per second in a single shard (fully connected, P2P mesh of nodes). The latency is measured in seconds. The following are the characteristics of hedera hashgraph.

- *Performance*: Energy intensive proof-of-work is avoided in hasgraphs. Custom mining rigs are not purchased rather they can run the available hardware. Hashgrap is extremely fast and efficient, it adds only a small overhead beyond the amount the bandwidth usage available.
- *Security*: The transactions are digitally signed, the communication is encrypted with TLS 1.2, and cryptographic hashes are used in the construction of hashgraph. The hashgraph is asynchronous byzantine fault tolerant. Community consensus cannot be hindered by any member. The consensus once reached cannot be changed. Hashgraph is resistant to attacks, like

344 CHAPTER 13 CRYPTOCURRENCY MECHANISMS FOR BLOCKCHAINS

distributed DoS against the consensus algorithm, and attains the theoretical limits of security defined by aBFT.

- *Governance*: A council of up to 39 leading global enterprises govern the hedera network. These enterprises are hugely responsible for introducing innovative experience in process and business expertise eliminating its absence in previous public ledger platforms. This governance model protect network users by reducing the risk of forks, ensuring the integrity of the codebase, and providing open access for reviewing the underlying software code. To prevent centralized control, all the governing members exercise equal voting rights and are required to serve a limited period of time.

   This governance model is permissioned with permissionless or open consensus.
- *Stability*: The stability of this platform is ensured by both legal and technical controls.
- *Regulatory compliance*: Users and developers are provided with a choice of binding certified identities to otherwise pseudonymous Hedera network accounts in the hedera technical framework which includes an opt-in verified identity mechanism.

Hedra hashgraph has several network services. It is used for deploying smart cards that would be immuable and also presents an optional mechanism to enable binding arbitration. Hedera also supports cryptocurrency. The hedera cryptocurrency API provides with tools to manage accounts and ensure scalable, low-latency transactions fee that costs a fraction of a cent of using native network coin, HBR [44–46].

Fig. 13.15, describes the chain in hedera hashgraph (left) and its characteristic behavior, as described above. Individual node (right) of the hashgraph is shown with various components, including timestamp, transactions, and hashes.

13.4 FURTHER ADVANCEMENTS

Blockchains have shown immense potential in, both, the industry and academia. The scope of further advancement in the application and research pertaining to the blockchain technology is huge. Truly, blockchain technology could revolutionize countless sectors—AI, Insurance, Voting Systems, Smart Contracts, Big Data Analytics, and many others.

The financial sector, particularly cryptocurrencies, has had its own advancements—in terms of efficiency, security, and accessibility. To achieve this, many aspects of cryptocurrency mechanisms are updated. New consensus algorithms are devised aiming to solve the many limitations of their predecessors. Many algorithms including PeerCensus, Kraft, and Chepurnoy have been introduced with this objective. Along with this, cryptocurrencies, like Postcoin, are being introduced that can withstand volatility. Since a major drawback for many cryptocurrencies is their high exchange rate volatility, Postcoin resolves this by fully or partially backing the cryptocurrencies with other assets, or by involving trusted parties. In addition, the concept of cryptocurrencies and blockchain technology is being consistently used to innovate accounting and assurance. Blockchain has provided real-time, transparent, and verifiable accounting environment with the potential to radically transform the current practices leading to a uniform, transparent and timely automated system. Numerous other advancements including Post-Quantum Blockchain, that can resist quantum computing attacks, are changing the scope and direction of blockchain technology for cryptocurrencies. They

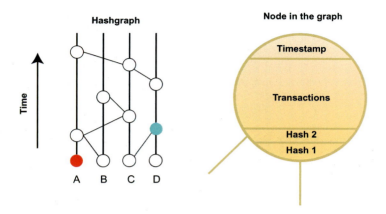

FIGURE 13.15

Hedra hashgraph chain (left) and individual node structure (right).

are improving the security and efficiency of the system, while bringing about innovation and stability.

Hence, these advancements in blockchain technology for cryptocurrencies have significantly impacted, both the applications and research sphere [47–50].

13.5 CONCLUSION

The blockchain technology provided a revolution in the way transactions were handled and is now being used in various fields. The transparent nature and decentralized infrastructure of the mechanism is appraised by many. The various cryptocurrencies present at the moment are working toward providing the users with faster and secure transactions. Even though, currently, blockchain is being used mostly with cryptocurrency, it's potential for transforming any industry is now being recognized. It is evident that even through the technical challenges, rapid change, and lack of acceptance in various countries, the technology is here to stay. There have been a development of various other distributed ledger such as IOTA Tangle and Hedera hashgraph, going beyond blockchain and alleviating all its limitations. These are more energy-efficient and transparent form of technologies. The cryptocurrencies based on the above distributed network are supported with features of management, fair ordering, secure services, and *ad hoc* transactions.

REFERENCES

[1] J. Yli-Huumo, D. Ko, S. Choi, S. Park, K. Smolander, Where is current research on blockchain technology?—a systematic review, PLoS One 11 (10) (2016) e0163477.

[2] D.K.C. Lee, L. Guo, Y. Wang, Cryptocurrency: a new investment opportunity? J. Alternat. Invest. 20 (3) (2018) 16–40. Available from: https://doi.org/10.3905/jai.2018.20.3.016.

346 **CHAPTER 13** CRYPTOCURRENCY MECHANISMS FOR BLOCKCHAINS

[3] M.H. Miraz, M. Ali, Applications of blockchain technology beyond cryptocurrency, Ann. Emerg. Technol. Comput. (AETiC) 2 (1) (2018) 1−6.

[4] E. Kazan, C.-W. Tan, E.T.K. Lim, Value creation in cryptocurrency networks: towards a taxonomy of digital business models for bitcoin companies, PACIS 2015 Proceedings, 2015, p. 34.

[5] A. Gervais, G.O. Karame, K. Wüst, V. Glykantzis, H. Ritzdorf, S. Capkun, On the security and performance of proof of work blockchains, ACM SIGSAC Conference on Computer and Communications Security (2016), pp. 3−16.

[6] M. Vukolić, The quest for scalable blockchain fabric: proof-of-work vs. BFT replication, IBM Research, Zurich, 2015.

[7] D.D. Wood, Ethereum: A Secure Decentralized Generalized Transaction Ledger, 2014, http://gavwood. com/paper.pdf.

[8] C. Natoli, V. Gramoli, The Balance Attack Against Proof-of-Work Blockchains: The R3 Testbed as an Example, 2016, arXiv: 1612.09426.

[9] I. Eyal, A.E. Gencer, E.G. Sirer, R. van Renesse, Bitcoin-ng: A Scalable Blockchain Protocol, 2015, arXiv preprint arXiv:1510.02037.

[10] F. Saleh, Blockchain Without Waste: Proof-of-Stake, 2019, SSRN Electronic Journal, https://doi.org/10.2139/ssrn.3183935.

[11] B. David, P. Gaži, A. Kiayias, A. Russell, Ouroboros Praos: An Adaptively-Secure, Semi-Synchronous Proof-of-Stake Blockchain, in: J. Nielsen, V. Rijmen (Eds.), Advances in Cryptology − EUROCRYPT. Lecture Notes in Computer Science, vol. 10821. Springer, Cham, 2017.

[12] L. Chen, L. Xu, N. Shah, Z. Gao, Y. Lu, W. Shi, On security analysis of proof-of-elapsed-time (PoET, in: P. Spirakis, P. Tsigas (Eds.), Stabilization, Safety, and Security of Distributed Systems. SSS 2017. Lecture Notes Computer Science, vol. 10616, Springer, Cham, Switzerland, 2017.

[13] P. Coelho, T.C. Junior, A. Bessani, F. Dotti, F. Pedone, Byzantine fault-tolerant atomic multicast, Dependable Systems and Networks (DSN) 2018 48th Annual IEEE/IFIP International Conference, 2018, pp. 39−50.

[14] E. Androulaki, A. Barger, V. Bortnikov, et al., Hyperledger Fabric: A Distributed Operating System for Permissioned Blockchains, 2018, Thirteenth EuroSys Conference Article No. 30.

[15] Q. Nasir, I.A. Qasse, M.A. Talib, A.B. Nassif, Performance analysis of hyperledger fabric platforms, Sec. Commun. Netw. 2018 (2018) 14. Available from: https://doi.org/10.1155/2018/3976093.

[16] C. Berger, H.P. Reiser, Scaling byzantine consensus: a broad analysis, in: Proceedings of the 2nd Workshop on Scalable and Resilient Infrastructures for Distributed Ledgers, 2018, pp. 13−18, https://doi.org/10.1145/3284764.3284767.

[17] J. Innerbichler, V. Damjanovic-Behrendt, Federated byzantine agreement to ensure trustworthiness of digital manufacturing platforms, in: Proceedings of the 1st Workshop on Cryptocurrencies and Blockchains for Distributed Systems, 2018, pp. 111−116, https://doi.org/10.1145/3211933.3211953.

[18] A. Baliga, Understanding Blockchain Consensus Models − Semantic Scholar, Persistent Systems Ltd., 2017.

[19] P. Tasatanattakool, C. Techapanupreeda, Blockchain: challenges and applications, in: International Conference on Information Networking (ICOIN), 2018.

[20] J. Bonneau, A. Miller, J. Clark, et al., SoK: Research Perspectives and Challenges for Bitcoin and Cryptocurrencies, 2015, IEEE Symposium on Security and Privacy, San Jose, CA, USA.

[21] Z. Zheng, S. Xie, H. Dai, H. Wang, An overview of blockchain technology: architecture, consensus, and future trends, in: Proceedings of IEEE International Congress on Big Data (BigData Congress), Honolulu, HI, 2017, pp. 557−564.

REFERENCES 347

[22] B. Koteska, E. Karafiloski, A. Mishev, Blockchain implementation quality challenges: a literature review, in: Z. Budimac (Ed.), Proceedings of the SQAMIA 2017: 6th Workshop of Software Quality, Analysis, Monitoring, Improvement, and Applications, Belgrade, Serbia, 2017.

[23] J. Lindman, V. Tuunainen, M. Rossi, Opportunities and risks of blockchain technologies: a research agenda, in: Proceedings of the 50th Hawaii International Conference on System Sciences, 2017. Available from: <https://doi.org/10.24251/HICSS.2017.185>.

[24] R. Houben, A. Snyers, Cryptocurrencies and Blockchain: Legal Context and Implications for Financial Crime, Money Laundering and Tax Evasion, 2018, Policy Department for Economic, Scientific and Quality of Life Policies.

[25] S. Kethineni, Y. Cao, The rise in popularity of cryptocurrency and associated criminal activity, Int. Criminal Justice Rev., 2019.

[26] C. Brenig, R. Accorsi, G. Günter, Economic analysis of cryptocurrency backed money laundering. ECIS 2015 Completed Research Papers, Paper 20, 2015.

[27] G. Papadopoulos, Blockchain and digital payments: an institutionalist analysis of cryptocurrencies. Chapter 7, 2015. https://doi.org/10.1016/B978-0-12-802117-0.00007-2.

[28] V. Dostov, P. Shust, Cryptocurrencies: an unconventional challenge to the AML/CFT regulators? J. Financ. Crime 21 (2014). Available from: https://doi.org/10.1108/JFC-06-2013-0043.

[29] B. Donato Masciandaro, Financial instruments and markets, world banking abstracts, 35 (6) (2019) 409–420. Wiley Online Library, https://doi.org/10.1111/1467-8462.12304.

[30] Z. Zheng, S. Xie, H.-N. Dai, X. Chen, H. Wang, Blockchain challenges and opportunities: a survey, Int. J. Web and Grid Services (2017).

[31] E.B. Abeer, A. Laura, K. Anne, P.-S. Romualdo, B. Andrea, Evolutionary dynamics of the cryptocurrency market. 4R. Soc. Open Sci., 4 (11), 2017. https://doi.org/10.1098/rsos.170623.

[32] G. Hileman, M. Rauchs, Global Cryptocurrency Benchmarking Study, Cambridge Centre for Alternative Finance, 2017.

[33] J. Christian, C. Bach, Blockchain Technology and Cryptocurrencies: Opportunities for Postal Financial Services, Working Papers 0056, Swiss Economics, 2016.

[34] W. Raymaekers, Cryptocurrency Bitcoin: disruption, challenges and opportunities, J. Payments Strategy Syst. 9 (1) (2015) 30–46.

[35] K.A. Sontakke, A. Ghaisas, Cryptocurrencies: a developing asset class, Int. J. Business Insights Transform 10 (2) (2017) 10–17.

[36] P. Serguei, The Tangle – Calgary, 2017. Available from: <http://pages.cpsc.ucalgary.ca/~joel.reardon/blockchain/readings/IOTA_Whitepaper.pdf>.

[37] B. Kusmierz, The First Glance at the Simulation of the Tangle: Discrete Model, IOTA Foundation, 2017.

[38] R. Alexander, IOTA – Introduction to The Tangle Technology: Everything You Need to Know About the Revolutionary Blockchain Alternative, ACM, 2018.

[39] S. Popov, O. Saa, P. Finardi, Equilibria in the tangle, Comput. Ind. Eng. 136 (2019) 160–172.

[40] P. Ferraro, C.R. King, R. Shorten, IOTA-Based Directed Acyclic Graphs Without Orphans, 2018. Available: arXiv:1901.07302.

[41] Q. Bramas, The Stability and the Security of the Tangle, 2018. hal-01716111v2f.

[42] L. Tennant, Improving the Anonymity of the IOTA Cryptocurrency, 2017. Available from: http://iota-feed.com/wpcontent/uploads/2017/08/anonymityiota.pdf.

[43] G. Bu, Ö. Gürcan, M. Potop-Butucaru, G-IOTA: Fair and Confidence Aware Tangle, 2019. Available from: <https://arxiv.org/abs/1902.09472v1>.

[44] C. Myongsu, Open Hashgraph: An Ultimate Blockchain Engine, Whitepaper, 2018.

[45] L. Baird, M. Harmon, P. Madsen, Hedera: A Governing Council & Public Hashgraph Network: The Trust Layer of the Internet, Hedera Hashgraph, LLC, 2018.

[46] I.V. Mayer, LoRaWan-Hyperledger robust network integrity on IoT devices, Proceedings of SPIE 11013, Disruptive Technologies in Information Sciences II, 110130R, 10 May 2019.

[47] B. Scott, How Can Cryptocurrency and Blockchain Technology Play a Role in Building Social and Solidarity Finance? UNRISD, 2016.

[48] J. Herbert, A. Litchfield, A novel method for decentralised peer-to-peer software license validation using cryptocurrency blockchain technology, in: Proceedings of the 38th Australasian Computer Science Conference (ACSC 2015), Sydney, Australia, 27−30 January 2015.

[49] R.A. Memon, J.P. Li, J. Ahmed, Simulation model for blockchain systems using queuing theory, Electronics 8 (2) (2019) 234.

[50] J.C. Mendoza-Tello, H. Mora, F.A. Pujol-Lopez, M.D. Lytras, Social commerce as a driver to enhance trust and intention to use cryptocurrencies for electronic payments, IEEE Access 6 (2018).

CHAPTER

OVERVIEW OF BLOCKCHAIN
TECHNOLOGY CONCEPTS

14

C. Komalavalli, Deepika Saxena and Chetna Laroiya
Jagan Institute of Management Studies, Rohini, New Delhi, India

14.1 INTRODUCTION

The need for modernization in current century has pushed rapid development of technologies in the past decade. Technologies such as Internet Of Things (IOT), Augmented reality, Machine learning, and Blockchain become integral part of modern day-to-day lives. Blockchain is a decentralized database which helps in information sharing between domains that do not trust each other. Blockchain participates in decision-making process in a cooperative, collaborate, and coordinated manner. Blockchain technology (BCT) created a major change in the business world and transformed the industry in every aspect. It brings new innovations in supply chain, healthcare, agriculture, banking, etc. by providing trust, security, and transparency.

A blockchain is a peer-to-peer distributed ledger forged by consensus, combined with a system for "smart contracts," and other assistive technologies—hyperledger.org

Let us understand the blockchain concept with an example.

Mr. X wants to transfer some amount to Mr. Y. In the traditional system, first X will send the request to the bank for initiating the transaction. Bank first checks X account for the given amount and credentials. If account is verified and finds correct, given amount gets transferred to Mr. Y. This process raises the issues like third-party intervention such as bank, cost, and time also.

BCT coined into the presence of resolving these issues with the support of a block. If this transaction is initiated over the Internet and included in the block, the transaction is validated with the support of members of the network and the amount gets transferred to Mr. Y without any hurdles. Validated transactions cannot be altered in future.

Characteristics of BCT attracted different communities for exploring and building new applications in decentralized, distributed network. New solutions are proposed, whereas some applications are changed into blockchain from their traditional system.

14.1.1 NETWORK ARCHITECTURES

14.1.1.1 Client/server systems

Let us compare centralized and decentralized systems in the following topics.

Handbook of Research on Blockchain Technology. DOI: https://doi.org/10.1016/B978-0-12-819816-2.00014-9
© 2020 Elsevier Inc. All rights reserved.

349

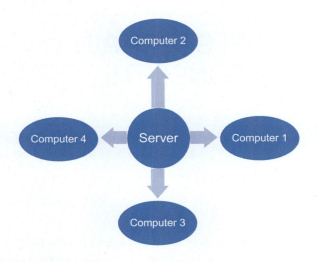

FIGURE 14.1

Client/server architecture.

The following two architectures are used in the various domains of IT industry.

1. Client/server
2. Peer-to-peer (P2P) network

Client/server systems are widely used in the enterprise applications. It is computing systems wherein server computer manages all resources and client computers are connected to the server. It follows the request/response model and their services are distributed over the network. In this architecture, client posts the request to the central computer known as server, server receives the request and replies back to the client (Fig. 14.1).

14.1.1.2 *Peer-to-peer network*

P2P network is a distributed architecture that widely used in the applications which facilitates online demand delivery such as content distribution, etc. In this computing model, each and every computer in the network has the same capabilities and responsibilities. This architecture is a decentralized network and does not support sever concept. They are coordinated and communicated by message passing to one another. Distributed architecture is the enabler for the BCT and provides the decentralized environment, that is distributed environment. Each and every node of the P2P network can send and receive transaction within a network. Every peer of the network is acting as publisher as well as subscriber (Fig. 14.2).

14.2 MOTIVATIONS BEHIND BLOCKCHAIN

All new technological innovations are almost always brought about by the need to solve a problem. The problem could be a new one that arose during the course of running a process or an existing problem left unresolved or partially resolved due to the limitations of existing technologies.

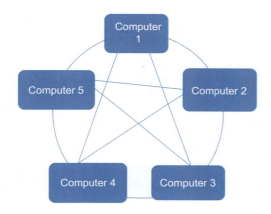

FIGURE 14.2

Peer-to-peer network.

Blockchain was an innovation born as a result of attempts to address the uncertainty existing in the financial transactions. Though uncertainty could not be eliminated, it can certainly be lowered. Third-party institutions have long played the role of adjudicator/lawmakers between the parties making an agreement that lower the uncertainty or act as the bridge that covers the trust deficiency during a transaction.

A good example would be an e-commerce transaction. The buyer expects fair goods/services that ensure value for his money. The seller expects to receive the agreed payment once he delivers the promised goods/services as per the agreement.

Obviously, there would be a lack of trust or trust deficiency between the parties intending to enter into the agreement. Hence, a need for a third party arises, who in this case could be organizations like e-Bay/Amazon that provides the trust platform for both the parties. This mediating party assures the first and second party or the seller and buyer in this case assured legitimate trade. As noted earlier, the uncertainty or trust deficiency is not completely eliminated, as the mediating party has to be "trusted."

Trusting an institution however requires a lot of research and knowledge. Blockchain aimed to overcome this uncertainty by implementing the applications in a secure and decentralized way, thereby providing an assurance better certainty. BCT is gaining more and more acceptance and adoption, in the trustless society, precisely due to this reason.

The key to the success of BCT lies in the implementation of other protocols like decentralization and distributed data ledger. The blockchain is assisted by these protocols in making it robust and much resilient. The implementation of blockchain in trustless environment is made possible due to the decentralization of computation in dense P2P networks.

The maintenance of secure and publicly distributed ledger provides complete transparency over the entire transactions, mutations that take place in the blockchain. The information regarding the latest state of blockchain is held by every node, being made sure by P2P protocol.

We have seen that the decentralization is an essential motivating factor contributing to the developing and success of BCT. The decentralization is achieved by virtue of distributing the computation tasks to all the nodes. Now the decentralization resolves two or some of the most critical problems associated with a typical centralized system. Single-point failure is one such problem. In large centralized networks like a bank or railways, the centralized server having a few backup servers faces the risk of failure at any instant.

Even in a very well-conceived and architected network, cent percentage availability of the central server cannot be guaranteed, where surge in traffic can overload the central and backup servers. Such a situation results in slowdown or complete breakdown, resulting in crashing and consequent shutdown of the server. Thus downtimes cannot be averted completely in a centralized system.

In a decentralized network, all the transactions are stored in every node, and hence, each node acts as a backup server. This is possible due to maintaining distributed data ledger which is an essential feature of blockchain. The immutability feature of blockchain which is a key factor in making an integrity. Blockchain "trustable" ensures the integrity of the publicly accessible ledger as well.

14.2.1 FOUNDATION OF BLOCKCHAIN

BCT is built on the foundation of distributed computing, software engineering, cryptography, and game theory (Fig. 14.3).

14.2.2 NEED FOR BLOCKCHAIN

BCT is integrated with the applications of different domains. The need of this technology for the following reasons:

- *Faster settlements*
    Traditional banking systems are very time-consuming process for the settlement of transactions. But, blockchain reduced the time significantly.
- *Security*
    Cryptography functions and consensus protocols enable the secured transactions.
- *Immutable*
    Since blocks are immutable, tampering of the block is very difficult.
- *Transparent*
    Since blockchain is a decentralized system, third-party invention is not needed at all. All stockholders of the network can participate in the network, thus ensures transparency.

FIGURE 14.3

Foundation of blockchain.

14.3 **BLOCKCHAIN CONCEPTS**

In order to know the BCT in depth, the basic understanding of BCT concepts and key terms are to be discussed in this chapter. Blockchain is considered as a digital concept of storing data. The data such as list of transactions, document, or any other related data come in blocks. Block can be considered as a container for the data BCT consists of chain of consecutive blocks of transactions. But, BCT is a decentralized, P2P network and distributed in nature. Each and every participant of the network can control over the network. The integrity of the chain is achieved by the iterative process from last block to genesis block.

Blockchain network is defined as interconnection of many computers, and each and every computer holds the copy of the ledger. It can be observed as continuously budding chain of blocks, and blocks are interconnected with the support of hash function. Validating of new blocks is followed by a set of protocols and consensus from every participant of the network. The records are kept and arranged in linear fashion chain.

Pointers and linked list data structures are used in blockchain for the block representation. Pointers are variables that are used to point the location of the next block. Blocks are arranged in sequence and lined with each other using linked list. Each block has data and connection to the next block with the aid of pointers.

14.3.1 **HISTORY OF BLOCKCHAIN**

The first blockchain was conceptualized by Satoshi Nakamoto in 2008. Invention of Bitcoin in 2009 by Satoshi Nakamoto paved a way to the evolution of new digital technology called blockchain. First cryptocurrency Bitcoin made substantial contribution for the concepts such as blockchain and smart contracts. Blockchain, the underlying technology of the Bitcoin can be explored for other applications or domains. Followed by blockchain, smart contract was devised in 1994 by Nick Szabo. A set of executable codes are Smart Contracts that can directly run on top of the blockchain systems. Agreement between untrusted parties without the requirement of a third party is enforced by this technology. Smart contracts are allowed in financial transactions than Bitcoin. Proof-of-Stake (PoS) is the new innovation for replacing Proof-of-Work (PoW) was introduced later. Nowadays, every transaction is processed by every computer of the network. Speed of the operation can be reduced drastically. By scaling-up the blockchain, process can be accelerated and validated the transaction can be divided efficiently.

14.3.2 **NETWORK VIEW**

See Fig. 14.4.

14.3.3 **DATABASE VERSUS BLOCKCHAIN**

Relational databases support CRUD operation that is Create, Read, Update, and Delete operation model. In the database, centralized administrator is managing the database and assigning access control to the users. It follows client/server model. Databases are fast and do not contain details about the history of records. Authenticated users only read or write in the database. It is not completely robust technology.

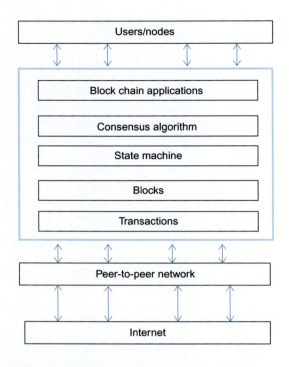

FIGURE 14.4

Network view.

But, blockchain methodology is read and append only by the users. Blockchain blocks are restricted for updating and deleting operation. New blocks can be attached to the end of the chain. There is no single administrator for managing the access control. All participants of the network have equal access rights for reading and storing the block. It follows distributed model. Blockchain is slow compared to database and contains history and ownership of records. Users who have Proof of Work (PoW) rights can write on the blockchain and completely robust technology.

14.3.4 DRIVING CONCEPTS FOR BUSINESS
- *Shared Ledger*
   Shared Ledger is an immutable ledger of all transactions of the network. Transactions are recorded once and shared among the members by replication, Every network participants authorized to view the transactions of the network.
- *Permissions*
   Permissions are rights to be given to the participant of the network for accessing the transactions. It can be categorized as permissioned or permissionless by considering application requirement. Participants are identified with a unique identity for enabling their permissions.

With the help of permissions, some users can view only certain transactions and some users have rights to view all transactions of the network. This ensures the transaction's security and authentication.

- *Smart Contract*

  These are a set of rules defined for the transaction and execute automatically. They are considered as a part of the transaction and embedded into that.

- *Consensus*

  It is defined as an agreement between parties of the network for validating the transactions. There are playing a crucial role for maintaining the Integrity and Security in blockchain network.

14.4 KEY TERMS RELATED TO BLOCKCHAIN

The following section describes the key terms associated with the blockchain architecture.

- *Node*

  Computers in the blockchain architecture are called nodes and each node has the copy of the ledger. Every node participates in the transactions and checks against validity of the transaction.

- *Transaction*

  These are smallest unit of blockchain which maintains the records, data, information, etc. They generally have sender address details, receiver details, and value of the transaction. Value gets transferred by digitally signing the hash from the sender side and the transaction is distributed over the network. Time stamped transactions are stored in a block.

- *Addresses*

  These are unique identifiers for denoting the senders and recipients of the blockchain network. An address is a public key.

- *Block*

  Block is a collection of data that stores the transaction details such as timestamp, linkage to the preceding block which is generated by secure hash algorithm. We can understand the hash as a fingerprint which is distinctive to each block. Every block contains two parts: block header and block body. Block header and body shall be discussed in detail in later part of the chapter.

- *Chain*

  Blocks arranged in a sequence order creates the chain of the architecture.

- *Nonce*

  It is a random number that miners of the network are trying to find a valid hash value. This value helps in producing a valid hash with the combination of other information from that block. Finding suitable nonce value is called Mining. Miners start nonce value from 0 and increment the value by 1 till the valid hash is found. The number of valid hashes decreases with the difficulty increases. Mining takes more computational power for finding a valid hash. The number of leading zeros determines difficulty level of the block.

- *Mining*

  It involves the creation of hash for the block of transaction. The transaction cannot be forged, thus the integrity of the chain is protected. Complex mathematical calculations are

356 **CHAPTER 14** OVERVIEW OF BLOCKCHAIN TECHNOLOGY CONCEPTS

involved in the mining for verifying the transaction. Miners are verifying the validity of the transaction, and then transaction is included in secure block. Miners get awarded for creating secure hash of the block.

- *Consensus*
  It is defined as a set of rules to be followed for adding block into the structure.
- *Hard fork*
  It is a rule change for validating the block. All nodes should work in accordance with the new rules. A fork is formed when two miners are finding the block at the same time.
- *Merkle Tree*
  It is also called as hash tree. In this tree, every leaf node is considered with the hash of a block and every nonleaf is considered with the hash of the labels of its child nodes.

14.4.1 **BLOCK STRUCTURE**

Block header stores the metadata about the block and body which stores all the information about the block. Metadata comprises of version of Block, Merkle Tree Root Hash, Time Stamp, n Bits, Nonce, and Parent Block hash. Block Version indicates the validation rules to be followed in the network. Merkle Tree Hash stores the hash value of all transactions in a block (Fig. 14.5).

Timestamp refers the current date and time. Transactions are action performed by the users of the network. Mining includes the formation of hash for the block of transaction that cannot be easily forged, thus protects reliability.

Block body consists of transaction counter and transactions. Block size and size of each transaction determine the number of transactions in a single block. Asymmetric cryptography is used to confirm the authentication of transaction (Figs. 14.6 and 14.7).

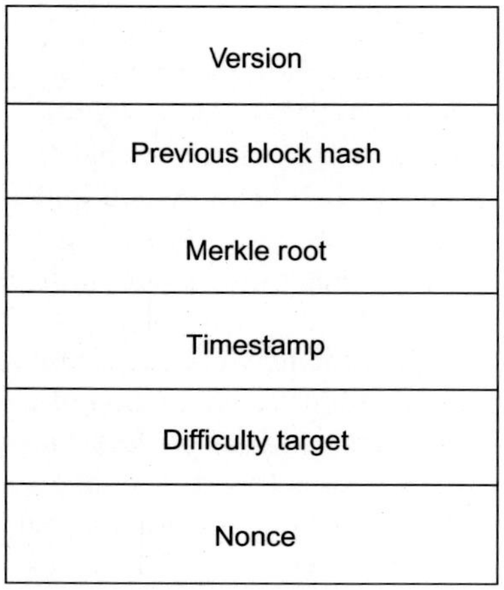

Version
Previous block hash
Merkle root
Timestamp
Difficulty target
Nonce

FIGURE 14.5

Block structure.

14.4.2 GENESIS BLOCK

The beginning block of blockchain and the block in which no previous block is Genesis block. Consensus protocols are the rules for adding block to the blockchain. Protocol decides how a block is processed and how to calculate the hash of the block. If any block is tampered, the block followed by that block becomes invalid (Table 14.1).

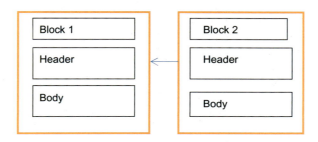

FIGURE 14.6

Chain of block.

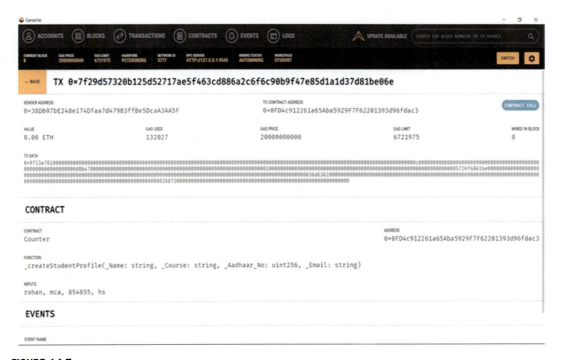

FIGURE 14.7

Example of a block.

358 CHAPTER 14 OVERVIEW OF BLOCKCHAIN TECHNOLOGY CONCEPTS

Table 14.1 Block Content	
Previous hash	0
Timestamp	Mon, 06 Aug 2019, 02.00 GMT
Data	Welcome to Blockchain
Hash	00003456ad2345566. . .
Nonce	45673

14.4.3 BLOCK TIME

It is denoted by the average time taken for generating block in the network. When miners complete block, the included data become testable and valid. Validating of new blocks is followed by a set of protocols and consensus from every participant of the network.

14.4.4 BLOCK SIZE

Block size can be defined as the maximum limit of a block that can be filled up with transaction. Block size is a crucial factor for preventing denial-of-service attacks in the network. Network rejects the block when the block size exceeds limit. Block size limitation raises the issues of slowing of the network and high processing. When the number of users transacting is increasing, fixed block size results in slowdown of the network. Network speed can be increased by increasing the block size and dynamic allocation of block size. When the transaction volume increases above the limit, block size dynamically changes by itself, thus avoids the slowing down of the network. Soft fork called Segregated Witness is another way of increasing the block capacity, by removing signature data from transactions. The implementation of Segregated Witness in Bitcoin application, results a significant increase in block capacity.

14.4.5 CRYPTOGRAPHY IN BLOCKCHAIN

Cryptography is method for protecting information through the use of encrypting and decrypting the data. Hashing is mainly used in linking of blocks and in consensus algorithms. Asymmetric key cryptography is driving the blockchain applications for identifying the contributors of the network and proof of their ownership. So cryptography is an excellent way for replacing the third parties and provides trustless environment in network.

- *Hashing*

  Hash value denotes a numeric value of a fixed length that will be generated using cryptographic hash algorithm. It identifies the data uniquely and Blockchain state is represented by hash function SHA256. Hashing generates a fixed length hash value that uniquely represents the contents of an arbitrary length string. Identical strings are generating the same hash value. Retrieving the original string from hashed values is not possible, since it is a one-way function.

  Genesis block hash is calculated using initial transactions. Index of the block, previous block hash, timestamp, block data, and nonce are used for calculating the hash value of the consecutive blocks. Example of a hash function

> Data: Blockchain Concepts Book Chapter

> Hash: 06ecd9a034556c403064a9114d26e2d227324520e4c2d5b330cf5f881564ac9b

Even a very small change in input string results a new hash value. In the above-mentioned string, only b changed in the book string. It creates a completely new hash value for the new string.

> Data: Blockchain Concepts book Chapter

> Hash: def6a36ca079b54d3004cadfe14068c54880b6a1873e6da467f20ac9f44ba5b5

- *Digital signature*
  Digital signature implements an asymmetric cryptography. It uses public and private key pairs for encryption and decryption of the data by the user, thus ensuring the privacy and security. Both the keys are used in blockchain network for authentication. A pair of private and public keys are possessed by every user. The transactions are signed by private key and the transactions are broadcasted in the whole network. The transactions are accessed by public key by the users of the network. Signing and verification phases are two phases which are involved in blockchain.

14.5 FORMATION OF CHAIN IN BLOCKCHAIN

Blocks are interconnected in blockchain by referencing hash value of the preceding block. Each node of the network will have the complete chain and attach a new block whenever required. When a new block is inserted to the chain, index of the block must be greater than the latest block. Block hash meets difficulty requirement. Each node will be verified before appending to the network. The computed hash value is the combination of the previous block hash and its own block data. All items are hashed and the block stores the timestamp also. It helps in creating an ordered chain of blocks. The block will be added by matching the hash value of the preceding block and approval of all members of the network (Fig. 14.8).

- *Merkle Tree*
  The integrity and validity of data are provided by Merkle Tree. They play a vital role for saving the memory or disk space. The following example shows simple Merkle Binary Tree and

360 CHAPTER 14 OVERVIEW OF BLOCKCHAIN TECHNOLOGY CONCEPTS

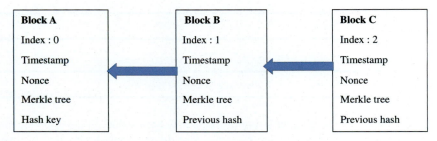

FIGURE 14.8

Formation of chain.

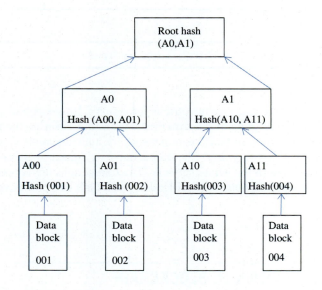

FIGURE 14.9

Merkle Tree.

top hash that is the hash of the entire tree, known as the Root Hash. Each of these is repeatedly hashed, and stored in each leaf node (Fig. 14.9).

- *New node synchronization*

When a new node is connected in the chain, block synchronization is needed for maintaining the updated copy of the blockchain in the new node. By exchanging the message with the peers in the network, it updates its local copy. Initially, new node sends a chain length for identifying an index of the last block. If received block can be appended win the chain after checking the index, node completes its process and block is appended to the network. In case of hash mismatch, new node requests for all blocks from its peer for ensuring the longest chain and updated chain.

In P2P network, nodes communicate with each other, because no single peer is trustable. By broadcasting information to all peers of the network retains the integrity of the network.

14.6 CONSENSUS ALGORITHM

It is a mechanism through which blocks are added into the structure. In blockchain network, all nodes reach a common contract for adding a new block. This ensures the trust between unknown peers of the network.

14.6.1 PROOF OF WORK

First algorithm PoW is developed and used in Bitcoin cryptocurrency. This algorithm is widely used in mining process. To add a block, miners are attempting to solve the puzzle. Miners are trying to find out the hash value for the next block. In this algorithm, the nodes/computers in the network agree on the hash of the minor and block will be added in the network. First miner will get the reward for solving puzzle. But, it requires high computational power for solving the puzzle.

14.6.2 PROOF OF STAKE

This algorithm validates the block according to the stake of participants. Validation is determined by the investment of the currency. Miners are not rewarded with the money.

14.6.3 PRACTICAL BYZANTINE FAULT TOLERANCE

This algorithm is based on reaching agreement even when the failure of the nodes happened in the network. This algorithm is focused on the node failure by considering both faulty and working nodes. Byzantine Generals' Problem is the base for this algorithm. Fault tolerance can be achieved by taking the correct values of working nodes and assigning the default vote value for the faulty nodes. Thus network reaches an agreement on correct values.

14.6.4 PROOF OF ACTIVITY

It is a combined approach of PoW and PoS. It provides assurance that all transactions are genuine in nature and users reach at a consensus on the status of the ledger. In first phase, miners are trying to find out the new block using PoW consensus. When new block is identified, process changes into PoS.

362 CHAPTER 14 OVERVIEW OF BLOCKCHAIN TECHNOLOGY CONCEPTS

14.6.5 PROOF OF BURN TIME

The principle behind this consensus algorithm is burning or destroying coins detained by the miners. It ensures the agreement of all participating nodes by valid state of network, thus avoiding cryptocoin double spending. Energy consumption of PoW is addressed in this algorithm. Miners are allowed to write block proportion to the destroying or burning the number of tokens.

14.6.6 PROOF OF CAPACITY

This algorithm is based on hard disk space instead of computational power of the node. Hard disk space of the nodes is utilized for mining the cryptocoins. Miners can store the possible solutions in the hard drive before mining, thus avoiding changes in the header value rapidly. Miners can match the required hash value from the list for winning the reward.

14.6.7 PROOF OF IMPORTANCE

The algorithm is based on PoS. It is working in a concept of harvesting and vesting. Harvesting determines the node's eligibility for adding block into the network and node vests transaction fees in turn within that block.

14.7 BLOCKCHAIN VERSIONS

There are three versions existing in the system.

14.7.1 BLOCKCHAIN 1.0 CURRENCY

Cryptocurrencies are termed as Blockchain version 1.0. This permits financial transactions based on BCT. Bitcoin is the most prevalent example of this version.

14.7.2 BLOCKCHAIN 2.0 SMART CONTRACTS

Second version of blockchain is Smart Contracts. They are introduced and the replacement for traditional contract. These are a small set of code running on the top of the blockchain and simplest form of decentralized. This code facilitates, verifies, and enforces the cooperation of a contract or transaction. They reduce the incurred cost for verification, execution, and fraud prevention, thus allow transparent contract definition.

14.7.3 BLOCKCHAIN 3.0 DAPPS

DApps are the third version of blockchain network. It is a decentralized application and can have front end code and user interface.

14.7.4 BLOCKCHAIN 4.0

This version makes blockchain usable in the industry. It defines approaches and solutions to cater the needs of business demands.

14.7.5 DIFFERENT USERS OF BLOCKCHAIN TECHNOLOGY

- *User*

    All participants or members joining the network are blockchain users. They are not aware of the working of blockchain and its knowledge. They are novice user and can conduct transactions in the network.
- *Developer*

    They are only playing a major role for the successful creation of the blockchain network. Programmers those are having vast knowledge of blockchain network become developers and develop various applications with the support of different platforms. They decide on the type of blockchain network also and create smart contracts for the network.
- *Regulators*

    Regulators are authorities who can monitor the translation within the network and provide rules for the transaction
- *Network operator*

    They are responsible for managing the network and define permission for various types of participants of the network. Every business application associates with the network operator.
- *Certificate authority*

    They are managing and issuing certificates for permissioned blockchain.
- *Application interfaces*

    Blockchain network offers two interfaces: HTTP API and WebSocket. HTTP API interface used to provide rights for accessing and manipulating blockchain information. WebSocket interface is designed for bidirectional communication by peers of the network.

14.8 BENEFITS OF BLOCKCHAIN TECHNOLOGY

The distributed nature of blockchain offers a lot of benefits:

- *Distributed*

    Since the blockchain data are stored in thousands of nodes, reducing the possibility of single-point failure of node. Each and every node is able to replicate and store the data itself.
- *Transparency*

    Since blockchain is a distributed ledger, all participants of the network share the same copy of the transaction. Thus it ensures the most accurate, consistent, and transparent data for all participants.
- *Speed*

    Real-time settlement of transactions raises the security threat. But, blockchain is an enabler for settlement of transactions in real time by reducing risk and increasing speed.

364 CHAPTER 14 OVERVIEW OF BLOCKCHAIN TECHNOLOGY CONCEPTS

- *Reduced cost*
  Any two parties can transact without the involvement of third party and bring the cost efficiency in the system. It brings the drastic reduction in cost of the transaction in the system.
- *Reduced fraud*
  Since the transactions are immutable and irreversible, all transactions can be easily verified by participants of the network; thus minimizes fraud in the system.
- *Security*
  Data are protected by cryptography functions. So blockchain provides enhanced security for the block.
- *Collaboration*
  Parties can directly execute transactions with each other without the involvement of third party.

14.9 TYPES OF BLOCKCHAIN

Blockchain is characterized by consensus, distributed computation, immutability, and authentication. The blockchain type varied on the domain and users of the application. Authentication of users in the network also plays a vital role for the type.

14.9.1 PUBLIC

Public chains are regarded as open source system and known as permissionless ledger and any participant can join in the network. No permission is needed to join the network is required. They have full rights for reading and writing into the network and having same copy of the ledger. Public blockchain work seamlessly in trustless networks due to the immutable nature of the records. Examples of public blockchain are Bitcoin, Ethereum, etc.

14.9.2 PRIVATE

They are permissioned ledger and restricts access to only selected participants of the network. Each contributor in the network will have the whole record of the transactions and the associated blocks. All blocks are encrypted by a private key and cannot be interpreted by anyone. They are controlled only by authorized users of a specific organization. Businesses can only support trusted nodes to join in the network. Users can be validated with the support of identity management system. It involves trusted network and suitable for sharing the ledger internally. Examples of private blockchain are Hyperledger fabric, Corda, and Quorum.

14.9.3 HYBRID

It denotes a combination of public and private blockchains and also called as consortium blockchain. This blockchain is semidecentralized and semiprivate structure. Characteristics of both public and private blockchains are supported by hybrid network. It has a controlled user group but works across different organizations.

Table 14.2 Comparison of Blockchain Types			
Property	**Public**	**Private**	**Consortium**
Consensus algorithm	Permissionless	Permissioned. Only authorized users are allowed	Permissioned. Only
Efficiency	Low	High	High
Determination of consensus	All miners of the block will be participating	Users of one organization will be participating	Selected set of nodes only will be participating
Read	Open to public, anyone can read	Can be public or restricted	Can be public or restricted
Immutability	Tampering is not passible	Possibility of tampering	Possibility of tampering

The following table provides a detailed comparison among these three blockchain systems (Table 14.2).

14.10 WORKING OF BLOCKCHAIN

The following steps are involved in the working of blockchain.

1. Users initiate a transaction in the network. The transaction could be contracts, cryptocurrency or records of other information.
2. The request for the transaction has to be represented as block in the network.
3. The block is created first and then disseminated to the participants of the network.
4. All participants analyze the received block from the network and validate that block.
5. The block validation is done with the support of consensus algorithm.
6. Members of the network validate the block for attaching the block to the network.
7. The new block is attached to the network and transaction gets completed.
8. Block added with the consent of members of the network becomes permanent and immutable (Fig. 14.10).

14.11 PLATFORMS

Blockchain is a platform with the scripting language that is used to solve the problems in different domains other than cryptocurrency. There are many blockchain platforms are existing based on different consensus algorithms, developing tools, and programming languages. Some of them are briefly discussed in this chapter.

14.11.1 CRITERIA FOR SELECTING BLOCKCHAIN PLATFORM

The following parameters are considered for selecting the platform for developing applications.

- *Type of network*: Depends on application, choice has to be chosen for public or private network.

366 CHAPTER 14 OVERVIEW OF BLOCKCHAIN TECHNOLOGY CONCEPTS

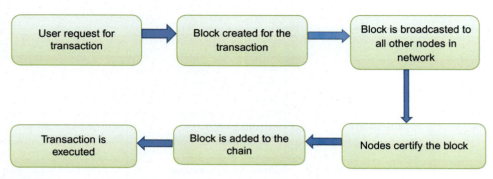

FIGURE 14.10

Working of blockchain.

- *Activity*: For selection of platform, active development of the platform is also playing a role.
- *Price*: Pricing of the platform is a crucial factor for the decision. It is a most important factor for decision.
- *Programming languages*: Programming languages used for writing the code in the platform also have significant role for the selection.
- *Popularity*: Popularity of the platform among the communities is also one of the main factors for the selection.

14.11.2 DIFFERENT PLATFORMS

- *Bitcoin*

    Bitcoin is the first well-known blockchain network based on cryptocurrency. It follows PoW consensus algorithm. This is a single-chain architecture and C++ is used for programming.
- *Ethereum*

    Ethereum is most popular platform used by various communities. It is an open source platform and used widely for testing purpose. It is designed for smart contracts and large variety of decentralized applications. It is a public network and open source network. Ethereum supports Solidity language for programming. Ethereum offers faster processing and private transactions within a permissioned group of participants within a network.
- *Hyperledger*

    Hyperledger was introduced by the Linux Foundation and open source, crossindustry BCT. This platform is effort of crossindustry global collaboration by developing distributed legers. It includes various industries such as finance, banking, manufacturing, etc. It consists of Hyperledger Burrow, Fabric, Grid, Indy, Iroha, and Sawtooth frameworks.
    - *Hyperledger Burrow*

        It provides permissioned modular blockchain client. It is developed with the description of the Ethereum Virtual Machine.
    - *Hyperledger Fabric*

        This is intended for developing applications with the modular architecture.

14.11 PLATFORMS 367

- *Hyperledger Grid*

  Grid is a framework which is designed for accelerating the development of ledger based solutions for all types of crossindustry supply chain.
- *Hyperledger Indy*

  It is a distributed ledger and designed for decentralized entity. The tools and libraries provided by Indy are interoperable across administrative domains.
- *Hyperledger Iroha*

  It is blockchain platform and C++ language used in this. It incorporates the unique chain-based Byzantine Fault tolerant consensus algorithm.
- *Hyperledger Sawtooth*

  This is a platform for building, deploying, and running distributed ledgers.
- *HydraChain*

  It is a private network and supports parallel chain architecture. It is the extension of Etherum platform. It is used for permissioned distributed ledger. Both private and consortium networks are supported in Hydrachain.

- *Corda*

  It is an open source and single-chain architecture. The supported language for this framework is Java.
- *IOTA*

  IOTA is a distributed ledger technology was developed by IOTA foundation for enabling fee-less transaction for IOT. Tangle, the innovation of IOTA is a decentralized platform and P2P network. Python is used in this IOTA.
- *Chain Core*

  This platform for issuing and transferring financial assets. It is a permissioned, private architecture. It supports parallel chain architecture and enables the transactions between entities directly. These transactions help in issuing new assets, transfer assets between parties or release the assets.
- *Azure*

  Developed by Microsoft and provides blockchain as a service. It provides services and capabilities for creating and deploying blockchain applications. Business process and data can be shared with other organizations.
- *AChain*

  It is a public blockchain platform which enables smart contract creation and issue of tokens. It provides platform for developing decentralized application systems. It is a public, global blockchain network for exchanging the information and transactions. Programming language Glua is used in Achian for developing blockchain network. It supports parallel chain architecture.
- *EOSIO*

  It is an open source platform used for developing private and public blockchain networks. It supports single-chain architecture. This platform provides authentication, database, asynchronous communication, and better scalability. Millions of transactions are performed within a second, thus eliminate user fees. It is very much user friendly for quick and easy deployment of applications. It uses C++ as smart contract programming language.

368 CHAPTER 14 OVERVIEW OF BLOCKCHAIN TECHNOLOGY CONCEPTS

14.12 CHALLENGES IN BLOCKCHAIN TECHNOLOGY

Since the technology is growing rapidly, challenges are also raised with the same pace.

- *Scalability*

  Though blockchain is gaining popularity in the banking sectors, it cannot handle the financial transactions that happen each and every day. The modification of data is not allowed in blockchain systems. New nodes should be created for the modification and multiple nodes are validating each and every transaction. It leads to the decrease in the transaction speed and increased the fee per transaction. Scalability problem is quite tough for adopting BCT. Scalability can be addressed by storage optimization and redesigning the block size. Growing block size is a major problem to be addressed. Blockchain must be redesigned for making trade-off between block size and optimization.

- *Public perception*

  Understanding of this technology other than Bitcoin application is a challenge. Widespread lack of knowledge of people about this technology other than Bitcoin, hampers the exploration of ideas in other fields.

- *Cultural shift*

  Blockchain transforms the traditional means of doing things to digital technologies with the concept of decentralized network. Trust and security of the blockchain are not centralized.

- *Initial cost*

  The adoption of BCT affords promising key benefits in productivity, efficiency, and reduced cost for the P2P transaction. But, the initial blockchain infrastructure and exchanging existing legacy systems are cost effective. Banks will be responsible for the initial aggregate cost of the blockchain implementation.

- *Integration with legacy systems*

  It may be very difficult to replace the existing legacy systems with the blockchain systems completely. Blockchain systems need to be integrated with the legacy systems. Initially, for facilitating the smooth transition from legacy to blockchain systems. This process takes a significant amount of time and extensive planning also.

- *Security and privacy*

  Customers' trust is the driving force in the growth of banking and financial transactions. The transactions in the blockchain are available publicly raises the question of privacy in the systems. Banks and financial institutions are needed to be implementing private blockchain.

- *Legal regulations*

  Legal regulations are mandatory for implementing the BCT in the banking and finance institutions.

- *Energy consumption*

  One of the major challenges is energy consumption. Miners are using PoW as consensus algorithm and it takes more electricity for mining. If the technology grows, computational power of the system and energy consumption also raises a big challenge.

- *Lack of skill set*

  Blockchain is an evolving and tough technology to understand the concepts and to provide support for that. This needs highly skilled personnel, which is very tough to find out. Handling the complexity of the network raises the challenge.

14.13 STRENGTHS OF TECHNOLOGY

The following are described as key characteristics of BCT.

- *Trust*

    Blockchain offers full control for the users. Participants are able to authenticate the data without an intervention of third party. All transactions are digitally signed for maintaining and validation of the ledger.

- *Distributed*

    Blockchain offers distributed computing. All members of the networks have same copy of the ledger.

- *Digital*

    All information are stored in digitized form, thus reduce the manual documentation.

- *Decentralization*

    Blockchain offers the concept of distributed ledger. Every node in the network has the copy of the information and can validate without the interference of centralized server. It provides transparency, trust, and data security.

- *Chronological and time stamped*

    Blockchain stores the chain of blocks and each block stores the information of the transaction and the link points to the previous block. All blocks are verified with the timestamp and arranged in chronological order.

- *Robustness*

    Every transaction is replicated and stored across all nodes of the network. This characteristic offers greater robustness to the BCT.

- *Immutable*

    Each transaction has to be validated before added into the network. It cannot be reversible and impossible to tamper the transaction. All blocks are cryptographically sealed, thus avoid any manipulation. It ensures the robustness.

- *Auditability*

    The transactions are validated and stored into the network with the timestamp. User can verify the chain of transaction at certain point of time.

- *Consensus driven*

    Each block of blockchain network is verified and appended to the network with the approval of the participating nodes of the network. Various consensus algorithms are existing for approval of the participating nodes ensures the trust verification.

- *Anonymity*

    Each user can communicate with the network using their generated address. User's privacy information is not stored in centralized server. It ensures the privacy of the users in the blockchain network.

14.14 APPLICATIONS

The underlying technology of Bitcoin application had brought a major transformation in society by its immutable and distributed nature. They impacted the life from the way we transact, managing

370 CHAPTER 14 OVERVIEW OF BLOCKCHAIN TECHNOLOGY CONCEPTS

assets to identity management, voting, etc. Wide range of applications in different industries is benefitted by implementing BCT. Companies such as Amazon, IBM, Microsoft, Samsung, etc. are discovering the potential of BCT for various types of applications. They transformed the public and private sectors also. It is applied in fields such as identity management, land registration, storing musical records, maintaining healthcare records, etc.

Cryptocurrencies are prime application of blockchain. Smart Contacts are regular contracts but they are in digital format and enforced in real time. The intervention of any third party is avoided in these types of contracts. Crossborder payments are simplified with the support of BCT. Applications in various industries are discussed later chapters of this book.

14.15 SUMMARY

In this chapter, we have provided an insight to the BCT which is fast evolving world's favorite technology for secure and transparent one-to-one transactions. Ideas about the distributed systems, various key terms were also discussed in this chapter. Then, we explained the history of blockchain and types of blockchain also. Supported platforms, strengths of the technology, and challenges were also deliberated in this chapter. After reading this chapter, readers can understand the insights of BCT in detail.

FURTHER READING

R. Aggarwal, Blockchain and Financial Inclusion, Digital Currency Group, 2017. Available from: <http://finpolicy.georgetown.edu/sites/finpolicy.georgetown.edu/files/Blockchain%20and%20Financial%20Inclusion%20120417.pdf>.

A. Kharpal, Blockchain Revolution: Everything You Need to Know About the Blockchain, CNBC, June 2018.

M. Gebert, Application of Blockchain Technology in Crowdfunding, New European, 2017. Retrieved March 2019, from: <https://www.researchgate.net/publication/318307115_application_of_blockchain_technology_in_crowdfunding>.

S.K. Johansen, Working Paper on a Comprehensive Literature Review on the Blockchain Technology as a Technological Enabler for Innovation, 2018. Retrieved March 2019, from: <https://www.researchgate.net/publication/312592741>.

D.E. Krause, V. Velamuri, T. Burghardt, D. Nack, M. Schmidt, T.M. Treder, Blockchain Technology and the Financial Services Market—State of the Art Analysis. A Joint Report by Infosys Consulting and HHL Leipzig Graduate School of Management, Infosys, November 2016. Retrieved from: <https://www.infosysconsultinginsights.com/insights/blockchain-technology-and-the-financial-services-market/>.

A. Krishnan, 24 Industries That Blockchain Will Radically Transform, February 2018. Retrieved from Invest in Blockchain: <https://www.investinblockchain.com/blockchain-transform-industries/>.

K. Sulthan, U. Ruhi, R. Lakhani, Conceptualizing blockchains: characteristics and applications, in: IADIS International Conference Information Systems, 2018.

A. Narayanan, J. Bonneau, E. Felten, A. Miller, S. Goldfeder, Blockchain technology in India I opportunities and challenges, 3 Princeton University Report on "Bitcoin and Cryptocurrency Technologies" by Arvind Narayanan, Joseph Bonneau, Edward Felten, Andrew Miller, Steven Goldfeder. Deloitte, 2017. Retrieved from: <https://www2.deloitte.com/content/dam/Deloitte/in/Documents/strategy/in-strategy-innovation-blockchain-technology-india-opportunities-challenges-noexp.pdf>.

M. Gupta, IBM Blockchain for Dummies, A Wiley brand, M. Young, The Technical Writer's Handbook, second ed., University Science, Mill Valley, CA, 1989.

M. Peck, Freelance technology writer, a white paper on "Reinforcing the links of Blockchain", in: IEEE Spectrum Magazine Special Edition "Blockchain World", November 2017.

M. Swan, Blockchain: Blueprint for a New Economy, vol. 3, no. 3, O'Reilly Media, February 2015, pp. 38−72.

J. Naughton, Is Blockchain the Most Important IT Invention of Our Age? 2016. Available from: <http://www.theguardian.com/commentisfree/2016/jan/24/blockchain-bitcoin-technology-most-important-tech-invention-of-our-age-sir-mark-walport?CMP = share_btn_fb>.

S. Seebacher, R. Schüritz, Blockchain technology as an enabler of service systems: a structured literature review, in: International Conference on Exploring Services Science, Springer International Publishing AG, Italy, 2017.Available from: http://dx.doi.org/10.1007/978-3-319-56925-3_2.

T. Shah, S. Jani, Applications of Blockchain Tehnology in Banking and Finance, 2018, Available from: http://dx.doi.org/10.13140/rg.2.2.35237.96489.

L.J. Trautman, Is Disruptive Blockchain Technology the Future of Financial Services? 69 The Consumer Finance Law Quarterly Report 232, 2016. Retrieved from: <https://papers.ssrn.com/sol3/papers.cfm?abstract_id = 2786186>.

G.R. White, K. Brown, Future applications of blockchain: toward a value-based society, in: Conference: INCITE, at Amity University, India, 2016. Retrieved from: <https://www.researchgate.net/publication/308916112_Future_Applications_of_Blockchain_toward_a_value-based_society>.

H.T. Vo, A. Kundu, M. Mohania, Research directions in blockchain data management and analytics, in: 21st International Conference on Extending Database Technology (EDBT), OpenProceedings.org, March 26−29, 2018, pp. 445−448.

Y. Guo, C. Liang, Blockchain Application and Outlook in the Banking Industry, n.d.

M.H. Miraz, M. Ali, Applications of blockchain technology beyond cryptocurrency, Ann. Emerg. Technol. Comput. 2 (1) (2018).

Z. Zheng, et al., An overview of blockchain technology: architecture, consensus, and future trends, IEEE 6th International Congress on BigData (2017) 557−564.

Z. Zheng, S. Xie, H. Dai, X. Chen, H. Wang, An overview of blockchain technology: architecture, consensus, and future trends, in: Proceedings of the 2017 IEEE BigData Congress, Honolulu, HI, 2017, pp. 557−564.

J. Yli-Huumo, D. Ko, S. Choi, S. Park, K. Smolander, Where is current research on blockchain technology? A systematic review, PLoS One 11 (10) (2016) e0163477. Available from: https://doi.org/10.1371/journal.pone.0163477.

S. Nakamoto, Bitcoin: A Peer-to-Peer Electronic Cash System, 2008. <https://bitcoin.org/bitcoin.pdf>.

Hyperledger Sawtooth. <https://www.hyperledger.org/projects/sawtooth>.

Proof of Authority. <https://github.com/paritytech/parity/wiki/Proof-of-Authority-Chains>.

Blockchain Applications. <https://blockgeeks.com/guides/blockchain-applications-real-world/>.

Blockchain Technology. <https://www.blockchaintechnologies.com/applications/>.

https://www.coindesk.com/information/applications-use-cases-blockchains

https://www.investinblockchain.com/blockchain-transform-industries/

Introduction to Blockchain. <https://www.javatpoint.com/blockchain-introduction>.

CHAPTER

SCALABILITY IN BLOCKCHAIN: CHALLENGES AND SOLUTIONS

15

Gagandeep Kaur and Charu Gandhi
Department of CSE&IT, Jaypee Institute of Information Technology, Noida, India

15.1 INTRODUCTION

Blockchain has been considered as the fifth disruptive technology after mainframes, personnel computers, the internet, mobile communications, and the social media. It is referred to as a distributed peer-to-peer (P2P) database enabler for systems decentralization. Although it itself does not provide decentralization, it provides distribution of data storage and transactions. Blockchain consists of blocks in a chronological order. A block has a block header and the transactions in that block. The block has a link to the previous block through a hash value and the new transaction that is to be stored.

From facilitating Bitcoin-supported cryptocurrency to Ethereum's smart contracts, the concept of blockchain has found takers in various applications like currency, voting, health data, predictions, food, electricity distribution, data storage, and so on. The growth of blockchain technologies has led to scalability issues. When Nakamoto introduced blockchain, its block size for data storage had a limit of 1 MB [1]. Since then, it has become famous technology to provide secure transactions without the need for third-party validations. This growth, however, has had its consequences. Now a new participant of a blockchain network has to download whole of the long chain to learn about the chain and to be able to validate transactions. This not only requires vast memory but also has become time-consuming as well.

Bitcoin offers P2P electronic cash transfer, whereas Ethereum supports virtually any kind of decentralized application. With popularity of cryptocurrencies, the number of transactions being carried out both in Bitcoin and Ethereum are growing exponentially [2]. The whole concept of freedom in blockchain from third-party validations is based on verification of each and every transaction by mining nodes. The average waiting time for a transaction has increased to 29 minutes. It is here that the blockchain suffers scalability. It therefore has become the need of the hour to look out for solutions which can minimize mining latency and provide better scalability in blockchain.

There are several solutions to handle blockchain scalability. These have either been or will be implemented. Some important ones are block size, sharding, proof-of-stake (POS), lightning network, and Segwit [3–5].

Blockchain technology has three main components, namely, application, protocol, and cryptographic solution. A transaction in blockchain describes the transfer of ownership from seller to buyer and blockchain is used to document this transfer. Transaction consists of an id, input, output,

---

Handbook of Research on Blockchain Technology. DOI: https://doi.org/10.1016/B978-0-12-819816-2.00015-0
© 2020 Elsevier Inc. All rights reserved.

373

374 CHAPTER 15 SCALABILITY IN BLOCKCHAIN

and timestamp. The sender signs a transaction which is then broadcasted to the nodes for verification and validation. Afterwards, the buyer receives ownership over sent asset. During verification, the sender and receiver nodes are checked for legitimacy. Validation is done to achieve consensus. When a blockchain system receives the transaction from user, it carries out the transaction and writes down the results into ledger. In single chain, blockchain assets are transferred among accounts in the system. However, handling transactions of assets between different blockchains are challenging. For example, an institution may want the arrival of student fee to trigger a corresponding transfer of funds to hostel mess. The source system needs to know how to make a transaction reach the target chain system. Involved chains must keep the same results after finishing cross-chain transaction. Global consensus mechanism in blockchain brings that the speed of dealing with the transaction cannot be improved by adding extra nodes. So a single chain has limited performance. The performance can be improved by executing transactions in parallel. Gideon proposed configurable multichain which is easy to configure and can work with different blockchains. Parallel consensus protocols for performance scalability can be divided into on-chain, off-chain, side-chain, child-chain, and interchain techniques, sharding for scale-out throughput and nonlinear block organization. Pegged sidechains enable Bitcoins and other ledger assets to be transferred between multiple blockchains. Under on-chain solutions are available for Big-block, MAST, Segwit, and sharding. Off-chain provides solutions for lightning network and Raiden Network. Child-chain provides plasma and interchain provides atomic-swap solutions. Other multiblockchain architectures like Cosmos, Polkadot, Multichain private blockchain, and so on, are also been worked upon to handle scalability.

Bitcoin developers have proposed lightning protocol to reduce transaction latency, and Ethereum developers are working on partitioning schemes like sharding [6]. In distributed environment, if partitioning leads to handling of all of the application requests in a single shard and balanced load among shards, then performance scales up. But practically, there are very few applications that can be optimally partitioned. Therefore the system must be able to handle requests spanning multiple shards. In Ethereum, concept of directed acyclic graphs (DAG) is used where nodes represent transactions and edges represent direction of confirmation. Although the problem of balanced graph partitioning is Non-deterministic Polynomial (NP) complete, methods have been devised to partition Ethereum blockchain graphs. These methods have been categorized as hashing methods, Kernighan−Lin (KL) method, METIS, R-METIS, and TR-METIS.

Segwit or segregated witness provides scalability solution by increasing the number of transactions in a block but keeping the block size same. A segregated witness creates space for new transaction by removing the signature data from the Bitcoin transaction. This signature data are stored outside the chain in a base transaction block. This separates the validation part, and more transactions are allowed without increasing block size.

This chapter aims to present the current state of the art knowledge available in blockchain technology with respect to scaling challenges being faced by the blockchain users.

The works referred to are mainly conference and journal papers from reputed sources like Institute of Electrical and Electronics Engineers (IEEE), Association for Computing Machinery (ACM), and so on. Since not all is available in these repositories, therefore articles published as white papers and related reports available at github, and so on have also been studied. The study presented here is of recent years only.

Rest of the chapter covers core concepts in Section 15.2, scalability challenges in blockchain Section 15.3, scalability solutions in blockchain in Section 15.4, future directions in Section 15.5, followed by conclusion and references.

15.2 **CORE CONCEPTS**

In this section, we have discussed key terms and basic concepts related to understanding of blockchain in general and for understanding scalability in blockchain in specific. Firstly, key terms have been discussed.

1. Blockchain data structure: A blockchain consists of three important things; blocks are used for holding data, hashes for block authentication and security, hash functions for generating hashes [7]. As shown in Fig. 15.1, a block contains a block header and the set of transactions done by the user in the block. In a blockchain, the $n - 1$th block computes its hash and records it. It also records the hash value of its previous block, that is, $n - 2$th block hash is stored in $n - 1$th block as well. The previous block is called the parent block, and the first block in a blockchain is called genesis block, because it does not have a parent block. A hash is generated based on SHA256 cryptographic algorithm and is difficult to revert. This makes the blockchains resistant to collusion attacks by forming chain of immutable data blocks. In addition to hash value, the block also stores timestamp when block was created, additional information depending upon the application of the said blockchain and payload. In a block, self-payload and the payload of previous blocks are stored. Blocks are organized in a chronological order to form blockchain. These blocks are like a ledger by recording all the transactions performed in a block and facilitating processing of transactions in sequential manner according to the order of transactions within a block. The record of all the transactions within a block is kept through Merkle tree. Merkle tree root hash is the hash of all the transactions carried out in the block.
2. P2P network: In blockchain, information exchange among the network entities is done using Gossip Protocols. Blockchain networks are P2P networks because there is no central authority like in client−server models, and third-party intervention is not required for security. Gossip protocol allows exchange of request and response messages among the peers.
3. Distributed ledger: In a distributed ledger (DL), record of transactions of a user is maintained in a log called ledger [9]. These transactions are mainly online financial transactions. New transaction is appended in the individual ledger. This ledger is distributed among the participating nodes, and its copy is maintained. Ledgers are of two types, namely, public and private. In public DL, anybody can join the P2P network whereas in private, only members of a company or an organization are allowed.

    As shown in Fig. 15.2, functioning of the blockchain ledger can be described based on sending and receiving nodes, consensus algorithm validation, and so on. The node interested in doing a transaction with another node forms a block and records the data in the block. This information is broadcast in the network. The network nodes receive the block hash information and verify if the block is authentic. If block is found correct, payload is added to the block. Consensus algorithms like Proof-of-Work (POW), POS, and so on are used to verify the block. Once verified by member nodes, the block is added to the blockchain. Blockchain uses private

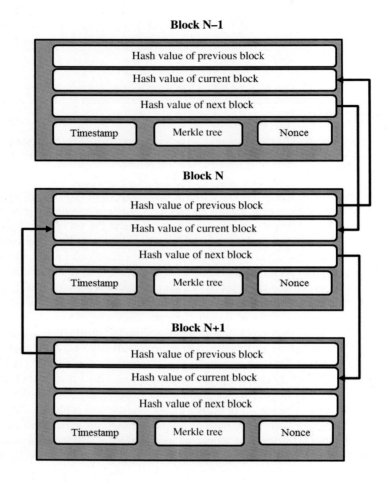

FIGURE 15.1

Block structure of blockchain.

Adapted from I. Acharjamayum, R. Patgiri, D. Devi, Blockchain: a tale of peer to peer security. in: IEEE Symposium Series on Computational Intelligence (SSCI), Bangalore, India, 2018, pp. 609–617 [8].

and public keys to provide security. A user's address is a public key and through the private key user gets their Bitcoin or access to other digital asset. Transactions can be done in blockchain without involving third party, and this functionality has made blockchain popular.

4. Consensus protocol: Blockchain does not involve third party for ensuring trust in the transactions. The interacting parties also do not trust each other. A fault tolerant consensus protocol is therefore required in blockchain. Consensus protocols ensure that all involved nodes agree to the new block, and it gets added to the chain only if all agree. P2P networks like blockchain mainly suffer from two kinds of problems. Firstly, double-spend where a user can do two parallel transactions with same amount of money and secondly suffer Byzantine

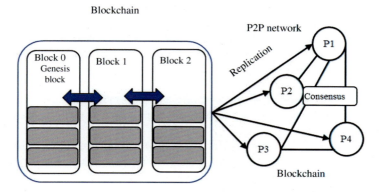

FIGURE 15.2

Basic layout of blockchain distributed ledger.

Adapted from H. Dang, et al., Towards scaling blockchain systems via sharding. Available from: <comp.nus.edu.sg>.

Generals Problem of handling malicious nodes. Two popular consensus protocols are used in blockchain, namely, POW and POS. Newer consensus protocols are also available depending upon user needs in the blockchain. Here we have briefly explained the above two only.

POW: POW requires two components. A *block proposal* is required to enable the nodes to propose to add a new block in the network. Because at any instant of time, proposal for adding new block can come for more than one node; therefore many branches can occur. To handle multiple branches, *branch selection* is done. POW by Satoshi Nakamoto requires nodes to solve cryptographic puzzle. Combined with it is the length of the chain. Multiple blocks with valid proofs could be generated and then candidate with longest chain is given preference.

POS: POW is discouraged nowadays due to very long blockchains due to which computational costs have increased and have bad impact on the environment as well. POS requires its users to commit a percentage of their shares in order to be able to forge a new block. This way unnecessary participant will not join in block creation, and less computational efforts are needed.

Some of the other consensus algorithms are delayed PoW (dPoW), proof-of-burn (PoB), proof-of-capacity (PoC), proof-of-activity (PoA), proof-of-existence (PoE), proof-of-intelligence (PoI), proof-of-luck (PoL), proof-of-property (PoP), ripple ledger, lightning network, and cross blockchains.

15.3 SCALABILITY CHALLENGES IN BLOCKCHAIN

Block size in Bitcoin is of 1 MB fixed size. Small block size of Bitcoin has become a causing concern for delays in transactions. To make things worse, popularity of freedom from third-party validations has led to exponential growth of the blockchain. But now, there is price to pay for its success [11,12]. In terms of energy cost involved in mining Bitcoin, the estimates are

982 MWh/day which in terms of hard currency is equivalent to 15 million US dollars. This energy cost is increasing every year. As per the modeled estimates by Long Future Foundation done in 2015, Bitcoin mining would end up consuming 60% of the world's energy. This means heavy carbon footing by Bitcoin mining. A human being on an average has carbon footing of 5K kg in a year whereas Bitcoin mining has carbon footing of 4K kg per mined Bitcoin. Combined with the ever-increasing size of its blockchain (125 GB in 2017), the search for solutions is becoming tedious. Through newer technology both in hardware and chip making, Bitcoin mining is going to triple its efficiency, but the rate at which the blockchain is growing, there is need for better solutions in terms of algorithms.

Due to increased mining time in Bitcoin, its network suffers under requirements of high-frequency trading, and the numbers of transactions per second remain low. The primary reason behind this is that every transaction has to be verified before it gets mined. With increase in the size of the network, verification times are increasing, and a number of blocks end up waiting in the queues for verification. Handling more than 130,000 transactions per day, the average waiting time for Bitcoin transaction is now at 29 minutes, which if compared with services like Paypal and Visa with 193 and 1670 transaction verifications per second respectively, is very high. It can therefore be seen that mining of blocks has become main bottleneck in the Bitcoin blockchain.

Ethereum, which was once being popularized as an alternative to Bitcoin, is suffering from scalability as well. Due to constraint of gas limit of 20 transactions per second only, Ethereum is unable to handle increased load of more than 1000 transactions per second.

According to the rules of blockchain consensus [13], any block being added to the blockchain is not allowed to have weight more than 4 million. As Bitcoin blockchain was in its initial phase of growth, in order to ensure decentralized network and to allow nodes with low bandwidth and limited computational powers to join the network, the constraint on block size was introduced.

But with current size of the Bitcoin network and increased transaction times, one needs to look out for solutions to handle scalability issues.

The simplest method could be to remove the size limit. This, however, may result into hard fork of the chain because due to tremendous size of the network, it is immensely difficult to upgrade all its nodes. By design, Bitcoin is not backward compatible and individually upgrading all its nodes is not feasible. Fallout from hard fork could result into two different Bitcoin currencies.

Another possible solution being worked upon is replacing the old Nakamoto consensus with more evolved Practical Byzantine Fault Tolerance (PBFT). This would provide increased throughput and reduce the commit time of the transactions. However, the PBFT consensus algorithms do require validators to validate the transactions, and with the ever-increasing size of the blockchain network, it would still be cause for reduced speed.

Among the other possible solutions, vertical and horizontal scaling may involve partitioning the blockchain. Horizontal partitioning or sharding can be helpful for DLs in improving their scalability. Sharding could lead to subsets to be validated proportionally, and with increase in number of participating nodes, the processing time would improve. This, however, dilutes the strongest feature of serializability of blockchain responsible for security, and hence, new mechanisms for ensuring security of the participating nodes need to be looked into.

Bitcoin stores three important values, namely, input data, output data, and signatures to validate the transactions. It is the signatures that consume significant amount of space. Segwit is a modification to Bitcoin, whereby the signatures are being saved onto a separate block rather than

15.4 SCALABILITY SOLUTIONS IN BLOCKCHAIN

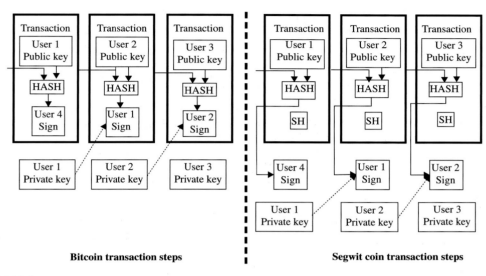

FIGURE 15.3

Bitcoin versus Segwit coin transaction.

Adapted from Accessed bitcoin cash (BCH) debunking the FUD. Available from: <https://steemit.com/cryptocurrency/@cryptoshoppe/bitcoin-cash-bch-debunking-the-fud-V2.0>.

maintaining it on the chain [14]. This speeds up the processing of Bitcoin chain. Comparative differences between Bitcoin and Segwit have been shown in Fig. 15.3.

15.4 SCALABILITY SOLUTIONS IN BLOCKCHAIN

Blockchain scalability problems can be divided into three categories: storage space, cost-of-scalability, and achieved throughput-with-scalability. At present, limited block size of the blockchain blocks has become biggest bottleneck. Blockchains are facing increase in waiting time required per transaction due to small block size. Moreover for blocks with short block generation time, many forks get created in the waiting time period of the previous block. It means that block generation time period cannot be reduced arbitrarily. As a result, blockchain throughput does not grow well. Secondly, the blockchain users have to pay transaction cost, even for microtransactions. It becomes burdensome for micropayments and impacts blockchain. Thirdly, as with any growing application with addition of blocks, the chain grows and requires storage, and hence, memory becomes an issue. Current blockchain sizes of Bitcoin and Ethereum are 163.34 GB and 667.10 GB, respectively. Scholars have suggested several methods to solve the above issues. We have covered Blockchain scalability solutions under three categories:

1. Chain partitioning-based scalability
2. DAGs-based scalability

3. Horizontal scalability through sharding
1. Chain partitioning-based scalability: As shown in Fig. 15.4, five main techniques being practiced under this category are on-chain, off-chain, side-chain, child-chain, and interchain solution [15].
   a. *On-Chain blockchain scalability:* On-chain scalability solutions use techniques, whereby the elements of the blockchain like blocks are modified. One such solution is increasing the size of the block to a very large size like "Big Block." Bitcoin Unlimited [16] uses this method to grow its blockchains. It provides higher throughput due to increased transmission limit, and transmission cost in terms of block size is decreased. The increase in size, however, affects the network propagation delay. Propagation delay increases with bigger block sizes. This increases maintenance cost of the blockchain, and centralized mining for grown blockchain slows down. It therefore requires better techniques other than increasing the block size only.

   Example: Bitcoin's BIP-144 uses combination of Merkle Tree and Abstract Syntax Tree (MAST) and is also known as Merkelized Abstract Syntax Tree. Merkle Tree is binary tree in which pairs of blockchain nodes are repeatedly hashed.

   The process is iteratively done until only one hash node is left, and this node is called root node. Nodes use individual transaction IDs, and tree grows from bottom-up. Each leaf

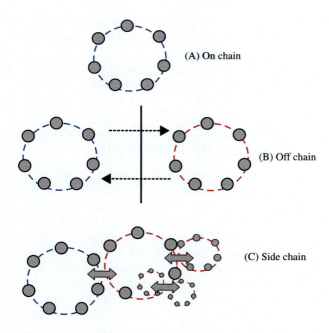

FIGURE 15.4

Basic idea of on-chain, off-chain, and side chain.

Adapted from M. Pärssinen, M. Kotila, R. Cuevas Rumin, A. Phansalkar, J. Manner, Is blockchain ready to revolutionize online advertising? IEEE Access 6 (2018) 54884–54899 [17].

15.4 SCALABILITY SOLUTIONS IN BLOCKCHAIN 381

node therefore contains hash of transaction, and nonleaf node contains hash of previous hashes. Abstract Syntax Tree (AST) is also a tree representation, but it is used to construct tree of source code in programming. So MAST converts or stores bitcoin script into Merkle Tree format. The benefit of using MAST is that unused encumbrance data responsible for increasing the transaction size can be removed, and only the required data can be kept in the block. Fig. 15.5 shows encumbrance script specifying the policy under which Alice can spend her bitcoins whenever she wishes, or allow her brothers Bob and Charlie to use her coins provided they mutually agree and Alice has not spent her bitcoins in some period of time. Fig. 15.5 shows the script and Fig. 15.6 shows its MAST. Depending upon either of the two transactions where either Alice uses her bitcoin or her brothers are allowed to use her bitcoins based on time condition, only one node can be traced in MAST. Fig. 15.7 shows both conditions, where case 1 MAST shows hash of previous blocks and unused block for first condition and case 2 MAST shows hash of previous blocks and unused block for second condition.

The Merkle proof can be visualized depending upon the subscript to be used. MAST grows logarithmically and hence reduces size of blockchain. It, however, compromises privacy because other nodes of the blockchain are not hidden completely.

b. *Off-Chain blockchain scalability:* Off-chain Scalability solutions are also known as layer-2 solutions for blockchain scalability. Off-chain scalability technique provides transaction processing outside the blockchain. In this, intermediate states are not broadcast. Only the last state is broadcast and therefore also called as state-channel solution for blockchains. Instead of a global consensus local consensus is used in the off-chains.

Example: Bitcoin's Lightning Network supports off-chain scalability. Lightning network reduces the transaction fee paid on micropayments. A channel is first opened between Alice and Bob. Both Alice and Bob use the payment channel and do payment transaction for multisig address. Opening this channel is on-chain transaction and fee from the main chain is charged. Once the channel is opened further transactions between Alice and Bob can happen on the off-channel. Off-channel charges zero transaction fee and chains face zero waiting times. After the transactions, the channel is closed, and main channel is notified. The user is charged only for the last notified transaction. This reduces cost of the blockchain.

```
OP_IF
                <Alice's Public-Key> OP_CHECKSIG
OP_ELSE
                <timedelay> OP_CHECKSEQUENCEVERIFY
OP_DROP
                <Bob's Public-Key> OP_CHECKSIG
OP_ENDIF
```

FIGURE 15.5

Bitcoin script example.

Adapted from S. Kim, Y. Kwon, S. Cho, A Survey of Scalability Solutions on Blockchain, ICTC, 2018.

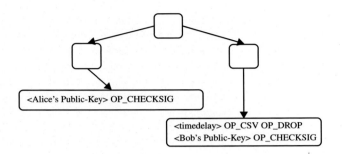

FIGURE 15.6

MAST example of Code 1.

Adapted from S. Kim, Y. Kwon, S. Cho, A Survey of Scalability Solutions on Blockchain, ICTC, 2018.

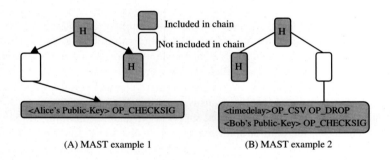

FIGURE 15.7

MAST examples.

Adapted from S. Kim, Y. Kwon, S. Cho, A Survey of Scalability Solutions on Blockchain, ICTC, 2018.

Moreover a third person, Carol can also make use of this method. Say Alice knows Bob and Bob know Carol. Then transactions between Alice and Carol can be done off-channel through Bob.

Raiden Network from Ethereum is also based on the same concept except that it allows other types of transaction like in IoT apps, Smart apps, advertisements, and so on, whereas Bitcoin only provides cryptocurrency.

c. *Side-Chain blockchain scalability*: A pegged sidechain is a sidechain in which the users can transfer assets from parent chain to the side chain and vice versa [19]. Two-way peg means a mechanism to ensure transfer of bitcoins between sidechains at some prefixed rate or current exchange rate. Two-way pegged sidechains use simplified payment verification (SPV) proof for validation. A SPV proof is a Dynamic Membership Multi-Party Signature (DMMS) that ensures that an event was performed on Bitcoin and proof was verified using POW consensus. Generally, the main components of an SPV proof are list of blockheaders and cryptographic proof. Block headers store POW, and cryptographic proof stores hash to

15.4 SCALABILITY SOLUTIONS IN BLOCKCHAIN   383

ensure that some output was generated in one of the blocks of the parent blockchain. This is required by the validators to know that work was done to carry out the asset transfer.

Working of two-way pegged chains is simple. In order to transfer bitcoins from parent chain to the sidechain, a special output is formed where coins are sent. This special output can be unlocked using SPV proof of the parent chain only. In a symmetric two-way pegged chain, both the parent and the side chain use SPV proof for unlocking the coin. Synchronization between the chains is done for proper communication.

Before the coins can be taken by the sidechain from the parent chain, there is confirmation period of one or two days. The coin remains locked in the parent chain during the confirmation period. This period ensures that denial of service (DoS) attacks are avoided. This would be recipient of the coin in the side chain creates a transaction on the sidechain referencing the parents output. An SPV proof created on the recipient's sidechain is provided as a proof that work was done in creating it. After this, the user waits for the duration of a *contest* period. This period is the duration before which the coin cannot be spent on the sidechain. It ensures that blockchain does not have problem of double spending. After the contest period, the coin is transferred to the side chain. The length of this period is often one to two days. But these days atomic swaps are being used which are faster. To transfer the coins from sidechain to the parent chain, same process is done with a difference that now SPV proof is first created by the sidechain and coin is sent to an output point. Fig. 15.8 shows main steps followed in two-way pegged chains.

Another two-way pegged type is asymmetric two-way pegged chains. Its functioning is similar to symmetric two-way pegged sidechains but instead of SPV proofs used by parent and sidechains, validators are required for transfer of coins from parent chain to sidechain. It is assumed that all validators in the sidechain are aware of the parent chain's state. But to transfer coins from sidechain to parent chain, SPV proof is required. It is because parent chain cannot be made aware of all the validators of the sidechains.

Side chain is a blockchain that runs alongside parent blockchain. It allows transactions of different blockchains. For example, if a user already using Bitcoin wants to use it on Ethereum's smart contract, then side chains are helpful. The user interested in doing the transaction firstly transfers the Bitcoin amount to a special account and freezes it. Then the equivalent of the Bitcoin value in Ethereum is created, and trade is done through it. After all the transactions on Ethereum blockchain are finished, the remaining amount after the freeze is released and provided to Bitcoin chain. This way interoperability between different blockchains has been made feasible.

Care needs to be taken where cryptocurrency prices between different blockchains vary heavily. Rootstock and Blockstream provide side-chain scalability to their users. Rootstock is advanced version of Turing-complete cryptocurrency named QixCoin. It is connected to Bitcoin blockchain through two-way peg. Recently, Lumino Transaction Compression Protocol (LTCP) has been proposed for Rootstock. LTCP has been used for "Delta Compression," a way of reducing amount of data. It is based on the concept that initially when a user gets a transaction it fills its details at payment channel. When it runs out, the details are refilled. LTCP looks at the transaction it was previously linked to and prunes signature data with old transactions. It only keeps the first and last transaction. This way middle signatures and data can be removed. Using the old details reduces transaction time

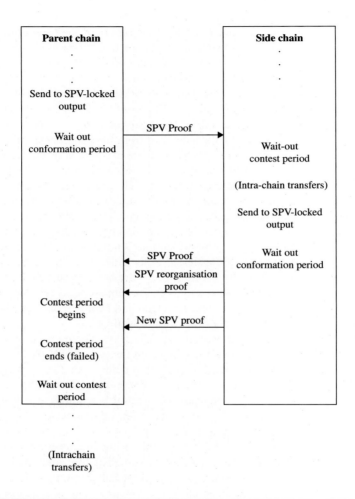

FIGURE 15.8

Example of two-way pegged protocol.

Adapted from A. Back, et al., Enabling blockchain innovations with pegged sidechains. Available from: <http://www.opensciencereview.com/papers/123/enablingblockchaininnovations-with-pegged-sidechains>, 2014.

and 2000 transactions per second are possible in contrast to —two to three transactions per second in normal Bitcoin.

    d. *Child-chain blockchain scalability:* In child-chain blockchain network of child blockchains run through common root blockchain known as parent blockchain. These child chains are connected to the parent node in tree-like structure. Individual child chains have their independent blockchains with consensus mechanisms, whereby the transactions are done in the child chain, and corresponding outcomes are recorded in parent blockchain. These child

15.4 SCALABILITY SOLUTIONS IN BLOCKCHAIN

chains can create more child chains for themselves. Security to child chains is provided by parent chain. Fig. 15.9 shows different child chains connected to parent blockchain.

Example: Ethereum plasma is example of child-chain—based blockchain scalability [20,21]. Fig. 15.10 shows example of plasma child chain. Transactions are processed by the child-chain nodes. All the transactions in the child chain—generated block are given single hash for root node in the parent chain. When a new child chain is created, it joins the parent chain by giving token deposit to the parent chain. Token is given by parent chain to be used by child chain. In practice, all the transactions are done in the child chains only, but if some fraud inclination is reported, then parent chain is given responsibility to resolve. The withdrawal by fraud node is slowed down, and its owner node is allowed to connect to another plasma chain. These chains are successful because any number of child chains can be created with better services like Ethereum, which is considered slow can have Paypal services and can provide better services to its users. The challenges, however, are encountered when fraud has to be detected. Recursive traceback by the root chain becomes time consuming at that stage.

Ardor Platform is also providing child chains for business purposes. Ignis, Aeur, Bitswift, and Max Property Group are the child chains provided by Arbor Platform for different business needs.

e. *Interchain blockchain scalability:* Interchain blockchain has its inspiration in the Internet—the network of networks [22]. Before internet, there were local area networks with their set of protocols. But with success of World Wide Web, it became eminent to make these networks join together through some standard protocols, and this led to internet. Interchain blockchain is an effort for the same, that is, to be able to have different blockchains communicate among themselves. This requires common protocol standards and policies.

Fig. 15.11 shows architecture of Interchain. It requires subchains, interchains, validating nodes, gateway nodes, and so on. Subchains would join the interchain through gateway nodes [23]. Gateway nodes register requests of the subchain to the interchain node. SPV method could be used to testify subchain to the interchain. Subchains can maintain individual sending contracts and exchange contracts. Cosmos blockchain is providing interchain solutions using Tendermint Core Protocol and PBFT with scaling through POS consensus.

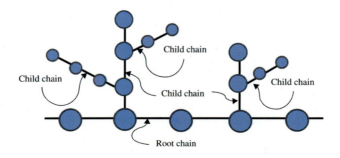

FIGURE 15.9

Child chain.

386 CHAPTER 15 SCALABILITY IN BLOCKCHAIN

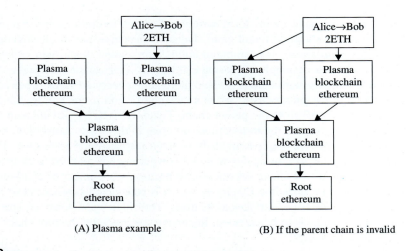

FIGURE 15.10

Plasma chain example.

Adapted from S. Kim, Y. Kwon, S. Cho, A Survey of Scalability Solutions on Blockchain, ICTC, 2018.

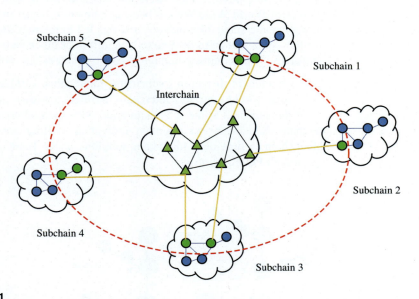

FIGURE 15.11

Interchain architecture.

Adapted from D. Ding, et al., InterChain: a framework to support blockchain interoperability.

15.4 SCALABILITY SOLUTIONS IN BLOCKCHAIN   387

Example: Concept of atomic swap can be used for interchain blockchain. For two users Alice and Bob with Alice using Bitcoin and Bob using Ethereum can be allowed to do transaction without involvement of third party for verification. Let's say Bob generates 1 Ethereum Classic (ETC) transaction for Alice using Alice's Signature and number n known to Bob and Alice only. Bob creates public key for Alice and sends to Alice along with the message. Similarly, Alice can send Bob her BTC amount by using shared public key. Alice uses shorter time than Bob. Since Alice knows number n and therefore can do the transaction.

The advantage is that any two different types of blockchains can interact with each other atomically, but this requires PoW and dependence on time period when exchange can be done and its expiry.

1. DAGs-based scalability: A graph in blockchain has vertices to represent contracts or accounts and edges to represent interactions or transactions [18,24]. By definition "A good partitioning is defined as one that minimizes the edges connecting two partitions (edgecut) while maintaining each partition balanced." A user can have accounts or contracts and can do transactions. Edges originating in accounts are single transactions; edges originating in contracts can perform different calls in the same transaction, both to accounts and other contracts. The weight in each edge denotes the number of times the interaction happened; when no weight is specified, the interaction happened once. Ethereum blockchain partitioning graphs [25] can be constructed as Hashing, Kernighan-Lin, METIS, R-METIS, TR-METIS, as explained next:

   a. *Hashing:* The simplest way to divide the DL via graph is hashing. Depending upon the number of required partitions or shards (k) generate hash for the vertex v as unique identifier modulo partitions number. For uniform hash distribution, optimal static balance between partitions is achievable. Hashing, however, fails when number of edge cuts increase exponentially.

   b. *Kernighan—Lin algorithm*: KL method partitions the graph based on transactions in period of time. The blockchain may exist in one of the many partitions and based on the transactions carried out at that instant in the blockchain new hashing value can be computed. New hash is computed only if partitioning leads to minimization in edge cuts. It is decided which vertices will go to which partition, and central node is informed. Central node checks the balancing of the nodes after partitioning and then sends matrix to all the participating nodes. The benefit of this technique is that very large blockchains can be partitioned dynamically on time-based manner.

   c. *METIS:* METIS is algorithm package used for partitioning very large irregular graphs and can be used in blockchain. METIS takes the current graph as an input and based on the number of partitions required provides the vertices as output. It is multilevel partitioning algorithm and tries to minimize edge cuts while balancing the partitions. If a new contract is to be added, then METIS check the partitions and finds out the partition which maintains the maximum balance. If new addition leads to unbalance state, then METIS partitions the graph further and maintains balance. It is recursive algorithm and cost of MENTIS in blockchain with thousands of transactions is heavy.

   d. *R-METIS:* METIS is costlier for very large blockchains and therefore a modified form R-METIS can be used. It takes a reduced graph as an input. This graph consists of user contracts, accounts, and transactions. R-METIS works in fixed window span and maintains

388 CHAPTER 15 SCALABILITY IN BLOCKCHAIN

partitioning as per the window size. So, for example, if window size is two weeks, then it will work on only the partitions with transactions in two-week period only. This way whole of the blockchain will not need to be repartitioned every time.

e. TR-METIS: TR-METIS does not used fixed interval of time for partitioning. Rather it considers a set threshold value for partitioning the graph. A new partition is computed only if threshold value is equal or higher than the number of edges generated in partitions and repartitioning is done to reduce the edge cuts.

Some of the famous DAG based Blockchains have been explained here:

(i). IOTA: IOTA has been designed for the IoT industry supporting permission-less DL [26]. IOTA is based on DL, Tangle, and promises to solve the scalability problems of traditional blockchain. Tangle uses DAG that acts as core of IOTA. Tangle does not use the concept of transaction blocks stored in chain format. As shown in Fig. 15.12, it uses stream of transactions bundled or tangled together. Tangle uses ledger for storing the user transactions through tangle protocol. Transactions in tangle are connected as edges of

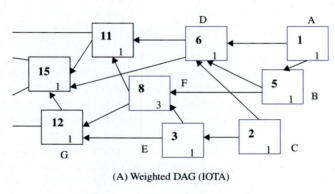

(A) Weighted DAG (IOTA)

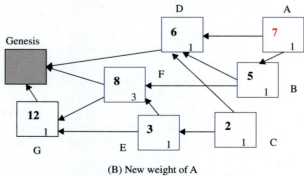

(B) New weight of A

FIGURE 15.12

Weighted DAG (IOTA).

Adapted from S. Popov, The tangle, April 30, 2018, White paper. Available from: <https://assets.ctfassets.net/r1dr6vzfxhev/2t4uxvslqk0EUau6g2sw0g/45eae33637ca92f85dd9f4a3a218e1ec/iota1_4_3.pdf> [27].

graph. These edges are used for validation of transactions in the network. Tangle is based on a very simple rule of validation. A new transaction requires the validation of two previous transactions in the network. This way Tangle is fast in validating new transactions. Unlike Bitcoin miners are not required in IOTA. Every user that can do transactions in the Tangle network can act as miner. For example, if Alice sends five IOTAs to Bob on her mobile phone, then POW calculations are carried out on her mobile phone only. IOTA has simplified hashcash algorithm for POW and therefore requires less computing power. IOTA would enable any computing machine transact with each other.

(ii). *Nano block lattice:* Nano creates blockchain lattice where users maintain account-based blockchains [28]. The nano network does not provide shared DL, rather it is group of independent nonshared blockchains. Signatures are used by the account owners to update the blockchains. At the beginning, genesis block is created by account user providing account amount. The initial genesis account balance can never be exceeded by sum of all of the chain accounts. Nano makes use of send and receive transactions to do transactions. Send transaction deducts the sending balance from the sender's account and encodes it in the recent block of the chain. Validation of correctness is not done in Nano. The receive transaction adds the received amount to the receiver's account and encodes the balance in the receivers recent block. Funds are deducted from the account chain when send transaction is sent and remain pending till the receiving account receives the transaction, and balance is added to the receiver's chain. Nano employs POW. Fig. 15.13 shows send

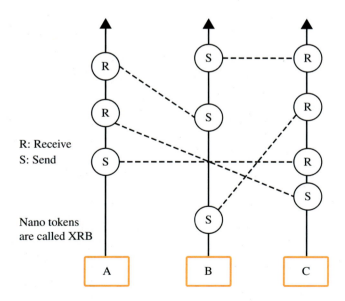

FIGURE 15.13

Nano block lattice.

Adapted from C. LeMahieu, Nano: A Feeless Distributed Cryptocurrency Network, White Paper. Available from: <nano.org/en/whitepaper>.

and receive transactions in nano chains. Nano is based on DAG and maintains consensus using DPoS using balance-weighted vote to handle conflicting transactions.

ByteBall: As shown in Figs. 15.14—15.16, Byteball is free from blocks and fixed block sizes [29]. It uses DAG and byte is storage size and hence, the name ByteBall. Byteball currency is Bytes. On ByteBall DAG, once a new transaction is added, it becomes visible to its peers. Child transactions can be added on the newly created transaction by peers. These child transactions further form their own child transactions. This way a ball of transaction is formed. Double spends in ByteBall are handled by special nodes called "witnesses."

These witnesses maintain partial order. Nonconflicting transactions can be done without witnesses through the DAG chain called "Main Chain." For every new transaction, Main Chain gets updated. If there is double spend, then the transaction appearing in Main Chain is retained and other is returned void. ByteBall finds its use in voting, insurance products, betting, payments via text messaging, etc. For each 16 BTC, you acquire 0.1 GB (1 GB = 1 billion bytes). For each 1 GB, you acquire extra 0.1 GB.

(iii). *Orumesh:* Orumesh is *DAG*-based and free from blocks [30]. It uses "Tips" as unapproved transactions. These transactions are linked to each other and every "tip" has hash value of previous tip. The users generate new transaction and add its hash in the tip. When new tips are inserted, then previous tips having completed specified level in DAG insert their weight to the new tip. The new transaction is added when the user solves POW cryptographic puzzle and selects and authorizes two unauthorized transactions. Contradictory unauthorized transactions are not allowed. OruMesh uses attention credit model whereby nodes calculate a percentage of latest transactions performed by neighbors. Participating nodes earn "Orus." For OruMesh protocol the OruMesh system, attention credit model is used to give transactions an incentive. Every Oru creates a new structure based on this transaction, calling it knot. Each knot consists of information about all its forefather knots (via parents). Flag in the knot is used to check if it is not a void node. A knot is built only if the corresponding transactions are found stable. Fig. 15.17 shows snapshot of knot.

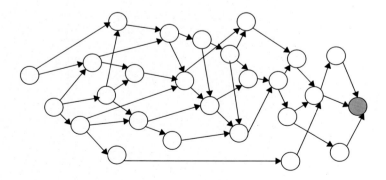

FIGURE 15.14

Single child transaction.

15.4 SCALABILITY SOLUTIONS IN BLOCKCHAIN

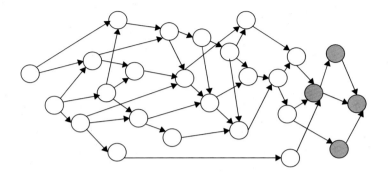

FIGURE 15.15

Single child transaction creates its child transactions.

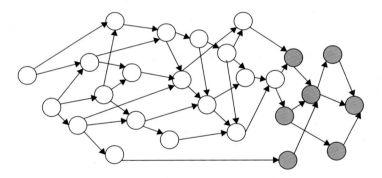

FIGURE 15.16

Child transactions further create child transactions.

```
Knot: {
        Transaction: "hash of transaction",
        Parent_knots: [array of hashes of knot
based on parent knots],

        Is_nonserial:true, //include this field
only if the transaction is non-serial
        Skiplist_knots: [array of earlier knots
used to build skiplist]
}
```

FIGURE 15.17

Knot structure.

Adapted from J. Ahmed, Orumesh. Available from: <https://orumesh.com/whitepaper2.0.pdf>, 2018.

(iv). *HashGraph:* HashGraph is Gossip about Gossip. It is DAG-based consensus algorithm that uses Gossip Protocol. In Gossip Protocol, members in the network inform about updates to other members [31]. Take for example, a node A chooses a node B randomly and informs node B about all the information it has. It further chooses another node say node C randomly and informs about all the information it has. Similarly, node B also chooses another node say node D and informs all the information it has. This way all nodes in the network come to know of latest updates. The passing of information is in itself exponential and therefore fast.

Fig. 15.18 shows HashGraph data structure. Node A creates an event (red) recording occurrence of Node B doing a gossip sync to her and telling her everything it knows. The event contains a hash of two-parent events (blue): the self-parent (dark blue) by the same creator A, and the other parent (light blue) B. It also contains payload of any new transactions that A chooses to create at that moment, and a digital signature by A. The other ancestor events (gray) are not stored in the red event, but they are determined by all the hashes. The other self-ancestors (dark gray) are those reachable by sequence of self-parent links and the others (light gray) are not. All the *events*, represented as vertices, are stored as byte-sequences in memory. These *events* are signed by their creator nodes. As shown in Fig. 15.18, an *event* created by node A (colored red) records the information that node B performed a *gossip sync* with Node A and sent all the information it had to node A. The *event* thus created and signed by node A also has two other hashes, one hash for the previous *event* by node A and a hash for node B's previous *event* before recent *sync* by node B. The *event* by node A can also record other information like transaction data, time, and date of the transaction by node A.

HashGraphs-based communication allows faster convergence of information at all nodes in a network due to very little communication overhead. In a network if instead of

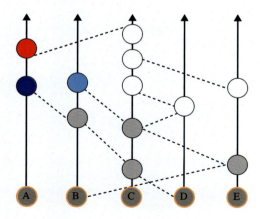

FIGURE 15.18

HashGraph data structure.

Adapted from L. Baird, The swirlds hashgraph consensus algorithm: fair, fast, Byzantine Fault Tolerance, in: Swirlds Tech Report Swirlds-TR-2016-01, 2016.

gossiping, the full transaction information only signed hashgraph information is gossiped and if this transaction contains enough *events* then in a single-hashed transaction information can be passed. Another advantage of HashGraph is that it does not require voting as in Byzantine Fault Tolerant Consensus Algorithms. If, for example, two nodes in a network say node A and node B have same hashgraph then using a deterministic function the total order can be computed by both of them individually at their nodes itself. Since hashgraph for both of them was same therefore the value computed will be same, and they will not have to send vote message. Since these hashgraphs cannot be same for all the nodes at all the times, especially when a new transaction has taken place therefore *virtual voting* is done.

(v). *Lachesis DAG:* Fantom blockchain is based on Lachesis Protocol-based DAG. Fantom proposed to have 300,000 transactions per second using Lachesis-based Consensus. Lachesis has two parts, the *OPERA* chain and the *Main* chain. *OPERA* chain is used for consensus algorithm whereby the network nodes can create, send, and receive messages in asynchronous communication mode [32]. Lachesis allows creation of *event* blocks after the node has informed nodes in the network about recent activity. In OPERA, *events* are the vertices of the graph and *edges* represent communication between the blocks. For consensus, the *event* needs to have been shared between at least 2n/3 nodes where *n* is the number of nodes in the network.

The second chain is the *Main* chain. This chain helps to speed up the convergence of consensus algorithm. A local hash table is maintained to find out nearest root to an *event* block. OPERA introduced concept of *root* which is an *event* block with 2/3*n* linked roots. Using *root* events the "*vital*" blocks in the network can be tracked. In Lachesis, a *Clotho* is a root if it is known to at least 2n/3 nodes. An *Atropos* consists of *Clotho* nodes and is used to decide order of *events* in the network. *Atropos* blocks form *Main* chain. Fig shows example of *Main* chain consisting of *Atropos* blocks. *Main* chain provides fast access to past transaction because members of the *Main* chain maintain individual copy of the *Main* chain and can use it search their order in the consensus. The new *event* block can compute its consensus order from the nearest contributing *Atropos* block. This makes the consensus fast.

(vi). *Avalanche DAG*: Perlin uses Avalanche DAG for its consensus building [33]. It is strictly append-only type of DAG. Avalanche uses Snowball instances with multiple degrees and new nodes are added to previously existing parent nodes. The first DAG link is known as *genesis vertex*. When a new transaction is created by the user, it can have two or more *parents* that are appended to the DAG chain. *Ancestors* are the transactions that the child can reach for a parent through its history. *Progeny* are the children transactions and may include their offspring as well.

Information in the system which is *t*-generations old is considered to have high degree score and is assumed to have been accepted by all the nodes. To handle conflicting transactions like double spends, concept of "chits" is used. A chit is given to a new transaction if a certain set of nodes decided in the past vote to confirm that the said transactions as well as its parent transactions are preferred by these chosen nodes. The numbers of chits for a transaction are then used to decide the order of transaction. If the numbers of chits also come out to be equal then nodes decide based on the transaction seen first.

Fig. 15.19 shows transaction T2 with larger confidence than transaction T3, and therefore it can be said that its children transactions T4, T5, T8, T9 have better chances of getting forward going chits than those of T3.

2. *Horizontal scalability through sharding*: The biggest challenge in blockchain is verification of transaction as fast as possible. Each blockchain node stores entire chain and verification becomes cumbersome [35–37]. Sharding provides a scalable solution for this. Fig. 15.20 shows sharding in traditional distributed databases versus blockchains.

A transaction can be broken down into *shards* which can be further spread in the network. Network nodes work on shards simultaneously and result in reduced processing time. A block in blockchain consists of a header and the data part. Header block records the Merkle root for all the transactions. Via sharding recording of transaction hashes in Merkle tree can be divided into two levels, namely, transactions group and normal blockchain.

Transactions group consists of transaction groups, header and body part. When sharding is done each of the *shards* gets its own set of transactions without redundancy. The transactions group header in each *shard* is further divided into two parts, left-shard partition and right-shard partition. The *shards* are identified through their IDs, and therefore *shard-id* can be stored in left-shard partition. Left-shard partition also stores prestate root which is the state of the *shard root* at the beginning, that is, before transactions are applied. Poststate root is third component of the left-shard partition, which maintains the state of the *shard root* after the transaction has been applied. Fourth component of left-shard partition is receipt root, which is used to maintain receipt after all the transactions in the *shard* have been applied. The right-shard partition is simpler and is related to set of validators only. These validators are randomly selected and are

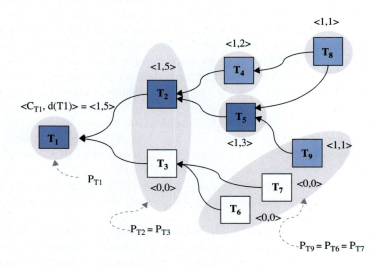

FIGURE 15.19

A sample set of transactions highlighting their different (chit, confidence) values.

Adapted from Protocol Spotlight: Avalanche (Part 2). Available from: <https://hackernoon.com/protocol-spotlight-avalanche-part-2-80cd4a530b6b> [34].

15.4 SCALABILITY SOLUTIONS IN BLOCKCHAIN

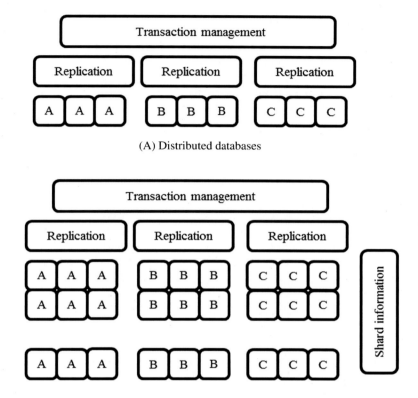

FIGURE 15.20

Sharding protocols in distributed databases versus blockchain.

Adapted from H. Dang, T.T.A. Dinh, et al., Towards scaling blockchain systems via sharding, in: Proceedings of the 2019 International Conference on Management of Data, 2019, pp. 123–140 [38].

used to verify the transactions. The body of the transactions group is used to store the *transaction ids* of all the transactions in the individual *shards*.

Second level in *shards* consists of normal blockchain. This blockchain has two primary roots, namely, state root and transaction root. The state root is used to store the state of the entire blockchain broken down into *shards*. Transaction group root on the other hand is used to maintain record of all the transaction groups in the block.

Sharding provides scalability because many transactions can be done in parallel by using multiple Merkle trees constructed from Merkle root of the transactions group. Distributed memory can be used to store the receipts with read/write privileges by *shards*.

Sharding in blockchain is very different from sharding in distributed databases due to some important reasons. Firstly, the high-performance consensus protocol that is used for consensus

396 CHAPTER 15 SCALABILITY IN BLOCKCHAIN

in distributed databases cannot be applied as it is in blockchain. Blockchain relies on Byzantine Fault Tolerance (BFT) consensus protocols which have proven to be scalability problem. Secondly, distributed databases can have any number of *shards* and any node can belong to any of the *shards*. But in blockchain, *shards* have to be assigned securely so as to protect them from any attack from the malicious user. Thirdly, distributed databases have atomicity of transactions ensured by transaction coordinators, but in blockchain, the coordinators or validators can be malicious, and therefore honest nodes are required.

The number of shards grows linearly with respect to the total computational power of the network; hence, the transaction rate also grows linearly. The limitation of sharding is that it is only optimal if transactions stay in the same shard.

a. Elastico: Elastico is the first sharding-based consensus protocol for public blockchains [10]. Elastico automatically parallelizes available computation power by dividing it into smaller groups called committees. Each committee handles set of transactions. All committees have disjoint set of transactions. The number of committees grows proportionally to the total computation power in the network. Elastico uses byzantine consensus protocol to decide number of members per committee. Consensus committee is used for combining the shard outputs by all committees. Consensus committee also computes PoW puzzle and broadcasts it to all the blockchain. Final committee also generates a set of shared public random bit strings. This randomness is used to ensure the security for the shards such that the attacker could not predict the next set of shards. Elastico works in epochs where for every epoch all committees have to rebuild and reestablish their identities. This leads to latency. The final committee broadcasts only one value. This value is formed by combining signatures from 51% of the members per *shard*. This value consists of two to power s subvalues received from different committees and is verified by one validator. At the end of an epoch, each user has received a bit string of specified length with sufficient randomness. However, the attacker can guess the randomness because the final committee broadcasts the random string in the network and can therefore damage the blockchain. In Elastico, the number of blocks per epoch increase linearly as number of nodes is increased in a network but after 100 nodes epoch time increases heavily because relatively large time is needed to find sufficient PoWs for the committees to fill up.

b. OmniLedger: OmniLedger supports an inter-shard protocol for transactions [39]. Although deciding the appropriate size of the shard to support the transaction is difficult. It provides bias-resistant public randomness-based security and correctness. Crosshard commit protocol ensures atomicity of the transactions when dealing with multiple shards. Omniledger provides parallel processing of intra-shard transactions and enables optimized performance. The throughput in Omniledger scales linearly with increase in size of the blockchain.

Randomness is critical in Omniledger and RandHound protocol is used for the same. RandHound provides unbiased decentralized randomness in Byzantine environment. For this, RandHound splits the shards into smaller shards and creates a protocol that works on commit-then-reveal concept. It is based on having confidence of at least one honest validator for allowing perfect randomness. To select subset of validators, cryptographic sortation is used. Validators are assigned weights and use public/private key pair for authentication to a set of verifiable random function (VRF) has been used to provide sortition.

15.4 SCALABILITY SOLUTIONS IN BLOCKCHAIN   397

Validators process the transactions and provide consistent state to the blockchain. Validators have unique private/public key pair which is used as *validator-id*. The *shards* have uniform distribution of validators. A *shard-policy* file is used to maintain configuration settings of the parameters. After every epoch new assignment of validators to the shards is done. Each shard gets set of transactions which are processed in every round. To participate in an epoch the validators have to register in the previous epoch.

Concept of *Atomic Commit (Atomix)* protocol based on Unspent Transaction Output (UTXO) state model has been used for *atomically* processing transactions across shards. Atomic Commit ensures that either the transaction is committed or it is aborted when multiple shards are involved. In the beginning clients create cross-shard transactions and gossips the cross transactions to all input-shards. Each input-shard validates the transactions within its set. If found valid the leader marks it spent, logs the transaction in shards ledger and gossips proof-of-acceptance (POA).

If rejected proof-of-rejection is created. After locking the client has proofs and can now either commit or abort the transaction. If all input shards issued POA then transaction is committed. Each output shard then validates the transaction and includes it in the next block of its ledger in order to update the state and enable the expenditure of the new funds. But if POR is issued even by one input-shard then transaction is aborted. This transaction is unlocked by gossiping unlock-to-abort transaction. Upon receiving this request the input-shards mark the original UTXO as spendable again. Fig. 15.21 shows an example of the same.

c. RapidChain: RapidChain is the first sharding-based public blockchain protocol [40]. It provides full-sharding without the use of any trust setup. It uses intra-committee consensus algorithm to achieve high throughput. It allows pipelining for reducing transaction processing time and uses gossip protocol for cross-shard transactions. RapidChain can process 7300 transactions per second within latency of 8.7 seconds in a blockchain network of 4000 nodes.

The protocol begins with Bootstrap whereby a set of clients run committee election protocol and form root group. The root group helps in generating reference committee. Reference committee generates k different committees of long size. After Bootstrap phase which is run only once there is consensus phase. Multiple epochs are run in consensus phase followed by Reconfiguration phase.

Users submit their transactions in the consensus phase to a subset of nodes. An intra-consensus protocol is run to check the transactions which are further added to the ledger. Reconfiguration is done to provide identities to new nodes and allow them to join already existing committees. Cuckoo rule is used to maintain committees when new nodes join instead of regenerating the committees from scratch. This reduces network latency for transaction processing.

Each committee picks one leader based on epoch randomness. The leader collects all the transactions and forms a block. It creates block header having iteration id and stores root of the Merkle tree. It uses IDA-gossip protocol to create the header and store root information. The block is saved using public key. Leader gossips the header messages. Other nodes also gossip the received headers. If more than one header is listened by honest nodes then it is

398 CHAPTER 15 SCALABILITY IN BLOCKCHAIN

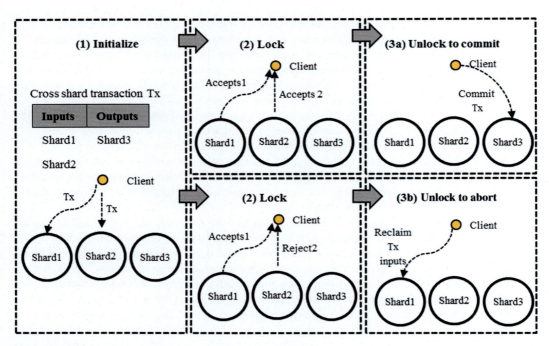

FIGURE 15.21

Atomix protocol in OmniLedger.

Adapted from K.K. Eleftherios, et al., OmniLedger: a secure, scale-out, decentralized ledger via sharding, in: IEEE Symposium on Security and Privacy, 2018.

considered corrupt. If the honest node receives 51% echoes of the header for iteration it is accepted. If honest node accepts the header then other nodes accept the header as well.

Transactions have unique id, list of inputs and a list of outputs. All inputs to a transaction are unspent UTXOs from previous transaction. Received transactions are verified by nodes and valid transactions are added to the next block. Transactions in RapidChain are partitioned based on their IDs by the committees which maintain the database. Cross-shared transactions are done in same way as in OmniLedger.

d. Harmony: Harmony uses PoS-based full sharding. Harmony provides linear scalability and uses beacon chains and multiple-shard chains. Beacon chains ensure randomness and maintain identity register. Shard chains are used to store blockchains. Transactions are processed concurrently in the shard chains [41].

Harmony uses unique algorithm for computing randomness which is a mix of VRF and Verifiable Delay Function (VDF). Harmony uses vote shares (staking-based sharding) for deciding number of shards.

The nodes interested in becoming validators have to show certain amount of tokens as a stake. Voting shares are then assigned to the stakers depending upon the stake amount. Each voting share counts as one vote. The amount of tokens required is dynamically adjusted.

Initially, new validators are assigned to shards randomly. The consensus in a shard is given by validators possessing 2f + 1 vote shares. Harmony's adaptive thresholded PoS guarantees that malicious stakers fail to claim the transactions. At the start of the epoch, random permutation on all the voting shares is done, and the list of voting shares is divided into buckets.

Buckets are equal to the number of shards. The voting shares in the ith bucket are allocated to i^{th} shard. The shard leader is selected as validator if it possesses first voting share in the shard. Adaptive threshold PoS is used to set price of share. Fig. 15.22 shows example of stake-based sharding.

e. CHECO: In CHECO, each node has its own genesis block and a hash chain. Hash chain stores the transactions involved at a particular period of time. All the transactions of the nodes are not stored in hash chains. There are transaction blocks used to store transactions and at a time only one transaction can be stored in a block. For a transaction to happen between sender node and a receiver node, there can only be two blocks and their hash chains. CHECO uses block of checkpoints (CP). CP is used to represent state of hash chain. Collection of all the CP blocks thus constructed then ensures state representation of the network. Fig. 15.23 shows the visual representation of the same.

CHECO has consensus protocol, transaction protocol and validation protocol to interact with the hash chain. The consensus protocol based on HoneyBadger BFT runs iteratively, starting a new execution immediately after the previous one is finished. After the execution ends, the blockchain nodes create new CP block to reach consensus on new values proposed by the

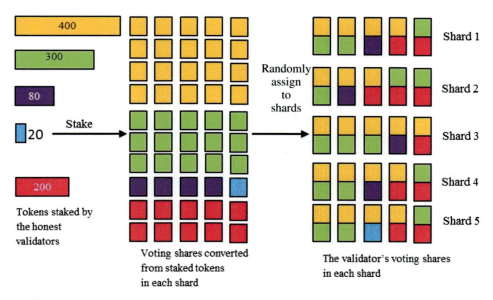

FIGURE 15.22

Harmony sharding.

Adapted from Harmony Team, Harmony, Technical White Paper. Available from: <https://harmony.one/whitepaper>.

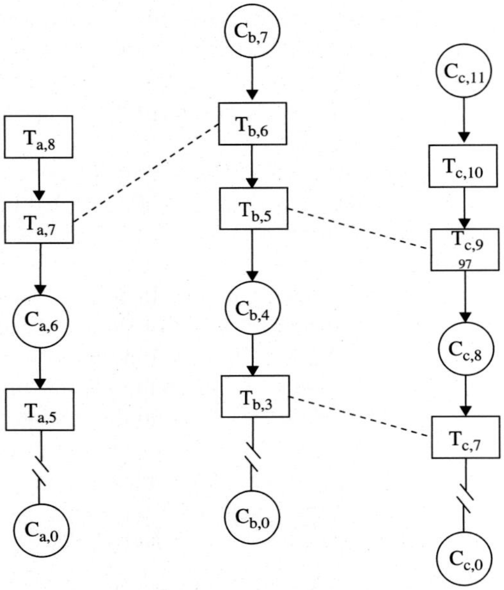

400 CHAPTER 15 SCALABILITY IN BLOCKCHAIN

$T_{u,i}$ represents a TX block on u's chain with a sequence number i, $c_{v,j}$ represents a CP block on v's chain with a sequence number j. The blocks at the ends of the dotted lines are pairs of each other. Blocks of sequence number 0 (e.g., $C_{c,0}$) are genesis blocks

FIGURE 15.23

CHECO example.

Adapted from K. Cong, Z. Ren, J. Pouwelse, A Blockchain Consensus Protocol With Horizontal Scalability.

nodes. The transaction protocol is used by nodes to exchange messages. New transaction blocks are creates by nodes in their individual nodes, and with transaction protocol, the request and response messages are shared. For validation of the transaction, the sender node asks for transaction agreement from the receiver. Receiver node responds and the receiver node checks if the sent CP blocks in consensus are same. Transaction is valid if conditions are met. This is also called horizontal scalability because transaction and validation protocols only make point-to-point communication.

3. Heterogeneous multichain scalability

   a. *Cosmos:* Cosmos is a network of independent heterogeneous blockchains, called zones. Tendermint Core provides stake-of-proof PBFT-based consensus [42]. Cosmos provides interoperability between zones with different policies and hence allows permmisionless blockchains. Success of Cosmos is also devoted to facilitating secure mobile−client transactions. Using Tendermint's language independent Application Blockchain Interface (ABCI) and freedom from strict-fork smartphones can provide Cosmos blockchain transaction through mobile phones as well. This use of low-power and limited-memory independence of Cosmos has great use for Internet of Things applications.

The primary zone in Cosmos is known as the Cosmos Hub. It provides multiple facilities and easies upgradation of blockchains to newer versions. Cosmos uses "atomic crosschain" (AXC) transactions and inter-blockchain communication (IBC) is done based on virtual User Datagram Protocol (UDP) or TCP-based communication protocols. Tokens are used by transactions for transferring currency securely from one chain to another. The transfers happen through Cosmos Hub. IBC ensures transfer of these tokens from one zone to another and are known as "coin packets." Since Hub is the central authority and therefore is responsible for maintain global state of the blockchain amount. IBC coin packet transactions are therefore committed by all three, that is, sending-client, hub, and the receiving client. Transactions are committed with the help of validators. Staking-tokens can be used by nonvalidators to earn and use rewards. Fig. 15.24 shows basic layout of different zones connected to Cosmos Hub and communicating through IBC.

The Hub keeps track of states of all zones and zones post status of their recently committed blocks. Similarly zones also maintain state of the Hub. However, these zones themselves do not maintain state of each of the blockchains. After commit, communicating packets provide proof of sent and done transactions through Merkle-tree proofs. Cosmos allows zones to further form zones like in acyclic graph form. A privileged zone called "bridge-zone" is used for transactions between blockchains having different cryptocurrencies, like, Ethereum and Bitcoin.

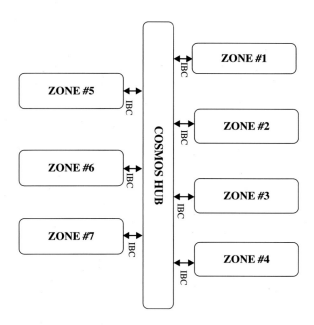

FIGURE 15.24

Cosmos blockchain.

Adapted from J. Kwon, E. Buchman, Cosmos: A Network of Distributed Ledgers.

Consider an example in Cosmos network with three heterogeneous blockchains: ZoneA, ZoneB, and the Hub. A user in ZoneA wishing to do transaction with user in ZoneB would create packet with destination ZoneB but passing through the Hub. The packet is generated in the users, blockchain ZoneA and its proof is posted on the blockchain ZoneB. Receiving blockchain ZoneB keeps senders block header. The IBC protocol used for communication provides two communication types, namely, IBCBlockCommitTx and IBCPacketTx. Blockchain provides its latest block-hash to its observers through IBCBlockCommitTx. IBCPacketTx transaction is used to proof that the sender had actually published the said packet. In the example shown in Fig. 15.25, IBCBlockCommitTx transaction is used to update through posting the block-hash value of ZoneA on Hub. Same is done for posting block-hash of ZoneB on Hub.

b. *Polkadot blockchain network:* Polkadot blockchain network consists of heterogeneous blockchains just like Cosmos [43,44]. It uses Tendermint's HoneyBadger BFT consensus algorithm. It provides stake through Proof-of-Authority. First version of Polkadot supported around 100 parachains. Parachains are like individual heterogeneous blockchains. These parachains are built upon main chain called Relay-chain. There are four roles defined in a Polkadot network, namely, collator, fisherman, nominator, and validator. Validator ensures adding of new blocks in the chain. Validator allows chains to ratify new block on parachain in relay-chain client communication form. The validators take help from third party to provide full-synchronization among blockchain databases. This third party is called collator. Nominator consists of parties holding stake-claim and ensures security through bonds in blockchain. Nominators risk the inflow and outflow of capital in the blockchain and therefore have vote in nominating the validators. Collators gather and execute the transactions and use zero-proof of knowledge to store them on special blocks called unsealed blocks. These blocks are used by validators to propose new blockchain in the network. Collators ensure valid parachain blocks.

Fishermen are like bounty-hunters who look out for rewards by finding out illegal blockchains. Collators collect and propagate client transactions. Block candidates are also propagated by collators to fishermen as well as validators. Relay chain is used to transfer amount from an account in

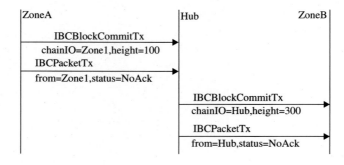

FIGURE 15.25

Example of transactions between zones and hub.

Adapted from J. Kwon, E. Buchman, Cosmos: A Network of Distributed Ledgers.

a parachain to another parachain where its transaction is interpreted. Relay chain therefore handles parachain queues and I/O as well as outbound transactions, second-order relay chains, interchain-transactions and parachain bridge between two different cryptocurrencies. Polkdot is being sold as the next generation of blockchain for World Wide Web.

15.5 FUTURE DIRECTIONS

Blockchain technology has great prospectus of growth in industry. The contributions from the academia hold big scope for improvement of blockchain networks. Decentralization, security, and scalability are three main pillars of blockchain development. Possible future directions with respect to scalability in blockchain can be worked upon in four areas, namely, On-Chain scalability solutions, Off-Chain scalability solutions, DLs-based scalability solutions, and consensus algorithm-based scalability solutions.

1. *On-Chain scalability solutions:* On-Chain would require changes to be made in the codebase of the chain [45]. This involves changing the features of the blockchain like block size. The block size can be increased from 1 MB to variable larger fixed size blocks like 10 MB. This, however, requires hard fork and upgrading all the nodes of the long chain is difficult. Another solution could be using segwit. The signature data are known to consume 70% of the block chain space. Removing signature data would enable more transactions to be carried out. Litecoin is using segwit. However, security would be a compromise in this case. Third possible solution is sharding. Each of the *shards* can be processed independently and hence, partitioning of the chains into *shards* can enhance scalability. Ethereum will be using PoS-based *sharding* in the coming years.

2. *Off-Chain scalability solutions:* Off-Chain solutions involve building chains in the second layer and transactions can be shifted to side chains. This can cause reduction in network congestion. Lightning network, Ethereum's Plasma, and Raiden Network are models worked on in this area. Using Lightning network, TPS of 1,000,000 + is achievable. Plasma claims to have infinite TPS based on this model.

3. *DLs-based scalability solutions:* Serial blockchain transactions are becoming time consuming and are no longer a viable solution. DAGs with asynchronous transaction processing can be used to make blockchain scalable. IOTA with approximately 140 + TPS, Byteball with 20 TPS, HashGraph with 100,000 + TPS, Nano Lattice with approximately 700 + TPS are some of the blockchains supporting DAGs.

4. *Consensus algorithms-based scalability solutions*: New Consensus algorithms like PoS, Delegated PoS (DPoS), PBFT, Proof-of-Authority (PoA) can be used. These algorithms provide efficient scalability to blockchain networks. Blockchains like Seemit with 24 TPS, EOS with 1000 TPS, Bitshares with 17.5 TPS, Lisk with 25 TPS are known to have been using DPoS. Hyperledger with 3500 TPS, Zilliqa with 2000 + TPS are known to be using PBFT. Stellar with 1500 TPS, Ripple with 1500 + TPS are based on Federated Byzantine Agreement (FBA).

404 CHAPTER 15 SCALABILITY IN BLOCKCHAIN

In addition to above lightweight cryptography, solutions can be looked into in order to reduce the processing time for signature and space required to store the signature. IoT devices with constrained resources can immensely benefit from lightweight cryptography.

15.6 CONCLUSION

Blockchain is based on decentralization and P2P networking. Use of Bitcoin for online transactions without involving third-party security has made Blockchain popular. However, there are various fields where blockchain can be used. The biggest challenge being faced by blockchain providers is its scalability. In this chapter, we have given overview of the challenges in scalability of blockchain and solutions for solving the problem of scalability. We have also discussed future directions for achieving the same.

REFERENCES

[1] S. Nakamoto, Bitcoin: a peer-to-peer electronic cash system. Available from: <bitcoin.org>, 2008.

[2] M. Bez, G. Fornari, T. Vardanega, The scalability challenge of ethereum: an initial quantitative analysis, in: IEEE International Conference on Service-Oriented System Engineering (SOSE), 2019.

[3] C. Worley, A. Skjellum, Blockchain tradeoffs and challenges for current and emerging applications: generalization, fragmentation, sidechains, and scalability, in: IEEE Confs on Internet of Things, Green Computing and Communications, Cyber, Physical and Social Computing, Smart Data, Blockchain, Computer and Information Technology, Congress on Cybermatics, 2018.

[4] K. Luo, W. Yu, et al., A multiple blockchains architecture on inter-blockchain communication, in: IEEE International Conference on Software Quality, Reliability and Security Companion, 2018.

[5] D. Burkhardt, M. Werling, H. Lasi, Distributed ledger: definition & demarcation, in: IEEE International Conference on Engineering, Technology and Innovation (ICE/ITMC), 2018.

[6] J. Poon, T. Dryja, The bitcoin lightning network. Available from: <https://lightning.network/lightning-network-paper.pdf>, 2016.

[7] Z. Zheng, et al., An overview of blockchain technology: architecture, consensus, and future trends, in: IEEE 6th International Congress on Big Data, 2017.

[8] I. Acharjamayum, R. Patgiri, D. Devi, Blockchain: a tale of peer to peer security. in: IEEE Symposium Series on Computational Intelligence (SSCI), Bangalore, India, 2018, pp. 609−617.

[9] K. Zhang, H. Jacobsen, "Towards Dependable, Scalable, and Pervasive Distributed Ledgers with Blockchains," 2018 IEEE 38th International Conference on Distributed Computing Systems (ICDCS), Vienna, 2018, pp. 1337−1346.

[10] H. Dang, et al., Towards scaling blockchain systems via sharding. Available from: <comp.nus.edu.sg>.

[11] Z. Zheng, et al., Blockchain challenges and opportunities: a survey, Int. J. Web and Grid Serv. 14 (4) (2018).

[12] A. Chauhan, et al., Blockchain and scalability, in: IEEE International Conference on Software Quality, Reliability and Security Companion, 2018.

[13] W. Wang, et al., A survey on consensus mechanisms and mining strategy management in blockchain networks, IEEE Access 7 (2019).

[14] Accessed bitcoin cash (BCH) debunking the FUD. Available from: <https://steemit.com/cryptocurrency/@cryptoshoppe/bitcoin-cash-bch-debunking-the-fud-V2.0>.

REFERENCES 405

[15] H.T. Vo, et al., Internet of blockchains: techniques and challenges ahead, in: IEEE Confs on Internet of Things, Green Computing and Communications, Cyber, Physical and Social Computing, Smart Data, Blockchain, Computer and Information Technology, Congress on Cybermatics, 2018.

[16] CoinMarketCap, Cryptocurrency market capitalizations. Available from: <https://coinmarketcap.com/currencies/bitcoin/>, 2017.

[17] M. Pärssinen, M. Kotila, R. Cuevas Rumin, A. Phansalkar, J. Manner, Is blockchain ready to revolutionize online advertising? IEEE Access 6 (2018) 54884−54899.

[18] S. Kim, Y. Kwon, S. Cho, A Survey of Scalability Solutions on Blockchain, ICTC, 2018.

[19] A. Back, et al., Enabling blockchain innovations with pegged sidechains. Available from: <http://www.opensciencereview.com/papers/123/enablingblockchaininnovations-with-pegged-sidechains>, 2014.

[20] J. Poon, V. Buterin, Plasma: Scalable Autonomous Smart Contracts, White paper, 2017.

[21] J. Poon, V. Buterin, Plasma: scalable autonomous smart contracts. Available from: <https://plasma.io/plasma.pdf>.

[22] D. Ding, et al., InterChain: A Framework to Support Blockchain Interoperability in Conference Proceedings of apnet, 2018.

[23] P.R. Rizun, Subchains: a technique to scale bitcoin and improve the user experience. Available from: <http://ledgerjournal.org/ojs/index.php/ledger/article/view/40>, 2016.

[24] H. Pervez, et al., A comparative analysis of DAG-based blockchain architectures, in: International Conference on Open Source Systems and Technologies (ICOSST), 2018.

[25] E. Fynn, F. Pedone, Challenges and pitfalls of partitioning blockchains, in: 48th Annual IEEE/IFIP International Conference on Dependable Systems and Networks Workshops, 2018.

[26] B. Shabandri, P. Maheshwari, Enhancing IoT security and privacy using distributed ledgers with IOTA and the tangle, in: 6th International Conference on Signal Processing and Integrated Networks (SPIN), 2019.

[27] S. Popov, The tangle, April 30, 2018, White paper. Available from: <https://assets.ctfassets.net/r1dr6vzfxhev/2t4uxvsIqk0EUau6g2sw0g/45eae33637ca92f85dd9f4a3a218e1ec/iota1_4_3.pdf>.

[28] C. LeMahieu, Nano: a feeless distributed cryptocurrency network, White paper. Available from: <nano.org/en/whitepaper>.

[29] A. Churyumov, Byteball: a decentralized system for storage and transfer of value. Available from: <https://byteball.org/Byteball.pdf>.

[30] J. Ahmed, Orumesh. Available from: <https://orumesh.com/whitepaper2.0.pdf>, 2018.

[31] L. Baird, The swirlds hashgraph consensus algorithm: fair, fast, Byzantine Fault Tolerance, in: Swirlds Tech Report Swirlds-TR-2016-01, 2016.

[32] S.-M. Choi, et al., OPERA: Reasoning About Continuous Common Knowledge in Asynchronous Distributed Systems, ArXiv e-prints, October 2018.

[33] Team Rocket, Snowflake to avalanche: a metastable protocol family for cryptocurrencies. Available from: <https://www.dropbox.com/s/t5h2weaws1vo3c7/paper.pdf>, 2018.

[34] Protocol Spotlight: Avalanche (Part 2). Available from: <https://hackernoon.com/protocol-spotlight-avalanche-part-2-80cd4a530b6b>.

[35] K. Cong, Z. Ren, J. Pouwelse, A Blockchain Consensus Protocol With Horizontal Scalability, in IFIP Networking Conference (IFIP Networking) and Workshops, pp. 1−9, 2018.

[36] Z. Ren, et al., A scale-out blockchain for value transfer with spontaneous sharding, in: Crypto Valley Conference on Blockchain Technology, 2018.

[37] H.K. Yoo, et al., A blockchain for domain based static sharding, in: 17th IEEE international conference on trust, security and privacy in computing and communications, 2018.

[38] H. Dang, T.T.A. Dinh, et al., Towards scaling blockchain systems via sharding, in: Proceedings of the 2019 International Conference on Management of Data, 2019, pp. 123−140.

406 CHAPTER 15 SCALABILITY IN BLOCKCHAIN

[39] K.K. Eleftherios, et al., OmniLedger: a secure, scale-out, decentralized ledger via sharding, in: IEEE Symposium on Security and Privacy, 2018.

[40] M. Zamani, et al., RapidChain: scaling blockchain via full sharding. Available from: <https://eprint.iacr.org/2018/460.pdf>, 2018.

[41] Harmony Team, Harmony, Technical White Paper. Available from: <https://harmony.one/whitepaper>.

[42] J. Kwon, E. Buchman, Cosmos: A Network of Distributed Ledgers, pp. 1−41, published online by Cosmos Network, 2018.

[43] S. Deshpande, Analysis of Polkadot. Available from: <https://medium.com/@saurabh.s.deshpandey>.

[44] G. Wood, POLKADOT: Vision For Heterogeneous Multi-Chain Framework, available at https://github.com/w3f/polkadot-white-paper/raw/master/PolkaDotPaper.pdf, Nov 2016.

[45] J. Poon, T. Dryja, The Bitcoin Lightning Network: Scalable Off-Chain Instant Payments, available at http://lightning.network/lightning-network-paper.pdf, 2015.

FURTHER READING

L. Luu, et al., A secure sharding protocol for open blockchains. Available from: <https://people.cs.georgetown.edu/teaching/cosc841-spring19/papers/new>, 2016.

CHAPTER

A VITAL ROLE OF BLOCKCHAIN TECHNOLOGY TOWARD INTERNET OF VEHICLES

16

Vikram Puri[1], Raghvendra Kumar[2], Chung Van Le[1], Rohit Sharma[3] and Ishaani Priyadarshini[4]

[1]*Duy Tan University, Danang, Vietnam* [2]*Department of Computer Science and Engineering, GIET University, Gunupur, India* [3]*Department of Electronics & Communication Engineering, SRM Institute of Science and Technology, Ghaziabad, India* [4]*University of Delaware, Newark, DE, United States*

16.1 INTRODUCTION

The Internet of Things (IoT) is a rapidly growing concept that is being transpired with the amalgamate of innumerate smart objects and things [1]. Modernistic technologies have used through the faraway devices and interconnected network, which become a vital part of the Internet. IoT is a system that integrates the computing devices, machinery, things or objects, living creatures that have the capability to exchange data or information over the network with their associated devices without any human interference. In addition, IoT systems are capable enough to collect the data and process it either locally or send to the cloud server through the gateway and vice versa [2]. Advancement in IoT technology maintains proper trust regarding framework, architecture, and security. Nowadays, IoT extends its limits and merges with various technologies such as blockchain, artificial intelligence (AI), virtual reality, augmented reality to enhance and unlock its potential in different domains, namely, Internet of Vehicles (IoV), Internet of Medical Things, Internet of Nano Things, Internet of Cloud, Internet of Industrial Things, and mainly Internet of Everything [3]. Fig. 16.1 shows IoT as honey comb and other related technologies as a part of IoT.

Nowadays, blockchain technology rapidly spreads in all domains of IoT technology [4,5] and also enticed attraction toward the security and peer-to-peer technologies. Blockchain [6] is a decentralized and scattered database or digital public ledger that is used to store the transaction executed in the across the connected participant's computers. Every transaction is checked and verified by the concord of lion's share of the participant. In the blockchain, information is unerasable, it means if information is entered in the system, it can never be erased. To explain blockchain, consider an example of stealing a biscuit from the box inside the home is more easier as compared to stealing a biscuit from the box placed in market because hundreds of people are being noticed in the market.

In the past decades, central servers are gestated with digital cash to avoid the double-spending prevention [7]. Double spending is a flaw in the digital payment system in which a single token or coin can be used more than one time [8]. Blockchain technologies are able to overcome the double-spending flaw with the aid of cryptography key technique, whereas every participant allocated a private key that work as password and public key distributed with all other participants [9].

Handbook of Research on Blockchain Technology. DOI: https://doi.org/10.1016/B978-0-12-819816-2.00016-2
© 2020 Elsevier Inc. All rights reserved.

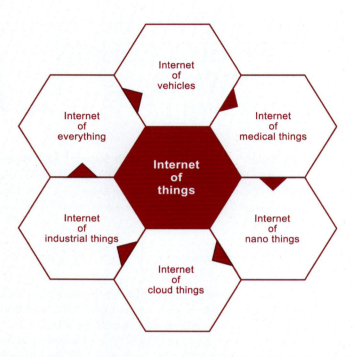

FIGURE 16.1

Different domains of IoT.

Blockchain authorizes to access shared data that are distributed across the connected participant on the basis of trust [10]. Integration of connected cars, IoT, AI, and blockchain not only unlock the potential of many smart applications but also change the futuristic way of research toward the connected vehicles. Nowadays, blockchain is the essential part of the connected cars or "IoV" to manage the vehicle management, store the lifetime data for car, autonomous transaction executions, efficiency of service provider management.

IoV is based on the extension of ad hoc network that can improve the efficiency of traffic [11]. IoV is subdomain of IoT and their characteristics are almost the same but mostly IoT works on static position and IoV works on the mobile. It deployed on the open wireless sensor network that increases complexion during computing and processing the data because of rapidly changing topology networking due to mobility. Simultaneously, it increases the security and privacy issues such as malware attacks, hack confidential information, as well as data tampering.

Blockchain is an appropriate technology for decentralized applications with distributed agreement features especially in the case of composite traffic where mobile nodes or vehicles cannot setup proper connection due to the lack of trust on each other. Under the protection shield of blockchain technology, hackers cannot collapse the shield through the usage of encryption feature. Moreover, this feature helps to maintain the replica account data from more than one service provider. On the other side, energy consumption of IoV network is less as compared to other IoT applications because every vehicle provides energy by itself. Vehicle privacy is another challenge in the IoV network that includes location coordinates as well as identities. Blockchain helps to maintain the security level as well as hides the information from the neighbor vehicles. In this chapter, we discuss blockchain architecture for the IoV network, elucidate security and privacy issues analysis, and also illustrate applications of blockchain-enabled IoV network as well as compared to the tradition IoV network. Section 16.2 discusses the related studies, Section 16.3 proposes the blockchain-enabled IoV architecture, Section 16.4 illustrates the application of blockchain-enabled IoV network. Section 16.5 discusses the analysis of security and privacy issues, and Section 16.6 concludes the proposed study.

16.2 RELATED STUDIES

Many studies have been proposed in the last few years regarding the integration of blockchain and IoV technologies. Sharma [10] proposed a vehicular-based blockchain network for the smart cities. This architecture operated in a distributed way to construct transport management system based on the distributed network. It also illustrated vehicle-connected network on the basis of vehicular data as well as networking. Dorri [12] presented blockchain-based architecture for the users' privacy protection as well as increased the security for the vehicle in the IoV network. They also illustrated blockchain architecture on the remotely software updation as well as other services such as pay vehicle insurance fee. Puthal [13] discussed about the blockchain technology for the perception of security across the network. Moreover, it also explained the role of blockchain technology among different domains such as financial services, IoT, healthcare services, government smart services. In Ref. [14], blockchain-based data coins and energy coin are presented to enable the security in electric vehicles cloud and edge computing. To overcome the challenges in vehicular network, some security solutions are also proposed. Kang [15] proposed data sharing scheme based on the reputation to share high-quality data across the IoV network. In addition, edge computing is combined with vehicular network to form vehicular edge computing and network that help to achieve high efficiency and high level of security in the data sharing between the vehicles. Zhang [16] discussed the edge information system include edge cache, compute, as well as AI for the IoV network. Moreover, design issues, methodologies, as well as hardware-supported platform for IoV are also illustrated. In Ref. [17], the author proposed secured authentication protocol for the IoV and also overcome the limitation of Ying and Nayak's [18] authentication protocol regarding location spoofing attack, identity guessing attack, and replay attack. Moreover, this algorithm also improved the authentication time. Kong [19] proposed scheme based on the data sharing that helps to collect and distribute data through the sensors integrated on the vehicles. This scheme enabled the resistance between vehicle-to-vehicle collision through the sensors and privacy preserving. Kandah [20] presented scheme based on the multitier that enabled the trust building/distribution framework designed to improve the security and privacy of information exchanged within IoV network.

16.3 ARCHITECTURE

Rapid growth of blockchain technology attracts many researchers toward fruitful outcome from collaboration of blockchain technology with other related technologies such as healthcare, IoT,

Industry 4.0. Nowadays, blockchain works as a backbone for the IoV network. Combination of blockchain technology with IoV helps to improve the big data storage, decentralization architecture, picky privacy, as well as quality of service. Many limitations already overcome regarding the security and privacy due to the decentralized nature of blockchain technology. Architectural design of blockchain-enabled IoV network is categorized into different components as follows (see Fig. 16.2): (1) data classification, (2) blockchain nodes, and (3) transmission.

1. *Data classification*: With reference to different blockchain application purposes, storage of data is categorized into different blockchain components. It is not mandatory to setup

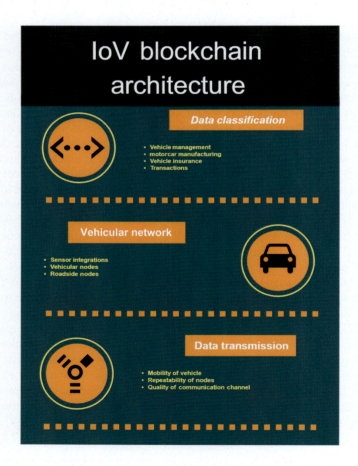

FIGURE 16.2

Blockchain-based IoV architecture.

communication between every component. Different component data in the blockchain are as follows:

a. *Vehicle management component*: In this component, vehicle parameters such as speed, braking, and way of driving are mandatory, but personal data storage is not mandatory in the components such as media scrolling or playing include audio and video both.

b. *Moto-car manufacturing component*: Embed sensors include tyre pressure, temperature, humidity, engine status provide a technical data to the vehicle manufacture through the aid of mutual contract between them. This technical data is used for the better improvement.

c. *Vehicle insurance component*: With the help of blockchain technology, vehicle insurance will automatically choose according to the driver skills, habits, conditions of cars through the internal sensors and related data to the vehicle insurance.

d. *Normal transaction component*: Normal user transaction is done through the tokens for daily to daily needs.

2. *Vehicular network*: In the IoV network, communication is independent of different blockchain components. Due to the sensor integration, vehicles are capable enough to fetch, store, and send data to the nodes. In addition, these vehicles can completely control the information. Vehicles are monitored through the two sides. On the one side, sensors embedded inside the vehicle collect the information concerning about fuel consumption, engine condition, pressure inside tyres, and radiator temperature, and on the other side, external sensors are deployed along the roadside to monitor motion speed of vehicle, and both sides are responsible to make a communication between the neighbor nodes, vehicles, and roadside nodes to expand the network.

3. *Data transmission*: On the roads, vehicle speed varies drastically due to obstacles, traffic congestion that makes challenge for the wireless connectivity between vehicles to vehicles, vehicles to roadside nodes. Blockchain-based IoV network is to consider the mobility of vehicles, repeatability of nodes, quality of communication channel, and areas covered for the data transmission. As an example, in every regular interval of time, vehicles generated particular data that required neighbor nodes or roadside nodes for setup communication network. However, there are a good number of roadside nodes available, but vehicles have to choose the right one to avoid transmission failure. The main reason behind the transmission failure is lack of quality of signal between two nodes.

16.4 APPLICATION OF BLOCKCHAIN WITH INTERNET OF VEHICLES

Blockchain technology is a key to unlock the potential of IoV. In the near future, vehicles will recognize from digital identity on the blockchain with vehicle registration number to vehicle left from manufacturer. In this section, we discuss about the various applications related to the blockchain-enabled IoV network. Fig. 16.2 represents the application of the blockchain-enabled IoV network. Table 16.1 represents the application comparison between the traditional and blockchain-enabled IoV network (Fig. 16.3).

Table 16.1 IoV Application With the Traditional Method and Blockchain-Based

S. No.	Applications	Traditional Method	Blockchain-Enabled Benefits
1.	Wireless-enabled software	• Based on the centralized structure • Lack of privacy • Easy to interface in the history of installed updation as well as easily downloadable • Only original manufacturers are able to verify the communication between them	• The presence of blockchain means the introduction of decentralized structure in the IoV network • Capable to maintain the security regarding the user as well as vehicle • Software updation history as well as any authentication can be publicly verified
2.	Smart contacts	• Based on the centralized, contracts and private information are insecure • Vulnerability of the system is high	• Blockchain-based IoV network used analytical modeling tracking technique that allows system to record outcomes executed at end layer or edge layer of the network • With the use of hashes, authentication security is very high as compared to traditional systems
3.	Vehicle insurance	• Due to lack of privacy, system is insecure • Control over data is very less • Data sharing between the insurance company and user is totally continuous	• Data exchange is totally secured, decentralized, distributed, and privacy protected • User has full control on data transferred between the nodes • Data sharing system is on demand that makes it more secure when compared with the continuous data sharing
4.	Vehicle-sharing scheme	• Due to centralized nature, payments are done through centralized structure • Tracking of vehicle done via nonchangeable identities	• Payment done through the tokens, hashes that make it more secure and private • Vehicle tracking is done through changeable identities

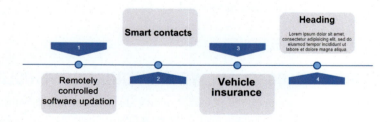

FIGURE 16.3

Application of IoV with blockchain technology.

16.5 SECURITY AND PRIVACY ISSUES

As illustrated in Fig. 16.4, Section 16.5 discussed about the security and privacy issues in the blockchain-enabled IoV network.

16.5.1 SECURITY

Security in the blockchain-based network referred as indemnity of transaction information from the internal or external threats. This indemnity includes detection of threat as well as some action against threats according to their mutual policies. Important indemnities are as follows (Fig. 16.5):

- *Fortification in penetration testing:* In this technique, multilayers are used for indemnity of data instead of single layer of security.
- *Analysis and manage the risk:* In deploying or deployed IoV network, its mandatory to identify, selection for authentication, action against any risk, which can be unshield the network.

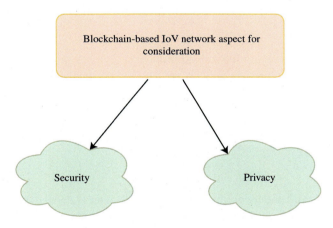

FIGURE 16.4

Aspects of IoV-enabled blockchain technology.

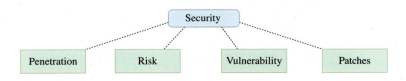

FIGURE 16.5

Indemnity in the security.

- *Vulnerability management*: Checking the vulnerability in the IoV network and manage from identify to patching.
- *Analysis and modify patches*: Identify the error part of the code, algorithm, or application and modify according to the need.

To achieve security in the transaction of information or block information, blockchain technology plays a major role and used different techniques. Bitcoin technology is one of the techniques used for information safety that is combination of the two different keys: one is public and the other one is private. Both the keys are useful for the encryption or decryption of information. True and reliable blockchain system is the longest blockchain network and also one of the most secure systems. In addition, it helps to reduce the security threats due to fork problem as well as long chain eliminate authentication based attacks.

16.5.2 PRIVACY

In terms of blockchain, privacy is referred as transaction done through the network without losing any user-related information. The main goal behind the privacy in the blockchain-enabled IoV network is to attenuate the copying information of other nodes or vehicles in the same network. The privacy policies vary from personal use to organization data. Some of the common characteristics are as follows (Fig. 16.6):

- *Data sorting:* In the IoV network, blockchain provides a flexible structure to collect and store the information or data. The privacy policies vary for the personal and organization data. Moreover, privacy policies are designed for personal user data, but for the organization data, it is more strict.
- *Distribution regarding storage:* Available node present in the network that is capable of storing or replicating the information is referred to as full node. This node leads to data redundancy that supports two key features:
  - Transparency
  - Verification
- *Blockchain domains:* Authentication privacy in the blockchain is dependent on the public or private domain. In the restricted data, only authenticated users can access the data as well as maintain copy of access data in the blockchain network. In the public domain, all users are permitted to modify the data.

FIGURE 16.6

Privacy characteristics.

16.6 CONCLUSION

Amalgamation of IoT and wireless connectivity has led to open new opportunities toward IoV development. This amalgamation encompasses IoV services such as vehicle insurance, vehicle management, and road and life safety. Well-organized communication between two nodes can prevent many incidents and can save life before time. However, security and privacy is a major issue in the IoV regarding the personal information, payment, as well as smart contacts. This chapter discussed about the blockchain-based IoV architecture to overcome security and privacy issues of traditional IoV network. The proposed architecture also focused on blockchain-enabled IoV applications as well as compared with the traditional IoV networks. Moreover, analysis of security and privacy issues is also illustrated in this chapter.

REFERENCES

[1] A. Al-Fuqaha, M. Guizani, M. Mohammadi, M. Aledhari, M. Ayyash, Internet of Things: a survey on enabling technologies, protocols, and applications, IEEE Commun. Surv. Tut. 17 (4) (2015) 2347−2376.

[2] A. Bahga, V. Madisetti, Internet of Things: A Hands-on Approach, Vpt, 2014.

[3] J. Gubbi, R. Buyya, S. Marusic, M. Palaniswami, Internet of Things (IoT): a vision, architectural elements, and future directions, Fut. Gen. Comp. Syst. 29 (7) (2013) 1645−1660.

[4] J. Robinson, BitCon: the naked truth about bitcoin, Amazon Dig. Serv. 26 (2014) 149.

[5] M.A. Ferrag, M. Derdour, M. Mukherjee, A. Derhab, L. Maglaras, H. Janicke, Blockchain technologies for the Internet of Things: research issues and challenges, IEEE Internet Things J. 6 (2) (2018) 2188−2204.

[6] M. Crosby, P. Pattanayak, S. Verma, V. Kalyanaraman, Blockchain technology: beyond bitcoin, Appl. Innov. 2 (6−10) (2016) 71.

[7] D. Chaum, Blind signatures for untraceable payments, Advances in Cryptology, Springer, Boston, MA, 1983, pp. 199−203.

[8] G.O. Karame, E. Androulaki, S. Capkun, Double-spending fast payments in bitcoin, Proceedings of the 2012 ACM Conference on Computer and Communications Security, ACM, 2012, pp. 906−917.

[9] F.X. Olleros, M. Zhegu (Eds.), Research Handbook on Digital Transformations, Edward Elgar Publishing, 2016.

[10] V. Sharma, An energy-efficient transaction model for the blockchain-enabled Internet of Vehicles (IoV), IEEE Commun. Lett. 23 (2) (2018) 246−249.

[11] T. Qiu, X. Liu, K. Li, Q. Hu, A.K. Sangaiah, N. Chen, Community-aware data propagation with small world feature for Internet of Vehicles, IEEE Commun. Mag. 56 (1) (2018) 86−91.

[12] A. Dorri, M. Steger, S.S. Kanhere, R. Jurdak, Blockchain: a distributed solution to automotive security and privacy, IEEE Commun. Mag. 55 (12) (2017) 119−125.

[13] D. Puthal, N. Malik, S.P. Mohanty, E. Kougianos, C. Yang, The blockchain as a decentralized security framework [future directions], IEEE Consum. Electron. Mag. 7 (2) (2018) 18−21.

[14] H. Liu, Y. Zhang, T. Yang, Blockchain-enabled security in electric vehicles cloud and edge computing, IEEE Netw. 32 (3) (2018) 78−83.

[15] J. Kang, R. Yu, X. Huang, M. Wu, S. Maharjan, S. Xie, et al., Blockchain for secure and efficient data sharing in vehicular edge computing and networks, IEEE Internet Things J. 6 (3) (2018) 4660−4670.

[16] J. Zhang, K.B. Letaief, Mobile Edge Intelligence and Computing for the Internet of Vehicles, 2019. arXiv preprint arXiv:1906.00400.

[17] C.M. Chen, B. Xiang, Y. Liu, K.H. Wang, A secure authentication protocol for Internet of Vehicles, IEEE Access, 7, 2019, pp. 12047−12057.

[18] B. Ying, A. Nayak, Anonymous and lightweight authentication for secure vehicular networks, IEEE Trans. Veh. Technol. 66 (12) (2017) 10626−10636.

[19] Q. Kong, R. Lu, M. Ma, H. Bao, A privacy-preserving sensory data sharing scheme in Internet of Vehicles, Fut. Gen. Comp. Syst. 92 (2019) 644−655.

[20] F. Kandah, B. Huber, A. Skjellum, A. Altarawneh, A blockchain-based trust management approach for connected autonomous vehicles in smart cities, 2019 IEEE 9th Annual Computing and Communication Workshop and Conference (CCWC), IEEE, 2019, pp. 0544−0549.

CHAPTER

NEED TO KNOW ABOUT COMBINED TECHNOLOGIES OF BLOCKCHAIN AND MACHINE LEARNING

17

X. Alphonse Inbaraj[1] and T. Rama Chaitanya[2]

[1]Research Scholar, Department of Information Engineering, I-Shou University, Taiwan, R.O.C

[2]Professor, Department of Computer Science and Engineering, PACE Institute of Technology and Sciences, Ongole, India

17.1 INTRODUCTION

Initially, we must know about what is blockchain, bitcoin, and combination of these technologies (artificial intelligence and machine learning combination, AL/ML), how these work together. Only one thing that refers about blockchain is incorruptible decentralized digital ledger. That will be programmed to store important data and transaction. AI/ML enables the machine to solve the human problems to have the cognitive human functions. This can be done through vast amount of data in order to train the models and fit. Blockchain is technological advances that are already making some waves. According to 2018 analysis, International Data Corp predicts that artificial intelligence spending will reach $ 46 billion before 2020. Machine learning is a subset of AI. Samuel coined this term and defined that computer is able to learn without explicit programming knowledge. This technology increases efficiency within the accounting profession to an unprecedented degree, which in turn will affect our future workflow process and how we interact with clients.

These combinations of two technologies, the technological revolution is not going to be short. With help of these, can envision a world where accounting and auditing can happen in real time. Accounting and transaction firms will not only start learning about how these technologies but also debugging the new ways of working internally with clients and with their teams.

17.2 LITERATURE SURVEY

According to network infrastructure design and its methods, some users are considers as malignants [1], but not all time. Currently, these users are performing some financial transactions deceitfully under some scope of financial network. But these are finishing up [1]. The solution to that is blockchain technology. According to Xu [2], blockchain technology will prevent some activities such as double-spending and record hacking. Some transactions will be affected by delayed network and/or propagation delay, and this is often realistically attributable to the delay in the broadcasting

Handbook of Research on Blockchain Technology. DOI: https://doi.org/10.1016/B978-0-12-819816-2.00017-4
© 2020 Elsevier Inc. All rights reserved.

418 CHAPTER 17 COMBINED TECHNOLOGIES OF BLOCKCHAIN AND ML

pending payment through network that is being unvalidated at different times [2]. According to Refs. [1,3], these papers, some studies, and analysis solution were given to the anomaly detection problems that can even be generalized or applied to alternative networks.

In this chapter [4], some cluster methods are discussed for preventing some malicious activities according to Smith et al. Here, K-means clustering and Self-Organizing Map (SOM) provide solution to those problems.

Some ML studies give solution for threats according to Refs. [1,3] and called as K-means clustering metric [5]. Pham and Lee [3] tried to provide solution for some abnormal conditions in transactional techniques, but those refer to the unsupervised ML techniques. Moreover by using ML-supervised approach [6], one of unique technique was introduced in order to reduce the anonymous transaction in Bitcoin according to Harlev et al. [7]. Carunana and Niculescu [6] exposed ML samples concerning with 200 million group action blocks.

According to Hirshman et al. [8], ML techniques is used to find abnormal conducts in Bitcoin network dealing with unsupervised activities. Furthermore, their ML approach was applied on dataset of Bitcoin network transactions to find obscurity guarantees within the Bitcoin system through data set clump, and Monamo et al. [9] examined some deceitful behaviors in Bitcoin network, namely, trimmed k-means and also in both local and global perceptivities of trimmed k means, its trees were found and identified.

According to Zhdanova et al. [10], mobile money transfer system was introduced to develop some novel techniques for fraud chain detection, and predictive security analysis was used for event-driven approach. Additionally, Yin and Vatrapu [11] determined some criminal nodes in the Bitcoin network and classified them into 12 categories. Again, Yin and Vatrapu [11] classified observations through some supervised ML techniques, and its classifiers called Bagging and Gradient Boosting was about 29.81% and 10.95%, respectively. Zambre and Shah [12] took some promised clients account to trace back the row of money. They used different techniques to trace that account, three different methods called Stone Man Loss (SML), All In Vain (AIV), and Mass MyBitcoin Theft (MMT). These are mainly used for robberies/thieves-related reports. In order to utilize stylometry to differentiate the advertisement, ML-based classifiers presented by Portnff et al. [13] were used. They explored Bitcoin memory leakage mechanism related to some matched transactions. Public Ethereum blockchain was proposed by Bonger [14] and discussed about some on-chain decentralized marketplace by Buterin et al. [15]. Kurtulmus and Daniel [16] presented a novel protocol on top of the Ethereum blockchain such that reputation and identity system are not imposed in order to elaborate transactions of the marketplace. However, some majority attacks [17] based on the Bitcoin and Litcoin [18] are not security threats. In that way, Dey [19] proposed a new way to monitor the clients who are in the Bitcoin networks. This method used supervised ML approach.

17.3 BITCOIN METHODOLOGY

The Bitcoin term was coined and introduced in 2008, which is an immense alternative to the conclusion of transactions entirely. White paper was initiated by Satoshi Nakamoto. Moreover, that article was the first to consider an authorized and trust transaction on the blockchain. It leads to instant bitcoin transactions around the world and also for secured bitcoin transactions end to end with trustworthy.

17.3.1 A DECENTRALIZED CHAIN

A blockchain protocol is used for decentralized payment network in bitcoin transaction. However users can get encrypted and transaction history in form of ledger that will be seen on both sides and realize that no more data are lost. That cipher value of transaction minimizes the verification of user cost majorly. So users can find the authorized transaction with fulfilment of their satisfaction. To verify that some authenticated code will be arrived to strengthen the transaction. That will add some worth values for each end of users' side of transaction and that it is not at all necessary to find the users' conclusion [20].

So here the definition about blockchain transaction is that decentralized way of digital database of independent authority and also referred as public ledger that stores all types of numerous blocks of connected transaction structure. Blockchain characteristics include:

- Transparent transaction for all users
- Unchangeable of transaction history (nonbreakable)
- Secure data protection high security (all personal data are highly protected)
- Cost-efficiency
- Decentralization
- Ease of use

To ensure high-level reliability, cryptocurrency transfers between parties must satisfy all the required credentials that are ensured by smart contract [20]. Smart contract is the main role in blockchain technology because it derives some methodology and protocols. A client can execute his/her transaction conditions on any of the blockchain platforms in which there is no more third party. This is one of smart and speed response processes and moreover is a highly trustworthy one. Intruders could not identify this because of secure transaction.

17.4 BLOCKCHAIN IMPLEMENTATIONS IN VARIOUS INDUSTRIES

17.4.1 BLOCKCHAIN TRANSACTION IN BANKING

Uses of blockchain are banking, FinTech, and credit history include:

- One of secure and fast process among parties
- Currency exchange information reveals properly
- Very less audit effort
- Ledger details will be shared among parties

17.4.2 EDUCATION

In education, blockchain is implemented to

- Sharable credits score among institution
- Academic hassle-free grade system

420   CHAPTER 17 COMBINED TECHNOLOGIES OF BLOCKCHAIN AND ML

- Smart and simply coordinates among students and parents
- Issuing sharable knowledge with proper delivery to students

17.4.3 ENERGY

In energy, blockchain is implemented to

- Peer to peer transaction for commercial and sensex demands

17.4.4 LEGAL, GOVERNMENT, AND INSURANCE

Blockchain is implemented to legal, government, and insurance to

- Hassle free and without any scandal procedure
- Secure policies and procedures in all state governance
- Recognizes proof for each party
- Issuing authorized and authenticated user data
- Transparent and more hidden transaction
- Easy accessibility in data transaction

17.4.5 INTERNET OF THINGS AND ARTIFICIAL INTELLIGENCE

In Internet of things and AI, the blockchain is used to

- Access the data through blockchain boundary
- Conductivity–chain process in the production world
- ML and ANN are growing technologies

17.4.6 HEALTH CARE AND MEDICINE

Uses of the blockchain in health care and medicine are to

- Patient history sharing among hospital
- Proper medicine supply management
- Efficient and authenticated medicine issue. Maintain the patient history among doctors and among hospital

17.5 PROCESS STAGES OF IMPLEMENTATION OF BLOCKCHAIN TECHNOLOGY

By using virtual keys, bitcoin allows to encrypt and decrypt the confidential users' information and transaction history without any problem. The fact that this is decentralized one leads to transparent process between users and [20] it enables the sender to substantiate the legitimacy of the bitcoin group action.

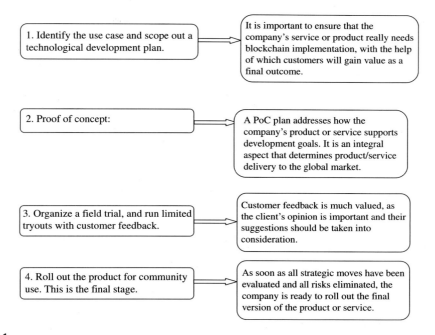

FIGURE 17.1

Process stages of blockchain technology.

On the other hand, distributed ledgers are log files of transactions that will be added while buying or creating the bitcoin that you engage in consensus. This is known as mining of transactions.

One of the main characteristics of bitcoin is a freelance currency rather than a ruled payment system. It leads to a new era in all types of transaction especially that is decentralized and fully digitalized currency rather than the liquid. Bitcoin is commonly known as "digital gold." The reason for this is the similarity of the following features:

- Very less container
- Autonomous
- Ledger of transaction

Users can send and receive funds with the help of bitcoin with legal condition. Because of higher level of security ensured in supply chain, this contributes beneficially to the anonymity feature. See Fig. 17.1.

17.6 CREATION OF BITCOIN TRANSACTION AND MINING

There are three steps to buy a BITCOIN (see Fig. 17.2)

Step 1: First, install a digital wallet on PC or mobile device, and it will keep track of your Bitcoin balance and transactions.

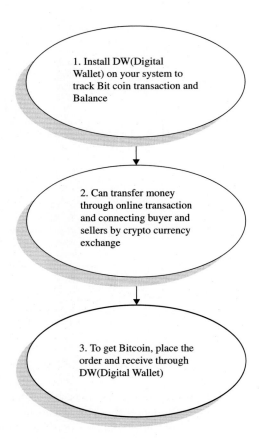

FIGURE 17.2

Bitcoin resources 1.

Step 2: Deposit money through an online payment company or transfer directly from a bank account to a crypto currency exchange connecting buyers and sellers.
Step 3: Place an order for a Bitcoin purchased and receive it in digital wallet.

The following steps are used to get a Bitcoin (see Fig. 17.3)

Step 1: A Crypto currency exchange: You can exchange "regular" coins for Bitcoins.
Step 2: A Bitcoin ATM: Can change Bitcoins or cash for another crypto currency.
Step 3: Classified service where we can find a seller who will help trade of Bitcoins for cash.
Step 4: Can sell the product or service for Bitcoins

This is one of decentralized unbreakable public ledger group of maintaining the secure chain. In this group, action also can be allowable. In order to verify the process, group action is necessary to confirm so that it will enable bitcoin to authenticate. Then only next or new transaction process will update in that ledger. A group action may be a definition for price transfer from one notecase to a different one. The wallet keeps track of the user's addresses and personal details. Actually the

17.6 CREATION OF BITCOIN TRANSACTION AND MINING

FIGURE 17.3

Bitcoin resources 2.

FIGURE 17.4

Beginning of transactions mode.

address will be the combination of all such that numbers and letters in order to specify the public key. Mining is one of consensus system that is mainly used to confirm parties' transaction request and response in all over the decentralized database system.

- The purpose of mining system is for neutrality and proper transaction which is hassle free. That enables different nodes to reach out many agreements on the system state without discard of transaction history ledger.

17.6.1 TRANSACTION OF DATA

This is exclusive series of transactions in the Bitcoin transaction and one of the attractive tracking history ledger and can find the transaction log files [21] (Refer Fig. 17.4).

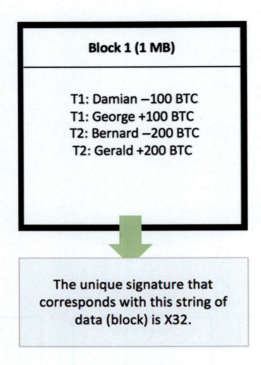

FIGURE 17.5
Unique signature for the corresponding block.

17.6.2 CHAINING THE BLOCKS (WITH HASH)

All users get corresponding unique ID in that linked chain transaction and ledger details. This is much more strengthened by signature. If anything changes, that will affect or intimate to whoever in this track of the transaction and give the chance to create or allow new signature (see Fig. 17.5) [21].

17.6.3 CHAINING

In this track of transaction, all blocks are together linked and hence allow to create ledger. Some block does not allow signature to create and other block allows without signature involvement because of string of data in another block 2 (see Fig. 17.6).

17.6.4 FINAL CHAINING PROCESS

17.6 CREATION OF BITCOIN TRANSACTION AND MINING

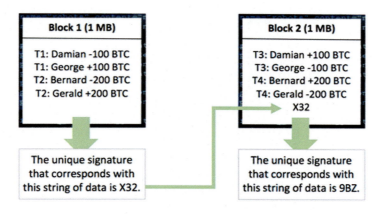

FIGURE 17.6

Chaining with each block.

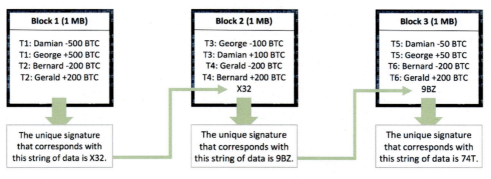

FIGURE 17.7

Final Bitcoins transaction.

17.6.5 ALTERATION IN FIRST BLOCK

The signature W10 does not match the signature that was previously added to block 2 anymore. Blocks 1 and 2 are now considered no longer chained to each other [21].

From the above figures, we can understand that Blocks 1 and 2 are no longer chained together. If any changes occur, that will spread entire blocks. Blockchain is immutable that cannot be changed in order to perform any alternate in block, that will affect entire transaction and changes in log files also (see Figs. 17.7 and 17.8). This is showing the transparent transaction on both parties with secure.).

17.6.6 HASH SIGNATURE CREATED

This specific string of data now requires a signature. In blockchain, this signature is created by a *cryptographic hash function*. A cryptographic hash function is created that is very important to

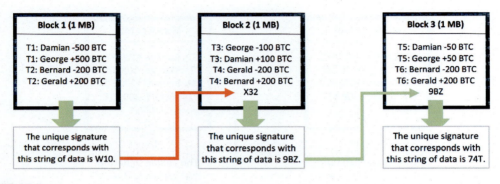

FIGURE 17.8

Modification in first block.

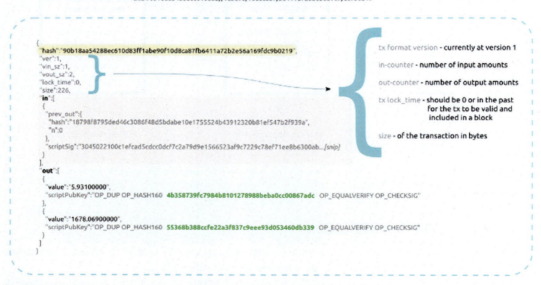

FIGURE 17.9

Bitcoin transaction sample.

secure transaction. It is taking some string of input and creates some unique 64 digital bit as output [20]. For example, creating hash function for the word "ROSES." This is used to allow the hash function. After the processing, you will get the following string of data.

"761A7DD9CAFE34C7EYZPO6C1270E17F773025A61E511A56F700D415F0D3453E199868"

In the example, you can find a transaction ID highlighted in yellow. Meta-data along, with a description, is on the right [20]. Inputs and outputs are highlighted in pink and green (see Fig. 17.9).

17.7 COMBINING BLOCKCHAIN WITH ARTIFICIAL INTELLIGENCE/ MACHINE LEARNING

17.7.1 ENERGY CONSUMPTION AND EFFICIENCY

One of the inherent challenges of blockchain is massive energy use for transactions verification. Artificial intelligence/machine learning can be used to optimize energy consumption through efficient node allocation based on need, type of transaction, and therefore taking the most efficient path. This would also make blockchains more efficient and faster in verifying transactions. This is currently not possible through blockchain alone and also because it is nascent, and the technology needs to mature. Furthermore, blockchains store all the data, and transactions are tied to each other. To reduce the burden of data storage, AI can help verify the safest and fastest node to store data instead of having to go through all the nodes.

17.7.2 BLOCKCHAIN CAN HELP ARTIFICIAL INTELLIGENCE/MACHINE-LEARNING SYSTEMS TO BETTER EXPLAIN THEMSELVES

Artificial intelligence/machine learning is complicated and can move at fast speeds that normal human brain cannot comprehend. Blockchain can help AI/ML to better explain itself. If two are linked together, it is easier to track the decisions made by AI on the blockchain, since all transactions are recorded. In case of dispute, these can be executed on the blockchain.

17.7.3 MORE DATA

Artificial intelligence/machine learning works better with more data, not less. However, the more data that are added to the blockchain, it becomes harder to process and slows the whole architecture. However, if AI/ML is linked with blockchain, more data would be better, since decisions about the use of data would be made by AI and therefore increase the applications of the blockchain.

17.7.4 SMART CONTRACTS

Smart contracts are promising to be the heart of business transactions in the near future. They will enable transfer of value and making of business deals on the blockchain in a transparent and immutable way across vast geographical distances in an unprecedented way. However, the operationalization of smart contract still requires human/computer nodes verification process. With ML, verification of financial transactions on smart contracts will not require "human hand" as clearing agent, since they will be automated. This would enable them to not only function faster but also more efficient. The data on smart contracts can be used by AI to make better decisions at faster

428 CHAPTER 17 COMBINED TECHNOLOGIES OF BLOCKCHAIN AND ML

speeds than the current cases. Also, the way in which most of the smart contracts are structured is such that they do not collect data from outside. Therefore, smart contracts may not have the latest data about the current changes in the external environment such as price changes or weather patterns. These aspects are important for accuracy in the execution of smart contracts with AI/ML such that data can easily be collected and verified. This would make smart contracts more "smarter" in making near-instantaneous business and value engagements.

17.7.5 REDUCE INSTANCES OF 51% ATTACKS

Currently, 51% attack is difficult on more established blockchains such as Bitcoin and Ethereum. However, we have witnessed instances of 51% attack on smaller blockchains such as Tether has been hacked. Artificial intelligence/machine learning can help in training nodes, the characteristics of hacks, and therefore anticipate them. Since AI can make decisions faster than the current blockchain architecture, it can help identify and even prevent hacks before they happen.

17.8 STARTUP WITH BLOCKCHAIN AND MACHINE LEARNING

A new crop of startups has popped up that are looking at interesting ways to utilize blockchain with AI. Many of these startups have similar business plans that hope to use blockchain to give people control over their data in the hope that they will share more data with ML algorithms and AI apps. See Tables 17.1 and 17.2. Some examples of startups are:

1. TraneAi
2. SingurityNET
3. OpenMinded
4. BurstIQ
5. Neural

17.9 CENSUS TYPES AND PROS AND CONS

Blockchain census provide many solution to various problems because of peer-to-peer decentralized distributed ledger. Here we discussed about some Pros and Cons. Refer Table 17.2.

17.10 ARTIFICIAL INTELLIGENCE VERSUS BLOCKCHAIN

- Decisions created by AIs will generally be arduous for humans to grasp. This is as a result of capable of assessing an outsized variety of variables that are not dependent on every alternative and "learning" which of them are necessary to the task it is trying to achieve.

17.10 ARTIFICIAL INTELLIGENCE VERSUS BLOCKCHAIN 429

Table 17.1 Startup and Their Stages

Startup	Working Scenario
TraneAi	Wants to use the blockchain to streamline data tasks, such as the tagging of training data, in a crowdsourced manner. Those who help with the artificial intelligence (AI) training would work with the company's Transaction Protocol for Artificial Intelligence (TPAI) and be compensated with TPAI tokens.
SingurityNET	The goal is to create a "full-stack" blockchain-powered AI solution that can democratize access to machine learning capabilities. "The AI code is there, free and open source, on the Internet right now," says Ben Goertzel, the company's CEO and Chief Scientist, in a video on the company's website. "It's just hard to find and hard to use. So SingularityNET can complete that connection between AI developer and AI users."
OpenMinded	It uses a variety of technologies, including federated learning and blockchain, to create an anonymous and secure grid for the training of machine-learning models. The OpenMinded grid give data scientists and developers access to data supplied by "miners" who help train the models and are rewarded.
BurstIQ	Its goal is to facilitate the sharing of health-care data among businesses, researchers, and users. It uses blockchain to maintain the security and privacy of the data and to adhere with regulations such as Health Insurance Portability and Accountability Act (HIPAA) and General Data Protection Regulation (GDPR). "The BurstIQ platform allows businesses to get more from their own data, share it with the right stakeholders at the right time, and access new data streams through a global network of people, partners and products," the company says.
Neural	It is using blockchain to build a peer-to-peer framework to harness idle computing power for big data analytics. The company, which was founded in 2014, boasts that its network is 1 million times more powerful than the world's biggest supercomputer. It even claims that neither Facebook nor Google can "harness the massive and raw amounts of data needed to surpass what Neural's architecture allows."
Namahe	One of the major goals in most industries is to *increase automation* to increase efficiency and reduce costs, whether this be on the hardware side (e.g., Robotics) or on the *software and logistics side via AI.*
DeepBrainChain	One of the major challenges for new, or small AI companies is access to enough computing power to run their AI-training simulations, slowing the rate of progress and leading to a high financial barrier to entry.

- As an example, AI algorithms are expected to be more and more utilized in creating choices concerning whether or not monetary transactions are deceitful and will be blocked or investigated.
- For some time though, it will still be necessary to have these decisions audited for accuracy by humans. And given the large quantity of knowledge that will be taken into thought, this could be a posh task. Walmart, as an example, feeds a months' price of transactional data across all of its stores into its AI systems that build choices on what product ought to be equipped and where.
- No matter however clearly we see that AI offers vast benefits in several fields, if it is not trusted by the public, then its usefulness will be severely limited. Recording the decision-making method on blockchains can be a step toward achieving the extent of transparency and insight into mechanism minds which will be required so as to achieve public trust.

430 CHAPTER 17 COMBINED TECHNOLOGIES OF BLOCKCHAIN AND ML

Table 17.2 Consensus Types

Consensus Type	Definitions	Advantages	Disadvantages	Implementations
Proof of Work (PoW)	Miners solve hard mathematical problems to create blocks. The miner who solves the problem first gets a reward.	It works	Slow throughput; expensive	Bitcoin, Ethereum, Litecoin, Dogecoin, etc.
Proof of Stake (PoS)	In PoS, blocks aren't created by miners doing work, but by minters staking their tokens to "bet" on which blocks are valid. The one with more wealth has a more valued vote. Those who vote wrong will "lose their stake."	More decentralized, energy efficient	More attacks, more expensive, nothing at stake	Decred Ethereum, Peercoin, and others
Delegated PoS (DPoS)	In DPoS, token holders don't vote on the validity of the blocks themselves but vote to elect delegates to do the validation on their behalf. There are generally between 21 and 100 elected delegates in a DPoS system.	Cheap transactions, scalable, and energy efficient	Partially centralized	Steemit, EOS, and BitShares
Proof of Weight (PoWeight)	PoWeight is a broad classification of consensus algorithms based upon the Algorand consensus model. The general idea is that in PoS, your percentage of tokens owned in the network represents your probability of "discovering" the next block. In a PoWeight system, some other relatively weighted value is used.	Customizable and scalable	Incentivization can be a challenge	Algorand, Filecoin, and Chia
Byzantine Fault Tolerance (BFT)	The idea of a BFT is that there are several validators that have to come to a common decision that a certain transaction is valid (true). Only with the approval of all validators can the transaction go through.	High throughput, low cost, scalable	Semi-trusted	Hyperledger, Stellar, Dispatch, and Ripple

- Traditionally, computers are very fast. Without specific directions on how to perform a task, computers cannot get them done. This means that because of their encrypted nature, operational with blockchain data on "stupid" laptops needs giant amounts of computer process power. As an example, the hashing algorithms that want to mine blocks on the Bitcoin blockchain take a "brute force" approach—effectively attempting each combination of characters till they realize one which fits to verify a transaction.
- Artificial intelligence is an effort to pass over far from this brute force approach and manage tasks in a very additional intelligent, thoughtful manner. Consider, however, an individual's professional on cracking codes can, if they are sensible, become higher and additionally economical at code breaking, as they succeed with cracking additional and additional codes throughout their career. An ML-powered mining algorithm would tackle its job in a similar

way—although rather than having to take a lifetime to become an expert, it could almost instantaneously sharpen its skills, if it is fed the correct coaching knowledge.

17.11 FUTURE DIRECTIONS

By considering the altcoin concept environment, we developed flexibility in Bitcoin and blockchain technologies, and ML also improved and developed in its aspects of security such as cryptocurrency system.

- *Protocol of blockchain*
    In order to improve and holding of blocks of transactions, prove the usefulness of gametheoretic techniques that need some mining pools and resources to modeling the various issues and delivering efficient solutions.
- *Techniques for clustering*
    Specific threshold are designed in order to address the broad range of threats such that specifying cluster and obtaining extra signature to enable read and write on some cryptographic data [10]. Here some string searching filters are used (i.e., Bloom filter).
- *Miner nodes in crypto currency*
    The miner nodes must be addressed to settle currency in this cryptocurrency network because some fraud miners are acquiring the extra awarding coins by getting Bitcoin allowances, whether it is constant and inconstant.
- *Avoiding backtracks*
    Handling contracts by third parties should be conditioned and restricted in financial networks such as Bitcoin infrastructures.

17.12 CONCLUSION

The Bitcoin blockchain is a free, open technology. It operates in a peer-to-peer network without a central authority (financial institution). This allows the exchange of objects (BTC) and records each transaction in a large, unchangeable ledger. In this chapter, we addressed what is Bitcoin, blockchain technology, and its transactional process in a step-by-step detailed fundamental explanation and trends of blockchain technology. Then, we explained about combination technology that ML and blockchain solves the common security threats and explained about some startups that will be familiarized in these combinations. Finally, we detailed some open-research questions and future research directions followed by concluding remarks.

REFERENCES

[1] T. Pham, S. Lee, Anomaly detection in the Bitcoin system A network perspective, 2016. [Online]. Available from: https://arxiv.org/abs/1611.03942.

[2] J.J. Xu, Are blockchains immune to all malicious attacks? Financial Innov. 2 (1) (2016) 25.

432 **CHAPTER 17** COMBINED TECHNOLOGIES OF BLOCKCHAIN AND ML

[3] T. Pham, S. Lee, Anomaly detection in Bitcoin network using unsupervised learning methods, 2016. [Online]. Available from: https://arxiv.org/abs/1611.03941.

[4] R. Smith, A. Bivens, M. Embrechts, C. Palagiri, B. Szymanski, Clustering approaches for anomaly based intrusion detection, in: Proc. Intell. Eng. Syst. Through Artif. Neural Netw., 2002, pp. 579−584.

[5] H. Xiong, J. Wu, J. Chen, K-means clustering versus validation measures:A data-distribution perspective, IEEE Trans. Syst. Man Cybern. B. Cybern. 39 (2) (2009) 318−331.

[6] R. Caruana, A. Niculescu-Mizil, An empirical comparison of supervisedlearning algorithms, in: Proc. Int. Conf. Mach. Learn. (ICML), 2006, pp. 161−168.

[7] M.A. Harlev, H.S. Yin, K.C. Langenheldt, R. Mukkamala, R. Vatrapu, Breaking bad: De-anonymising entity types on the Bitcoin blockchain using supervised machine learning, in: Proc. 51st Hawaii Int. Conf. Syst. Sci., 2018, pp. 1−10.

[8] J. Hirshman, Y. Huang, S. Macke, Unsupervised approaches to detecting anomalous behavior in the Bitcoin transaction network, Stanford Univ., Stanford, CA, USA, Tech. Rep., 2013.

[9] P. Monamo, V. Marivate, B. Twala, Unsupervised learning for robust Bitcoin fraud detection, in: Proc. Inf. Secur. South Africa (ISSA), Aug. 2016, pp. 129−134.

[10] M. Zhdanova, J. Repp, R. Rieke, C. Gaber, B. Hemery, No smurfs: revealing fraud chains in mobile money transfers, in: Proc. Int. Conf. Availability, Rel. Secur. (ARES), Sep. 2014, pp. 11−20.

[11] H.S. Yin, R. Vatrapu, A first estimation of the proportion of cybercriminal entities in the Bitcoin ecosystem using supervised machine learning, in: Proc. Int. Conf. Big Data (Big Data), Dec. 2017, pp. 3690−3699.

[12] D. Zambre, A. Shah, Analysis of Bitcoin network dataset for fraud, Stanford CS 224W Project Final Rep.-Group 30, Dec. 2013.

[13] R.S. Portnoff, D.Y. Huang, P. Doerfier, S. Afroz, D. McCoy, Backpage and Bitcoin: uncovering human traffickers, in: Proc. Int. Conf. Knowl. Discovery Data Mining (SIGKDD), 2017, pp. 1595−1604.

[14] A. Bogner, Seeing is understanding: anomaly detection in blockchains with visualized features, in: Proc. Int. Joint Conf. Pervasive UbiquitousComput., Int. Symp. Wearable Comput., 2017, pp. 5−8.

[15] V. Buterin et al., A next-generation smart contract and decentralized application platform, White Paper, 2014.

[16] A.B. Kurtulmus, K. Daniel, Trustless machine learning contracts; evaluating and exchanging machine learning models on the ethereum blockchain, 2018. [Online]. Available from: https://arxiv.org/abs/1802.10185.

[17] I. Eyal, E.G. Sirer, Majority is not enough: Bitcoin mining is vulnerable, Commun. ACM 61 (7) (2018.) 95−102.

[18] Litecoin. Accessed: Jul. 2018. [Online]. Available from: https://litecoin.org/.

[19] S. Dey, Securing majority-attack in blockchain using machinelearning and algorithmic game theory: a proof of work, 2018. [Online]. Available from: https://arxiv.org/abs/1806.05477.

[20] https://applicature.com/blog/crypto-enthusiasts/.

[21] https://blog.goodaudience.com/blockchain-for-beginners-what-is-blockchain-519db8c6677a.

Index

Note: Page numbers followed by "*f*" and "*t*" refer to figures and tables, respectively.

A

Abstract Syntax Tree (AST), 380–381
Access control policy, 44–45, 44*f*
Access policy (AP) tree, 80
AChain, 367
Active Network Management, 153
Additive manufacturing (AM), 19–20
Adoption rate, 121
Advanced Metering Infrastructure (AMI), 153
Alawite State, 6
Alliander, 162
Ambrosus, 317
Annan, Kofi, 13
Anonymity, of blockchain, 62–63, 174, 327–328
Apache HTTP server, 204
Application interfaces, 363
Applications of blockchain technology, 220–237, 275–277, 275*f*, 369–370
  in the existing music industry, 235
    use cases, 235–236
  in financial services, 220–223
    use cases, 222–223
  and financial technology, 225
  in healthcare, 225–229
    use cases, 227–229
  in identity management, 236–237
    use cases, 236–237
  in insurance sector, 223–225
    use cases, 224–225
  in real estate, 230–232
    existing real estate sector, problems in, 231
    use cases, 231–232
  in supply chain, 232–235
    use cases, 233–235
  in voting, 229–230
Application-Specific Integrated Circuit (ASIC), 330
Approval Estimation and Editing (VEE), 158–160
Arab Socialist Baath Party, 6–7
Arab spring and the emergence of the Syrian civil war, 11
ArtemDuvanov, 26
Artificial intelligence (AI) techniques, 277, 296, 417, 427
  versus blockchain, 428–431
Artificial intelligence/machine learning and blockchain, combined technologies of, 427–428
  artificial intelligence versus blockchain, 428–431
  census types and pros and cons, 428
  energy consumption and efficiency, 427
  explaining artificial intelligence/machine learning, 427
  future directions, 431
  more data, 427
  reducing instances of 51% attacks, 428
  smart contracts, 427–428
  startup with blockchain and machine learning, 428
Assad government, 14–15
Astana, 14
Asymmetric cryptography, 356, 358–359
"Atomic crosschain" (AXC) transactions, 400–402
Attacks, types of, 104–107
  BGP hijacking attack, 106
  DAO attack, 105–106
  eclipse attack, 106
  liveness attack, 107
  selfish mining attack, 104–105
Attribute-based encryption (ABE) algorithm, 80–81
Auditability, of blockchain, 328
Authentication privacy in the blockchain, 414
Automated car, 298
Automatic Dependent Surveillance–Broadcast (ADS-B), 18
Automatic Voltage Control, 153
Avalanche DAG, 393
Avalanche effect, 89
Azure, 367

B

Backdoor attacks, 68
Background of blockchain, 215–220
  bitcoin, 215
  emergence of bitcoin, 216–217
  legal issues in bitcoin, 219–220
  risk in bitcoin, 218–219
  working of bitcoin, 217–218
Backtracks, avoiding, 431
Bagging, 418
Balfour Declaration, 6
Banking, blockchain transaction in, 419
Bashar al-Assad regime, Syria under, 8–10
BCPG, 164
Benefits of blockchain technology, 363–364
BGP hijacking attack, 106
BIP-144, 380–381
Bitcoin, 2–4, 25, 69, 100–101, 106, 113–114, 196, 215, 245, 273, 295, 366, 414
  case study, 339–340
  emergence of, 216–217
  energy consumption index, 114

433

434 Index

Bitcoin (*Continued*)
  legal issues in, 219–220
  methodology, 418–419
    decentralized chain, 419
  payment networks with, 310–311
    Bitcoin-NG, 310
    GHOST, 310–311
  risk in, 218–219
  script, 381*f*
  transaction, mechanism of, 274–275
  working of, 217–218
Bitcoin-NG, 310
Bitcoin transaction and mining, creation of, 421–427
  alteration in first block, 425
  chaining, 424
  final chaining process, 424
  hash signature created, 425–427
  transaction of data, 423
Bitcoin Unlimited, 380–381
Bitcoin versus Segwit coin transaction, 379*f*
B-learning management system, 36
Block, 326, 355
BlockApps, 16
Block body, 356
Blockchain 1.0 Currency, 362
Blockchain 2.0 Smart Contracts, 362
Blockchain 3.0 DApps, 362
Blockchain 4.0, 363
"Blockchain against Hunger", 16
Blockchain-based IoT (BIoT) system, 115–116, 118, 286–287
Blockchain Emergency ID (BE-ID), 16
Blockchaining, 172–173
Blockchain Technology-Based System (BTBS), 1–2, 4–5, 15–16, 21–22, 25, 29
Block size, 358
Block structure, 356, 356*f*, 357*f*
Block time, 358
Bouygues Immobilier, 162
Building Blocks, 16
Burstcoin, 333
BurstIQ, 429*t*
ByteBall DAG, 389–390
Byzantine agreement, 73
Byzantine Fault Tolerance algorithm, 73, 255, 430*t*
  and variants-hyperledger fabric, 333–334
Byzantine Fault Tolerant Consensus Algorithms, 392–393
Byzantine Generals (BG) Problem, 195–196

C

Capacity model, proof of, 332–333
Capital market operations, applications of blockchain technology in, 222

Carbon Chain, 145
Carbon Chain Project Token, 146
Carbon Chain Token (CCT), 146
Carbon credit, 142–144
Carbon Credit Exchange, 145–147
  carbon credits and carbon markets, 146–147
Carbon emission effect, 147–152
  blockchain application for distributed energy resources management in smart city, 151–152
Carbon Offsetting and Reduction Scheme for International Aviation, 144
Categories of blockchain development, 240–241
  application of blockchain technology
    as a development platform, 240
    as a marketplace, 240
    as a smart contract, 240
    as trusted service application, 241
Census types and pros and cons, 428
Center components of blockchain architecture, 141–142
Central authority, of blockchain, 327
Central servers, 407
Certificate authority (CA), 78, 363
Certificate-based security, 120
Certified emissions reductions (CERs), 143, 146–147
Chain, 355
Chain Core, 367
Chaining, 424
Chaining of blocks in blockchain, 255
Chain partitioning-based scalability, 380–385
ChainTutor application, 37
Challenges and concern, in blockchain, 237–238
Challenges in blockchain technology, 278–279, 335–339, 336*f*, 368
  Bitcoin (case study), 339–340
  implementation and acceptability, 337–338
  opportunities in cryptocurrencies, 338–339
  technical challenges, 336–337
Characteristics of BCT, 349, 369
CHECO, 399–402
Child-chain, 373–374
  blockchain scalability, 384–385
Ciphertext policy and hybridized weighted attribute-based encryption (CP–HWABE), 79–89
  experimental results, 81–85
    CP - HWABE decryption, 85
    CP - HWABE encryption, 83–84
    key pair generation, 81–82
    user registration, 81
  experimental setup, 80
  result analysis, 86–89
    Avalanche effect, 89
  working of ciphertext policy and hybridized weighted attribute-based encryption (ABE) algorithm, 80–81
Clean development mechanism (CDM), 143

Client/server systems, 349–350
Client side security threats, 99
Closed-door consultations, 14
Cloud-based blockchaining for enhanced security, 171
  anonymity, 174
  autonomy, 173
  blockchaining, 172–173
  block chain structure, 174–176
    blockchain security process, 175–176
    main data, 174
    timestamp, 174
    types of blockchain, 174
  cloud environment, incorporation of blockchaining with, 177–179
    access policy checking, 177–178
    sensitive and nonsensitive data separation and data package security checking process, 178–179
    virtual machine measurement security checking process, 178
  immutable, 173
  related works, 176–177
  transparent, 173
Cloud edge computing, 409
Cloud platform, IoT on, 289
Coin-base transaction, 3
Coin packets, 400–402
Collators, 402
Collision-free hash functions, 2–3
Combating Terrorism Technical Support Office (CTTSO), 18
Committees, 396
Communication channel attack, 67
Communication Network, 158
Computers, 430
Concepts of blockchain, 187–189, 353–355
  database versus blockchain, 353–354
  driving concepts for business, 354–355
  history of blockchain, 353
  network view, 354*f*
Connected car, 284, 408
Connected health, 284
Consensus, 195–198, 355–356
  in blockchain network, 55–57
  practical byzantine fault tolerance (PBFT), 197–198
  proof of stake (POS), 196–197
  proof of work, 196
Consensus algorithm in blockchains, 329–335
  Byzantine fault tolerance (BFT) and variants-hyperledger fabric, 333–334
  federated Byzantine agreement, 334–335
  proof of capacity model, 332–333
  proof-of-elapsed time (PoET) model, 331–332
    block verification, 332
    calculation of the waiting time, 332
    random waiting times, 331–332

proof-of-stake model, 331
proof-of-work model, 329–330
Consensus algorithms-based scalability solutions, 403
Consensus protocol, 324, 357, 376–377
  attack, 74–75
  and byzantine general problem, 72–73
Consensus types, 430*t*
ConsenSyswho, 16
Consenus algorithm, 361–362
  practical byzantine fault tolerance, 361
  proof of activity, 361
  proof of burn time, 362
  proof of capacity, 362
  proof of importance, 362
  proof of stake (PoS), 361
  proof of work (PoW), 361
Consortium accounts, monitoring of
  applications of blockchain technology in, 222–223
Consortium blockchain, 172, 193, 246, 364
Constituents of blockchain, 114–115
Constructive criticism, 8–9
Contemporaneous advancements in blockchains, 323–324
Content-addressed technique, access of transaction using, 51–54
  advantage of content addressed over location addressed, 54
  increases durability of data, 54
  maintains integrity, 54
  permanent storage of files, 54
Control Flow Graph (CFG) BUILDER, 109–110
Corda, 367
Cosmos, 400–402
Cosmos blockchain, 385, 400–402, 401*f*
Counterterrorism, 14
Creative thinking, 8–9
Criminal activity, 100–101
Cross border payments, applications of blockchain technology in, 222
Cross-fault tolerance (XFT), 334
Cryptocurrency, 120, 273, 296, 323, 362, 365, 370
  block, 326
  blockchain challenges and opportunities, 335–339, 336*f*
    Bitcoin (case study), 339–340
    implementation and acceptability, 337–338
    opportunities in cryptocurrencies, 338–339
    technical challenges, 336–337
  blockchain consensus model, 329–335
    Byzantine fault tolerance (BFT) and variants-hyperledger fabric, 333–334
    federated Byzantine agreement, 334–335
    proof of capacity model, 332–333
    proof-of-elapsed time (PoET) model, 331–332
    proof-of-stake model, 331
    proof-of-work model, 329–330
  blockchain features, 327–328

436 Index

Cryptocurrency (*Continued*)
 contemporaneous advancements in blockchains, 323–324
 digital signature, 327
 further advancements, 344–345
 Hedera Hashgraph, 342–344
 IOTA tangle, 340–342
 mechanisms, for blockchains, 324–325
 motivations for the use of blockchains in, 324
 taxonomy of blockchain system, 328–329
Cryptographic algorithms, 171, 177, 324–325
Cryptography, 3, 358–359
CTB-Locker, 100
Cuckoo rule, 397–398
Cultural shift, 368

D

DApps, 362
Database versus blockchain, 353–354
Data Collector Units (DCUs), 158
Data Concentrator Unit (DCU), 155–156
Data privacy, 186–187
Data structure, blockchain, 375
Data transaction, integrity of, 62
Decentralization, 62, 231
Decentralized Autonomous Organization (DAO), 74, 105–106
Decentralized chain, 419
Decentralized hash table–based storage approach, 40–41, 40*f*
DeepBrainChain, 429*t*
"De-escalation zones", ethical considerations and issues in for refugees and internally displaced persons, 14–15
Defense Advanced Research Projects Agency (DARPA), 17–19
Defense Science Board (DSB), 18
Delegated PoS (DPoS), 430*t*
Denial of service (DoS), 40
 attacks, 63, 67
Department of Defense (DoD), 17–18
Developer, 363
Development platform, application of blockchain technology as, 240
Device-Gateway-Cloud (DGC) model, 317
Digital gold, 421
Digital identity, 15–16
Digitalized globalization, 2
Digital signature, 327, 359
Directed acyclic graphs (DAG), 63–64, 374, 387–394
Disseminated quality sources, 134–135
Distributed denial of service (DDoS) attack, 40, 75, 299–300
Distributed ledger (DL), 375–376, 377*f*
 -based scalability solutions, 403
Distributed ledger technology (DLT), 134–135, 163–164
Distributed nature of blockchain, 363–364

Distributed peer-to-peer (P2P) database, 373
Distribution Transformers (DTs), 155
Divvi, 163
Double spending, 75, 98–99
Driving concepts for business, 354–355
Drug Supply Chain, 257
Dynamic Line Rating, 153

E

Eavesdropping, 186
Eclipse attack, 75, 106
Edge computing, 409
Education, blockchain implementations in, 419–420
Elastico, 396
Electronic Health Records (EHRs), 251, 257
Elements in a blockchain, 190–192
Elliptic Curve Digital Signature Algorithm (ECDSA), 327
End devices attacks, 68
Energo Labs, 163
Energy, blockchain implementations in, 420
Energy Block, 163
Energy conservation, blockchain implemented projects worldwide in, 161–164
Energy consumption, 368
Energy efficiency, 121–122
Energy-efficient blockchain for internet of things, 123–128
 energy consumption in attached machines, 126–128
 experimental setup, 125
Engineering Center Steyr (ECS), 18
EOSIO, 367
E-residency identity management, 237
ESP32, 203, 208*f*
Ethereum, 199–200, 203, 205, 330, 366, 373, 378, 384–385, 418
Etherium, 257, 259–260
Ethical considerations and issues of blockchain technology-based systems in war zones, 1
 Arab spring and the emergence of the Syrian civil war, 11
 in "de-escalation zones" for refugees and internally displaced persons, 14–15
 future of blockchain technology for global governance, 25–28
 issues and challenges, 21–25
 major actors in the conflict, 12–14
 methodology, 2
 process involved, 3–5
 refugees and internally displaced persons, blockchain technology for, 15–17
 refugees and internally displaced persons in Syria since 2011 (case study), 10–11
 Syria, 5–6
  beginning of modern nation-state, 6–8
  under Bashar al-Assad regime, 8–10

US defense forces and industry (case study), 17–21
European Defense Matters (EDA), 17
E-voting system prototype, 26
Existing technologies of blockchain, 68–75
  analysis of blockchain security, 74–75
    consensus protocol attack, 74–75
    distributed denial of service attack, 75
    double spending, 75
    eclipse attack, 75
    fraud in programming, 75
    private key leakage, 75
    smart contracts vulnerabilities, 74
  consensus protocol and byzantine general problem, 72–73
  data structure, 69–72

F

Fail2ban, 300
Fail stop failures, 73
Features of blockchain, 189
Federal Aviation Administration (FAA), 18
Federated Byzantine agreement, 334–335
51% attacks, 98, 428
Financial industry (FinTech), 95
Financial services, applications of blockchain technology in, 220–223
  use cases, 222–223
Financial technology, blockchain technology and, 225
Forking, 99–100
Forks and multichain management, 121
Formation of chain in blockchain, 359–361
Foundation as a Service (IaaS), 175
Foundation of blockchain, 352, 352*f*
Foundation of Defense of Democracies (FDD), 18
Fraud in programming, 75
Free Syrian Army (FSA), 12
Future cases, 238–240
  Internet of Things (IoT), blockchain technology in, 239–240
  military, blockchain technology in, 238–239
FX trading, applications of blockchain technology in, 222

G

General electric (GE), 279–280
General Packet Radio Service (GPRS), 155–156
General Syrian Congress, 6
Generation transaction, 3
Genesis block, 357
Genesis block hash, 358–359
Genesis vertex, 393
GHOST, 63–64, 310–311
Global consensus mechanism, 373–374
Globalization 2.0, 2
Global multinational corporations (MNCs), 16–17

Golan Heights, 7–8
Google home voice controller, 282
Gossip Protocols, 375
Government, 250–251
Government Accountability Office (GAO), 17–18
Gradient Boosting, 418
Greater Lebanon, 6
The Great Humanitarian Crisis, 10–11
Greedy Heaviest Observed Subtree (GHOST), 330
Greenhouse Gas (GHG) emanations, 143
Grid Singularity, 164–165
Guarantors, 14

H

Hard fork, 100, 356
Harmony sharding, 398–399, 399*f*
HashGraph, 392–393, 392*f*
Hashing, 358–359, 387
Hash tree. *See* Merkle Tree
Hawk, 110
HCI (Human–Computer Interaction), 318
Health care and medicine, blockchain implementations in, 420
Healthcare system, 245, 247–251
  application of blockchain in, 225–229
    use cases, 227–229
  blockchain applications in, 256–259
  blockchain technology, 253–256
  challenges to blockchain applications in, 259–261
  future research directions, 264–266
  government, 250–251
  insurance industry, 248–249
  integrated blockchain powered smart healthcare framework, 261–264
  issues and challenges for, 251–253
  patients, 250
  pharmaceutical industry, 249–250
  physicians, 250
Health Data Analytics, 259
Hedera Hashgraph, 342–344
Heterogeneous multichain scalability, 400–402
High Negotiations Committee (HNC), 13–14
History of blockchain, 353
Homomorphic encryption, 120
HoneyBadger BFT, 399, 402
Horizontal scalability through sharding, 394–400
Hybrid blockchains, 364–365
HydraChain, 367
Hyperledger, 198–199, 257, 260, 366–367

I

ID2020 Summit, 16–17
Identity management, application of blockchain technology in, 236–237

438 Index

Identity management, application of blockchain technology in (*Continued*)
    use cases, 236–237
Image Encryption Method, 171, 177
Immutable, 173
Implementations, blockchain
    in banking, 419
    in education, 419–420
    in energy, 420
    in health care and medicine, 420
    in Internet of things and artificial intelligence, 420
    in legal, government, and insurance, 420
    in power generation, transmission and distribution, 152–153
        Smart Energy Grids, 153
Industrial internet, 283–284
Industrial internet of things (IIoT), 285, 288–289, 298
Information storage and inference, 103–104
Initial cost, 368
Innovative combination of distributed consensus protocols, 3
Insurance industry, 248–249
Insurance sector, blockchain applications in, 223–225
    use cases, 224–225
Integration with legacy systems, 368
Inter-blockchain communication (IBC), 400–402
Interchain blockchain scalability, 385
Interim Smart Grid Pilot in activity, 155
Internally displaced persons (IDPs), 1, 14
The International Business Machines Corporation (IBM), 77
International Energy Agency (IEA), 147
International Syrian Support Group (ISSG), 13–14
Internet of Everything (IOE), 296, 318
Internet of Nano Things (IoNT), 318
Internet of things (IoT), 64–68, 271, 295, 407
    advantages of using, 285–286
    application of, 281–285, 281*f*, 297–298
        automated car, 298
        connected car, 284
        connected health, 284
        industrial internet of things, 298
        industrial internet, 283–284
        smart city, 283, 297
        smart farming, 285, 298
        smart grids, 283, 297
        smart home, 282, 297
        smart retail, 285, 298
        smart supply chain, 285, 298
        telemedicine, 298
        wearables, 282, 297
    architecture of, 280, 280*f*, 298–299
    and artificial intelligence
        blockchain implementations in, 420
    blockchain and, 239–240, 272, 303–305, 311–319
        applications of blockchain, 275–277, 275*f*

background of smart lock systems (use case), 316–318
    components, 313–314
    smart contracts (use case), 318–319
    smart home (use case), 315–316
    smart security (use case), 316
    static sensors and autonomous machines, 319
blockchain for IoT applications, 75–78, 198–201
    challenges in integrating blockchain with internet of things, 200–201
    Ethereum, 199–200
    hyperledger, 198–199
    integration of blockchain and IoT projects and applications, 77
    IOTA, 199
    malicious behaviors in IoT devices, 76
    risks in internet of things blockchain integration, 201
    sensor data correctness, 76–77
    structure of blockchain integrated with IoT applications, 77–78
challenges of, 286, 299–303, 308
challenges with blockchain, 278–279
characteristics, 64–66, 65*f*
    decentralization, 66
    high capacity and low-cost performance, 64–66
    large number of devices and huge internet of things data, 66
    mobility and stable topology, 66
    unpredictable and unstable connections, 66
convergence of blockchain, internet of things, and artificial intelligence, 319–320
crypto currency, 273
device becoming smart device, 281
different domains of, 408*f*
future work, 291
general architecture of blockchain in, 65*f*
integration of, 286–290
    benefits of integration, 288
    challenges in IoT and blockchain integration, 288
    cloud platform, IoT on, 289
    creating an environment, 289–290
    industrial IoT, 288–289
    need for integration of IoT and blockchain, 287
need for blockchain, 272–273
payment networks with bitcoin, 310–311
    Bitcoin-NG, 310
    GHOST, 310–311
potential areas of research in blockchain, 277–278
properties of blockchain, 305–307
    blockchains in data set sharing, 307
    immutable, 307
research on blockchain, 308–309
scope for research, 309–310
security analysis on applications of, 66–68
    communication channel attack, 67

Index **439**

denial of service attack, 67
end devices attacks, 68
network protocol attacks, 68
sensor data attack, 68
software attacks, 68
security and threats, 185–187
data privacy, 186–187
eavesdropping, 186
man-in-the-middle attacks (MitM), 186
types of blockchain, 273–275, 274f
mechanism of bitcoin transaction, 274–275
Internet of things (IoT), blockchain for, 113, 119–123
acceptance of blockchain, 115
Bitcoin, 113–114
Bitcoin energy consumption index, 114
challenges, 119–122
adoption rate, 121
energy efficiency, 121–122
forks and multichain management, 121
infrastructure, 122
privacy, 119–120
security, 120
smart contract administration, 121
throughput, 121
constituents of blockchain, 114–115
energy-efficient blockchain for internet of
things, 123–128
energy consumption in attached machines, 126–128
experimental setup, 125
literature work, 117–119
research opportunities, 122–123
results, 128–129
Internet of vehicles (IoV), blockchain-enabled, 407
application of, 411–412, 412t
architecture, 409–411, 410f
data classification, 410–411
data transmission, 411
vehicular network, 411
privacy, 414, 414f
related studies, 409
security, 413–414
Internet Routing in Space (IRIS) program, 297
Interplanetary file system (IPFS), 35, 45–57
adding peers into blockchain network, 54–55
adding transaction into blockchain network, 50–51
–based decentralized storage system, 39–41
decentralized hash table–based storage approach,
40–41, 40f
previously viewed content available off-line, 40
reliable and persistence, 39–40
secured against distributed denial of service-style
attacks, 40
commands, 45–46
connection and deployment, 46–47

content-addressed technique, access of transaction
using, 51–54
advantage of content addressed over location
addressed, 54
increases durability of data, 54
maintains integrity, 54
permanent storage of files, 54
installation, 45
uploading transactions to, 49–50
use of consensus in blockchain network, 55–57
validating the transaction using mining process, 50
Intrusion Detection System (IDS), 299–300
Intrusion Prevention System (IPS), 299–300
IOTA, 199, 367, 388–389
IOTA tangle, 340–342
Iran, 13
Islamic Front, 12
Islamic Revolutionary Guard Corps (IRGC), 13
Israel, 8

J

Jabal al-Druze State, 6
Jihadist groups, 12

K

Kernighan–Lin (KL) method, 374, 387
Key terms related to blockchain, 355–359
block size, 358
block structure, 356
block time, 358
cryptography in blockchain, 358–359
genesis block, 357
K-means clustering, 418
K-means clustering metric, 418
Know Your Customer (KYC), applications of blockchain
technology in, 223
Kurdish groups, 12
Kuri mobile robot, 282
al-Kuwatli, Shukri, 6–7
Kyoto Protocol, 143, 146–147

L

Lachesis DAG, 393
Lack of skill set, 368
LAMP server, installation of, 204–205
Layer Coordination/Cognitive Intelligence Smart Contracts
(LCSC), 263
Layer Functionality Smart Contracts (LFSC), 263
Layer Interoperability Smart Contracts (LISC), 264
Lebanon, 6
Ledger, 273
Legal, government, and insurance

440 Index

Legal, government, and insurance (*Continued*)
  blockchain implementations in, 420
Legal regulations, 368
"Legitimate" central supervisory authority, 3
LINUX, 204
Liveness attack, 107
LO3 Energy, 161
Low-Level Reader Protocol (LLRP), 300
Low-power and lossy network (LLN), 185
Low-power internet of things devices, blockchain integration
      with, 183
    blockchain for internet of things applications, 198−201
      challenges in integrating blockchain with internet of
          things, 200−201
      Ethereum, 199−200
      hyperledger, 198−199
      IOTA, 199
      risks in internet of things blockchain integration, 201
    concept of blockchain, 187−189
    consensus, 195−198
      practical byzantine fault tolerance (PBFT), 197−198
      proof of stake (POS), 196−197
      proof of work, 196
    elements in a blockchain, 190−192
    features of blockchain, 189
    internet of things security and threats, 185−187
      data privacy, 186−187
      eavesdropping, 186
      man-in-the-middle attacks (MitM), 186
    proposed architecture, 201−206
      deploying smart contract on network, 205
      installation of LAMP server, 204−205
      interfacing ESP32 with Ropsten network, 205−206
      smart contract, 203−204
    structure of blockchain, 190
    types of blockchain, 192−195
      consortium blockchain, 193
      private blockchain, 193−195
      public blockchain, 192−193
Low Power Radio (LPR), 158
Lumino Transaction Compression Protocol (LTCP), 384−385

M

Machine learning (ML), 417, 427
Macnee, Walt, 15−16
Malicious behaviors in internet of things devices, 76
Man-in-the-middle attacks (MitM), 186
Marketplace, application of blockchain technology as, 240
Markov Decision Processes, 108
Mediterranean Sea, 6
Members-Only Exchange, 145−146
Memory Hardness, 330
Merkle root, 3−4, 42−43, 43*f*

Merkle tree, 4, 4*f*, 5*f*, 71−72, 174, 179, 356, 359−360
Merkle Tree Hash, 356
Metadata, 356
Meter Data Acquisition System (MDAS), 158
Meter Data Management System (MDMS), 153, 158
METIS, 374, 387
Micropayments, 311
Military, blockchain technology in, 238−239
Miner nodes in crypto currency, 431
Miners, 3−4, 355−356, 422−423
Mining pool attacks, 99
MIStore, 259
Mobile Ad Hoc Network (MANET), 62
Motivations behind blockchain, 350−352
    foundation of blockchain, 352
    need for blockchain, 352
Motivations for the use of blockchains in cryptocurrencies,
      324
Multinational corporations (MNCs), 16−17
Music industry, application of blockchain in, 235
    use cases, 235−236
Muslim Brotherhood, 8−9
MySQL, 204

N

Namahe, 429*t*
Nano block lattice, 389−390, 389*f*
NASDAQ, 25−26
Nationally Determined Contributions (NDCs), 143
National Settlement Depository (NSD), 25−26
Nature of blockchain—eternal records, 104
Near-Field Communication, 64
Need for blockchain, 272−273, 352
Network architectures
    client/server systems, 349−350
    peer-to-peer (P2P) network, 350
Network Centric Operations (NCOs), 17
Network operator, 363
Network protocol attacks, 68
Network view, 354*f*
Neural, 429*t*
New node synchronization, 360−361
Nodes, 3−4, 355
Nonce, 355
North Atlantic Treaty Organization Communication and
      Information Agency, 18−19
Nuclear Threat Initiative (NTI) group, 20−21, 238
Nuclear weapons systems, 20−21

O

Off-Chain blockchain scalability, 381−382
Off-Chain scalability solutions, 403

Index **441**

OmniLedger, 396–397
  Atomix protocol in, 398*f*
On-Chain blockchain scalability, 380–381
On-Chain scalability solutions, 403
OneUp, 164
Open clouds, 176
Open key, 175
OpenMinded, 429*t*
OPERA chain, 393
Operation Euphrates Shield (OES), 12–13
Oracle Blockchain Cloud Service venture, 176
Original Equipment Manufacturers (OEMs), 20–21
OriginTrailis Protocol, 317
Orumesh protocol, 390–391
Ottoman Empire, 5–6
Outage Management System, 161
Ownership titles, managing, 231
Oyente, 108–110

P

Paris Accord, 144
Paris Agreement, 143–145
Patient-centric healthcare system, 249*f*
Patients, 250
PawanDuggal, 26
Payment networks with bitcoin, 310–311
  Bitcoin-NG, 310
  GHOST, 310–311
Peak Load Management, 160
Peers addition into blockchain network, 54–55
Peer-to-peer (P2P) control trading, 133–134
Peer-to-peer (P2P) electronic cash transfer, 373
Peer-to-peer (P2P) network, 350, 375
Peer-to-peer Bitcoin network, 3
Peer-to-peer scheme, blockchain-based framework for data
    storage in, 35
  blockchain structure to maintain availability, 41–45
    access control policy, 44–45
    availability of record, 44
    decentralized peer-to-peer structure over centralized
      structure of storage, 41
    immutability of record, 42–43
    integrity of record, 43
    transparency of record, 41
  implementation of framework using interplanetary file
    system and blockchain, 48–57
    access of transaction using content-addressed technique,
      51–54
    adding peers into blockchain network, 54–55
    adding transaction into blockchain network, 50–51
    uploading transactions, 49–50
    use of consensus in blockchain network, 55–57
    validating the transaction using mining process, 50

interplanetary file system, 45–47
  commands, 45–46
  connection and deployment, 46–47
  installation, 45
interplanetary file system–based decentralized storage
    system, 39–41
  decentralized hash table–based storage approach,
    40–41, 40*f*
  previously viewed content available off-line, 40
  reliable and persistence, 39–40
  secured against distributed denial of service-style
    attacks, 40
  motivation, 36
  related work, 36–37
  working model of framework, 37–38
Pereira, Frank Rausan, 27–28
PermaCoin, 333
Permissioned blockchain, 246, 255
Permissionless ledger, 364
Permissions, 354–355
Persistency, of blockchain, 327
Personal health information (PHI), 284
Pharmaceutical industry, 249–250
Phasor Measurement Unit, 153
PHP, 204
Physical–cyberspace boundary, 102–103
Physicians, 250
Pitts-Kiefer, Samantha, 20–21
Platforms, 365–367
  AChain, 367
  Azure, 367
  Bitcoin, 366
  Chain Core, 367
  Corda, 367
  criteria for selecting, 365–366
  EOSIO, 367
  Ethereum, 366
  HydraChain, 367
  Hyperledger, 366–367
  IOTA, 367
Poditok, 259
"Political transition" process, 14
Polkadot blockchain network, 402
Potential areas of research in blockchain, 277–278
Power-ID, 163–164
Power Ledger, 161–162
Power Line Carrier Communication (PLCC), 155–156
Power Quality Management, 160
PowerToShare, 164
Practical Byzantine Fault Tolerance (PBFT) protocol, 62,
    197–198, 361, 378
Price Waterhouse Cooper (PwC), 255–256
Privacy in the blockchain-enabled IoV network, 414
Privacy issues of blockchain technology, 97, 102–104

442 Index

Privacy issues of blockchain technology (*Continued*)
  information storage and inference, 103—104
  nature of blockchain—eternal records, 104
  physical—cyberspace boundary, 102—103
Private blockchain, 193—195, 364
Private clouds, 176
Private key leakage, 75
Private key security, 101
Private/permissioned and consortium, 139
Process stages of implementation of blockchain technology,
    420—421
Programming as a Service (SaaS), 175
Proof of activity, 361
Proof of burn time, 362
Proof of capacity (PoC), 324—325, 362
Proof of concept (PoC), 117
Proof-of-elapsed time (PoET) model, 324, 331—332
  block verification, 332
  random waiting times, 331—332
    registration, 332
  waiting time, calculation of, 332
Proof of importance, 362
Proof of shake (PoS), 72—73, 196—197, 324—325, 331, 353,
    361, 376—377, 430t
Proof of Weight (PoWeight), 430t
Proof of work (PoW), 62, 113—117, 119, 121—122, 187—188,
    196, 254—255, 324—325, 329—330, 330f, 353, 361,
    375—377, 430t
Properties of blockchain, 305—307
  transparent, 307
    blockchains in data set sharing, 307
    immutable, 307
Prosumers, 133—134
Protocol of blockchain, 431
ProvChain, 176—177
Public blockchain, 192—193, 246, 364
Public Ethereum blockchain, 418
Public key (PK), 62, 69—70, 422—423
Public ledger, 419
Public perception, 368
Public/permissionless blockchain, 138—139
Puducherry Electricity Distribution System, 154

Q

QixCoin, 384—385
Quantitative framework, 108
Quorum intersection, 335, 335f

R

Radio Frequency (RF), 155—156
Radio-Frequency Identification (RFID) tags, 64
Rajya Sabha Television Network (RSTV), 27—28

Raman, Mythili, 25
RandHound, 396—397
Random waiting times, 331—332
  registration, 332
RapidChain, 397—398
Reactive Power Compensation, 153
Real estate, application of blockchain in, 230—232
  existing real estate sector, problems in, 231
  use cases, 231—232
Real Estate Investment Trusts (REIT), 232
Real estate regulatory authority (RERA), 276
Record
  availability of, 44
  immutability of, 42—43
  integrity of, 43
  transparency of, 41
Reducing emissions from deforestation and forest
    degradation, 147
Refugee bank, 16
Refugees, 1
  and internally displaced persons
    blockchain technology for, 15—17
    ethical considerations and issues in "de-escalation zones"
      for, 14—15
Regulators, 363
Relay-chain, 402
Relevance of blockchain technology applications, 214—215
Remote Patient Monitoring, 257
Renewable energy certificates, 147
Renewable energy source (RES), 134—135
Research on blockchain, 277—278, 308—309
Research survey, 63—64
R-METIS, 374, 387—388
Root Hash, 359—360
Ropsten network, interfacing ESP32 with, 205—206
Royal Institute of International Affairs, 19

S

Satoshi Nakamoto, 3
Scalability in blockchain, 368, 373
  challenges, 377—379
  core concepts, 375—377
  future directions, 403—404
  scalability solutions, 379—403
Schumar, Chuck, 25
Secure Ad hoc On-Demand Distance Vector (SAODV)
    routing, 300
Security and privacy, 368
Security enhancement to blockchain systems, 107—110
  Hawk, 110
  Oyente, 108—110
  quantitative framework, 108
  SmartPool, 107—108

Security in the blockchain-based network, 413–414
Security issues of blockchain technology, 96–102
  client side security threats, 99
  criminal activity, 100–101
  double spending, 98–99
  51% vulnerability, 98
  forking, 99–100
  mining pool attacks, 99
  private key security, 101
  transaction privacy leakage, 101–102
Segregated Witness, 358
Segwit, 378–379
Selfish mining attack, 104–105
Self-Organizing Map (SOM), 418
Sensor data attack, 68
Sensor data correctness, 76–77
Service-style attacks, secured against distributed denial of, 40
SGX, 178
Sharding, 394–400
Shared Ledger, 354
Shia-affiliated militia groups, 13
ShoCard, 16
Side-Chain blockchain scalability, 382–384
Signature scheme, 2–3
Simplified payment verification (SPV), 382–384
Sinai Peninsula, 6
SingurityNET, 429t
Smart city, 283, 297
Smart contract, 74, 203–204, 353, 355, 362, 370, 419, 427–428
  administration, 121
  application of blockchain technology as, 240
  deployment on network, 205
Smart device, 281
Smart Energy Grids, 153
Smart farming, 285, 298
Smart grid, 283
Smart grid-based smart city, blockchain implementation using, 133
  background, 135–136
  blockchain concept, 136–138
  blockchain implementation in power generation, transmission and distribution, 152–153
    Smart Energy Grids, 153
  blockchain implemented projects worldwide in energy conservation, 161–164
  blockchain types, 138–140
  carbon credit, 142–143
  Carbon Credit Exchange, 145–147
    carbon credits and carbon markets, 146–147
  carbon emission effect, 147–152
    blockchain application for distributed energy resources management in smart city, 151–152
  center components of blockchain architecture, 141–142

Paris Agreement, 143–145
  smartgrid blockchain implementation in India, 153–161
Smart Grid components, 156–157
Smart Grid Pilot Project (Division-I) profile, 154–155
Smart grids, 297
Smart home, 282, 297
Smart meter, 156
SmartPool, 107–108
Smart retail, 298
  and supply chain, 285
Smart supply chain, 298
Social media, 277
Soft fork, 100
Software attacks, 68
Software Guard Extensions (SGXs), 331
Space and Terrestrial Communications Directorate of the US Army (S&TCD), 18–19
SpaceMint, 333
Spectral Energy, 162
Stage as a Service (PaaS), 175
Startup with blockchain and machine learning, 428
State of Aleppo, 6
State of Damascus, 6
Stellar system, 334–335
Stochastic Activity Networks (SAN), 300
Streamr, 317
Strengths of technology, 369
Structure of blockchain, 190
Supervised ML approach, 418
Supervisory Control and Data Acquisition Distribution Management System, 161
Supply chain, application of blockchain technology in, 232–235
  use cases, 233–235
Supply chain management (SCM), 276–277, 285
Surety, 2–3
Suricata, 300
Sykes–Picot Agreement, 6
Syria, 5–6
  beginning of modern nation-state, 6–8
  refugees and internally displaced persons in Syria since 2011 (case study), 10–11
  under Bashar al-Assad regime, 8–10
Syria–Lebanon region, 6
Syrian civil war, Arab spring and the emergence of, 11
Syrian Islamic Liberation Front (SILF), 12

T

Taurus Mountains, 6
Taxonomy of blockchain system, 328–329
Technical challenges faced while using blockchain technology, 336–337
Techniques for clustering, 431

444 Index

Telemedicine, 298
3D printing, 19−20
Time of Use (ToU) taxes, 164−165
Timestamp, 356
Time-Stamping Authority (TSA) framework, 3
Time-stamping service (TSS), 2−3
Tokenization of real estate platforms, 231−232
Top Hash, 4
Trade finance, applications of blockchain technology in, 222
TraneAi, 429*t*
Transaction, 355, 373−374
Transaction privacy leakage, 101−102
Transparency, of blockchain system, 173
TR-METIS, 374, 388
Trusted equipment units (TPM), 178
Trusted Execution Environment (TEE), 331
Trusted service application, application of blockchain
    technology as, 241
TrustZone, 178
Turkey, 12−13
Two-way pegged protocol, 384*f*
Types of blockchain, 192−195, 273−275, 274*f*, 328*f*,
    364−365
  bitcoin transaction, mechanism of, 274−275
  consortium blockchain, 193
  hybrid, 364−365
  private blockchain, 193−195, 364
    advantages of, 193−194
    disadvantages of, 194−195
  public blockchain, 192−193, 364

U

Ubuntu machine, 204−205
Ultra High-Frequency (UHF) RFID readers, 300
United Arab Republic (UAR), 6−7
United Nations Supervision Mission in Syria (UNSMIS), 13
UPort, 16

US Air Force (USAF), 17−18
US defense forces and industry (case study), 17−21
Users of blockchain technology, 363

V

VALIDATOR module, 109−110
Vector, 162
Vehicle privacy, 408−409
Vehicular Ad Hoc Network (VANET), 62, 66, 78
Verifiable Delay Function (VDF), 398−399
Verifiable random function (VRF), 396−397
Verified emission reduction (VERs), 143, 146
Versailles Peace Conference, 5−6
Virtual machine (VM), 171, 178
VISUALIZER module, 109−110
Voting, application of blockchain in, 229−230
Vulnerability, 98

W

Waiting time, calculation of, 332
WALMART, 274−275
Waltonchain, 317
Watson IoT Platform, 317
Weapons of mass destruction (WMD), 9
Wearable devices, 297
Wearables, 282
Weighted DAG, 388*f*
Working of blockchain, 365, 366*f*
Workspace ONE App, 177
World Bank Group (WBG), 143−144
World Food Programme (WFP), 16
WORM (write once read many) policy, 44−45

Z

Zero-knowledge proving methods, 120
ZoneB, 400−402

Printed in the United States
By Bookmasters